VOLUME TWO HUNDRED AND TWENTY-ONE

ADVANCES IN
IMAGING AND ELECTRON PHYSICS

VOLUME TWO HUNDRED AND TWENTY-ONE

ADVANCES IN
IMAGING AND ELECTRON PHYSICS

The Beginnings of Electron Microscopy - Part 2

Edited by

PETER W. HAWKES and MARTIN HŸTCH

CEMES-CNRS

Toulouse, France

ELSEVIER

ACADEMIC PRESS

An imprint of Elsevier

Cover photo credit:
From E. Ruska: The Early Development of Electron Lenses and Electron Microscopy (Hirzel, Stuttgart, 1980). Courtesy S. Hirzel Verlag.

Academic Press is an imprint of Elsevier
125 London Wall, London EC2Y 5AS, United Kingdom
525 B Street, Suite 1650, San Diego, CA 92101, United States
50 Hampshire Street, 5th Floor, Cambridge, MA 02139, United States
The Boulevard, Langford Lane, Kidlington, Oxford OX5 1GB, United Kingdom

ISBN: 978-0-323-98919-0
ISSN: 1076-5670

For information on all Academic Press publications
visit our website at https://www.elsevier.com/books-and-journals

Publisher: Zoe Kruze
Acquisitions Editor: Jason Mitchell
Developmental Editor: Jhon Michael Peñano
Production Project Manager: Vijayaraj Purushothaman
Designer: Greg Harris

Typeset by VTeX

Contents

Contributors *xiii*

Foreword *xv*

Foreword to first edition *xix*

Preface by Ernst Ruska *xxiii*

Afterword by Peter Hawkes xxv

References xxvi

1. Recollections from the early years: Canada–USA **1**

Cecil E. Hall and Peter Hawkes (Afterword)

1. Introduction 1
2. Initial work: University of Toronto (1935–1936) 3
3. Kodak Research Laboratories, Rochester, New York (1937–1941) 7
4. Massachusetts Institute of Technology 13
   - 4.1. Note on the war period 15
   - 4.2. Electron stains 16
   - 4.3. Paramyosin 17
5. Transition from the early years 19
   - 5.1. Shadowcasting 19
   - 5.2. Stigmator 20
   - 5.3. Microtomy 21
6. Afterword by Peter Hawkes 22

References 22

2. My recollection of the early history of our work on electron optics and the electron microscope **25**

Tadatosi Hibi and Peter Hawkes (Afterword)

1. Emission microscope 25
2. Studies of specimen techniques for electron microscopy 27
3. Electron-microscopic observations of ionic crystals 29
4. The pointed filament and its application 34
5. Interference and coherence 37
6. Afterword by Peter Hawkes 42

References 42

3. Walter Hoppe — X-ray crystallographer and visionary pioneer in electron microscopy **45**

Joachim Frank

1. Introduction 45
2. Life 46
3. Work in X-ray crystallography 47
4. Structure research using electron microscopy 48
5. Electron tomography 53
6. Walter Hoppe remembered 55

Acknowledgments 57

Appendix to "Walter Hoppe — X-ray crystallographer and visionary pioneer in electron microscopy" by Joachim Frank 58

References 66

4. Reminiscences of the development of electron optics and electron microscope instrumentation in Japan **71**

Koichi Kanaya and Peter Hawkes (Afterword)

1. Historical survey of the early stages of electron microscopy in Japan 72
2. Lens aberrations 80
   - 2.1. Fifth-order Seidel aberrations and third-order astigmatism 80
   - 2.2. Aberrations of a focused charged-particle beam caused by space charge 84
   - 2.3. Asymmetric aberrations and manufacture accuracy 88
3. Practical magnetic lens design 93
   - 3.1. Minimum aberration coefficients 93
   - 3.2. Aberrations of three-stage electron microscopes 96
   - 3.3. Minimum aberrations and achromatization 105
   - 3.4. Anastigmatic lens 109
4. Formation of caustic and shadow images 110
   - 4.1. Third-order caustic 110
   - 4.2. Caustic properties combined with second- and third-order astigmatism 113
   - 4.3. Shadow image 114
5. Stigmators 117
   - 5.1. Electrostatic stigmator used to correct second- and third-order astigmatisms 117
   - 5.2. Magnetic stigmator used to correct second- and third-order astigmatisms 120
   - 5.3. Anastigmatic alignment 122
   - 5.4. Anastigmatic deflection system for correcting astigmatism and its defocusing action 124

6. Electron phase microscope 127
7. High-voltage electron microscopes 130
8. Afterword by Peter Hawkes 131
References 133

5. Early electron microscopy in the Netherlands 139

J.B. Le Poole and Pieter Kruit (Afterword)

1. Afterword by Pieter Kruit: about Jan Bart Le Poole (1917–1993) 168
References 171

6. L. L. Marton, 1901–1979 173

Charles Süsskind and Peter Hawkes (Afterword)

1. Bibliographical appendix: publications of L. L. Marton 192
2. Afterword by Peter Hawkes 198
References 198

7. The invention of the electron Fresnel interference biprism 201

G. Möllenstedt and Peter Hawkes (Afterword)

1. Childhood years in Bielefeld 201
2. Gaining technical know-how as a student apprentice 202
3. A broad education in physics with Professor Walter Kossel 202
4. Influence of the work of Kikuchi and convergent beam diffraction 203
5. Electron-optical experiments with Brüche, Scherzer, and Mahl 205
6. Quantitative testing of the operation of the biprism 211
7. Measurement of the inner potential of solids 218
8. Electron interference microscope in the transmission mode 219
9. The intensity problem in electron interferometers 219
10. Atomic resolution electron holography 220
11. Editorial note 222
12. Afterword by Peter Hawkes 222
References 223

8. The industrial development of the electron microscope by the Metropolitan Vickers Electrical Company and AEI Limited 225

T. Mulvey and Peter Hawkes (Afterword)

1. Introduction 225
2. Cathode rays and electron images—Gabor and the ironclad concentration coil 226
3. The first commercial electron microscope—EM1 228

4. Wartime developments 232
5. The EM2 electron microscope 233
6. A three-stage electron microscope—EM3 234
7. A simple electron microscope—EM4 236
8. High-voltage electron microscopy—the EM5 237
9. Improving the resolving power—EM3A and the emergence of electron holography 238
10. High-resolution microscopes—EM6 243
11. The image intensifier 246
12. The electron microscope in biology 247
13. Very high-resolution microscopy 248
14. Afterword by Peter Hawkes 248
References 252

9. The development of the scanning electron microscope 255

C.W. Oatley, D. McMullan, K.C.A. Smith, and Peter Hawkes (Afterword)

1. Introduction 255
2. The initial aims of the three groups 256
3. The Cambridge microscope 260
4. Uses of the microscope 275
5. Afterword by Peter Hawkes 283
Bibliography including cited references 283

10. Some recollections of electron microscopy in Britain from 1943 to 1948 289

R. Reed and Peter Hawkes (Afterword)

1. Afterword by Peter Hawkes 307
References 309

11. Rudolf Rühle and the BOSCH electron microscope: another early commercial instrument 311

Hans R. Gelderblom and Heinz Schwarz

1. Origins of electron microscopy in Berlin 311
2. First steps for the S&H instrument 312
3. The commercial Siemens & Halske magnetic electron microscope 314
4. The electrostatic electron microscope of the AEG 317
5. Competition between concepts, characters, and companies 317
6. The gifted entrepreneur: Baron Manfred von Ardenne 321

7. Development of electron microscopy inside the Bosch company 324
8. Rudolf Rühle (1907-2003): personal background and development 325
9. Rudolf Rühle: the technical engineer at Bosch 327
10. Walter Rentschler and the Bosch electron microscope 331
11. Rudolf Rühle's groundwork for the Bosch electron microscope 332
12. Rudolf Rühle's thesis: on the future Bosch electron microscope 333
13. Rebuilding of Germany after WWII 337
14. Foundation of the German society for electron microscopy 340
15. The commercial Bosch electron microscope 340
16. Promotion for the Bosch electron microscope 345
17. The Bosch electron microscope in Hohenheim 347
18. The Bosch-EM in Tübingen 354
19. Giving up the commercial electron microscope 364
20. A list of Rudolf Rühle's patents 366
Acknowledgments 368
References 369

12. Otto Scherzer and his contributions to electron microscopy 377

Dieter Typke

1. Introduction 377
2. Hans Busch, "the father of electron optics" 378
3. Otto Scherzer 380
   3.1. Munich 1927-1931 380
   3.2. Berlin and Munich, 1932-1935 381
   3.3. Darmstadt 1935-1940, Scherzer's theorem 382
   3.4. The time during the Third Reich and the struggle against Deutsche Physik 386
   3.5. Peenemünde 1940-1945: radar investigations in war time 387
   3.6. Mosbach, Fort Monmouth, and again Mosbach 1946-1949; the start of aberration correction, "Scherzer Focus" 388
   3.7. Darmstadt, 1949-1982 390
   3.8. Scherzer's personality, other activities 400
   3.9. Scherzer anecdotes 401
Acknowledgments 403
References 403

13. 1950-1960: a decade from the viewpoint of an applications laboratory 407

Cilly Weichan and Maren Heinzerling (Afterword)

1. Introduction 407

2. Experience with testing a new microscope 409
2.1. The importance of a higher electron-optical magnification 409
2.2. Astigmatism and testing the resolving power 411
2.3. Contrast and aperture problems 413
2.4. Calibration and high-resolution test specimens 415
3. A microscope with higher resolving power 416
3.1. Contamination and an anticontamination device 418
3.2. Contrast improvement and focusing aids 420
3.3. I take up the aperture problem again 422
3.4. Extended application of microscopical techniques 423
3.5. One end point: 0.4-nm high resolution 423
4. Development and improvement of preparation methods 424
4.1. Investigation of rubber and filling material 424
4.2. Further applications: modifications of the replica technique 428
4.3. Shadowing and special techniques 432
5. Teaching duties in an applications laboratory for electron microscopy 435
6. Final remarks 438
7. Afterword by the late Dipl.-Ing. Maren Heinzerling (Deutscher Akademikerinnenbund e.V.) 439
References 442

14. From the cathode-ray oscillograph to the high-resolution electron microscope 443

Otto Wolff and Peter Hawkes (Afterword)

1. Introduction 443
2. Developments up to the end of World War II 444
2.1. Working toward Braun's tube 444
2.2. From Braun's tube to the cathode-ray oscillograph 445
2.3. Further development of the cathode-ray oscillograph into a technical instrument 447
2.4. The recording oscillograph 451
3. Period of transition 453
4. Improvements in the electron microscope after 1945 457
4.1. The task ahead 457
4.2. Improvements in the electron optics 458
4.3. Stability of the operating parameters 458
4.4. Replacement of the previous control circuits 460
4.5. Test equipment 460
4.6. A new high-voltage system 461
4.7. Improvement of resolution 464
5. Concluding remarks 465

6. Afterword by Peter Hawkes 467
References 467

15. Reminiscences **469**
R.W.G. Wyckoff and Peter Hawkes (Afterword)

1. Afterword by Peter Hawkes 473
References 473

16. Complementary accounts of the history of electron microscopy **475**
Peter Hawkes

1. Appendix 488
Acknowledgments 495
2. Afterword 496
Acknowledgments 498
References 498

Index *511*

Contributors

Joachim Frank
Department of Biochemistry and Molecular Biophysics, Columbia University Medical Center, New York, NY, United States

Hans R. Gelderblom
Retired from Robert Koch Institute, Berlin, Seddiner See, Germany

Cecil E. Hall
Formerly Jasper, AR, United States

Peter Hawkes
Laboratoire d'Optique Electronique du Centre, National de la Recherche Scientifique, Toulouse, France, now CEMES-CNRS, Toulouse, France

Maren Heinzerling
Formerly Berlin, Germany

Tadatosi Hibi
Formerly Sendai, Japan

Koichi Kanaya
Formerly Department of Electrical Engineering, Kogakuin University, Tokyo, Japan

Pieter Kruit
Department of Imaging Physics, Faculty of Applied Sciences, Delft University of Technology, Delft, The Netherlands

J.B. Le Poole
Formerly Technische Natuurkunde, Technische Hogeschool, Delft, The Netherlands

D. McMullan
Formerly Engineering Department, Cambridge University, Cambridge, United Kingdom

G. Möllenstedt
Formerly Institut für Angewandte Physik, Universität Tübingen, Tübingen, Germany

T. Mulvey
Formerly Department of Physics, University of Aston, Gosta Green, Birmingham, United Kingdom

C.W. Oatley
Formerly Engineering Department, Cambridge University, Cambridge, United Kingdom

R. Reed
Formerly Pudsey, West Yorkshire, United Kingdom

Heinz Schwarz
Retired from Max Planck Institute for Biology, Tübingen, Germany

K.C.A. Smith
Formerly Engineering Department, Cambridge University, Cambridge, United Kingdom

Charles Süsskind
Formerly College of Engineering, University of California, Berkeley, Berkeley, CA, United States

Dieter Typke
Formerly Munich, Germany

Cilly Weichan
Formerly Berlin, Germany

Otto Wolff
Formerly Auf dem Sand, Muehltal, Germany

R.W.G. Wyckoff
Formerly Duval Corporation, Tucson, AZ, United States

Foreword

With one exception, the chapter by Charles Süsskind on Ladislaus Marton, the contributions to the 1985 edition of 'Beginnings' were all written by the pioneers themselves. This inevitably left many major gaps. Early work on electron microscopy at the AEG Research Institute was entirely absent for Ernst Brüche, Hans Mahl, Carl Ramsauer, and Otto Scherzer had already died. (Scherzer had agreed to contribute but was finally unable to do so.) Hans Boersch was also too unwell at the time. In this second edition, I have filled in a few of the gaps. Boersch now appears in the form of a biographical article by Hans Niedrig, translated from the original in the house journal of the German Electron Microscopy Society (DGE), *Elektronenmikroskopie*. Raymond Castaing is now present: a full study by his former pupil Christian Colliex is reproduced in volume 220, accompanied by a tribute by Peter Duncumb and myself. The Royal Society has generously allowed us to reprint the long Memoir on Dennis Gabor by T.E. Allibone, also in volume 220. Joachim Frank, who began his scientific life in Walter Hoppe's laboratory, has composed an account of Hoppe's life and work.

I very much wanted to include the pioneers of electron holography in this new edition and am very pleased that Hannes Lichte has agreed to write an account of his own work in which that of Akira Tonomura is intertwined. This could not be completed in time to appear here but will be published in a regular volume of these Advances. For Gottfried Möllenstedt, we reprint his account of the genesis of the electron biprism, first published in *Advances in Optical and Electron Microscopy* but not hitherto available in electronic form. Another of Möllenstedt's historical articles appears in 'Growth of Electron Microscopy' (volume 96 of these Advances). It is not widely known that electron microscopes with magnetic lenses were produced by the Bosch company in Stuttgart in the early 1950s, at the instigation of Rudolf Rühle (author of *Das Elektronenmikroskop*, Curt E. Schwab, Stuttgart 1949). The life of Rudolf Rühle and the construction of these instruments are described by Hans Gelderblom and Heinz Schwarz. The last of the new articles is an account of Otto Scherzer's career by Dieter Typke, who prepared his doctoral thesis in Scherzer's laboratory in Darmstadt. Typke originally planned to cover the work performed at the AEG Research Institute in Berlin in the 1930s in this article. It was later decided to publish it as a separate chapter since Scherzer did not spend much of his career at the AEG Institute. It appears in volume 220.

Several other major figures of the early years are still missing but do appear in earlier volumes of these Advances. The widow of Bodo von Borries published a biography of her husband in volume 81 (1991) of *Advances in Electronics and Electron Physics*, though some sentences were found excessive by other survivors of the early years. His cousin Götz von Borries has also published a small book on *Bodo von Borries und das Elektronenmikroskop* (Fouqué Literaturverlag, Egelsbach 2001). In 'Helmut Ruska (1908–1973): His role in the evolution of electron microscopy in the life sciences, and especially virology' in volume 182 (2014), Hans Gelderblom and Detlev Krüger have retrieved Ernst Ruska's brother Helmut from oblivion and shown how important was his role in persuading Siemens to launch their first commercial instrument, the Übermikroskop. The article on Dennis Gabor by Tom Mulvey in volume 91 (1995) nicely complements the Royal Society Memoir printed here. The life of Ernst Ruska, again by Mulvey together with Lotte Lambert, in volume 95 (1996) should be read alongside Ruska's own historical volume. Jan Le Poole was present in the original 'Beginnings' and a fuller picture of the man emerges from the biographical article by Mulvey and D.J.J. van de Laak-Tijsen in volume 115 (2001).

I particularly regret the absence of a full biography of Walter Glaser, who contributed so much to the theory of electron optics. The references in his monumental *Grundlagen der Elektronenoptik* (Springer, Vienna 1952), which are very complete, make it clear that Glaser's own work is at the heart of almost every chapter. The composition of that book is evoked vividly by Glaser's research students Hans Grümm and Peter Schiske in volume 96 (1996).

Johannes Picht should also not be forgotten. His *Einfuhrung in die Theorie der Elektronenoptik* (Barth, Leipzig 1939) went into three editions and one of his doctoral students (M. Leitner) drafted the first full study of the aberrations of cylindrical electron lenses before being killed in World War II.

All the contributors to the 1985 edition of 'Beginnings' have since died. I have therefore added an Afterword to each of their chapters. In a few cases (Armin Delong, David Drummond, Jan Le Poole, Cilly Weichan), these were kindly provided by authors close to the subjects.

Nor have the historians and chroniclers of science been idle. In the final chapter, we provide full bibliographic details of several books and articles on various aspects of the history of the subject. To these we add a major publication by Falk Muller, *Jenseits des Lichts. Siemens, AEG und die Anfänge der Elektronenmikroskopie in Deutschland* (Wallstein Verlag, Göttingen 2021). Royal Society Memoirs are available on V. Ellis Cosslett and Sir

Charles Oatley, whose papers are housed in the archives of Churchill College, Cambridge. Obituary notices on many German electron microscopists are to be found in *Physikalische Blätter* and *Elektronenmikroskopie* (accessible via the website of the Deutsche Gesellschaft für Elektronenmikroskopie). Biographical papers on deceased members of the French Académie des Sciences can be found in their *Comptes Rendus*.

Peter W. Hawkes

Foreword to first edition

Electron microscopy is still a young subject and some of those who were active during its formative years, contributing to its development or implantation in their various countries, can still be met in their laboratories, while others, although retired, have not broken all contact with the subject. The idea of gathering recollections of the beginnings of electron optics and electron microscopy had been in my mind for some time but it was the publication of Professor E. Ruska's meticulous historical volume[1] that spurred me into action; that and the macabre reality that time was running out.

My initial approach to scientists in many countries met with a largely enthusiastic response, though the few refusals were particularly disappointing as they left whole areas of activity imperfectly represented. Also, several letters elicited no reply and I have no way of telling whether or not they failed to reach their targets. I mention these points here to explain the absence of a number of names, which might occasion surprise.

One country is, however, not represented for a quite different reason. Interest in electron optics and the associated instruments in Russia dates back to 1935 and the first commercial instrument appeared in 1946. Russian scientists have remained active in the field ever since. Despite innumerable letters and telegrams to members of the Russian electron microscope community, I have failed to obtain any response and I therefore address this more public appeal, to anyone concerned with the beginnings of electron microscopy in the USSR, to get in touch with me with a view to remedying this omission, perhaps in a future volume of these *Advances*.

This is not the only historical gap and I have therefore ventured to include a brief account of the whereabouts of further information on the history of the subject. Thus some of those who were not able to speak for themselves figure in these pages, if only at second hand.

I am sorry to have to record that a few of those who had planned to contribute, at least provisionally, have been prevented from doing so by ill health or by death. Helmut Johannson, who agreed to contribute, died in April, 1982. Otto Scherzer, who made no firm commitment but

[1] E. Ruska: Die Frühe Entwicklung der Elektronenlinsen und der Elektronenmikroskopie, *Acta Hist. Leopoldina* (1979), Nr 12, 136 pp.; translated into English by T. Mulvey as "The Early Development of Electron Lenses and Electron Microscopy" (Hirzel, Stuttgart, 1980).

from whom I still hoped for a short piece, perhaps on the first derivation of "Scherzer's Theorem," also died in 1982. Dr. R. W. G. Wyckoff sent me the short account published here but was prevented by illness from expanding it as he planned. Professor Hans Boersch has been obliged to withdraw for health reasons. Dr. A. C. van Dorsten has been delayed, but his contribution will appear in a future volume of the *Advances*.

It only remains for me to thank all the contributors for the time and hard work that they have devoted to preparing their articles for this book. From the outset, I made it clear that both formal history and more informal reminiscence would be welcome, best of all perhaps being a combination of the two. This explains the differences in length and tone between the various articles, and I trust that the reader will agree that this is not the least attractive feature of this collection. It also explains why the war figures so prominently in the articles by members of various occupied countries, where the first instruments were assembled in the grim conditions recalled here.

Finally, I am particularly grateful to Professor E. Ruska for agreeing to write a preface and for furnishing much invaluable information and numerous documents.

Peter W. Hawkes

Note on the references. The various European and International Conferences on electron microscopy are referred to so frequently that we merely give place and date in the individual lists of references. The full publishing details of these and the international high-voltage electron microscopy meetings are as follows:

Delft, 1949: *Proceedings of the Conference on Electron Microscopy*, Delft, 4–8 July, 1949, A. L. Houwink, J. B. Le Poole, and W. A. Le Rütte, eds. (Hoogland, Delft, 1950).

Paris, 1950: *Comptes Rendus du Premier Congrès International de Microscopie Electronique*, Paris, 14–22 September, 1950 (Editions de la Revue d'Optique Théorique et Instrumentale, Paris, 1953).

London, 1954: *The Proceedings of the Third International Conference on Electron Microscopy*, London, 1954, R. Ross, ed. (Royal Microscopical Society, London, 1956).

Gent, 1954: *Rapport Europees Congrès Toegepaste Electronenmicroscopie*, Gent, 7–10 April, 1954, edited and published by G. Vandermeersche (Uccle-Bruxelles, 1954).

Toulouse, 1955: *Les Techniques Récentes en Microscopie Electronique et Corpusculaire*, Toulouse, 4–8 April, 1955 (C.N.R.S., Paris, 1956).

Stockholm, 1956: *Electron Microscopy. Proceedings of the Stockholm Conference*, September, 1956, F. J. Sjöstrand and J. Rhodin, eds. (Almqvist and Wiksells, Stockholm, 1957).

Tokyo, 1956: *Electron Microscopy. Proceedings of the First Regional Conference in Asia and Oceania*, Tokyo, 1956 (Electrotechnical Laboratory, Tokyo, 1957).

Berlin, 1958: *Vierter Internationaler Kongress für Electronenmikroskopie*, Berlin, 10–17 September, 1958, *Verhandlungen*, W. Bargmann, G. Möllenstedt, H. Niehrs, D. Peters, E. Ruska, and C. Wolpers, eds. (Springer, Berlin, Göttingen, Heidelberg, 1960), 2 Vols.

Delft, 1960: *The Proceedings of the European Regional Conference on Electron Microscopy*, Delft, 1960, A. L. Houwink and B. J. Spit, eds. (Nederlandse Vereniging voor Elektronenmicroscopie, Delft n.d.), 2 Vols.

Philadelphia, 1962; *Electron Microscopy. Fifth International Congress for Electron Microscopy*, Philadelphia, Pennsylvania, 29 August to 5 September, 1962, S. S. Breese, ed. (Academic Press, New York, 1962), 2 Vols.

Prague, 1964: *Electron Microscopy 1964. Proceedings of the Third European Regional Conference*, Prague, M. Titlbach, ed. (Publishing House of the Czechoslovak Academy of Sciences, Prague, 1964), 2 Vols.

Kyoto, 1966: *Electron Microscopy 1966. Sixth International Congress for Electron Microscopy*, Kyoto, R. Uyeda, ed. (Maruzen, Tokyo, 1966), 2 Vols.

Rome, 1968: *Electron Microscopy 1968. Pre-Congress Abstracts of Papers Presented at the Fourth Regional Conference*, Rome, D. S. Bocciarelli, ed. (Tipografia Poliglotta Vaticana, Rome, 1968), 2 Vols.

HVEM Monroeville, 1969: *Current Developments in High Voltage Electron Microscopy (First National Conference)*, Monroeville, 17–19 June, 1969. Proceedings not published but *Micron* **1** (1969), 220–307, contains official reports of the meeting based on the session chairmen's notes.

Grenoble, 1970: *Microscopie Électronique 1970, Résumés des Communications Présentées au Septième Congrès International*, Grenoble, P. Favard, ed. (Société Française de Microscopie Electronique, Paris, 1970), 3 Vols.

HVEM Stockholm, 1971: The Proceedings of the Second International Conference on High-Voltage Electron Microscopy, Stockholm, 14–16 April, 1971; published as *Jernkontorets Annaler* **155** (1971), No. 8.

Manchester, 1972: *Electron Microscopy 1972. Proceedings of the Fifth European Congress on Electron Microscopy*, Manchester (Institute of Physics, London, 1972).

HVEM Oxford, 1973: *High Voltage Electron Microscopy. Proceedings of the Third International Conference*, Oxford, August, 1973, P. R. Swann, C. J. Humphreys, and M. J. Goringe, eds. (Academic Press, London and New York, 1974).

Canberra, 1974: *Electron Microscopy 1974. Abstracts of Papers Presented to the Eighth International Congress on Electron Microscopy*, Canberra, J. V. Sanders and D. J. Goodchild, eds. (Australian Academy of Science, Canberra, 1974), 2 Vols.

HVEM Toulouse, 1975: *Microscopie Electronique à Haute Tension. Textes des Communications Présentées au 4e Congrès International*, Toulouse, 1–4 Septembre, 1975, B. Jouffrey and P. Favard, eds. (SFME Paris, 1976).

Jerusalem, 1976: *Electron Microscopy 1976. Proceedings of the Sixth European Congress on Electron Microscopy*, Jerusalem, D. G. Brandon (Vol. I) and Y. Ben-Shaul (Vol. II), eds. (Tal International, Jerusalem, 1976), 2 Vols.

HVEM Kyoto, 1977: *High Voltage Electron Microscopy 1977. Proceedings of the Fifth International Conference on High Voltage Electron Microscopy*, Kyoto, 29 August to 1 September, 1977, T. Imura and H. Hashimoto, eds. (Japanese Society of Electron Microscopy, Tokyo, 1977); published as a supplement to *Journal of Electron Microscopy* **26** (1977).

Toronto, 1978: *Electron Microscopy 1978. Papers Presented at the Ninth International Congress on Electron Microscopy*, Toronto, J. M. Sturgess, ed. (Microscopical Society of Canada, Toronto, 1978), 3 Vols.

The Hague, 1980: *Electron Microscopy 1980. Proceedings of the Seventh European Congress on Electron Microscopy*, The Hague, P. Brederoo and G. Boom (Vol. I), P. Brederoo and W. de Priester (Vol. II), P. Brederoo and V. E. Cosslett (Vol. Ill), and P. Brederoo and J. van Landuyt (Vol. IV), eds. Vols. I and II contain the proceedings of the Seventh European Congress on Electron Microscopy, Vol. Ill those of the Ninth International Conference on X-Ray Optics and Microanalysis, and Vol. IV those of the Sixth International Conference on High Voltage Electron Microscopy (Seventh European Congress on Electron Microscopy Foundation, Leiden, 1980).

Hamburg, 1982: *Electron Microscopy 1982. Papers Presented at the Tenth International Congress on Electron Microscopy*, Hamburg (Deutsche Gesellschaft fur Elektronenmikroskopie, Frankfurt, 1982), 3 Vols.

HVEM Berkeley, 1983: *Proceedings of the Seventh International Conference on High Voltage Electron Microscopy*, Berkeley, 16–19 August, 1983, R. M. Fisher, R. Gronsky, and K. H. Westmacott, eds. Published as a Lawrence Berkeley Laboratory Report, LBL-16031, UC-25, CONF-830819.

Preface by Ernst Ruska

When electron microscopy came into being more than fifty years ago, light microscopy, already in existence for some 300 years, had long since attained a high degree of perfection. During those centuries, it had benefited numerous branches of science decisively while others had the light microscope to thank for their very existence. The scientists working in these fields were often conscious of the fundamental role of the microscope in their work. For some fifty years, the limit imposed on the resolution attainable with the light microscope by the wavelength of the light waves was likewise known. Considering that the waves associated with electrons would be some five orders of magnitude smaller than light waves, the development of an electron microscope seemed to offer the possibility at least of forming images of considerably better resolution than those the light microscope could provide. In a publication of 1932 it was already established that, even considering the technological difficulties to be expected, resolutions better than those of light microscope images could be anticipated.

With such a prospect, it must have seemed certain that the champions of microscope research in the various fields would welcome the development of the electron microscope so enthusiastically that the pioneers of the time in various countries would in turn receive generous financial assistance in a reasonably short time. But spontaneous support of this kind from microscope users scarcely materialized, and the financial means of the appropriate industries were not put at the disposal of the pioneers, for want of proof that there was adequate demand for such novel microscopes. The reasons for this lay certainly, to a considerable extent, in the serious reservations which obscured the real possibilities of electron microscopy for many years. Moreover, these reservations were perfectly comprehensible. An objection that was particularly difficult to refute was that the specimen preparation would be heated much too strongly by the electron energy absorbed within it. Nevertheless, it was found quite soon (1933/1934) that image contrast could arise not only from spatial variations in the absorption of electron energy but also from the difference in electron scattering from one point to another, scarcely any energy then being deposited in the specimen and converted into heat.

This knowledge was particularly important for the development of the electron microscope, which was just beginning, because it became reasonable to hope that specimens could be examined at high magnification but at a low enough temperature for there to be no danger of chemical modification. It also became clear, however, that if very little thermal energy was to be transferred to the specimen, and hence the image contrast created essentially by scattering, the thickness of the specimen would have to be almost a hundred times less than that attainable with the microtomes then used for light microscopy. The hope that it would one day be possible to cut such thin sections seemed as utopian as that of achieving electron microscope resolutions far better than that of the light microscope. Indeed, this problem severely limited the choice of objects for study in the electron microscope for some twenty years. Only with the successful development of the ultramicrotome, and also of other specimen-preparation techniques which permitted at least some kinds of objects to be obtained in sufficiently thin sections, did the electron microscope rapidly reveal itself to be useful in almost every discipline in which the light microscope had hitherto been an indispensible tool. Indeed, the widespread use of the latter had likewise been hampered, for some 200 years, by the lack of achromatic objective lenses. These did not come into use until 1825.

When writing this preface, I had not yet seen the material collected together here, contributed by those international pioneers of electron microscopy who are still living. I suspect, however, that in at least some of these recollections, as in my own conclusions, will figure the absence of encouragement on the part of experienced light microscopists and also, of course, the resulting doubts that tormented the pioneers themselves.

The technical development of the electron microscope consists almost exclusively of a struggle against the *bad* properties of electrons. Almost every property of these resembles, when we consider its influence on the development of the instrument, a coin, which notoriously possesses two sides. The property that is fundamental for microscopy, the shortness of the wavelength, is counterbalanced by the associated high energy. The latter also has advantages—for example, even highly magnified images remain visible on the fluorescent screen—but at the same time the decisive disadvantages of heating the specimen and generating X rays. The fact that electrons are deflected in magnetic fields is indispensible for the construction of good enough magnetic lenses but also created an obstacle to the achievement of the very highest resolution that has proved difficult to surmount. It was essential to hold the accelerating voltage and the lens currents extremely constant and to provide extremely good screening of the interior of the microscope against parasitic magnetic fields. Apart from the technical problems arising from the physical properties of electrons, other difficulties had to be overcome in the development of the electron microscope—the design of the specimen holder mechanism, for example. This and other mechanical difficulties resulted from the high magnification that must be attainable when the highest resolution is to be achieved.

The ever-increasing degree of perfection in the solutions of such problems as these, together with the laborious progress of specimen-preparation methods, has made the electron microscope a universal tool, taking it far beyond the imaginings of the pioneers of "the beginnings of electron microscopy."

Ernst Ruska

Afterword by Peter Hawkes

1937

1941

Figure 1 Ernst Ruska in 1937 and 1941. Courtesy of the Siemens Archives.

Ernst Ruska was born on Christmas Day 1906 in Heidelberg. I can add nothing to his book mentioned in the Foreword to the first edition and to the full biography by Ruska's secretary, Lotte Lambert and Tom Mulvey (Lambert & Mulvey, 1996). Frau Lambert's earlier article should also not be missed (Lambert, 1986). An account of the role of Ernst Ruska's brother Helmut by Hans Gelderblom and Detlev Krüger (2014) is also very relevant.

Ruska was awarded a half-share of the Nobel Prize in 1986, which produced the following reaction by Judith Reiffel, secretary of Elmar Zeitler, Ruska's successor at the Institute of Electron Microscopy in the Frtitz-Haber-Institute in Berlin and founder-editor of *Ultramicroscopy*:

> *"Just after your call the telephone rang again, and didn't stop ringing until five minutes ago. Because Ruska got the Nobel prize, sharing it with Binnig and Rohrer. Isn't that a gas? Ruska is on holiday, and the BZ* (Berliner Zeitung) *(yellow journal) sent a helicopter and found him taking a walk and schlepped him back to the hotel, where they interviewed him. Jesus, it's a good thing we've got a festschrift in press".*

Ernst Ruska's Nobel Lecture contains a wealth of details about his early years (Nobel Prize website and Ekspong, 1993); see Fig. 1.

Ernst Ruska died on 27 May 1988.

References

Ekspong, G. (Ed.). (1993). *Nobel Lectures in Physics 1981–1990*. Singapore: World Scientific.

Gelderblom, H. R., & Krüger, D. H. (2014). Helmut Ruska (1908–1973): His role in the evolution of electron microscopy in the life sciences, and especially virology. *Advances in Imaging and Electron Physics*, *182*, 1–94.

Lambert, L. (1986). Some anecdotes around and with Colonel Ernst August Ruska. *Ultramicroscopy*, *20*, 337–340.

Lambert, L., & Mulvey, T. (1996). Ernst Ruska (1906–1988). Designer Extraordinaire of the electron microscope: A Memoir. *Advances in Imaging and Electron Physics*, *95*, 2–62.

CHAPTER ONE

Recollections from the early years: Canada–USA

Cecil E. Hall[a] and Peter Hawkes (Afterword)[b,*]
[a]Formerly Jasper, AR, United States
[b]CEMES-CNRS, Toulouse, France
*Corresponding author. e-mail address: hawkes@cemes.fr

Contents

1. Introduction 1
2. Initial work: University of Toronto (1935–1936) 3
3. Kodak Research Laboratories, Rochester, New York (1937–1941) 7
4. Massachusetts Institute of Technology 13
   4.1 Note on the war period 15
   4.2 Electron stains 16
   4.3 Paramyosin 17
5. Transition from the early years 19
   5.1 Shadowcasting 19
   5.2 Stigmator 20
   5.3 Microtomy 21
6. Afterword by Peter Hawkes 22
References 22

1. Introduction

A student of physics entering graduate school in 1935 was fortunate in many ways. Advances in physical research would soon lead to enterprises of great consequence. There were disadvantages too. Economic conditions were in a severe state of stagnation. On the plus side, the quality of teachers and students in those days was excellent, probably never better. Furthermore, adversity was accepted cheerfully as a normality and on this scale morale was high.

At the University of Alberta we had good classical physics, optics, and spectroscopy, and in the Mathematics Department, a professor who covered theoretical physics at a whirlwind pace through the Dirac theory of the electron. The Electrical Engineering Department offered an interesting course on "Radio Engineering" that was a pioneering step into that

Advances in Imaging and Electron Physics, Volume 221
ISSN 1076-5670
https://doi.org/10.1016/bs.aiep.2022.03.001

technology at the time. My leaning for graduate work might have been described as toward "Physical Electronics" or more specifically as "Optics of Ultrashort Radio Waves," in other words, the field that would later simply be called "electronics." My first choice was to go to England and work with Professor Appleton, who was measuring distances by means of reflected radio waves. This was also the suggestion of the head of the department. I submitted an application for the only available scholarship for foreign study, giving this as the reason. It was not a good enough reason. The board of judges did not know what I was talking about. In retrospect, if other reasons had been advanced, which was possible, they might have worked and still achieved the desired result. (Professor Appleton was awarded the Nobel Prize in Physics in 1947 for his pioneer work in what became known as "radar.")

An alternate application was successful at the University of Toronto and I entered the Department of Physics there in September, 1935. For a few weeks I was assigned to a young instructor, Arnold Pitt, who was proficient in electronics. A few years later he made important and inventive contributions to military applications such as the proximity fuse and radar. At the time there was little or no interest or support for research in this area. Discussions with other faculty members followed.

Professor E. F. Burton, head of the department, brought up the subject of electron optics and the electron microscope. His interest had been stimulated by Dr. Walter H. Kohl, a German engineer–physicist who had come to Canada in 1930 and was engaged in developmental work with a radio manufacturer in Toronto. Dr. Kohl held a doctorate from the University of Dresden, where his work had been in photoelectricity and secondary emission. He became a guest lecturer in electronics at the University of Toronto and in a lecture on April 9, 1934, he showed a cathode-ray (CR) tube in which he had modified the electrodes so as to produce a magnified image of the oxide-coated cathode on the face of the fluorescent screen. This was no doubt the first demonstration of electron imaging in Canada and perhaps in North America. Although no experimental work had been done at the university, this kind of research attracted me.

The opinion of other graduate students was largely negative. They believed that electron optics was too ill-defined, although it might have some application in the future "if television ever became practical." This fitted my thinking too, except my outlook was more optimistic. I think the main aspect that frightened them away was the fact that Professor Burton voiced the strange conviction that an electron microscope could be built that could be used in biological and medical research. He had not mentioned this in our conversations but I could see how such a goal would discourage a budding physicist who wanted the results of his efforts to be a little more predictable. It seemed to me, however, that a transmission electron microscope could be built and used for the observation of colloidal particles and that the experience gained would be an asset in many other physical applications. The conservative attitude that prevailed at the time might have been a result of the hard economic conditions, or it might have been partly the cause. The development of new technologies eventually provided the forces that broke the economic stagnation.

My evaluation of the problem was as follows: An emission microscope could easily be built in a short time with very little financial outlay and a few interesting experiments could be performed. A high-resolution transmission electron microscope could also be built, but this would take time and a fairly large amount of cash for components and precision machine work. Results in the study of colloidal particles could be guaranteed. Biological substances were a different matter. If an application in medicine could be demonstrated even in a small way, we could reach a source of financial support in which our financial needs would be counted as trivial. First, we had to build a microscope and show results. We were physicists. That was our job. We should be able to do it ourselves. Another thing I liked about the venture was that the whole thing could be grasped by one person alone or at the most by two or three individuals working closely together. I decided to take up the task.

2. Initial work: University of Toronto (1935–1936)

The problem at Toronto was clear from the start. We would have to produce some results quickly with materials and facilities at hand in order to demonstrate a capability in electron optics. Then we could seek financial support for more substantial work. Finances are not the only requirement for a viable project in an academic environment. There must be a scientific base and it must be sufficiently promising to attract enterprising students in succession.

As a start, I made a demountable emission microscope with a three-electrode electrostatic lens close to the cathode. Such a lens is classified as an "immersion lens" because the object to be imaged (the cathode) is in a strongly refracting field. The system was enclosed in a horizontal metal cylinder 30 cm long and 15 cm in diameter. The vacuum was maintained

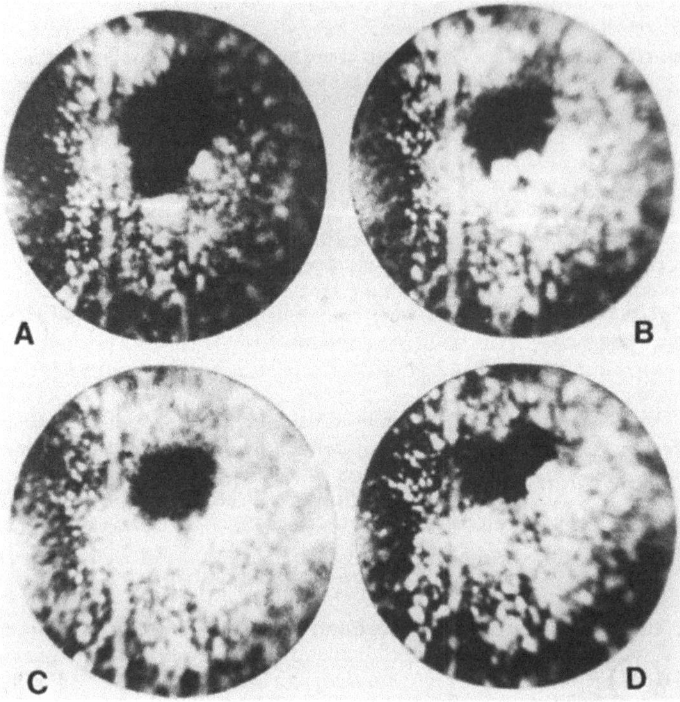

Figure 1 Emission electron micrographs of a BaO-coated cathode made with the Toronto microscope built in 1935. Sequence of pictures (A–D) shows effects of allowing the pressure of the vacuum to rise and produce positive-ion bombardment effects. Diameter of the field shown is 0.4 mm.

with a mercury-vapor pump and a conventional oil forepump. The first electron micrograph was made of the fluorescent screen with a 35-mm Leica camera in December, 1935.

This microscope was similar to one described previously by Johannson (1933, 1934). A wide range of electrode voltage ratios can produce focused images on the screen with various magnifications and useful image areas. Measurements were made of these characteristics. Electrode spacings could be varied externally and measured through a window with a traveling telescope. The work was recorded in a thesis at the university but was not submitted to a journal. Some of the data were published later in the first edition of my textbook (Hall, 1953). Some of the pictures are shown in Fig. 1 of a BaO-coated cathode after the vacuum pressure was allowed to rise purposely to produce positive-ion bombardment, which makes the dark crater. Emission varied actively as the internal pressure was raised and lowered and the reduced Ba migrated about the surface.

In May, 1936, it was time to evaluate the situation and decide on the next step. From a physicist's point of view there were several interesting avenues of research open, such as studies of thermionic, photoelectric, and secondary emission. For biological electron microscopy, an emission system seemed to be limited to ashed or carbonized specimens. It is worth noting that later (1939), Scott and Packer (1939), working in St. Louis, Missouri, built an emission microscope with magnetic magnifying lenses and were able to demonstrate the location of Ca and Mg naturally occurring in muscle and other tissues. Our goal had to be a high-resolution transmission electron microscope. We knew from the work of Knoll and Ruska (1932) and Ruska (1934) that such an instrument could be built, but we did not know whether it could be used for serious work in biology. Nobody seemed interested in its application to other materials, not at least to the point of providing substantial support. The money required could be estimated only roughly, mainly due to uncertainties as to what would be charged to existing facilities, student stipends, and so forth. My estimate came out at $3000 as the minimum to get moving and at $5000 as sensible. I did not bother to submit these figures to Professor Burton because I knew they were impossible to obtain from the government bureau with which he was negotiating. Then too, we had no hard supporting evidence.

In September, 1936, I decided to put together the simplest possible magnetic microscope for the sole purpose of experimenting with specimen technique and getting some pictures of biological material. I predicted that a resolution of 200 Å (2×10^{-6} cm) should be easy to achieve—and that is 10 times better than the best light microscope. It was a horizontal arrangement of components, essentially an optical bench. An electron gun produced a beam along a tubular brass casing of small diameter along which electron lenses could be moved externally. This led to a larger tube which ended in a fluorescent screen. The system was never completed as planned but some preliminary experiments were carried out. There was a good 5000-V dc generator, a pumping system, and lens current supply. With crude lenses a minimum focal length of about 5 mm could be obtained. The magnification was about 5000× at the screen. Opaque objects, fine wires with small particles sticking to the sides, were sharply imaged when the incident beam was well collimated. Estimates made on sharp projections and edges indicated a resolving power in the range 200–400 Å. Penetration of the beam through the thinnest specimens tried was negligible.

The plan was next to improve the components of the optical bench and most important of all to obtain a high-voltage supply which should reach at least 50 kV. Money would be needed for the latter. At this stage (December, 1936) Professor Burton could claim that we had constructed an electron microscope with magnetic lenses which could produce magnifications of about 5000× and had a resolving power superior to that of the light microscope.

During the Christmas vacation I went to Rochester, New York, to visit a friend who was employed in camera design with the Eastman Kodak Company, and he arranged a visit for me to see the research laboratories. There I talked with people who were working in optics and electronics and found there was interest in building an electron microscope. At the end of the afternoon I was offered a position on the research staff, which I accepted a week later, after returning to Toronto and talking with Professor Burton. I took residence in Rochester February 1, 1937.

I regretted leaving Professor Burton at this point but he was cheerful and indomitable and, I think, rather pleased that such a highly esteemed laboratory would want one of his students. Others of his students were to leave in succession after me, and his only comment was to say proudly, "They all obtained good appointments."

Although I had no further participation in the work at Toronto, it is pertinent to interject a few notes here on what happened there subsequently, because these events had important bearings on the course of my future work. My efforts while I was there did serve some purpose. They enabled Professor Burton to raise a pitifully small amount of money which was barely enough to keep the project alive. Yet, the first transmission electron microscope on the North American continent was completed at Toronto, in spite of the lack of funds, in a notably short period of time (Burton et al., 1939). The work was begun in the fall of 1937 by two graduate students, Albert Prebus and James Hillier (1939). The story of this achievement from conception to successful conclusion with behind-the-scenes commentaries has been recorded by U. M. Franklin, G. C. Weatherly, and G. T. Simon in a brochure commemorating the fortieth anniversary of the event (Franklin et al., 1978). This was in conjunction with the Ninth International Congress on Electron Microscopy at Toronto in August, 1978. Prebus and Hillier did not follow my scheme of working up to a good instrument gradually. They decided to go straight for a high-resolution microscope with whatever they could beg or borrow and with what they could do with their own hands, mostly the latter as it turned out. Their personal comments are recorded by Franklin (Franklin et al., 1978).

3. Kodak Research Laboratories, Rochester, New York (1937–1941)

Work at the Kodak Laboratories, mostly in optics and electronics, was interesting and enjoyable, but authorization to construct an electron microscope was slow in coming. Dr. C. E. K. Mees, vice president and director of the laboratories, was a pioneer in the science and arts of photography as well as the founder of the laboratories with George Eastman, the founder of the company. Dr. Mees, who was friendly with Professor Burton and apparently inspired from this direction, was in favor of getting the work started, but there was considerable skepticism and disbelief among his senior staff members concerning the practicality of such a device for the observation of photographic materials. An obvious objection was that sensitive materials would disintegrate under electron "bombardment," which certainly is the case. It was not until the group at Toronto had built their instrument and begun to show its capabilities that I was authorized to proceed.

It is interesting to recall how Dr. Mees, who was a master of administrative diplomacy as well as science, finally moved the project forward. He went directly to the board of directors and requested a personal allotment, separate from the general research laboratory appropriation. His stated purpose was to build an electron microscope which was to be used for the study of gelatin, an essential ingredient in photographic emulsions. This material has strange properties, depending on its origin, and its effects on photographic products were poorly understood. The possible use of the electron microscope for the study of the other components in the emulsion was not mentioned in the request. I knew nothing of these matters at high levels of policy and first heard about it when Dr. Mees told the story to a group of administrative people after the microscope had been completed and had been producing interesting results on silver halide crystals for some time. It is just as well that I did not know about the gelatin idea because what little knowledge I had of the substance was all unfavorable. It is made by breaking down the native structure of cowhide by drastic physical and chemical means to produce an extract containing very small organic molecules, or, rather, assorted pieces of molecules. It has some interesting aspects, but that came later.

Figure 2 Early transmission electron microscope, built in Kodak Research Laboratories, Rochester, New York. (Safety shields removed for photograph.)

The Kodak research electron microscope is shown in Fig. 2. It has the standard components and was designed and tested for operation well above 100 kV with a resolving power in the 10–20 Å range. The original intention was to shorten the column and install better manual control systems after testing, but it worked so well from the start that there was continued use and no time for refinements before I left for MIT. There are similarities between the instrument and those of Ruska and the Toronto group, but a few points in the design should be mentioned. The sections were stacked on their ground flat surfaces, which were sealed with high-vacuum lubricant so that they could be aligned laterally. The magnetic lenses were water cooled. The objective aperture could be centered independently. Specimens were inserted through a lock with a 1-to 2-min cycle. A plate camera held 20 2 × 2-in. photographic plates and required about 5 min for pump-down. An important improvement in the vacuum system was made possible by the invention of high-speed oil diffusion pumps by Dr. K. C. D. Hickman, also at the Kodak Laboratories. Backed by a fast forepump this type of diffusion pump came as a most welcome relief from the slow, troublesome, and poisonous mercury systems of the past.

Currents (dc) for the magnetic lenses were stabilized by feedback electron-tube circuits. High voltage (HT) was obtained from a 60-cycle transformer (donated by our X-ray department) with an output capacity far in excess of anything I would use. The HT from the transformer was fed through condenser input to the rectifier circuit. The ac input to the primary was from a line-voltage regulator. A number of commercial models were tested for performance with electronic equipment. Most were unsatisfactory. One was found with good waveform and other characteristics that was very good for the purpose. Large capacitors and low current drain were also an important factor in maintaining a steady HT for the EM beam.

The EM filament supply was a heavy-duty 4-V storage battery, housed in an isolated metal container and controlled through an insulated shaft. All this was rather bulky but could be housed in a room not much larger than is required for a "modern" instrument and maintenance problems were nil. The transformer and other ac components were at good distance from the EM, which was not always the case in later "office equipment style" instruments. While I was building, a new, high-permeability alloy, Mumetal, had become available and was used to shield the beam from stray fields along its path.

When the question of locating the EM arose, everyone seemed to assume that I would want to put it in a basement or sub-basement. That is where, traditionally, physicists are put or choose to go, often with good reason. It seemed to me that such isolation would set a bad precedent for electron microscopists. I chose a place in our own area, distant from the elevator shaft. Thus, one could remain a part of the community, with a "Danger High Voltage" sign on the door to discourage casual visitors. The location had another unexpected advantage. Dr. Mees could come from his office two flights up, down a remote unused stairwell, without traversing the busy traffic of the building.

The cost of the EM, including outside purchases and inside charges such as machine shop time but not including my time or the HT transformer, was very close to \$4000. The time from writing requisitions and making construction drawings to production of electron micrographs of the specified quality was about 4 months.

During the construction stage, an advertisement appeared for a new commercial product called "Lektromesh" screen (electroplated mesh), which is now a commonplace material, but at the time it was the answer to a prayer. Specimens for electron microscopy have to be supported on thin, fragile films and these in turn were supported, in the early technique, over small openings drilled in metal disks or over rather coarse woven mesh, both of which were troublesome and unsatisfactory. The electroplated material was the answer. It was flat and could be made with a wide variety of hole sizes. One specimen disk provided a large number of openings to be scanned so that experimentation with specimen materials was speeded up enormously.

The first test specimens were made of gold sols and silver halide crystals in gelatin. Halide crystals are decomposed, as expected, by the beam (Fig. 3). When they are exposed to light and put through the standard development procedures, the reduced silver appears as shown in Fig. 4. The unexpected filamentous texture of the silver produced by standard developers clarified the mechanism of developer action. Silver halide crystals are ionic conductors. Ag^+ ions may be freed to move through the crystal lattice. This release can occur with the help of energy from an absorbed photon. Such wandering ions tend to collect on imperfections at the surface of the lattice called "sensitivity spots." The ions of the developer solution can then penetrate the distorted electric field at these spots to donate electrons, which attract more silver ions until the entire crystal is reduced to silver. With fast development, the accumulation of silver at the spot is too rapid for growth inward, so it is extruded outward in ribbons. This is best illustrated in the case of very small AgBr crystals, for which there is, on the average, only one sensitivity spot per crystal (Fig. 5). Development results in a single ribbon of silver. Prior to these observations there were conflicting theories regarding the mechanism. One of these, favored by chemists, was that somehow the developer first dissolved the AgBr crystal and then precipitated the silver from solution. That concept is out for the standard type of development. There is, however, a humorous aspect to the situation. The process we have described is for conventional black and white development and is called "chemical development." It is really a physical picture and fits

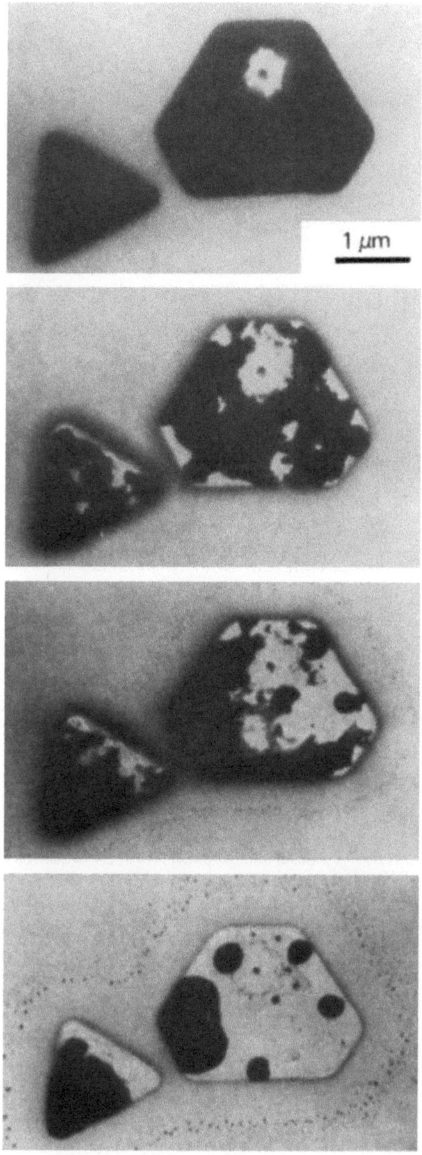

Figure 3 Crystals of AgBr in gelatin, from a photographic emulsion. Sequence from top to bottom shows effects of exposure to electron beam. HT = 75 kV.

perfectly into the physical theories of Dr. J. H. Webb (1940), who worked in the same laboratory and later became its director. There are other ways to develop a photographic image which do involve solution and reprecip-

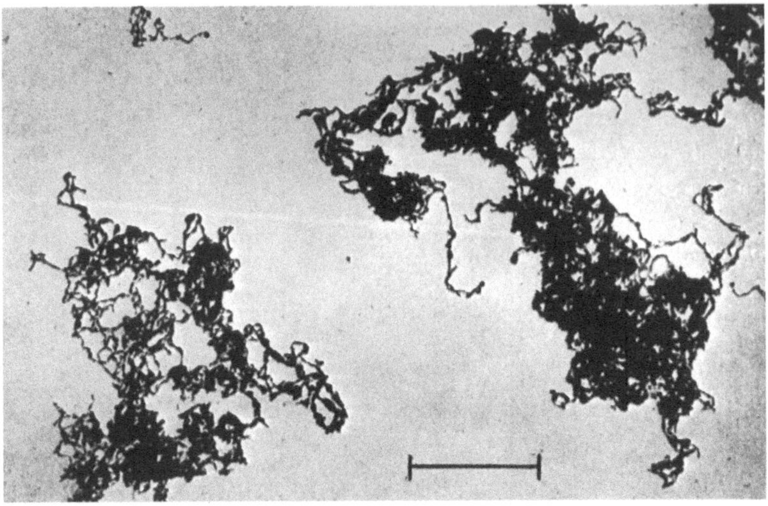

Figure 4 Filamentous silver reduced from silver halide crystals by the action of a standard photographic developer. Bar = 1 µm.

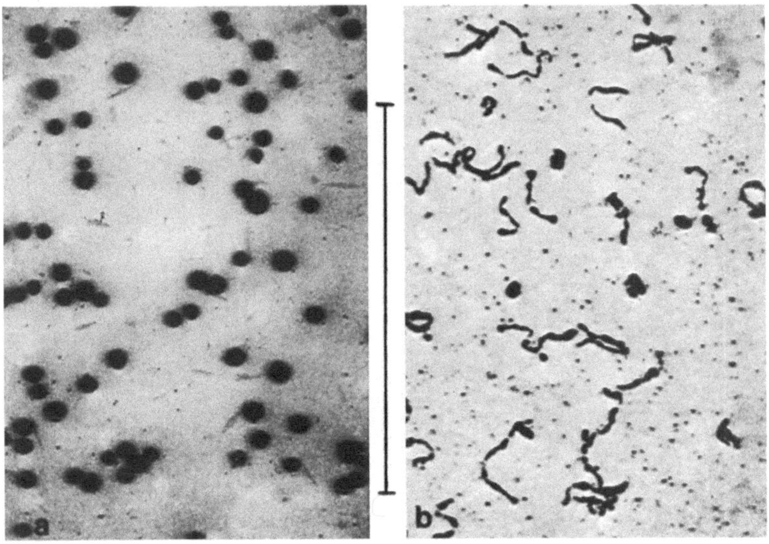

Figure 5 (a) Very small silver halide crystals. (b) Each crystal contains, on the average, only one sensitivity spot and so reduces to only one silver filament. Reductions of larger crystals as in Fig. 4 contain several sensitivity spots and reduce to clumps of filamentous silver. Bar = 1 µm.

itation of the silver. This is called "physical development." The deposit is amorphous or in crystalline clumps.

A summary of these applications of electron microscopy to the study of photographic materials was published (Hall & Schoen, 1941) with A. L. Schoen as coauthor. "Art" Schoen was the administrative leader of the small group within the Physics Department of which I was a member. He was originally a school teacher but had become a long-time member of the research laboratories, specializing in applications of scientific advances to problems in manufacturing. He had become very good at foreseeing such opportunities. Electron optics and the electron microscope had been on his list before my arrival. We already had two of V. K. Zworykin's first infrared image tubes (Zworykin & Morton, 1936) for observations in "photographically dark" situations. Thus, when the decision was made to build an electron microscope we were prepared to move quickly.

Dr. Mees made separate publications of these research results describing their significance for the broad subject of photographic phenomena. Noteworthy is his speech "Recent Advances in the Theory of the Photographic Process," which he presented before the Society of Motion Picture Engineers at their Spring Meeting at Rochester, New York, in 1941 (Mees, 1941). In this communication he describes with characteristic clarity the ways in which electron microscopy contributed to our understanding of the processes that occur between the absorption of photons and the development of an observable photographic image.

At about this time, spring of 1941, I was asked to join the faculty of The Massachusetts Institute of Technology and work with a group in the Department of Biology which was being organized to investigate the potential of the electron microscope for the observation of biological materials. I accepted and reported there in July, 1941.

4. Massachusetts Institute of Technology

In 1941 the Department of Biology at MIT was to be expanded in recognition of the growing relationships that were developing between the life sciences and the physical sciences, in light of the advances in physics, chemistry, and engineering. This reorientation toward the future biology was initiated by Dr. Karl T. Compton, president of MIT at the time. The Rockefeller Foundation for Medical Research awarded a grant in support of the general program and, in addition, a separate fund earmarked specifically for the exploration of the potentialities of electron microscopy as applied to biological materials. The EM was becoming known through published articles and a manufactured model (Hillier & Vance, 1941) was soon to become available for purchase in the United States at a fairly modest price (RCA; about $9500). The question as to whether this device could be used for a broad range of applications in biology had by no means been answered.

The new head of the Department of Biology was Professor Francis O. "Frank" Schmitt, who had early in his career become noted for his work on molecular (or macromolecular) structures in biological systems at Washington University in St. Louis. In addition to his other activities and duties, he was to be an active part of the EM trio, which also included Marie A. Jakus, who had been working as assistant and graduate student with Professor Schmitt in St. Louis. We received one of the first RCA Type B electron microscopes (Hillier & Vance, 1941) in September, 1941, and were soon at work.

Our initial class of materials chosen for investigation with the EM may be designated as "protoplasmic fibrils." Fibers and fibrillar structures occur in a wide variety of types, compositions, and functions throughout nearly all living substances. Common examples are tendons, muscle, cilia, flagella, chromosomal forms, and many others. A discussion of these materials and the outlook at the time is provided in a paper by Schmitt et al. (1943). Correlations were looked for between observations by X-ray diffraction methods and electron microscopy. The range of dimensions covered by the two methods overlaps. Professor Richard S. Bear, a chemist turned X-ray diffractionist, had also joined the department and was improving his X-ray diffraction techniques for observation at very small angles wherein the dimensions of the diffracting structures should be well within the resolving power of the EM.

The first material of this class chosen for study was collagen, the main substance of tendon, skin, and "connective tissue" in general. It is the binding and fabric stuff that holds us and many other living things together. Its fibrous nature may be broken down by heat to produce gelatin. Previous work had shown that the smallest visible fibrils could be dissolved in acetic acid to produce solutions which show strong flow birefringence. This indicates the presence of long molecules unresolved by light microscopy. Reassembly of the molecules into fibrils may be accomplished by addition of salt. The process is reversible. Our first efforts were to observe the unaggregated molecules and their reassembly into fibrils. We could not see the molecules. (This was in 1941, before we had effective staining or shadowcasting techniques.) We could easily see the salt-precipitated fibrils but they showed no significant structure. We drew a blank from this approach.

Meanwhile, Bear was examining intact collagen fibers with his improved low-angle X-ray camera. Previously, there had been reported long spacings of the order of 200 Å in the direction of the fiber axis, but Bear showed that these were all orders of a fundamental period of 640 Å which he resolved in the first order. This was quite an achievement when one considers that the X-ray wavelength is only 1.54 Å and there is a lot of diffuse scattering near the center of these patterns. It was also a challenge to the electron microscopists. So far, we had seen nothing, though we claimed resolving power at least 30 times better than 640 Å. Of course, the discrimination of structural irregularities by X-ray diffraction is not quite the same as the distinctions produced in electron micrographs by electron beam interactions. Still, one would expect to see *something* when the X-ray intensities showed up so well. In other words, the electron microscopist could find excuses but they all meant failure.

We had started this work on biological material with the intention that we would not follow the course of employing existing histological methods used in light microscopy. We would start at the physical and chemical base and work upward toward the larger, more complex systems, devising our own methods as we progressed. We were aware that we would probably create our own kinds of "artifacts" just as the classical microscopists had been doing for many years. We did not intend to borrow their concoctions (unless, of course, we found them useful). In the long run our approach proved correct but on this first encounter with collagen, our preoccupation with the physicochemical approach had caused us to skip an obvious, preliminary look at the untreated raw material that any rank amateur would have done at the beginning. This was now done. Marie took some small fragments of raw rat tail tendon in weak acetic acid and "teased" it with a pair of needles on an electron microscope specimen support. We saw mostly large, opaque areas as expected and, in between, we saw areas as shown in Fig. 6. There is a succession of light and dark bands in the direction of the fibril axis with an average distance between like bands of 640 Å. This was confirmed also for collagen from beef tendon and ligaments and human skin. Bear's results and ours were published early in 1942 (Bear, 1942; Hall et al., 1942; Schmitt et al., 1942).

4.1 Note on the war period

Shortly after we began the work in the fall of 1941, the United States was drawn actively into the war. Our activities were reoriented accordingly. Work on collagen and connective tissue, mostly of a biochemical nature,

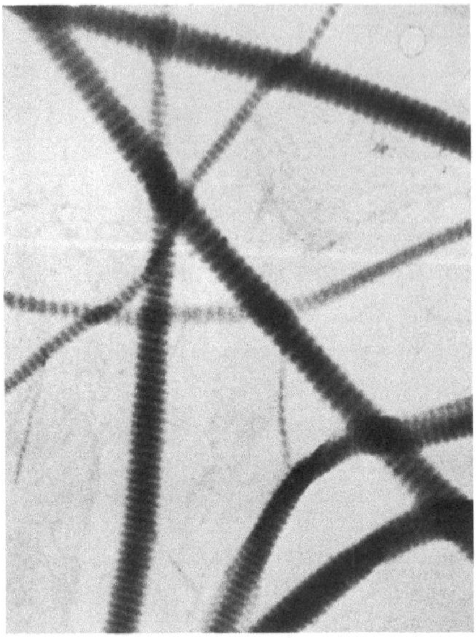

Figure 6 Fibrils of collagen from rat tail tendon, unstained. The length of the periodicity of light and dark bands is 640 Å (average).

had applications to a range of medical problems. Professor Schmitt organized a group to work on the medical aspects and spent much of his time on these efforts. The electron microscopy laboratory was employed in a variety of tasks, principally in the field of colloidal materials. A second RCA Type B EM was added to take care of extra work. The initial purpose of basic biological research was kept moving at a reduced rate.

4.2 Electron stains

The collagen results were gratifying in that we observed a molecular structure in unstained material that was confirmed by X-ray observations on intact dry or wet materials. We had a biological material with a characteristic molecular structure which we could disaggregate into its molecular constituents with mild physicochemical methods and reaggregate reversibly. We soon discovered that our initial failure to observe the characteristic period in salt-precipitated fibrils was due to interference by the salt. Other methods of reprecipitation revealed the period. The main problem that we faced now was that the X-ray pattern showed that there must be considerable fine structure within the 640-Å period that we could not see in our electron micrographs. We had found, however, a good test material on which to experiment with staining procedures which might render visible organic, low-contrast structures that were evidently within the resolving power of our microscope. Before we could begin these experiments with collagen, however, we were diverted to a study of a different fibrous protein which gave us an impressive demonstration of what we were seeking. This material, which we named "paramyosin," came from the adductor muscle of the clam, *Venus mercenaria*, which is the muscle that pulls the two halves of the shell together against the opening spring tension of the hinge.

4.3 Paramyosin

The adductor muscle of *Venus mercenaria* is a white translucent material which is easily dispersed in KCl solutions. A light centrifugation removes the heavy debris, leaving a suspension which exhibits a sheen when agitated and a pronounced flow birefringence when stirred between crossed Polaroid sheets. The long, suspended particles are brightly visible with the dark-field microscope even though the widths are below its resolving power. There are similarities between the solubility and optical behavior of these particles and the muscle protein myosin, but we doubted that the two were identical and named the clam muscle extract paramyosin.

Bear examined samples from the adductor muscle by X-ray diffraction and found a series of strong reflections on the meridian which indicated a spacing along the fiber axis of 145 Å. There were also reflections off the meridian in layer lines that indicated a longer repetition of 5 × 145 Å, or 725 Å. There was also a regular periodicity in a direction perpendicular to the axis. We observed paramyosin fibrils from the KCl suspensions with the EM. We could see them clearly but we could see no internal structure corresponding to the X-ray spacings. It was time to try attaching heavy atoms to enhance the scattering power of the structure which must be there.

One of the first reagents we tried was osmium tetroxide (OsO_4), which had often been used in biological reagents. We found some fibrils showing an axial spacing with a dimension that did not correspond to the diffraction pattern. This may have been another component (Jakus et al., 1944). We then began testing a list of compounds containing Pt, U, and so forth. There were three or four kinds of phosphotungstic acid on the shelf because these were sometimes used in classical techniques. The word "acid" was enough to steer us away from them in the beginning because acids are generally destructive. In buffered solutions, as we eventually used them, the reagents should probably be referred to as "phosphotungstates." The first specimen to be made with this kind of stain was prepared by Marie late one afternoon.

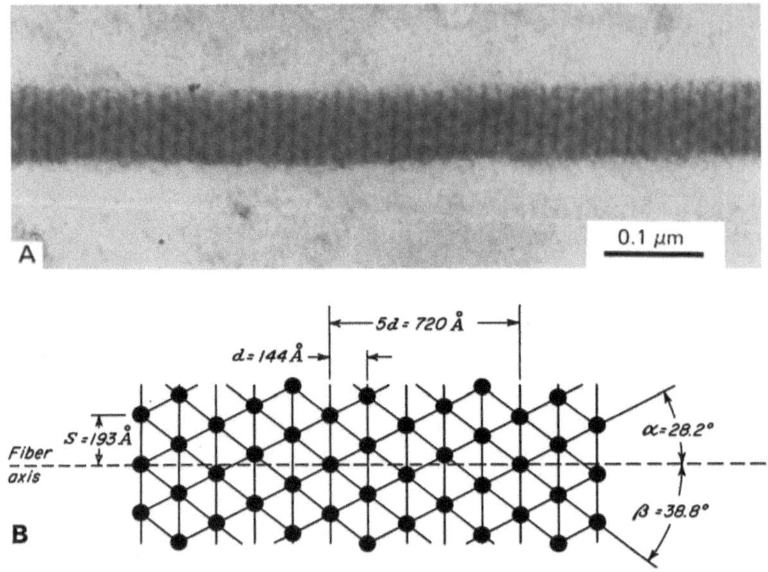

Figure 7 Fibrils of paramyosin from clam muscle (*Venus mercenaria*), stained with phosphotungstic acid (A). The geometry of the stained pattern is represented diagrammatically (B).

Micrographs were recorded through-focus, on a plate, and we never knew from observation of the fluorescent screen what we had until the plate was processed and could be observed with a magnifier. When the plate was processed the next morning, it took only a few seconds of observation with the magnifier to see that we had the answer. I had been looking at Bear's diffraction patterns (he was having difficulties in the structural interpretation) so often that the picture (Fig. 7) fell right into place. The 145-Å spacing is obvious. The dots are aligned at different angles to the left of the axis and to the right. A line drawn parallel to the axis through any dot will pass through another dot at a distance of 5 × 145 Å, thus explaining the 720-Å period. The difficulty in interpreting the diffraction pattern arises from the asymmetry. In an X-ray specimen the fibrils present both aspects to the beam and the superposition produces a symmetrical pattern. This work was presented at the Annual Meeting of the EMSA on November 17, 1944, and was published early in 1945 (Hall et al., 1945). Bear's X-ray studies of this and other materials were published separately (Bear, 1944a, 1944b).

At the same EMSA meeting we also presented the preliminary observations of phosphotungstic stain used on collagen fibrils (Schmitt et al.,

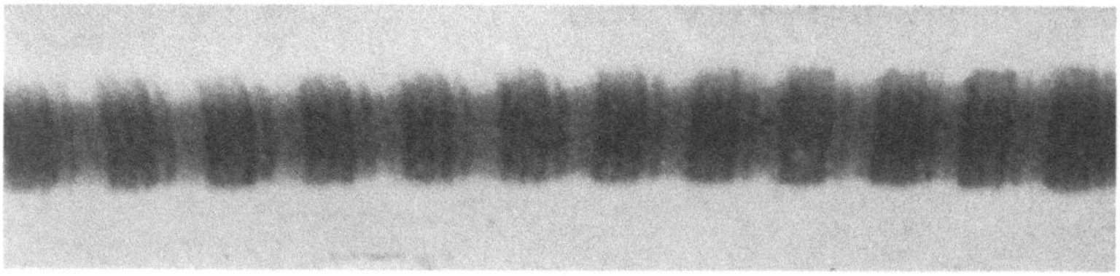

Figure 8 Part of a collagen fibril as in Fig. 6 but stained with phosphotungstic acid (buffered). The stain shows the presence of five stain-receptive finer bands within the fundamental period of 640 Å. The intensity and location of the bands have a directionality which indicates a lengthwise asymmetry in the structure of the collagen molecules.

1945). Within the 640-Å period, the stain showed five distinct sub-bands (Fig. 8). The position and intensity of the bands have a directionality that indicates a directionality in the arrangement of the atoms of which these long molecules are composed.

At about this time, when the war ended, electron microscopy entered a new era of growth. Research had been curtailed everywhere, though sufficient important results had been obtained to attract new investigators as they became available. The period between 1945 and 1950 seems to mark a transition between the "early years" and new times distinguished by better economic conditions, fresh outlooks, boldness in experimentation, and most importantly, enthusiasm in the vigorous band of recruits that had been attracted to this new branch of scientific endeavor.

5. Transition from the early years

For the present discussion we are regarding 1946 as about the end of the "early years." They did not come to an abrupt ending. There were still problems and much important new work to be done. There was a different attitude: we had good reasons for confidence and we knew where we were going. There was a short transition period characterized by successes in instrumentation and specimen techniques, and by an expanding body of competent electron microscopists. Some key accomplishments should be mentioned that extend beyond the personalized account that has been followed to this stage.

5.1 Shadowcasting

At about the same time that we were making our staining experiments, elsewhere (University of Michigan) another major advance was made on

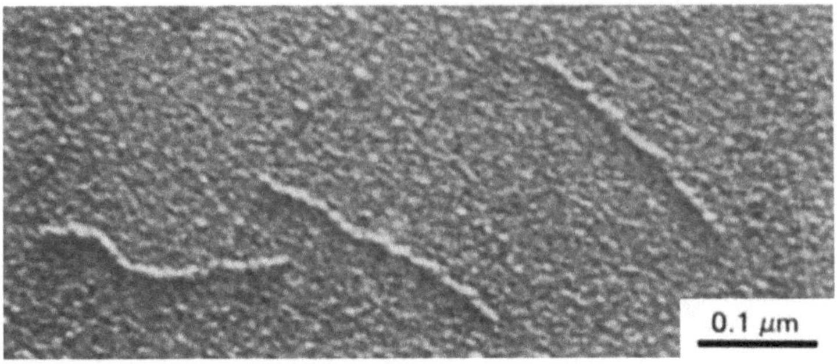

Figure 9 Individual molecules of collagen revealed by the shadowcasting technique. The molecules are 13 Å in diameter and about 1400 Å long.

the problem of enhancing the visibility of small or poorly scattering objects. This was the shadowcasting technique of Williams and Wyckoff (1944, 1945, 1946) and Williams and Backus (1949). This invention fitted in well with our efforts in macromolecular research. For example, the individual molecules of collagen, which are only about 13 Å in diameter, were rendered clearly visible by shadowcasting, as shown in Fig. 9. This example and other micrographs of macromolecules along with physicochemical measurements were included in a report by Hall and Doty (1958) at a later date, but the technique is based on work done in the 1940s. A review, "Electron Microscopy of Macromolecules," may be found in the volume edited by B. M. Siegel (Hall, 1964). A general reference for techniques and applications is my textbook (Hall, 1966).

5.2 Stigmator

The quality of electron micrographs produced by the average electron microscopist improved steadily due to improvements in skill and equipment. There was one improvement of a fundamental nature, however, that represents a breakthrough on a difficult problem that had limited attainable resolving power from the beginning. The problem is that with the highest precision in machining and the best quality of iron, manufactured magnetic lenses always contain a residual asymmetry in their fields (astigmatism). This problem was solved by Hillier and Ramberg (1947) by the invention of the "stigmator" or "compensator." This is a device which purposely applies a controllable asymmetry into the lens field which can be adjusted so as to oppose the residual asymmetry that the lenses carry with them from manufacture or acquire in use. With this device we could then expect to reach the resolution limits of electron microscopes as predicted by theory.

5.3 Microtomy

Ultramicrotomy was another major development that had barely started in the mid-1940s, but which moved rapidly forward and became an accomplished technique by the first years of the 1950s. Many workers and laboratories contributed to this achievement, which was not a victory over one problem but the solution to a bag of problems all tangled together. At MIT alone there were several contributors. We shall cite Latta and Hartman, both medical doctors who, frustrated by the chore of sharpening and using steel knives, reinvented the glass knife (Latta & Hartman, 1950) (previously used by prehistoric man). Our chief instrument maker, Mr. William Strovink, assisted them in finding civilized ways of breaking the glass. Also to be cited are Geren and McCulloch, who modified a standard screw-feed microtome (Minot) and produced excellent results (Geren & McCulloch, 1951). "Betty" Geren was also one of our group of doctors turned electron microscopist. David McCulloch was a regular biology graduate student. In all things mechanical, "Bill" Strovink could always be counted on.

Most of the graduate students and associates who came to work in electron microscopy at MIT had been diverted from their careers by involvement in war efforts. They were mature. Some came to study for an advanced degree and some to learn the new science in order to use it in their future research. Students in other departments as well as ours became interested in the subject. This prompted me to organize a series of lectures and laboratory sessions which resulted in a textbook on the subject (Hall, 1953). The time had arrived when electron microscopy could be described systematically from theoretical concepts to demonstrable practicalities. The subject could take its place in the academic curriculum. One could say that electron microscopy had come of age. If we reckon time from the original 1932 work of Knoll and Ruska (1932), the maturation comes out at about the age of many of our students—about 20 years.

The educational results of our enterprise at MIT were as important as the research accomplishments. Most of our students, as in other schools, participated in the experimental and inventive processes as well as such chores as maintenance of troublesome instruments and improvisations when instruments did not exist. After a year or two they went their way. We can say of them collectively, borrowing from Professor Burton, that "They all obtained good appointments."

6. Afterword by Peter Hawkes

Cecil Edwin Hall was born in Nottingham on 5 July 1912. By the 1930s, he was in Canada, passing his BSc at the University of Alberta in 1935, MSc at the University of Toronto in 1936. His article takes him up to 1946 and in 1947, he received his PhD from MIT. His book, *Introduction to Electron Microscopy* (McGraw-Hill, 1953, 2nd edition, 1966; the price of the first edition was $9.00) was widely read and an article in *Proc. Natl. Acad. Sci. USA* (**42**, 1956, 801–806) on the visualization of individual macromolecules with the electron microscope is another important landmark. For many years he was a consultant at RCA. In 1963, he was admitted to the American Academy of Arts and Sciences, emeritus in 1971. He died on 5 March 1991 in Jasper (Arkansas).

References

Bear, R. S. (1942). *Journal of the American Chemical Society, 64*, 727.
Bear, R. S. (1944a). *Journal of the American Chemical Society, 66*, 1297.
Bear, R. S. (1944b). *Journal of the American Chemical Society, 66*, 2043.
Burton, E. F., Hillier, J., & Prebus, A. (1939). *Physical Review, 56*, 1171.
Franklin, U. M., Weatherly, G. C., & Simon, G. T. (1978). *Toronto, 1978* Vol. 3 (Symposia). (pp. 4–17).
Geren, B. B., & McCulloch, D. (1951). *Experimental Cell Research, 2*, 97.
Hall, C. E. (1953). *Introduction to electron microscopy*. New York: McGraw-Hill.
Hall, C. E. (1964). In B. M. Siegel (Ed.), *Modern developments in electron microscopy* (pp. 395–415). New York: Academic Press.
Hall, C. E. (1966). *Introduction to electron microscopy* (2nd ed.). New York: McGraw-Hill (Reprint edition by Krieger Publ. Co. Inc., Melbourne, Florida, 1983).
Hall, C. E., & Doty, P. (1958). *Journal of the American Chemical Society, 80*, 1269.
Hall, C. E., & Schoen, A. L. (1941). *Journal of the Optical Society of America, 31*, 281.
Hall, C. E., Jakus, M. A., & Schmitt, F. O. (1942). *Journal of the American Chemical Society, 64*, 1234.
Hall, C. E., Jakus, M. A., & Schmitt, F. O. (1945). *Journal of Applied Physics, 16*, 459.
Hillier, J., & Ramberg, E. G. (1947). *Journal of Applied Physics, 18*, 48.
Hillier, J., & Vance, A. W. (1941). *Proceedings of the IRE, 29*, 167.
Jakus, M. A., Hall, C. E., & Schmitt, F. O. (1944). *Journal of the American Chemical Society, 66*, 313.
Johannson, H. (1933). *Annalen der Physik, 18*, 385.
Johannson, H. (1934). *Annalen der Physik, 21*, 274.
Knoll, M., & Ruska, E. (1932). *Zeitschrift für Physik, 78*, 318.
Latta, H., & Hartman, J. F. (1950). *Proceedings of the Society for Experimental Biology and Medicine, 74*, 436.
Mees, C. E. K. (1941). *Journal of the Society of Motion Picture Engineers, 37*, 10.
Prebus, A., & Hillier, J. (1939). *Canadian Journal of Research, 17*, 49.
Ruska, E. (1934). *Zeitschrift für Physik, 87*, 580.
Schmitt, F. O., Hall, C. E., & Jakus, M. A. (1942). *Journal of Cellular and Comparative Physiology, 20*, 11.
Schmitt, F. O., Hall, C. E., & Jakus, M. A. (1943). *Biological Symposia, 10*, 261.
Schmitt, F. O., Hall, C. E., & Jakus, M. A. (1945). *Journal of Applied Physics, 16*, 263.

Scott, G. H., & Packer, D. M. (1939). *The Anatomical Record, 74*, 17, 31.
Webb, J. H. (1940). *Journal of Applied Physics, 11*, 18.
Williams, R. C., & Backus, R. C. (1949). *Journal of Applied Physics, 20*, 98.
Williams, R. C., & Wyckoff, R. W. G. (1944). *Journal of Applied Physics, 15*, 712.
Williams, R. C., & Wyckoff, R. W. G. (1945). *Science, 101*, 594.
Williams, R. C., & Wyckoff, R. W. G. (1946). *Journal of Applied Physics, 17*, 23.
Zworykin, V. K., & Morton, G. A. (1936). *Journal of the Optical Society of America, 26*, 181.

CHAPTER TWO

My recollection of the early history of our work on electron optics and the electron microscope

Tadatosi Hibi[a] and Peter Hawkes (Afterword)[b,*]
[a]Formerly Sendai, Japan
[b]CEMES-CNRS, Toulouse, France
*Corresponding author. e-mail address: hawkes@cemes.fr

Contents

1. Emission microscope 25
2. Studies of specimen techniques for electron microscopy 27
3. Electron-microscopic observations of ionic crystals 29
4. The pointed filament and its application 34
5. Interference and coherence 37
6. Afterword by Peter Hawkes 42
References 42

1. Emission microscope

After my graduation from Tohoku University in 1933, I studied several works on ferromagnetism. One day at the beginning of 1936, Professor J. Okubo, my teacher, recommended that I attempt to build a magnetic-type emission microscope, which had not been tried in Japan at that time. The original work of Professors E. Brüche and O. Scherzer was at that time only known from one book on electron optics (1934). Therefore, I hesitated before deciding to embark on such a project. However, I started to work on the production of a magnetic-type emission microscope, as first made by Drs. M. Knoll, F. G. Houterman, and W. Schulze (1932), at the Department of Physics, Tohoku University. The construction (a) and its cross section (b) are shown in Fig. 1. The emission microscope consisted of two magnetic electron lenses each having a magnification of 10×. The focal point of each electron lens was adjusted by movement of the lens along the optical bench. The electron image obtained on a fluorescent screen was

Advances in Imaging and Electron Physics, Volume 221
ISSN 1076-5670
https://doi.org/10.1016/bs.aiep.2022.03.002

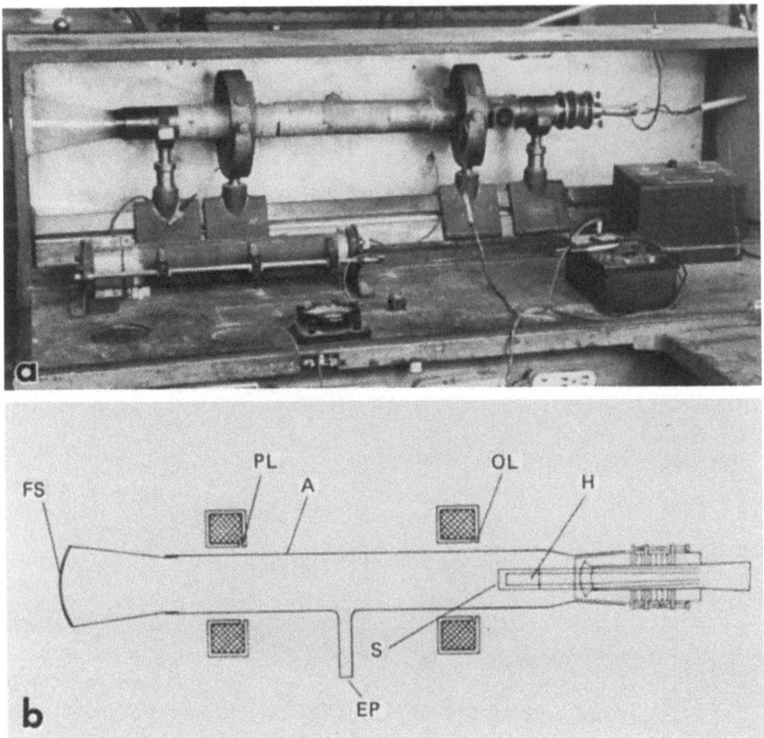

Figure 1 (a) The magnetic-type emission microscope built in 1936. (b) Its cross-sectional diagram: A, brass tube; FS, fluorescent screen; H, cathode; S, guard ring; PL, projector lens; OL, objective lens; EP, exit to vacuum pump.

recorded by a camera. The anode potential of 2 kV was obtained from an electric battery, and a vacuum of 10^{-5} mm Hg was obtained by means of an oil rotary pump and a mercury diffusion pump with a liquid-air trap. Mechanical construction of this emission microscope was finished only 1 year later, but we could not obtain a fine electron emission image for about 3 years. The main cause of this trouble was the existence of a disturbing magnetic field caused by electric currents in the electric circuits in our room. Therefore, magnetic shielding was introduced and consequently we could obtain an electron image at about 100× magnification. Fig. 2a shows the electron image obtained from barium azide coated on a nickel cathode. I believe that this is the first electron image obtained in Japan (Hibi, 1942a, 1942b). Fig. 2b is the reversed electron image obtained by raising the temperature of (a). Using this apparatus I made several electron-optical studies of the emission mechanism of oxide cathodes, for example, com-

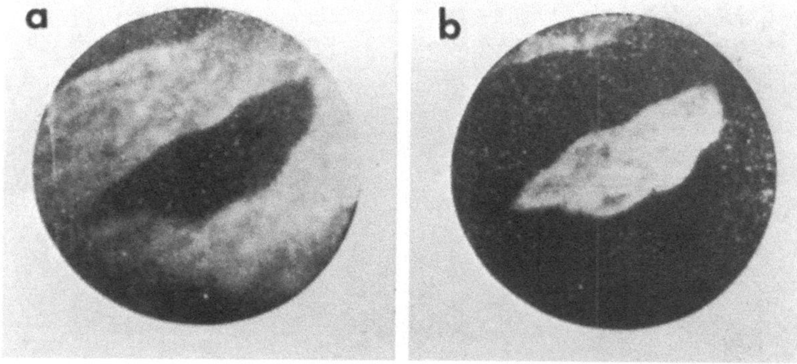

Figure 2 Electron images taken by the magnetic-type emission microscope. (a) Electron emission pattern of a nickel cathode coated with barium azide. Dark parts correspond to coated parts and bright parts correspond to the parts not coated. (b) The pattern is reversed by elevating the temperature.

parison of the thermionic emissions of BaN_6, BaO, and SrO (Hibi, 1949). Furthermore, the emission mechanism of oxide cathodes was continuously studied theoretically (Hibi, 1943, 1950; Hibi & Matsumura, 1951) and experimentally (Hibi & Ishikawa, 1951a, 1951b, 1952) by methods other than electron optics.

2. Studies of specimen techniques for electron microscopy

After the interruption of our study for 2 years due to my participation in World War II, I returned to Tohoku University. I then started trial production of an electrostatic-type conventional electron microscope as shown in Fig. 3. Two electrostatic lenses, one magnetic condenser lens, an optical bench, and a fluorescent screen from a Braun tube with a high-tension source were used in the apparatus. However, we could not obtain good results with this instrument, whereas good two-stage magnetic-type electron microscopes were coming on the market. Therefore, I gave up our trial production of an ordinary electrostatic electron microscope. In 1949, we bought an HU-6 made by Hitachi, having a resolving power of about 30 Å as shown in Fig. 4. This electron microscope was of the vertical type and had magnetic objective and projector lenses with one condenser lens. The anode potential was 50 kV and the magnification of the electron image in direct observation was 10,000×. It was with this apparatus that we mainly carried out the studies on specimen preparation techniques for electron

Figure 3 The electrostatic-type electron microscope built in 1948. Only the condenser lens is magnetic.

microscopy. Our first work was the study of metallic shadowcasting using a nozzle system to obtain a fine shadow, which was not much studied by makers of electron microscopes at that time. Here, a shutter plate (C) and a nozzle (D) with internal diameter of about 1 mm were located between the evaporation source (A) and specimen position (B), as shown in Fig. 5 (Hibi, 1952). By such a simple method we could obtain better shadowcasting than that obtained before (Williams & Bachus, 1949). This result was quoted in L. Holland's book (1956) and in an early book on electron microscopy published in the Soviet Union. For the vacuum evaporation of a metal such as tungsten, having a high melting point, we developed a vibrating-electrode method using the principle of an electric bell, as shown in Fig. 6, and obtained fine evaporation film (Hibi & Yada, 1954, 1956). In the case of a conical basket, material was not evaporated from the conical tip point of the basket, because the temperature of the material mounted was not highest in that region. To realize the highest temperature at the conical tip, a doublewinding basket was used, as shown in Fig. 7 (b and d) (Hibi & Yada, 1961). As the molecular beam from the thermal evaporation source normally had no charge, the molecular beam from evaporation source A was raised to unit charge by electron bombardment (electron path HN) and was then accelerated in an oblique direction with anode potential B and deposited on the specimen (E) through a nozzle (C), as shown in Fig. 8. Ultraviolet-light illumination (L) was used to obtain a much more highly ionized beam. We could obtain very fine shadowcasting and form

Figure 4 The Hitachi HU-6 electron microscope.

replicas for electron microscopy by using this apparatus (Hibi & Yada, 1954, 1956).

3. Electron-microscopic observations of ionic crystals

We also studied several kinds of materials with the electron microscope, including BaO, MgO, and alkali halide and silver halide crystals. Electron-optical observations on the crystal properties of magnesium oxide smokes (Hibi & Ogawa, 1954) and on the interference fringes in electron micrographs of magnesium oxide smokes (Hibi et al., 1955) aroused deep interest in many investigators of electron diffraction from the standpoint of the validity of a dynamical theory of electron diffraction and also from

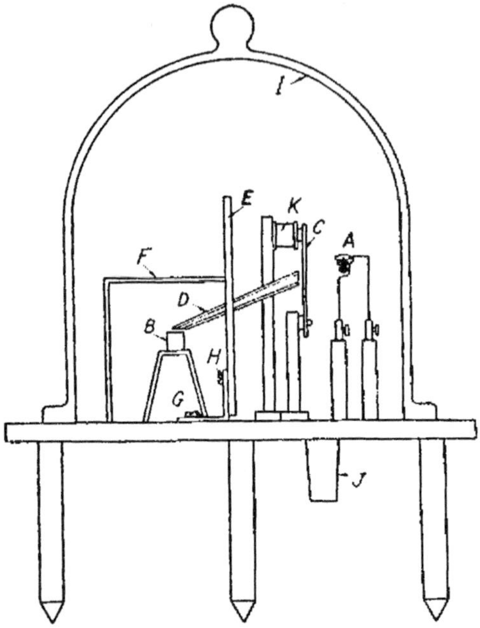

Figure 5 Metal shadowcasting apparatus using a nozzle system: A, tungsten evaporation basket; B, specimen supporter; I, bell jar; J, joint to a vacuum pump; C, shutter; D, glass nozzle; E, cover plate; F, glass vessel to cover the specimen supporter; G and H, rails to adjust the cover plate.

the standpoint of equal-thickness fringes, later demonstrated by Dr. R. D. Heidenreich (1967).

Coloring of KCl or NaCl by X-ray irradiation was studied by using replica techniques for electron microscopy. We found that with coloring of the ionic crystal by X-ray irradiation, growth of crystallites occurred clearly on the crystal surface or inner part of the crystal; with bleaching by holding the crystals in an atmosphere for a long time or at high temperature for a short time, the crystallites that had grown disappeared again (Hibi & Ishikawa, 1958). Confronted with such an unexpected and interesting phenomenon, violent objections arose from many investigators interested in the alkali halides who believed that this phenomenon might be only an artifact. We therefore made further studies to confirm the credibility of this phenomenon, namely, successive electron-microscopic observations of a definite field of a definite crystal were made by selected-area replica techniques (Hibi et al., 1956). Fig. 9 shows the result—many investigators recognized the credibility of this phenomenon (Hibi & Yada, 1959a). Seen

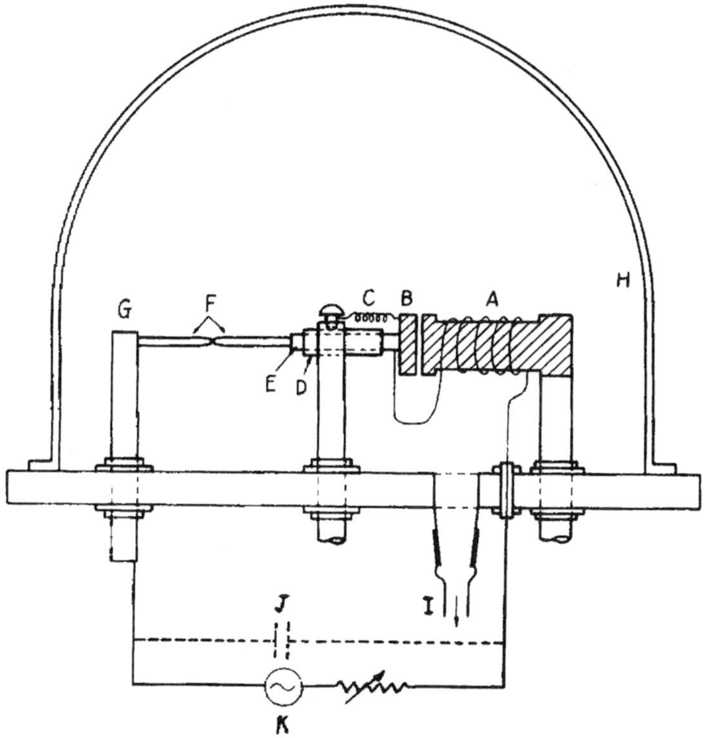

Figure 6 Apparatus for dynamical evaporation of metals of high melting point: A, electromagnet; B, polepiece; C, spring; D, silica tube; E, nickel wire; F, electrodes of a metal to be evaporated; H, bell jar; I, joint to a vacuum pump; J, condenser; K, alternating electric source.

in this figure (a–e) are the coloring process and the bleaching process (f–l). However, Professor F. Seitz, famous theorist of solid-state physics at the University of Illinois, commented that this phenomenon should not appear in the ionic crystal at low temperature. Therefore, we tried to make a low-temperature replica (Hibi & Yada, 1959b), quite independently of Dr. Steere's work (1957), and the effect of X-ray irradiation on a KCl crystal was then studied by electron microscopy. Fig. 10 shows the result for the crystal at liquid-air temperature. Growth of crystallites on the surface of KCl (a) and KBr (b) was clearly seen. Therefore, this phenomenon was recognized by many investigators of color centers in solid-state physics (e.g., Schulman & Compton, 1962). By using replica techniques, electron-microscopic observations on silver halide were also carried out (Iijima & Hibi, 1966a, 1966b).

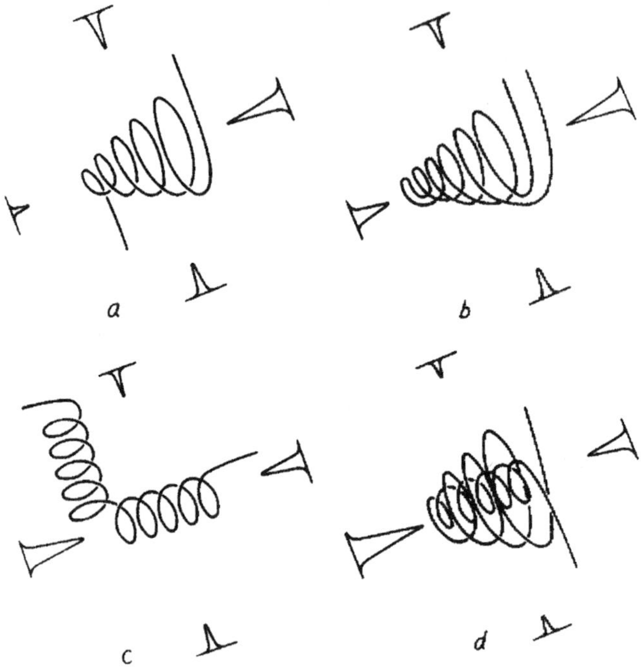

Figure 7 Evaporation baskets having different forms (a–d). Evaporation quantities for different directions are shown schematically.

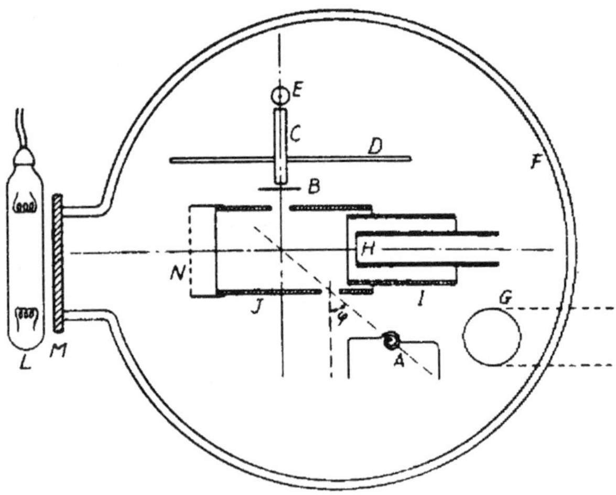

Figure 8 Apparatus for metal shadowcasting using metal vapor streams ionized by electron bombardment and by illumination with ultraviolet light and then accelerated to the specimen supporter kept at negative potential. A, Evaporation basket; B, shutter; C, nozzle; D, cover plate; E, specimen supporter; F, bell jar; G, joint to a vacuum pump; H, cathode; I, Wehnelt cylinder; J, anode; L, mercury-vapor lamp; M, quartz window; N, fluorescent mesh screen.

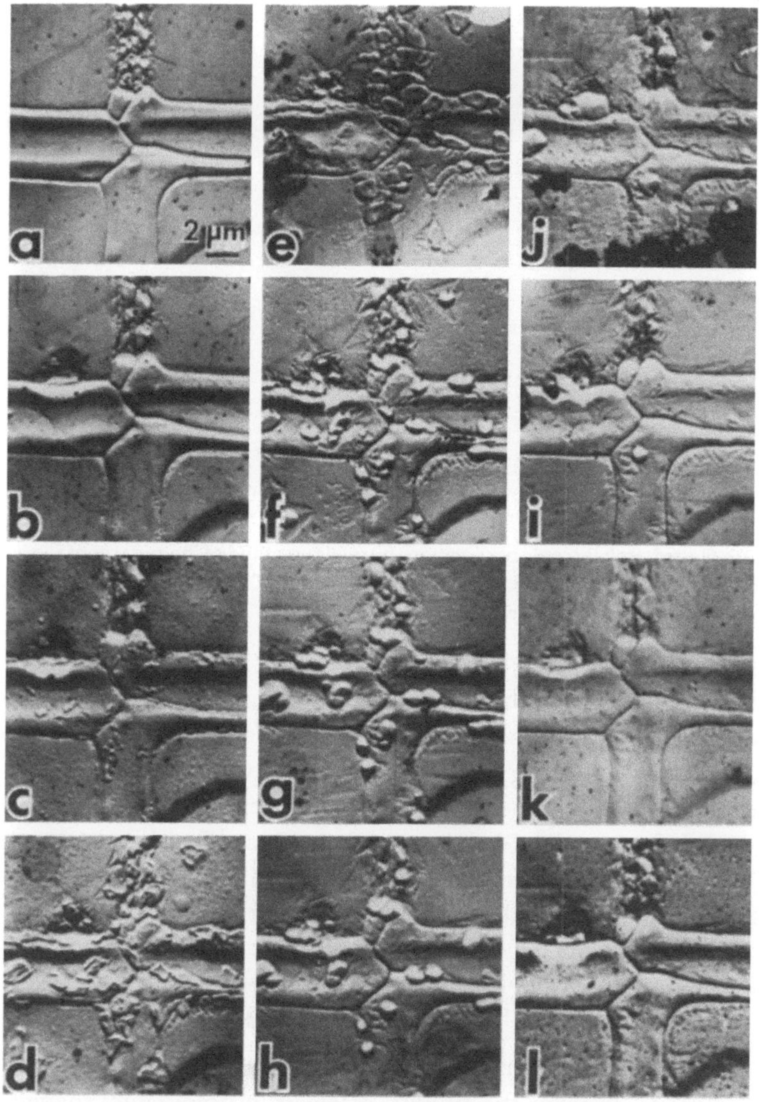

Figure 9 Electron-microscopic observations by replica method of surface changes in a designated area of a KCl crystal during coloration by X-ray irradiation and bleaching by natural-light illumination. (a) Before X-ray irradiation. (b–d) After 4, 7, and 11 hr of irradiation, respectively. (e–l) After bleaching for 30 min; 1, 2, 3, 23, 24, 29, and 53 days, respectively.

Direct observations of thin-metal specimens by electron microscopy were first attempted by Dr. W. Bollmann (1956). However, no one had attempted to make direct observations of thin alkali halide crystals. We

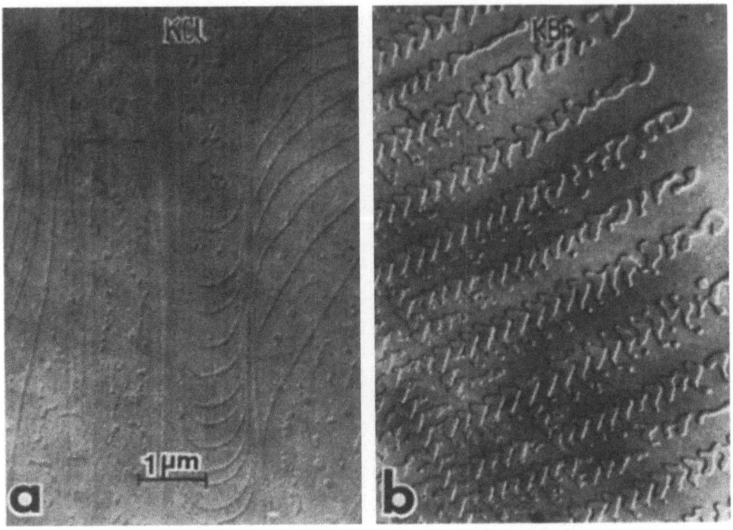

Figure 10 Electron-microscopic observations by replica method of the surfaces of alkali halides colored by X-ray irradiation at liquid-air temperature: (a) KCl; (b) KBr. Replicas were made at liquid-air temperature just after irradiation.

therefore tried to make thin crystals by drying one drop of KCl solution in vacuum and also by polishing bulk KCl crystals in water. We were the first to succeed in making a thin alkali halide crystal for direct observation in the electron microscope. By using these specimens, direct observations of crystal defects in alkali halide crystals produced by electron irradiation were actively made; Hibi and Yada (1962), Kawamata and Hibi (1965), and Sonder and Sibley referred to our results in their review (1972).

4. The pointed filament and its application

Doctors M. E. Haine and T. Mulvey (1952) performed an experiment to try to confirm the validity of electron holography suggested by Professor D. Gabor (1948, 1956) by using magnesium oxide smoke in the electron microscope. I was very interested in their work, though their attempt was not successful. Therefore, I tried repeatedly to carry out their experiment. In their case, a hairpin filament and a 2-μm rectangular slit, the size of which was adjustable from outside the vacuum, were used. In our case, however, such a fine slit could not be made at the workshop of our institute. To overcome this difficulty, I had the idea that a pointed filament should play the same role as the hairpin filament with a fine slit. I used a pointed filament having a tungsten tip as used in field emission microscopy by Professor E. W. Müller (1937). Professor Müller used an exceptionally good vacuum of 10^{-8}–10^{-9} mm Hg for the tip at room temperature, but we used a more usual vacuum of 10^{-4}–10^{-5} mm Hg for the tip at high temperature. As a result of this experiment, an unexpected result was obtained: a much brighter and more highly coherent electron beam than that available hitherto could be obtained. I also tried electron holography in the same way as Drs. Haine and Mulvey, but this trial did not succeed because no coherent light source such as laser light existed at that time. However, the Fresnel fringes observed by Drs. Haine and Mulvey at the edge of magnesium oxide smokes were only two or three in number, while we could obtain about 20 fringes by using a pointed filament in the electron microscope at 50 kV. This result was announced at the International Conference on Electron Microscopy, London (1954), by Dr. V. E. Cosslett, who read my paper, and a full paper on the pointed filament was published in 1956 (Hibi, 1956a, 1956b, 1962; Hibi, Yada, Kawamata, 1963). Many members of the audience, including Professor G. Möllenstedt, were deeply interested in this result. He asked me through Dr. Y. Sakaki (resident student in his laboratory) how to make pointed tips. Therefore, I taught him that, to grind the fine tungsten-pointed tip, an oilstone abrasive must be located on the tip, contrary to the ordinary manner of polishing. They made the fine tip by using our method and obtained the same good results as ours concerning Fresnel fringes (Maruse & Sakaki, 1958; Sakaki & Möllenstedt, 1956). Our photograph of Fresnel fringes created a sensation and was reproduced in the photographic book entitled "Thirteen Stages to the Atom" published in Sweden. Fig. 11 shows our micrograph of Fresnel fringes appearing on the edge of MgO; this photograph was used as the cover of a physics textbook for the introductory course at MIT.

The resolving power of the commercial two-stage electron microscope was then 30 Å (HU-6) and this could be raised to 15 Å by using a pointed filament and short-focal-length objective lens (Hibi & Takahashi, 1958). My pointed filament was called the "Hibische Katod" by some investigators of electron microscopy. In the German electron microscopy meeting held at Darmstadt in 1957, the high level of the Japanese electron microscope was first recognized by many members of the audience, including Professor H. Busch, through my lecture entitled "Two-Stage Electron Microscope Using a Pointed Filament" (Hibi & Takahashi, 1959a). Therefore, in the Fifth International Congress of Electron Microscopy held at Philadelphia in 1962, many participants met and actively discussed my lecture entitled

Figure 11 Electron image of Fresnel fringes that appeared at the edges of magnesium oxide smokes. These fringes were obtained by using a pointed filament. (This image was used on the cover of a book produced for internal use only at MIT, prepared by the Science Teaching Center, 1965; this revised edition was written by A. P. French and A. M. Hudson.)

"Point Cathode and Resolution of Electron Microscope." Afterward, in the case of the three-stage electron microscope (HU-11B), lattice images of 0.98-Å spacing of thin Pd film could be obtained by three-wave interference using a pointed filament. At that time, this lattice-fringe spacing was the smallest that had been observed anywhere in the world (Yada & Hibi, 1969).

Hitherto, to obtain a dark-field image, an objective aperture had been moved to one side, while we could obtain better contrast dark-field images by inclination of the illumination system and also by using a pointed filament as shown in Fig. 12 (Hibi & Takahashi, 1959b). In this figure, (a) is a bright-field image of MoO_2 and (b) is its dark-field image. The contrast of (b) is superior to that of (a). This dark-field method was used to obtain better contrast images by Professor G. Dupouy in the case of biological materials with his high-voltage electron microscope (1967).

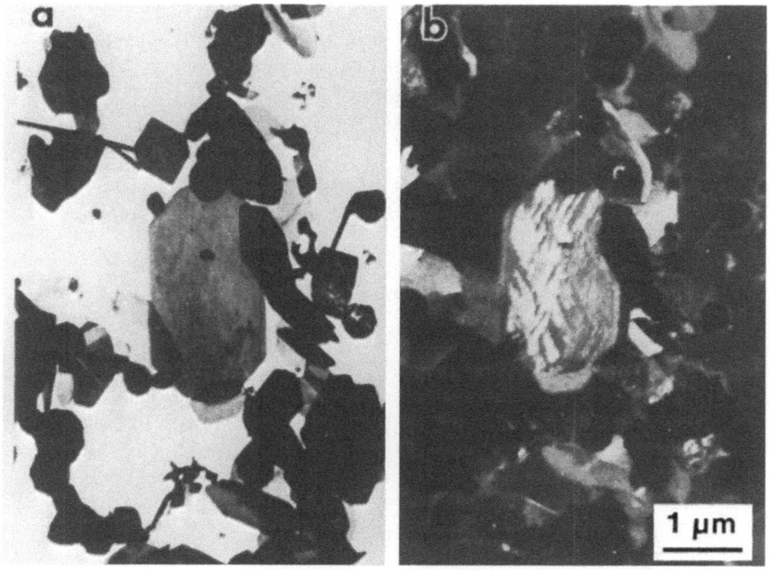

Figure 12 Electron images of molybdenum oxide smokes. (a) Bright-field image; (b) dark-field image. These images were obtained by using a pointed filament.

5. Interference and coherence

Studies on the quality of the electron beam have not been so active as in light optics. I found that the electron beam quality with a pointed filament was much superior, that is, more coherent than that of a hairpin filament. Therefore, a systematic study of electron beam quality was started (Hibi, Yada, Takahashi, 1963). The number of Fresnel fringes seen was at first selected as a measure of the degree of interference (Hibi & Yada, 1964). The quality of the electron beam was then determined from the degree of coherence obtained by using a Möllenstedt-type electron interferometer. Fig. 13 shows the Möllenstedt-type electron biprism produced by Hibi and Takahashi (1963). The central anode of the electron biprism was made from a very thin silica fiber about 0.8 μm in diameter on which chromium and gold were deposited under vacuum, following the technique of Professor G. Möllenstedt and Dr. H. Düker (1956). The outside cathode was made from nonmagnetic stainless steel. The distance between the central anode and the outside cathodes was 1.3 mm. Horizontal movements of the electron biprism were controlled by two Wilson seals (G and H). Fig. 14 illustrates the optical analog (a) of the electron interferometer produced by Hibi and Takahashi (1963) and its appearance (b). A pointed filament

Figure 13 Electron biprism produced on trial. A, Central anode; B, outer cathode; G and H, Wilson's seal.

having a radius of curvature of 1–2 μm was used as the electron source. The anode potential was 35 kV. A magnetic condenser lens was used as the condenser system. To obtain the interference fringe, an electric potential difference of 0–9 V between the central anode and the outer cathodes was obtained by using a battery. To obtain the magnified electron interference fringes, an ordinary electrostatic electron lens was used. Fig. 15 shows the electron interference fringes in the case of overfocus (a) and in the case of exact focus (b). In the case of exact focus (b), the central fringe has the maximum intensity. This electron interference microscope was used by us (Hibi & Takahashi, 1963) and afterward an attempt was made to set up a Möllenstedt-type biprism at the field-limiting aperture position of a high-resolution electron microscope, as shown in Fig. 16, so as to superimpose the electron interference fringes on the highly magnified images (Yada et al., 1973).

Although the degree of coherence defined by Professor C. Fert (1962) was correct, we could not determine the degree of coherence of an electron beam from it. Accordingly, Hibi and Takahashi (1969) determined the degree of coherence of the electron beam by using the visibility curve obtained from the electron interference fringe similar to that employed in the optical case (Nawata, 1964; Thompson & Wolf, 1957). The degree of

Figure 14 The electron interferometer (b) produced at Tohoku University. (a) Diagram of the optical analogy.

coherence γ_{12} can be determined not only theoretically but also experimentally from (b) of Fig. 15, namely, $\gamma_{12} = (h_1 - h_2)/(h_1 + h_2)$, where h_1 is the value of maximum intensity and h_2 is that of the first minimum intensity. By using such a definition of γ_{12}, we found that various values of γ_{12} are obtained by changing the temperature of the pointed filament. By using an electron interference microscope we found that the contrast of electron images of nonbiological and biological specimens is better in the case of large γ_{12} than that in the case of small γ_{12} (Hibi & Takahashi, 1966). The same result on contrast of electron diffraction was published at the International Conference of Electron Diffraction and Crystal Defects held at Melbourne in 1965. At that time, many investigators of electron diffraction did not agree with us because it was difficult to realize that γ_{12} is the lateral coherence rather than the usual longitudinal coherence. Recently, this result has interested a few investigators of electron diffraction and electron

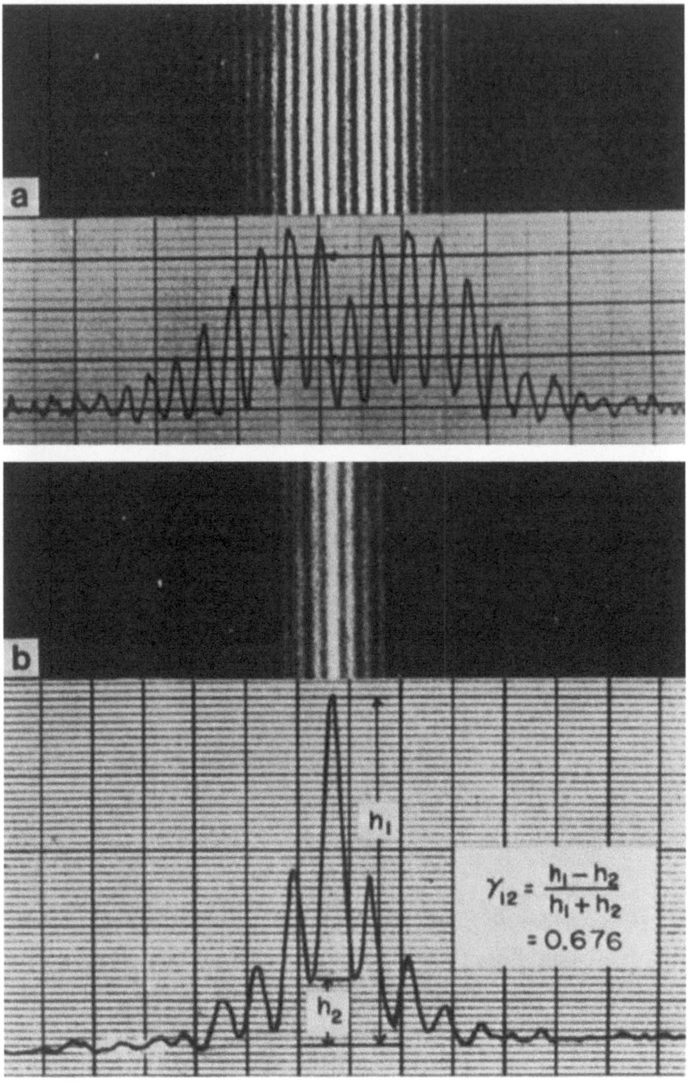

Figure 15 Electron interference fringes. (a) At overfocus; (b) at exact focus.

microscopy (e.g., Zeitler, 1979). Fig. 17 shows the result obtained from a copper phthalocyanine thin film by using the high-magnification electron interference microscope (Yada et al., 1973), for which γ_{12} values of the electron beams used were 0.16 (a) and 0.63 (b). The mean image contrast of (b) is twice as good as that of (a).

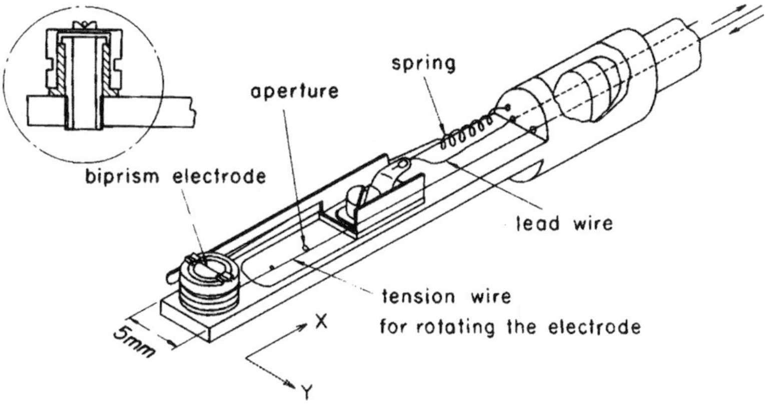

Figure 16 Perspective drawings of the electron biprism built in the aperture device of an electron microscope. The direction of the biprism can be rotated.

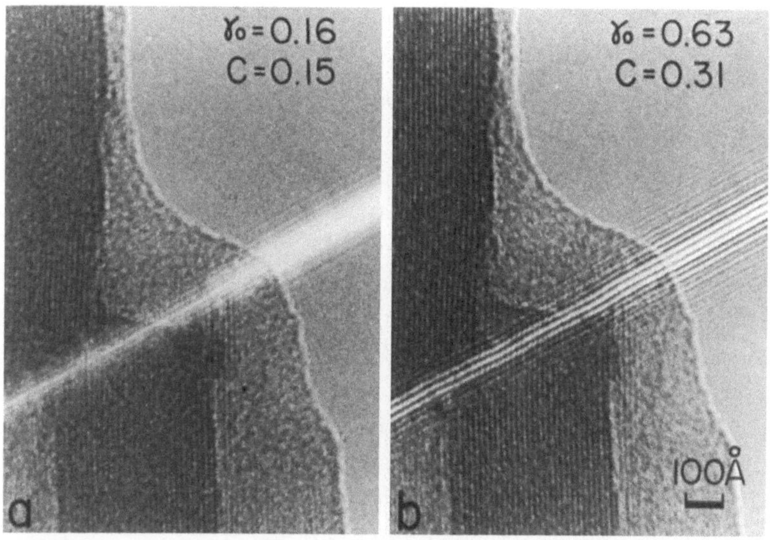

Figure 17 High-resolution electron interference micrographs showing the lattice fringes of copper phthalocyanine and the interference fringes produced by the biprism. Degree of coherence of the electron beam is 0.16 in (a) and 0.63 in (b). The image contrast in (b) is twice as good as in (a).

By using a high-magnification electron interference microscope, the mean inner potential of several materials was also measured (Yada et al., 1973). It is also capable of measuring the mean thickness of thick samples. At present, Dr. A. Tonomura et al. (1979) are actively pursuing interesting applications in electron holography by using a reference electron beam and laser light. For high-resolution electron microscopy such as direct observation of single atoms, a highly coherent electron beam is very important, as it is for electron holography (e.g., Missiroli et al., 1981).

By using a pointed filament with thermionic emission field emission, a comparatively high-coherence source is obtained and technological applications are being actively developed (see Hibi & Yada, 1976). However, the number of basic fundamental studies of these high-coherent sources is still very few. Accordingly, we cannot understand exactly why a pointed filament is a comparatively coherent source, though I think this may be due to the fact that many incoherent electron sources are very close together. It will be important to confirm this in the future by using a pumped electron beam and an electron mirror, to see whether an electron beam can be created in the same way as laser light.

6. Afterword by Peter Hawkes

Tadatoshi Hibi was born in 1909 in Kyoto. He graduated from the Tohoku Imperial University in 1933. At the Research Institute for Iron, Steel and other Metals, he was encouraged to build an electron microscope in order to observe the $\alpha \rightarrow \gamma$ transition in iron at high temperature (911 °C). A prototype emission microscope was constructed in 1936 and the first results, on the emissive properties of barium oxide, were described in 1940. In the same year, he and Junzo Ohkubo published the first book in Japanese on electron microscopy. He began construction of an electrostatic microscope but this project was halted by the war. He later continued to work in the field of electron microscopy at the Research Institute for Scientific Instruments at Tohoku University where he remained until his retirement in 1973. He died on 13 November 1994. An obituary appeared in the *Journal of Electron Microscopy* **44** (1995) 423–424.

References

Bollmann, W. (1956). *Physical Review, 103*, 1588.
Brüche, E., & Scherzer, O. (1934). *Geometrische Elektronenoptik*. Berlin and New York: Springer-Verlag.
Dupouy, G. (1967). *Journal of Electron Microscopy, 16*, 5.
Fert, C. (1962). *Journal of Electron Microscopy, 11*, 1.
Gabor, D. (1948). *Nature (London), 161*, 111.
Gabor, D. (1956). *Reviews of Modern Physics, 28*, 260.
Haine, M. E., & Mulvey, T. (1952). *Journal of the Optical Society of America, 42*, 763.
Heidenreich, R. D. (1967). *Journal of Electron Microscopy, 16*, 23.
Hibi, T. (1942a). *The Science Reports of the Tohoku Imperial University, 30*, 372.

Hibi, T. (1942b). *The Science Reports of the Tohoku Imperial University, 30*, 384.
Hibi, T. (1943). *Nihon Sugaku Butsurigaku Kaishi, 17*, 270 (in Japanese).
Hibi, T. (1949). *Science Reports of the Research Institutes, Tohoku University, Series A: Physics, Chemistry, and Metallurgy, 1*, 231.
Hibi, T. (1950). *Science Reports of the Research Institutes, Tohoku University, Series A: Physics, Chemistry, and Metallurgy, 2*, 157.
Hibi, T. (1952). *Review of Scientific Instruments, 23*, 383.
Hibi, T. (1956a). *London, 1954* Vol. 1. (p. 636).
Hibi, T. (1956b). *Journal of Electron Microscopy, 4*, 10.
Hibi, T. (1962). *Philadelphia, 1962* Vol. 1. p. KK-1.
Hibi, T., & Ishikawa, K. (1951a). *Physical Review, 83*, 659.
Hibi, T., & Ishikawa, K. (1951b). *Physical Review, 84*, 1254.
Hibi, T., & Ishikawa, K. (1952). *Physical Review, 87*, 673.
Hibi, T., & Ishikawa, K. (1958). *Journal of the Physical Society of Japan, 13*, 709.
Hibi, T., & Matsumura, T. (1951). *Physical Review, 81*, 884.
Hibi, T., & Ogawa, T. (1954). *Journal of the Physical Society of Japan, 9*, 627.
Hibi, T., & Takahashi, S. (1958). *Berlin, 1958*. (p. 169).
Hibi, T., & Takahashi, S. (1959a). *Journal of Electron Microscopy, 7*, 15.
Hibi, T., & Takahashi, S. (1959b). *Science Reports of the Research Institutes, Tohoku University, Series A: Physics, Chemistry, and Metallurgy, 11*, 490.
Hibi, T., & Takahashi, S. (1963). *Journal of Electron Microscopy, 12*, 129.
Hibi, T., & Takahashi, S. (1966). *Kyoto, 1966* Vol. 1. (p. 71).
Hibi, T., & Takahashi, S. (1969). *Zeitschrift für Angewandte Physik, 27*, 132.
Hibi, T., & Yada, K. (1954). *Journal of Applied Physics, 25*, 712.
Hibi, T., & Yada, K. (1956). *London, 1954* Vol. 1. (p. 460).
Hibi, T., & Yada, K. (1959a). *Journal of the Physical Society of Japan, 14*, 445.
Hibi, T., & Yada, K. (1959b). *Journal of Electron Microscopy, 7*, 21.
Hibi, T., & Yada, K. (1961). *Science Reports of the Research Institutes, Tohoku University, Series A: Physics, Chemistry, and Metallurgy, 13*, 105.
Hibi, T., & Yada, K. (1962). *Journal of the Physical Society of Japan, 17*(Suppl. B-II), 212.
Hibi, T., & Yada, K. (1964). *Journal of Electron Microscopy, 13*, 94.
Hibi, T., & Yada, K. (1976). In M. A. Hayat (Ed.), *Principles and techniques of electron microscopy: Vol. 6* (p. 312). Princeton, New Jersey: Van Nostrand-Reinhold.
Hibi, T., Kambe, K., & Honjo, G. (1955). *Journal of Electron Microscopy, 3*, 15.
Hibi, T., Ishikawa, K., & Yada, K. (1956). *Tokyo, 1956*. (p. 302).
Hibi, T., Yada, K., & Kawamata, Y. (1963). *Journal of the Physical Society of Japan, 18*(Suppl. III), 345.
Hibi, T., Yada, K., & Takahashi, S. (1963). *Journal of Electron Microscopy, 11*, 244.
Holland, L. (1956). *Vacuum deposition of thin films*. London: Chapman & Hall.
Iijima, S., & Hibi, T. (1966a). *Kyoto, 1966* Vol. 1. (p. 387).
Iijima, S., & Hibi, T. (1966b). *Japanese Journal of Applied Physics, 5*, 1183.
Kawamata, Y., & Hibi, T. (1965). *Journal of the Physical Society of Japan, 20*, 242.
Knoll, M., Houtermans, F. G., & Schulze, W. (1932). *Zeitschrift für Physik, 78*, 340.
Maruse, S., & Sakaki, Y. (1958). *Optik, 15*, 485.
Missiroli, G. F., Pozzi, G., & Valdrè, U. (1981). *Journal of Physics E, 14*, 649.
Möllenstedt, G., & Düker, H. (1956). *Zeitschrift für Physik, 145*, 377.
Müller, E. W. (1937). *Zeitschrift für Physik, 106*, 541.
Nawata, S. (1964). *Science Reports of the Research Institutes, Tohoku University, Series A: Physics, Chemistry, and Metallurgy, 16*, 54.
Sakaki, Y., & Möllenstedt, G. (1956). *Optik, 13*, 193.
Schulman, J. H., & Compton, V. S. (1962). *Color centers in solids*. Oxford: Pergamon.
Sonder, E., & Sibley, W. A. (1972). In J. H. Crawford Jr., & L. M. Slifkin (Eds.), *Point defects in solids: Vol. 1* (p. 201). New York: Plenum.

Steere, R. L. (1957). *The Journal of Biophysical and Biochemical Cytology*, *3*, 45.
Thompson, B. J., & Wolf, E. (1957). *Journal of the Optical Society of America*, *37*, 895.
Tonomura, A., Matsuda, T., & Endo, J. (1979). *Japanese Journal of Applied Physics*, *18*, 9.
Williams, R. C., & Bachus, R. C. (1949). *Journal of Applied Physics*, *20*, 98.
Yada, K., & Hibi, T. (1969). *Journal of Electron Microscopy*, *18*, 266.
Yada, K., Shibata, K., & Hibi, T. (1973). *Journal of Electron Microscopy*, *22*, 233.
Zeitler, E. (1979). *Proceedings — Annual Meeting, Electron Microscopy Society of America*, *37*, 376.

CHAPTER THREE

Walter Hoppe — X-ray crystallographer and visionary pioneer in electron microscopy

Joachim Frank
Department of Biochemistry and Molecular Biophysics, Columbia University Medical Center, New York, NY, United States
e-mail address: jf2192@columbia.edu

Contents

1. Introduction 45
2. Life 46
3. Work in X-ray crystallography 47
4. Structure research using electron microscopy 48
5. Electron tomography 53
6. Walter Hoppe remembered 55
Acknowledgments 57
Appendix to "Walter Hoppe — X-ray crystallographer and visionary pioneer in electron microscopy" by Joachim Frank 58
References 66

1. Introduction

Walter Hoppe (March 21, 1917 – November 3, 1986) (Fig. 1) was a German physicist who started as an accomplished X-ray crystallographer with both experimental and theoretical contributions to the field, and then moved into electron microscopy (EM) to become one of the early pioneers of quantitative structure research using the electron microscope. His laboratory has spawned two Chemistry Nobel Prizes, one in each area: to Robert Huber in 1988 and to the author in 2017. A third one, to Johann Deisenhofer jointly with his mentor Huber, is also in Hoppe's direct lineage. In my brief account I will mix personal recollections from the time I spent as a student in Hoppe's lab with a summary of his life and research accomplishments in the EM field. I will also draw from published and unpublished accounts of several of his former students (Huber & Holmes, 1987; Huber, 2019; Steigemann, 2013, 2021a; Typke, 2019).

Advances in Imaging and Electron Physics, Volume 221
ISSN 1076-5670
https://doi.org/10.1016/bs.aiep.2022.03.003

Figure 1 Walter Hoppe. Courtesy Archiv der Max-Planck-Gesellschaft, Berlin-Dahlem.

2. Life

Walter Hoppe was born in Wallsee/Donau, Austria and obtained his doctorate in chemistry at the German University in Prague under Professor J. Boehm in 1941. He joined Günter Scheibe's Institute for Physical Chemistry of the Technische Hochschule (TH) Munich (later renamed Technische Universität München, TUM) where he completed his *Habilitation* in 1944 on X-ray crystallographic methods. After working several years in industry he returned to Scheibe's institute as *Privatdozent* in 1953 and became professor of the TH in 1958. In 1959 he became head of the department for X-ray structure analysis of the *Max-Planck-Institut für Eiweiss* und *Lederforschung* (i.e., Protein and Leather Research), and in 1964 he became director. This institute, located in Munich's Schillerstrasse, was where both Robert Huber and I prepared our dissertations, to finish our Ph.D.s in 1963 and 1970, respectively (Fig. 2). It was later subsumed into the Abteilung für Strukturforschung I (Department for Structural Research I) – still under Hoppe's direction – of the newly founded Max Planck Institute for Biochemistry in Martinsried, which was inaugurated in 1972. At that time Huber assumed the directorship of an independent protein crystallography group in Martinsried that focused on large membrane protein complexes. In the fall of 1972, on my return from a two-year postdoctoral fellowship in the USA, I re-joined Hoppe's group as a visitor for a brief

Figure 2 Max-Planck Institute for Protein and Leather Research, Schillerstrasse 46, Munich, after the opening on April 4, 1957. Courtesy Archiv der Max-Planck-Gesellschaft, Berlin-Dahlem.

period, until June 1973. Hoppe remained in his position as Director of the Structural Research Department until his retirement in 1985.

3. Work in X-ray crystallography

As an outsider of the X-ray crystallography field I will let others speak about Hoppe's work in this area of structure research. In the tribute by the *Journal of Applied Crystallography* (King, 1987), Alfred Gieren, one of Hoppe's former students, is quoted as saying that "it would be impossible to cite all the contributions made by Hoppe to X-ray and neutron diffraction in detail, but his main fields of interest were thermal diffuse scattering, the convolution molecule method (i.e., *Faltmolekülmethode*, see below), phase determination, the shift product method, direct methods, the phase correction method, phase determination of protein structures, and last but not least the development of computer programs. He also developed the label triangulation method for structure determination of aperiodic objects by neutron scattering."

In Robert Huber's words (Huber, 2019), "Hoppe was an extraordinarily multifaceted researcher in focusing on problems in structural chemistry and on theoretical methods in X-ray crystallography. His '*Faltmolekülmethode*,' (Hoppe, 1957a, 1957b) today known as the *molecular replacement method*, has become the most important method for solving protein structures . . . " In their review of macromolecular X-ray crystallography, which also allotted Hoppe a significant position in a genealogical tree of X-ray crystallography, Jaskolski et al. (2014) explain the significance of this invention: "Practically all novel crystal structures, that is those not expected to be similar to any known atomic model, must be solved by some variant of the 'special atom' method. However, currently the Protein Data Bank contains close to 100,000 models of macromolecules, thus often a similar structural analogue is available and can be used as a search model in the method of Molecular Replacement. The '*Faltmolekülmethode*' suggested early on by Walter Hoppe did not acquire popularity, perhaps because it was published too early for its time and only as an abstract in German language. The practical use of the Molecular Replacement Method was pioneered by Michael Rossmann and David Blow (1962). This approach has also evolved significantly and presently a majority of protein crystal structures are solved by several powerful molecular replacement programs" (Jaskolski et al., 2014).

It is noteworthy in this context that Hoppe spent half a year as a visiting scientist in the lab of Max Perutz and John Kendrew at the Laboratory of Molecular Biology of the Medical Research Council in Cambridge in 1957/1958. Here as he became acquainted with the new field of protein crystallography he tried out his *Faltmolekülmethode* on myoglobin. In 1962 Perutz and Kendrew won the Nobel Prize in Chemistry for solving the structure of hemoglobin.

Further commenting on Hoppe's multifaceted interests, Huber (2019) cites his active participation in the design and construction of precision instruments: "A mechanical Fourier synthetizer (Fourier transformation, today easily accomplished on a small laptop, was a formidable problem of data processing at the time), a microdensitometer for photographical films, which served as detectors for both X-ray crystallography and electron microscopy, and a single-crystal diffractometer (Fig. 3), all posing complex problems of engineering that he brilliantly mastered."

4. Structure research using electron microscopy

Hoppe's group, along with Aaron Klug's at the Laboratory of Molecular Biology in Cambridge and Eduard Kellenberger's in Basel, was among the early pioneers in molecular structure research using the electron microscope (Fig. 4). According to Huber (2019), "a credible legend cites Hoppe with the pronouncement that X-ray crystallography had ceased to be a challenge for him, but that electron microscopy still offered plenty of it".

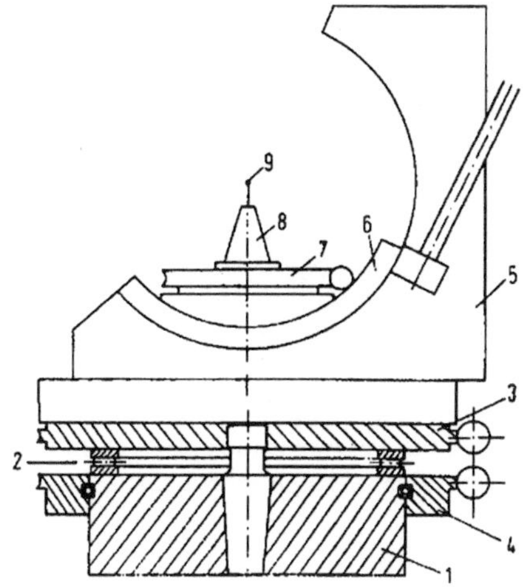

Figure 3 Schematic of the automated single-crystal diffractometer. Reproduced with permission from Hoppe (1965).

Figure 4 Walter Hoppe with the Siemens Elmiskop I electron microscope. Courtesy Archiv der Max-Planck-Gesellschaft, Berlin-Dahlem.

Hoppe's forays into quantitative methods in EM clearly benefited from the close similarity of the theoretical frameworks in the two fields, with Fourier space as the common meeting ground and reference system intimately connected with the data collection geometries in both fields. In Hoppe's words, the electron microscope was a diffraction instrument capable of phasing; that is, the tedious business of phase determination that X-ray crystallography was plagued with simply did not exist (Hoppe, 1970). On the other

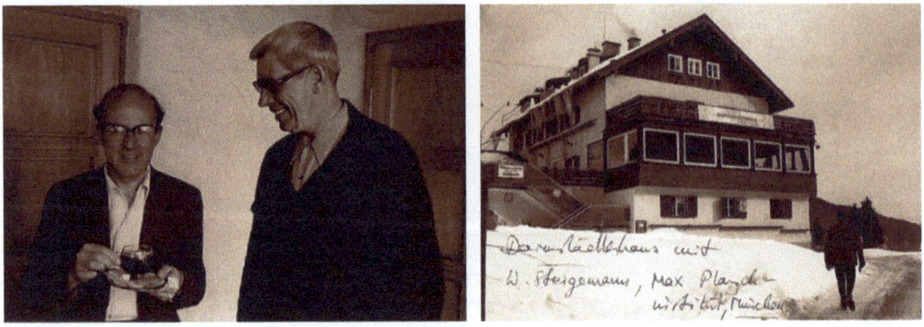

Figure 5 Workshop on Protein Crystallography in Hirschegg in 1968. Left: Walter Hoppe with Max Perutz. Right: Hotel where the workshops took place. From Hoppe's family album (handwritten note by Hoppe). Courtesy Franz Hoppe and Wolfgang Steigemann, 2013.

hand, the electron microscope could also be used in the diffraction mode, leading to methods of electron crystallography in strict analogy to the established X-ray methods but with the advantage that very small crystals or even two-dimensional crystals could be analyzed.

Hoppe's vision expanded on both approaches of EM structure research, as I will elaborate further below, but also went beyond: he developed the theoretical framework of *ptychography*, a method of electron crystallography in which phases could be determined computationally from patterns generated by interference of adjacent, illumination-expanded diffraction spots. The important attractive feature of this method is that it is insensitive to the wave aberrations of the instrument. Hoppe's invention, much ahead of his time, was made a reality by John Rodenburg (see review by Rodenburg, 2008) and has recently led to the highest resolution ever achieved by EM, 0.39Å (Jiang et al., 2018).

Already in the 1960s the exchange of ideas between the community of X-ray crystallography and the nascent community of structural biologists in electron microscopy was fostered by meetings in Hirschegg and Alpbach in the Austrian Alps that were co-hosted by Hoppe and Max Perutz (Fig. 5). In a way we can see these as early precursors of the "hybrid meetings" that were later organized starting 2000 in Granlibakken, Lake Tahoe with the intent to seek common ground between the conceptual foundations and working practices of cryo-EM and X-ray crystallography. "[The] alternation of research and leisure time – reminding me of the format of the 'Gordon Research Conferences' – also has been one the successful recipes in the famous 'Hirschegg meetings' ... These meetings really have been highlights in my scientific apprenticeship" (W. Steigemann, pers. commun., 2021a).

Because of the historical significance of these meetings for the origins of protein crystallography as well as the beginnings of molecular structure research with the electron microscope, the complete list of participants, an unpublished compilation by Richard Dickerson (W. Steigemann, 2021b), is included here as an appendix.

Although he started with far-reaching visions in EM, Hoppe's practical moves to put them into practice could be unrealistic. He approached the challenges in developing EM image processing programs with the idea that they could be met by modifying existing computer programs that had been developed for crystallography. In reality, as I found out when I joined his lab in 1967, the properties of typically noise-ridden electron micrographs required an entirely different approach, drawing on literature from fields like statistical optics and signal processing. Programs had to be written from scratch. The beginning of image processing was also handicapped by technical limitations: The image scanner dumped the data on yards and yards of punched tape. The computer I had to work with, the IBM1130, had a memory of 4k and was programmed by punched cards. The only way to visualize processed images was by overprinting characters on the line printer.

In the year after I joined Hoppe's group, the first 3D reconstruction of a biological object from electron micrographs was published by Aaron Klug's group at the MRC Laboratory of Molecular Biology in Cambridge (DeRosier & Klug, 1968). As Hawkes and Kasper (1994) pointed out, in view of some controversies later on, the principle of 3D reconstruction from projections collected in the electron microscope as tilt series or through use of known symmetries – giving rise so the coverage of 3D Fourier space on central sections – was spelled out clearly in articles by both labs (DeRosier & Klug, 1968; Hoppe et al., 1968). In fact it goes back to a theorem formulated much earlier by Radon (1917). The difference was that Hoppe's article was solely conceptual, while the one by the LMB/MRC group showed results for an object of biological interest, a bacteriophage tail, taking advantage of its helical symmetry. As we will see, the first reconstruction by the Hoppe group would not appear until six years later (Hoppe et al., 1974).

Reconstructing an object from images obtained in the electron microscope also requires an understanding of how these images are related to the object. This relationship is not at all straightforward; it is complicated by the presence of lens aberrations and its formulation, by recourse to the earlier work by Otto Scherzer (1949), is usually governed by a number of simplifying assumptions (e.g., weak phase object in bright-field mode). The most important lens aberration is the spherical aberration, which severely distorts the image in a manner described by the contrast transfer function in bright-field imaging. Over the years much thought has gone into describing the effects of the aberrations on image formation and compensating for them. One far-flung idea at the time I joined the lab (1967) was the use of zone plates, as proposed earlier by Hoppe (1961, 1963) and Friedrich Lenz (1963). These would be designed such as to cut out contributions from those ring-shaped zones in the back focal plane of the objective lens that were transferred with the 'wrong' polarity. The computation of image formation in the presence of a correctly designed and positioned zone plate at different focus settings, as well as the tolerances acceptable, kept the computers of the institute busy. The small dimensions of the zone plate – in the range of 50 microns diameter – created a challenge for both the manufacture and the exact positioning of the device in the back focal plane of the objective lens. Several of these tiny plates were actually made by Hoppe's collaborator Gottfried Möllenstedt, a physicist known for his biprism electron interference experiment (Möllenstedt & Düker, 1956) confirming de Broglie's wave-particle duality theorem for electrons. Möllenstedt was one of the pioneers of microfabrication. These efforts (Hoppe, 1971) were eventually abandoned in the face of technical difficulties, and the prospect of correcting the contrast transfer function computationally (Frank et al., 1970).

Another ambitious initiative was the design of an electron microscope with intrinsic 3D imaging capabilities (Hoppe, 1972; Plies & Typke, 1978; Typke, 1981). The idea was to create an illumination and imaging system that allowed the creation of separate incident beams on a cone with sufficiently high angle (up to about 50 degrees). Successive images taken with differently switched incident beams would then serve as a conical tilt series, thus abrogating the need for a mechanical tilt stage. A large amount of effort went into computations designing the electron optics and examining the imaging characteristics of such a device before the project was eventually abandoned, mainly because of the lack of funding.

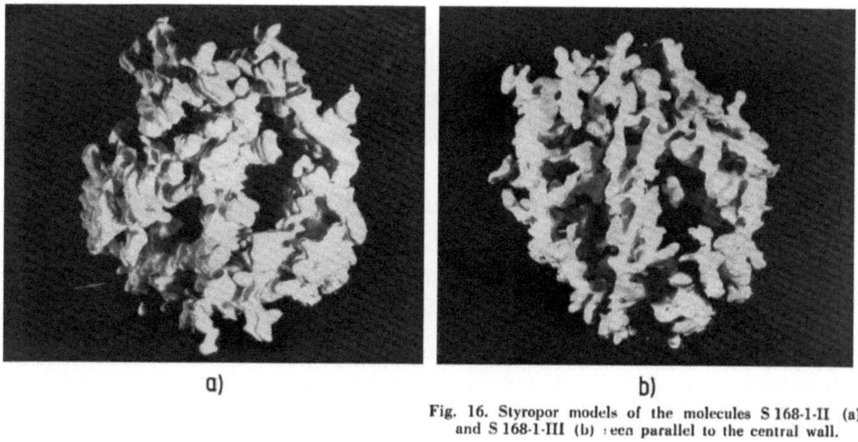

Figure 6 Styropor models of tomographic reconstructions of two fatty acid synthetase molecules. (Reproduced with permission from Hoppe et al., 1976c).

5. Electron tomography

Undoubtedly, Hoppe's most important achievement in electron microscopy was the early development of electron tomography. This method of three-dimensional imaging using a tilt series has universal applications as it does not rely on the presence of symmetries. Except for the first reconstruction (Hoppe et al., 1974), which employed a Fourier-Bessel approach (Cormack, 1973; Smith et al., 1973), the mathematical approach Hoppe's group pursued in the implementation of tomography was weighted back-projection (Lewitt & Bates, 1978; Radermacher, 1980, 1988; Radermacher & Hoppe, 1978), which had the advantage of mathematical tractability and high computational speed, and gained popularity in medical applications such as CAT-scan. Still, the large size of the volumes posed enormous computational challenges at the time. In the application to biological macromolecules, a portion of the EM grid containing several molecules supported on carbon film was imaged by a tilt series and reconstructed in 3D. From the resulting reconstruction, the 3D arrays containing the molecules were carved out. The entire workflow was described in (Hoppe et al., 1976a, 1976b; Hoppe and Hegerl, 1980).

In the first biological application his lab focused on the structure of fatty acid synthetase, resulting in the first 3D EM reconstruction of a single molecule (Hoppe et al., 1974, 1976c). Unfortunately, the results of these pioneering efforts (Fig. 6) were of limited value for biology and were marred by unrealistic resolution claims. Some of the problems were pointed out in a paper by Baumeister and Hahn (1975) that was later rebutted by Hoppe, Gassmann, et al. (1976). Without trying to go through the detailed arguments, here is a list of concerns: (1) the sample was prepared by negative staining, a sample preparation technique plagued with many artifacts including lack of information from the interior, partial collapse of the molecule and other effects of dehydration, partial immersion in stain, and granularity of the stain; (2) the limitation of the tilt range, leading to artifacts ("missing wedge" in Fourier space and the resulting resolution anisotropy); (3) the 3D structure of an individual molecule is extremely noisy, so many copies would need to be averaged before a common structure could be reliably ascertained and modeled; (4) radiation damage from repeated exposure throughout the tilt series leads to obliteration of the molecule during the experiment. In fact, the degree of similarity vs. dissimilarity between two individual molecules can be appreciated by comparing the two styropor models in Fig. 6.

Aware of the artifacts introduced by the inaccessible tilt range in the single-axis data collection – item 2 in the above list – Hoppe explored the use of the conical tilt geometry as a practical alternative (Radermacher, 1980), but this route was not realized until much later. Hoppe addressed the important problem (item 4), the radiation damage limitation, by developing and proving the *theorem of dose fractionation* (Hegerl & Hoppe, 1976; Hoppe & Hegerl, 1981; Hoppe et al., 1977), whose wide-ranging implications for the electron tomography field have been rightly stressed later on (McEwen et al., 1995). The theorem states that the electron dose required to define an element (a voxel) in the 3D density map with the necessary significance is the same as the dose required to obtain an image of that voxel at the same significance level. Practically this means that the total dose for 3D imaging can be distributed among all the projections contributing to the reconstruction. Dose fractionation was subsequently exploited in the reconstructions of the ribosomal subunits performed in his lab (Knauer et al., 1983; Oettl et al., 1983), and has become part of the canon of electron tomography.

However, the other two obstacles – sample preparation by negative staining and lack of statistical significance in the final averaged reconstruction – still persisted and limited the impact of Hoppe's early pioneering work.

He did not live to see the advances of frozen, hydrated sample preparation in electron tomography of molecules (Nitsch et al., 1998) or frozen sections (Al-Amoudi et al., 2004; Hsieh et al., 2002; Marko et al., 2007).

Both required automation of the data collection, first reported in 1991 (Typke et al., 1991). In the case of electron tomography of molecules, the ultimate solution for both the statistical significance problem and the missing wedge problem would later be found through development of subtomogram averaging (Förster & Hegerl, 2007; Renken et al., 2008).

As far as structure determination of biological molecules was concerned, Hoppe envisioned in far-reaching articles a way to trace the events of radiation damage for individual atoms across successive experiments, under the name 'Trace Structure Analysis' (Hoppe, 1975, 1982). However, in their scope these ideas were so far outside the instrumentational and computational realities of the time that the impact was negligible. An entirely different approach – single-particle reconstruction based on the collection of information from thousands of individual molecule projections – gained traction in the end (Frank, 1996; Radermacher et al., 1987). Significantly, both the author and Radermacher received their training in Hoppe's lab and benefited from its sophisticated expertise stemming from the confluence of ideas from X-ray crystallography and electron microscopy. Interestingly, the single-particle approach was discounted by Hoppe in the same year it was proposed: "[But] one of the final aspects of the new electron microscope methods is the study of structure laws of aperiodic objects. For these problems the use of periodicity is obviously useless. One could think of averaging over many identical molecules in an amorphous specimen. But this is dangerous, as a structure depends on the environment, which is – in contrast to the environment in a crystal lattice – different for every molecule" (Hoppe, 1975). Hoppe's dictum was right insofar as molecules in the cellular context are in different states, and perhaps he had no way of foreseeing the power of a method that precluded or minimized interactions between molecules, by suspending them in solution.

6. Walter Hoppe remembered

Hoppe was tall and, at the time I met him, full of energy, which he kept up by playing tennis with Johann Gaßmann (Fig. 7), who assisted his lab with IT. "Hoppe had thick glasses that dominated his face. I still see him, standing up in the discussion of a lecture, not with a question but with a comment – invariably referring to a footnote in an article he had published years before … I find it plausible that his habit was a major reason why the use of footnotes has been barred in most of today's scientific journals" (Frank, 2016).

Figure 7 Walter Hoppe with Johann Gaßmann (1967). From Hoppe's family album; courtesy Franz Hoppe and Wolfgang Steigemann, 2013.

Many accounts attest to the fact that Hoppe was a man of strong opinions, and this certainly jives with my own experience. "All who have known Walter Hoppe will remember his strong personality. We will miss his stimulating and critical remarks in discussions, which he usually started very innocently with the remark 'I want to make only a short comment'. In fact this was the signal for a long scientific discussion" (King, 1987).

"I saw first saw Walter Hoppe at EMBL in 1979. He and John Kendrew were evidently good friends and he most likely pushed Kendrew's enthusiasm towards developing EM in structural biology. EMBL certainly encouraged a number of EM groups to grow in several different directions. Hoppe's talk I recall, because of his wild dream of a computer/electron microscope team. His eyesight must have been rather poor as some of the slides he showed were side-ways. Like most in the audience I looked with my head tilted – I presume he was equally unaware of his audience's behavior." (John Berriman, personal communication 2016).

"The style of management with combination of scientific discussions in local seminars and non-scientific get-togethers certainly strengthened the feeling of cohesion in the research group. I well remember Hoppe's invitation to a party on the occasion of his 50th birthday in a very relaxed environment at his home close to the English Garden in Munich where he asked his 13-year-old son Franz to look for some additional champagne. Franz asked: 'von dem besseren oder dem einfacheren?' (from the better one or from the more ordinary one?). At another invitation at a later time in his new home in Gräfelfing closer to the new Max Planck Institute

Kurzmitteilung

aus dem Department für Elektronenmikroskopie ....

Im Hoppe'schen Strahlengang zur Messung des Hoppe-Winkels wird eine Hoppe-Platte im Langer'schen Tubus mit Hilfe einer Frank-Halterung so gehaltert, daß das 3. Zwischenbild (Hoppe'sches Zwischenbild) um 5 Hirt defokussiert

▪▪▪▪▪

Hoppe-Reiter möglich.) Dieser Gedanke wurde bereits 1948 in einem Briefwechsel von Hoppe mit Hoppe geäußert.

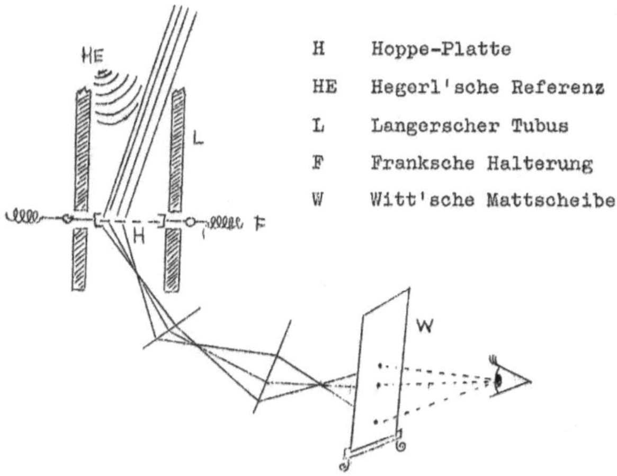

Figure 8 Excerpt from the internally produced Christmas journal of the Department of Crystallography of the MPG for Protein and Leather Research in 1967. "Short communication from the Department of Electron Microscopy. In Hoppe's electron optical path for the measurement of the Hoppe angle, a Hoppe plate is fastened with the aid of a Frank positioning device in a Langer tube in such a way that the third intermediate image (i.e., the Hoppe intermediate image) is defocussed by 5 Hirt... This idea was already mentioned in 1948 in an exchange of letters between Hoppe and Hoppe." Courtesy Wolfgang: Steigemann, 2013.

of Biochemistry in Martinsried (merger of three institutes) I have been deeply impressed not only by the hospitality of Hoppe's family, but also by his performance in piano playing and by his local setup of high fidelity audio equipment. Not to forget the yearly Christmas parties where institute work has been presented in cabaret like papers (Fig. 8)" (Steigemann, pers. commun., 2021a).

Acknowledgments

I would like to thank Peter Hawkes for inviting me to contribute a biography of my mentor. The correspondence with him added richness to the product as memory is fading away. I'm grateful to John Berriman, Reiner Hegerl, Franz Hoppe, Robert Huber, Michael Radermacher, Wolfgang Steigemann, and Dieter Typke for valuable recollections and suggestions, for pointing me to sources, for critical reading of the draft, and letting me use some unpublished material.

Appendix to "Walter Hoppe — X-ray crystallographer and visionary pioneer in electron microscopy" by Joachim Frank

Hitherto unpublished letter by Richard E. Dickerson, now Professor Emeritus at UCLA, to the Alumni of the Austrian Workshops, 1955-1972. The letter, which was kindly made available by Wolfgang Steigemann, contains a complete list of participants of the workshops in Hirschegg and Alpbach.

TO ALUMNI OF THE AUSTRIAN WORKSHOPS, 1955-1972:

This message brings to a successful conclusion the collection of data on the four Austrian protein structure workshops: 1966, 1968 and 1970 in Hirschegg, 1972 in Alpbach. Thanks to Anders Liljas, Robert Huber, Jan Drenth, Johann Deisenhofer, Brian Matthews, Fred Richards, and most recently Wolfang Steigemann, I have been able to assemble a complete set of both scientific programs and attendance lists for all four meetings, along with quite a bit of peripheral material: letters, memos, travel schedules, etc. If any of this material is of interest to you, I would be happy to send you xeroxes.

The enclosed table (of which you received an incomplete draft earlier) lists all the workshop participants by year, and whether they were principal speakers (P), coauthors (2) or audience (*). The table could be completed after Wolfgang Steigemann sent me the missing bit of data: the attendance list for the 1970 workshop. In all, 290 people attended one workshop or another, and 19 hardy souls showed up for all four.

My book, "Present at the Flood: How Structural Molecular Biology Came About", has just been published, and more than one person has asked whether all this data-gathering about the Austrian workshops was preparation for a sequel of some kind. No, it isn't. The period covered by the book, from 1934 to 1964, was an exciting one marking a genuine revolution in our knowledge of biological structure. It is fascinating to me to actually read the original words of the pioneers who created that revolution, in the various scientific papers reprinted in the book. I hope that current students will find this to be equally interesting reading. But from 1964 on the story is one of the growth and evolution of a new field of science, in dozens of individual laboratories. The sense of drama—the feeling of "Will we, or won't we succeed?"— is missing. It is replaced by "Will we, or won't we, get our grants and get a big enough computer?" It's not the same thing.

So thanks for your help. Let me know if you would like copies of any of the Austrian workshop materials.

With best regards,
Dick Dickerson

Workshop Attendees 5 July 2005 1

PROTEIN WORKSHOPS: HIRSCHEGG (1966, 1968, 1970) AND ALPBACH (1972)

P = Principal speaker; 2 = Coauthor; * = Audience

Note: Where multiple authors are listed for a talk, the first name among those authors actually attending the meeting is assumed to be the speaker (P), with others listed as coauthors (2). This may be incorrect, but is impossible to put right at this late date. In some cases, a person listed as first author does not show up among the list of attendees, and these are indicated by (P?). Either the attendee list is in error, or another of the authors really attended and presented the paper. Your corrections would be appreciated; as would the first names of many attendees.

Year:		66	68	70	72
Horst G. Abel	Max Planck, Munich	*	*	2	
Margaret J. Adams	Oxford	P	*		
	Purdue			*	2
Traudl Aidinian	Max Planck, Munich			*	
R. Allmann	Marburg	*			
A. Amit	Max Planck, Munich	*			
N. Amit	Max Planck, Munich	*			
L. M. Amzel	Johns Hopkins				P
N. L. Anderson	Cambridge				*
Natalia Andreeva	Moscow		P	*	*
Uli W. Arndt	Cambridge	P	*		*
Arthur Arnone	Cambridge				P
Emil Arutyunyan	Moscow SU			*	
E. N. Baker	Oxford			*	
Joyce Baldwin	Cambridge			P	*
Len Banaszak	Wash. U., St. Louis		P	P	P
David W. Banner	Oxford			*	*
S. Banyard	Oxford			*	*
A. Barrett	Heidelburg			*	
Klaus Bartels	Max Planck, Munich				*
M. E. Baumber	London			*	
C. R. Beddel	Oxford		*		
Graheme Bentley	Oxford				*
Per-Chroster Bergstén	Uppsala			*	
(_?_) Berthon	Paris			*	
U. Biebl	Garching, GER				*
Kirsten Biederman	Heidelberg				*
Jens J. Birktoft	Cambridge		P		
	La Jolla			P	
Colin C. F. Blake	Oxford	P	*		
Ann C. Bloomer	Cambridge				2
	Oxford			2	

Workshop Attendees 5 July 2005 2

David M. Blow	Cambridge	P	2	*	2
Tom Blundell	Oxford		*	*	*
Wolfram Bode	Max Planck, Munich				*
Torne Boiwe	Uppsala			*	2
F. Borras	Oxford				*
Sir Lawrence Bragg	Royal Institution		*		
Carl Bränden	Uppsala	P	P	P	
Franz Brandl	Max Planck, Munich	*	*		
Norbert Brodherr	Max Planck, Munich			*	*
W. J. Browne	Oxford				*
T. N. Bryant	Bristol				*
Manfred Bühner	Purdue				*
C. W. Bunn	London			P	
Roger M. Burnett	Michigan				2
J. W. Campbell	Bristol				*
Charles W. Carter	UC San Diego				P
John N. Champness	Cambridge			*	*
Cyril Chothia	Cambridge				P
Peter M. Colman	Oregon				*
Michel Cotriat	Bristol			*	
Albert Cotton	MIT, Cambridge, MA			P	
J. Cox	Cambridge		*		
Thomas E. Creighton	Cambridge			*	*
R. A. Crowther	Cambridge			*	
(_?_) Cutfield	Oxford				*
J. Dargay	Oxford				*
David R. Davies	NIH		P	P	*
Johann Deisenhofer	Max Planck, Munich				*
Robert Diamond	Cambridge	P	P	P	P
Richard E. Dickerson	Caltech	P	P	P	P
Rodger M. Dittgen	Heidelberg			*	
M. Dobler	ETH, Zürich				*
Eleanor Dodson	Oxford	*	*		*
Guy Dodson	Oxford	P	*	P	*
S. D. Dover	ETH, Zürich			*	*
Jan Drenth	Groningen	P	P		P
Emile Duée	Grenoble			*	
	Bristol				*
M. Dupont	Max Planck, Heidelberg				*
Peter Eagles	Oxford			*	
Alan B. Edmundson	Argonne				*
B. F. P. Edwards	Harvard				2
David Eisenberg	UCLA			*	P
Hans Eklund	Uppsala			*	2
Jürgen Engel	Max Planck, Munich	*	*	*	
Hans Peter Englmeier	Max Planck, Munich	2			

Workshop Attendees 5 July 2005 3

Otto Epp	Max Planck, Munich	2	2	*	P
Birgitta Eriksson	Stockholm			*	*
David R. Evans	Cambridge			*	
Philip Evans	Oxford			*	2
Lynn Ten Eyck	Cambridge				P
Heinz Fehlhammer	Max Planck, Munich				*
R. E. Fenna	Oxford			*	*
K. Fleischmann	Max Planck, Munich	*			
Robert Fletterick	Yale				*
L. O. Ford	Oxford				*
Helmut Formanek	Max Planck, Munich	2	2	*	
Joachim Frank	Max Planck, Munich		*		
Hartmut Fuess	Grenoble				P
W. Fuller	King's College				*
Johann Gassmann	Max Planck, Munich	P	P	*	*
Rudolf Gatz	Max Planck, Munich			*	
M. J. Geisow	Oxford				P
Alfred Gieren	Max Planck, Munich	*	*	*	*
P. F. C. Gilbert	Cambridge		*	P	*
Beatrice Gorinsky	Birkbeck				*
S. Göttlicher	Tech. Hoch. Darmstadt	*			
David W. G. Green	Edinburgh	P	*		*
Jonathan Greer	Cambridge		*	P	*
Marvin L. Hackert	Purdue				2
Erich Hädicke	Max Planck, Munich	*	*	*	
B. Handford	Oxford		P		
Marjorie M. Harding	Edinburgh	P			
Karl D. Hardman	Argonne				*
Steve Harrison	Heidelberg/Harvard			P	*
Edward E. Hazen	MIT			2	*
Siegfried Hechtfischer	Max Planck, Munich	*			
Reiner Hegerl	Max Planck, Munich		*		
Elizabeth G. Heidner	Cambridge				*
D. M. van der Helm	Groningen			*	
Richard Henderson	Cambridge		2	*	
Wayne Hendrickson	Naval Res. Lab. Washington			2	*
Dorothy C. Hodgkin	Oxford		*		*
G. I. Hodgson	Bristol				*
Wim G. J. Hol	Groningen			*	
Ken C. Holmes	Cambridge	*	P		
	Max Planck, Heidelberg			P	P
Walter Hoppe	Max Planck, Munich	P	P	P	P
Robert Huber	Max Planck, Munich	P	P	P	P
E. M. C. Hutton	Edinburgh				*
Erwin Imsel	Max Planck, Munich			*	
A. Jack	Cambridge			*	*

Workshop Attendees 5 July 2005 4

B. Jacrot	Max Planck, Grenible				P
Uwe Jakubowski	Max Planck, Munich		P	*	2
J. Janin	Cambridge				*
Jan. N. Jansonius	Groningen	*		*	2
Lyle H. Jensen	U. Washington				*
Louise N. Johnson	Oxford		*		*
C. L. Jones	Edinburgh				*
T. Alwyn Jones	King's College				*
Michael A. Joynson	Oxford		*	P	
	Bristol				*
Wolfgang Kabsch	Max Planck, Heidenberg		*		*
K. K. Kannan	Uppsala		P	*	P
Gobinath N. Kartha	Roswell Park, Buffalo	P	P	*	*
Louia Katz				*	
P. F. C. Kierkegaard	Stockholm	*	*	*	*
Sung Hou Kim	MIT				*
Aaron Klug	Cambridge	P	P		
G. Knesch	Max Planck, Munich	*			
R. Koekoek	Groningen		2		
Don Koenig	Brookhaven		*		
Gerhard Kopfmann	Max Planck, Munich	*	P		
David Kotelchuck	Cornell			P	
C. Kratky	ETH, Zürich				*
Joe Kraut	UC San Diego		P		
Robert H. Kretsinger	U. Virginia			P	P
Dietmar Kukla	Max Planck, Munich			*	2
P. Laggner	Graz, Austria				*
R. Langer	Max Planck, Munich	*	2		
Robert Langridge	Princeton				*
L. O. Larsson	Stockholm			P	*
Eaton E. Lattman	Johns Hopkins			*	*
K. Laves	ETH, Zürich				*
R. Leberman	Max Planck, Heidelberg			*	*
M. Leijonmarck	Stockholm				*
Paul J. Lentz, Jr.	Uppsala			*	*
Michael Levine	York				*
Michael Levitt	Cambridge			P	
Anders Liljas	Uppsala	P	2	P	
Wm. N. Lipscomb	Harvard		P	2	2
W. Littke	Freiburg, GER				*
L. O . Lorsson	Stockholm			*	
Herman Lotter	Max Planck, Munich			*	
Warner E. Love	Johns Hopkins			P	P
S. Lövgren	Uppsala				2
Barbara Low	Columbia	P			
R. M. Lucas	Boston			*	

Workshop Attendees 5 July 2005 5

Martha L. Ludwig	Michigan				P
Penelope A. Machin	Oxford		*	*	
Peter Main	Manchester		*		
Eckhard Mandelkow	Max Planck, Heidelberg			*	*
P. Manor	Göttingen				*
H. Martin	Tech. Hoch. Munich	*			
F. Marx	Max Planck, Heidelberg				*
S. Mason	Oxford				*
F. Scott Mathews	Washington U. St. Louis				P
Brian Matthews	Cambridge	P			
	Oregon			P	P
A. Mayer	Tech. Hoch. Munich		*		
L. Mazzarella	Naples				*
Andrew D. McLachlan	Cambridge		*	*	
W. Duncan Mercer	Bristol			*	*
D. Mercola	Oxford				*
Edgar Meyer	TAMU				*
P. C. Moews	London			*	
J. Keith Moffat	Cambridge		*		
	Cornell			P	*
Shirley J. Morris	Max Planck, Heidelberg			*	*
J. Mosley	Oxford				*
John Moult	Oxford		*		
	Weizmann Institute				*
Hilary Muirhead	Cambridge	P	2		
	Bristol			*	*
P. E. Nixon	Oxford				*
Chris Nordman	Oxford				*
Bo Nordstrôm	Uppsala			*	P
A. C. T. North	Royal Institution, London	P			
	Oxford		P	2	P
S. C. Nyburg	Toronto		*		
R. E. Offord	Oxford			*	*
W. Palm	Max Planck, Munich				*
H. H. Paradies	Münster		*		
Fritz Parak	Tech. Hoch., Munich		P		
	Max Planck, Munich			*	*
Max F. Perutz	Cambridge	P	P	P	P
Greg A. Petsko	Oxford				2
David C. Phillips	Royal Institution, London	P			
	Oxford		P	2	P
Richard P. Phizackerley	Cambridge			P	
	ETH, Zürich				*
M. Pierrot	Groningen				*
Roberto Poljak	Baltimore			P	
H. Pollard	Oxford				*

Workshop Attendees 5 July 2005 6

P. del Pulsinelli	Cambridge				P
Florante Quiocho	Cambridge			*	
Dove Rabinovich	Weizmann Institute				*
C. Rérat	CNRS, Bellevue, France			*	2
Fred Richards	Yale		P		
David C. Richardson	Duke			*	*
A. Ripamonti	Naples				*
D. Rogers	Imperial College, London				*
Michael Röhrl	Max Planck, Munich	*	*	*	
O. Rönnquist	Stockholm				*
Gerold Rosenbaum	Max Planck, Heidelberg			*	2
David de Rosier	Cambridge		2		
Michael Rossmann	Purdue	P	P	P	P
B. Rubin	Duke				*
Alfred Rühlmann	Max Planck, Munich			*	2
M. N. Sabesan	Oxford				*
Wolfram Saenger	Max Planck, Göttingen			*	*
V. R. Sarma	Oxford		2		
L. Sawyer	Bristol				*
Christian Scheringer	Aachen	P			
P. Schlecht	Tech. Hoch. Munich		*		
Hans-Josef Schramm	Max Planck, Munich			*	*
Georg Schulz	Max Planck, Heidelberg			*	*
Peter Schwager	Max Planck, Munich			*	*
Carl Schwalbe	Max Planck, Göttingen			*	*
Benno Schönborn	Brookhaven			(P?)	
Helen Scouloudi	Oxford			P	
Karin Seige	Max Planck, Munich			*	
P. Seiler	ETH, Zürich				*
Peter v. Sengbusch	Tübingen	*			
	Heidelberg			*	*
David M. Shotton	Bristol			*	P
Paul Sigler	Cambridge	P			
	U. Chicago		*		*
J. D. G. Smit	Groningen				*
K. W. Snape	Oxford				*
Henry M. Sobell	Rochester				P
Bengt-Olof Söderberg	Uppsala		2	*	*
Gustav Söderlund	Uppsala			*	2
H.-J. Springer	Max Planck, Munich	*			
D. K. Stammers	Bristol				*
Wolfgang Steigemann	Max Planck, Munich			*	2
Tom A. Steitz	Cambridge		2		
	Yale				*
Bror Strandberg	Uppsala				*
Robert M. Stroud	Caltech				P

Workshop Attendees 5 July 2005 7

Gerald Stubbs	Oxford			*	
Manfred Sturm	Max Planck, Munich				*
D. Suck	Max Planck, Göttingen				*
Joel Sussmann	New York			*	
Ian D. A. Swan	Oxford		*	P	
Robert M. Sweet	Cambridge				*
H. M. Swen	Groningen				*
Ulrich-F. Thomanek	Munich				*
Robert Tijan	Oxford				*
B. Tilander	Uppsala	P			
Patrick Tollin	Dundee	*	*	P	*
Wolfie Traub	Weizmann Institute				*
D. Tsernoglou	Yale		*		
	Wayne State, Detroit				*
Alexander Tulinsky	Michigan State				*
Ivar Vaara	Uppsala				2
Boris K. Vainshtein	Moscow				P
M. E. Vijayan	Oxford		*	*	
Ivar Waara	Uppsala			*	
M. Wade	Max Planck, Heidelberg				*
K. Wallenfels	Freiburg, Germany				*
K. B. Ward	Johns Hopkins				*
Stephen G. Warren	Harvard			*	2
P. Wassarman	Cambridge		*		
Herman C. Watson	Cambridge	P	*		
	Bristol			*	P
H. Wawra	Graz, Austria			*	*
Jon Weinzierl	Cambridge			*	
P. Wendell	Bristol			*	*
P.-E. Werner	Stockholm				*
J. L. White	Purdue				2
Nickolas White	Bristol			*	
Don C. Wiley	Harvard			P	P
G. Will	Tech. Hoch. Darmstadt	*			
I. A. Wilson	Oxford				*
B. G. Wolthers	Groningen		2		
Alan J. Wonacott	Cambridge			*	*
K. F. Woods	Edinburgh				*
(_?_) Wootton				*	
Christine Wright	Cambridge			*	
H. T. Wright	Cambridge			P	
Hal W. Wyckoff	Yale	P		*	*
Ada Yonath	Weizmann Institute				*
Klaus Zechmeister	Max Planck, Munich	*			
Chr. Zelwer	CNRS, Gif-sur-Yvette, France				*
Eila Zeppezauer	Uppsala			*	2

Workshop Attendees **5 July 2005** **8**

M. Zeppezauer	Uppsala				2
Speakers and coauthors (P and 2):		32	39	41	610
Audience (*):		25	42	96	139
Total attendees:		57	81	137	199
Percent audience:		42%	52%	70%	70%

References

Al-Amoudi, A., Chang, J.-J., Leforestier, A., McDowall, A., Salamin, L. M., Norlén, L. P. O., Richter, K., Sartori Blanc, N., Studer, D., & Dubochet, J. (2004). Cryo-electron microscopy of vitreous sections. *EMBO Journal*, *23*, 3583–3588.

Baumeister, W., & Hahn, M. (1975). Relevance of three-dimensional reconstructions of stain distributions for structural analysis of biomolecules. *Hoppe-Seyler's Zeitschrift für Physiologische Chemie*, *356*, 1313–1316.

Cormack, A. M. (1973). Reconstruction of densities from their projections, with applications in radiological physics. *Physics in Medicine and Biology*, *18*, 195–207.

DeRosier, D., & Klug, A. (1968). Reconstruction of three-dimensional structures from electron micrographs. *Nature*, *217*, 130–134.

Förster, F., & Hegerl, R. (2007). Structure determination in situ by averaging of tomograms. In J. R. McIntosh (Ed.), *Methods in cell biology: Vol. 79* (pp. 741–767). New York, NY: Academic Press.

Frank, J. (1996). *Three-dimensional electron microscopy of macromolecular assemblies*. New York: Academic Press.

Frank, J. (2016). *Franklab blog 4/22/2016*. https://franklab.cpmc.columbia.edu/franklab/.

Frank, J. (2019). *Untersuchungen von elektronenmikroskopischen Aufnahmen hoher Auflösung mit Bilddifferenz- und Rekonstruktionsverfahren* [Investigations of high-resolution electron micrographs using image difference and reconstruction methods]. Technical University Munich Press, ISBN 9783958840218. Reprint of Frank, J. (1970) dissertation at TUM.

Frank, J., Bußler, P., Langer, R., & Hoppe, W. (1970). Einige Erfahrungen mit der rechnerischen Analyse und Synthese von elektronenmikroskopischen Bildern hoher Auflösung [Experience gained from digital analysis and synthesis of high-resolution electron micrographs]. *Berichte der Bunsengesellschaft für physikalische Chemie*, 74, 1105–1115.

Hawkes, P. W., & Kasper, E. (1994). *Principles of electron optics. Wave optics: Vol. 3*. London: Academic Press (pp. 1200–1202).

Hegerl, R., & Hoppe, W. (1976). Influence of electron noise on three-dimensional image reconstruction. *Zeitschrift für Naturforschung. A*, *31*, 1717–1721.

Hoppe, W. (1957a). Die 'Faltmolekülmethode' - eine neue Methode zur Bestimmung der Kristallstruktur bei ganz oder teilweise bekannter Molekülstruktur. *Acta Crystallographica*, *10*, 750–751.

Hoppe, W. (1957b). Die Faltmolekülmethode und ihre Anwendung in der roentgenographischen Konstitutionsanalyse von Biflorin ($C_{20}H_{20}O_4$). *Zeitschrift für Elektrochemie*, *61*, 1076–1083.

Hoppe, W. (1961). Ein neuer Weg zur Erhöhung des Auflösungsvermögens des Elektronemikroskops. *Naturwissenschaften*, *48*, 736–737.

Hoppe, W. (1963). Fresnelsche Zonenkorrekturplatten für das Elektronenmikroskop. *Optik*, *20*, 599–606.

Hoppe, W. (1965). Wege zur automatisierten Roentgenstrukturanalyse. Teil I. Automatisierung des Messvorgangs. *Angewandte Chemie*, 77, 484–492.

Hoppe, W. (1970). Principles of electron structure research at atomic resolution using conventional electron microscopes for the measurement of amplitudes and phases. *Acta Crystallographica. Section A*, *26*, 414.

Hoppe, W. (1971). Use of zone correction plates and other techniques for structure determination of aperiodic objects at atomic resolution using a conventional electron microscope. *Philosophical Transactions - Royal Society. B*, *261*, 7194.

Hoppe, W. (1972). Dreidimensional abbildende Elektronenmikroskope I. Prinzip der Geräte. *Zeitschrift für Naturforschung*, *27a*, 919–929.

Hoppe, W. (1975). Principles of trace structure analysis in electron microscopy. *Zeitschrift für Naturforschung*, *30a*, 1188–1199.

Hoppe, W. (1982). Trace structure analysis, ptychography, phase tomography. *Ultramicroscopy*, *10*, 187–198.

Hoppe, W., & Hegerl, R. (1980). Three-dimensional structure determination by electron microscopy. In P. W. Hawkes (Ed.), *Computer processing of electron microscope images* (pp. 127–186). Heidelberg: Springer-Verlag.

Hoppe, W., & Hegerl, R. (1981). Some remarks concerning the influence of electron noise on 3D reconstruction. *Ultramicroscopy*, *6*, 205–206.

Hoppe, W., Langer, R., Knesch, G., & Poppe, Ch. (1968). Protein-Kristallstrukturanalyse mit Elektronenstrahlen. *Naturwissenschaften*, *55*, 333–336.

Hoppe, W., Gassmann, J., Hunsmann, N., Schramm, H. J., & Sturm, M. (1974). Three-dimensional reconstruction of individual negatively-stained fatty-acid synthetase molecules from tilt series in the electron microscope. *Hoppe-Seyler's Zeitschrift für Physiologische Chemie*, *355*, 1483–1487.

Hoppe, W., Gassmann, J., Hunsmann, N., Schramm, H., & Sturm, M. (1976). Comments on the paper "Relevance of three-dimensional reconstructions of stain distributions for structural analysis of biomolecules.". *Hoppe-Seyler's Zeitschrift für Physiologische Chemie*, *356*, 1317–1320.

Hoppe, W., Schramm, H. J., Sturm, M., Hunsmann, N., & Gaßmann, J. (1976a). Three-dimensional electron microscopy of individual biological objects. I. Methods. *Zeitschrift für Naturforschung*, *31a*, 645–655.

Hoppe, W., Schramm, H. J., Sturm, M., Hunsmann, N., & Gaßmann, J. (1976b). Three-dimensional electron microscopy of individual biological objects. II. Test calculations. *Zeitschrift für Naturforschung*, *31a*, 1370–1379.

Hoppe, W., Schramm, H. J., Sturm, M., Hunsmann, N., & Gaßmann, J. (1976c). Three-dimensional electron microscopy of individual biological objects. III. Experimental results on yeast fatty acid synthetase. *Zeitschrift für Naturforschung*, *31a*, 1380–1390.

Hoppe, W., Wenzl, H. U., & Schramm, H. J. (1977). New three-dimensional reconstruction technique at minimal dose conditions in electron microscopy. *Hoppe-Seyler's Zeitschrift für Physiologische Chemie*, *358*, 1069–1076.

Hsieh, C.-E., Marko, M., Frank, J., & Mannella, C. A. (2002). Electron tomographic analysis of frozen-hydrated tissue sections. *Journal of Structural Biology*, *138*, 63–73.

Huber, H., & Holmes, K. (1987). Walter Hoppe 21.3.1917–3.11.1986. In Max-Planck-Gesellschaft (Ed.), *Jahresbericht 1986 und Jahresrechnung 1985: Nachrufe. Max-Planck-Gesellschaft, Berichte und Mitteilungen. 4/1987* (pp. 78–81).

Huber, R. (2019). In Frank (2019).

Jaskolski, M., Dauter, Z., & Wlodawer, A. (2014). A brief history of macromolecular crystallography, illustrated by a family tree and its Nobel fruits. *The FEBS Journal*, *281*, 3985–4009.

Jiang, Y., Chen, Z., Han, Y., Deb, P., Gao, H., Xie, S., Purohit, P., Tate, M. W., Park, J., Gruner, S. M., Elsner, V., & Muller, D. A. (2018). Electron ptychography of 2D materials to deep sub-ångström resolution. *Nature*, *559*, 343–349.

King, J. N. (1987). Crystallographers. *Journal of Applied Crystallography*, *20*, 324–325.

Knauer, V., Hegerl, R., & Hoppe, W. (1983). Three-dimensional reconstruction and averaging of 30S ribosomal subunits of *Escherichia coli* from electron micrographs. *Journal of Molecular Biology, 163*, 409–430.

Lenz, F. (1963). Zonenplatten zur Öffnungsfehlerkorrektur und zur Kontrasterhöhung. *Zeitschrift für Physik, 172*, 498.

Lewitt, R. M., & Bates, R. H. T. (1978). Image reconstruction from projections. II Modified back-projection methods. *Optik, 50*, 85–109.

Marko, M., Hsieh, C., Schalek, R., Frank, J., & Mannella, C. (2007). Focused-ion-beam thinning of frozen-hydrated biological specimens for cryo-electron microscopy. *Nature Methods, 4*, 215–217.

McEwen, B. F., Downing, K. H., & Glaeser, R. M. (1995). The relevance of dose-fractionation in tomography of radiation-sensitive specimens. *Ultramicroscopy, 60*, 357–373.

Möllenstedt, G., & Düker, H. (1956). Beobachtungen und Messungen an Biprisma-Interferenzen mit Elektronenwellen. *Zeitschrift für Physik, 145*, 377–397.

Nitsch, M., Walz, J., Typke, D., Klumpp, M., Essen, L.-O., & Baumeister, W. (1998). Archaeal group II chaperonin in an open conformation: The thermosome from *Thermoplasma acidophilum* examined by cryo-electron tomography. *Nature Structural Biology, 5*, 855–857.

Oettl, H., Hegerl, R., & Hoppe, W. (1983). Three-dimensional reconstruction and averaging of 50S ribosomal subunits of *Escherichia coli* from electron micrographs. *Journal of Molecular Biology, 163*, 431–450.

Plies, E., & Typke, D. (1978). Dreidimensional abbildende Elektronenmikroskope, II. Theorie elektronenoptischer Systeme mit gekrümmter Achse. *Zeitschrift für Naturforschung, 33a*, 1361–1377.

Radermacher, M. (1980). Dreidimensionale Rekonstruktion bei kegelförmiger Kippung im Elektronenmikroskop [Three-dimensional reconstruction in the electron microscope using conical tilt] (Dissertation). Technische Universität München.

Radermacher, M. (1988). Three-dimensional reconstruction of single particles from random and nonrandom tilt series. *Journal of Electron Microscopy Technique, 9*, 359–394.

Radermacher, M., & Hoppe, W. (1978). 3D reconstruction from conically tilted projections. In *Ninth international congress of electron microscopy, Toronto 1978: Vol. 1* (pp. 218–219).

Radermacher, M., Wagenknecht, T., Verschoor, A., & Frank, J. (1987). Three-dimensional structure of the large ribosomal subunit from *Escherichia coli*. *EMBO Journal, 6*, 1107–1114.

Radon, J. (1917). Über die Bestimmung von Funktionen durch ihre Integralwerte längs gewisser Mannigfaltigkeiten. *Berichte über die Verhandlungen der Sächsischen Akademie der Wissenschaften Zu Leipzig. Mathematisch-Physische Klasse, 69*, 262–277.

Renken, C., Hsieh, C. E., Marko, M., Rath, B., Leith, A., Wagenknecht, T., Frank, J., & Mannella, C. A. (2008). Structure of frozen-hydrated triad junctions: A case study in motif searching inside tomograms. *Journal of Structural Biology, 165*, 53–63.

Rodenburg, J. (2008). Ptychography and related diffractive imaging methods. *Advances in Imaging and Electron Physics, 150*, 87–184.

Rossmann, M. G., & Blow, D. M. (1962). The detection of subunits within the crystallographic asymmetric unit. *Acta Crystallographica, 15*, 24–31.

Scherzer, O. (1949). The theoretical resolution limit of the electron microscope. *Journal of Applied Physics, 20*, 20–29.

Smith, P. R., Peter, T. M., & Bates, R. H. T. (1973). Image reconstruction from a finite number of projections. *Journal of Physics. A. Mathematical Nuclear and General, 6*, 361–382.

Steigemann, W. (2013). *From "Walter Hoppe (1917-1986) - Leben und Wirken eines Pioniers"*. Talk given as part of the series "Dichter und Denker von Gräfelfing," on the occasion of the 1250th birthday of Gräfelfing, where Hoppe resided during very important years of his scientific life.

Steigemann, W. (2021a). Personal communication.

Steigemann, W. (2021b). The compilation by Richard Dickerson is contained in a letter he wrote in 2005 to Wolfgang Steigemann. Who made a copy available to the author.

Typke, D. (1981). Image aberrations of axially symmetric imaging systems with off-axial rays: Three-dimensionally imaging electron microscopes (part III). *Nuclear Instruments and Methods, 187*, 217–226.

Typke, D. (2019). In Frank (2019).

Typke, D., Dierksen, K., & Baumeister, W. (1991). Automatic electron tomography. In *Proc. 49th EMSA meeting, San Jose 1991* (pp. 544–545).

CHAPTER FOUR

Reminiscences of the development of electron optics and electron microscope instrumentation in Japan

Koichi Kanaya[a] **and Peter Hawkes (Afterword)**[b,*]
[a]Formerly Department of Electrical Engineering, Kogakuin University, Tokyo, Japan
[b]CEMES-CNRS, Toulouse, France
*Corresponding author. e-mail address: hawkes@cemes.fr

Contents

1. Historical survey of the early stages of electron microscopy in Japan 72
2. Lens aberrations 80
   2.1 Fifth-order Seidel aberrations and third-order astigmatism 80
   2.2 Aberrations of a focused charged-particle beam caused by space charge 84
   2.3 Asymmetric aberrations and manufacture accuracy 88
3. Practical magnetic lens design 93
   3.1 Minimum aberration coefficients 93
   3.2 Aberrations of three-stage electron microscopes 96
   3.3 Minimum aberrations and achromatization 105
   3.4 Anastigmatic lens 109
4. Formation of caustic and shadow images 110
   4.1 Third-order caustic 110
   4.2 Caustic properties combined with second- and third-order astigmatism 113
   4.3 Shadow image 114
5. Stigmators 117
   5.1 Electrostatic stigmator used to correct second- and third-order astigmatisms 117
   5.2 Magnetic stigmator used to correct second- and third-order astigmatisms 120
   5.3 Anastigmatic alignment 122
   5.4 Anastigmatic deflection system for correcting astigmatism and its defocusing action 124
6. Electron phase microscope 127
7. High-voltage electron microscopes 130
8. Afterword by Peter Hawkes 131
References 133

Advances in Imaging and Electron Physics, Volume 221
ISSN 1076-5670
https://doi.org/10.1016/bs.aiep.2022.03.004

1. Historical survey of the early stages of electron microscopy in Japan

According to a historical survey by L. Marton (1968), the transmission electron microscope was initially constructed by E. Ruska in Professor M. Knoll's laboratory at the Berlin Technische Hochschule in Germany in June, 1931, and was developed into the first Siemens microscope of B. von Borries and E. Ruska (1938) with an estimated resolving power of 100 Å. This transmission electron microscope and its results clearly shown in micrographs attracted the attention of several ambitious scientists and stimulated studies on electron microscopy in Japan, as was referred to in the book by Kanaya (1954a).

For instance, in 1936, E. Sugata constructed an electron microscope of the surface emission type. However, there was little development of electron microscope studies until a Cooperative Research Committee under the Ministry of Education in Japan was established in 1939 [see Sugata's book (1956)].

Under the leadership of S. Seto, this committee continued its activity during and after World War II with the following members: S. Seto, H. Yamashita, and Y. Tani (University of Tokyo); N. Kato, K. Sasagawa, and N. Higashi (Kyoto University); E. Sugata (Osaka University); Y. Sakaki (Nagoya University); S. Suzuki (Electrotechnical Laboratory); S. Asao and H. Kamogawa (Tokyo Shibaura Electric Works Ltd.); K. Kasai and B. Tadano (Hitachi Central Research Laboratory); and S. Shimadzu and T. Hori (Shimadzu Seisakusho Co. Ltd.). Since 1953, the committee has continued with additional members A. Fukami (University of Tokyo), K. Kanaya (Electrotechnical Laboratory), K. Kobayashi (Kyoto University), N. Morito (Hitachi Central Research Laboratory), and T. Hibi (Tohoku University).

During this earliest stage, in 1940 and 1941, a number of electron microscopes of various types were assembled: a magnetic type at the Electrotechnical Laboratory by Suzuki, at Osaka University by Sugata, at Kyoto University by Sasagawa and Higashi; and an electrostatic type at Toshiba by S. Asao, H. Inuzuka, H. Kamogawa, and H. Yako, in cooperation with Y. Tani.

Fig. 1.1 shows a preliminary electron microscope (a), as designed by S. Suzuki, and its micrograph (b) of butterfly ramentum taken at 50 kV in 1940. This was constructed using a system of three magnetic lenses which are located outside the column as in a cathode-ray oscillograph. In such

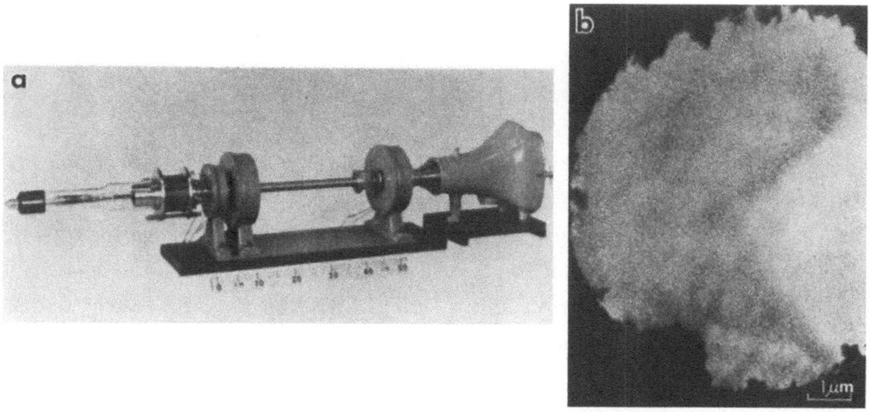

Figure 1.1 Prototype electron microscope: (a) column, (b) butterfly ramentum (S. Suzuki, 1940). Courtesy of E. Sugata (1968).

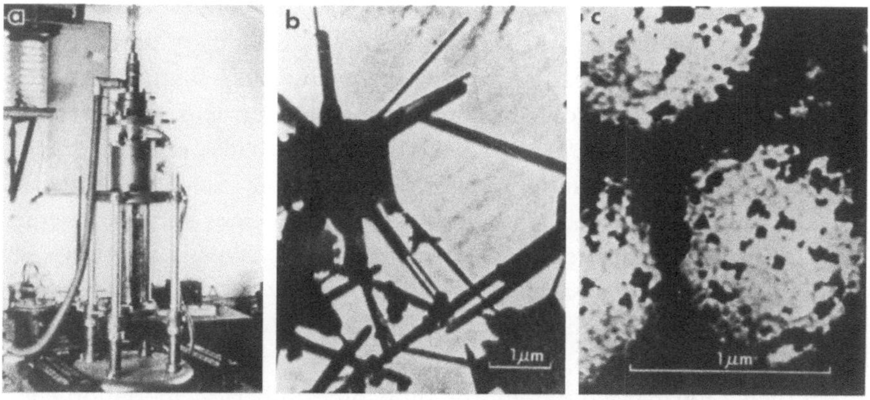

Figure 1.2 Electron microscope: (a) column, (b) zinc oxide, (c) carbon black (S. Suzuki, 1941). Courtesy of E. Sugata (1968).

early microscopes, the maintenance of high vacuum was the most important factor for achieving stable operating conditions.

Fig. 1.2 shows an improved electron microscope (a) and its micrographs (b and c) of zinc oxide and carbon black particles taken at 50 kV (S. Suzuki, 1941). The contrast stability can be improved by the self-biasing in the electron gun.

In 1941, the Hitachi and Shimadzu companies simultaneously started manufacturing commercial electron microscopes with the aid of the numerous constructional details given in von Ardenne's book (1940) and the fundamental theories on electron optics developed in Zworykin's book

(Zworykin et al., 1945). In 1947, Japan Electron Optics Laboratory (JEOL) began to manufacture a magnetic electron microscope; Akashi Seisakusho has also been manufacturing such instruments since 1950.

The Japanese Society for Electron Microscopy was organized in 1949. Under the auspices of the Society, an annual lecture meeting and symposium on recent progress in topics involving electron optics and instrumentation and their applications in the various scientific fields has been held every year and the *Journal of Electronmicroscopy* has been published quarterly, first edited by chief editor Y. Tani (1950), followed by K. Kanaya (1959), Y. Watanabe (1962), and so on, sponsored by the publication fund of the Ministry of Education.

Many national and government universities and their associated research laboratories were provided with continued research funds, enabling them to study basic applications using electron microscopes and especially the instrumentation of high-voltage microscopes. This is one of the most important factors in the remarkable progress and extraordinary production of electron microscopes (see, for example, the review by Y. Tani, 1956).

One of the earliest contributions on the electron lens was that of Y. Sugiura and S. Suzuki (1943a), who introduced the idea that the unipole magnetic lens gives minimum spherical aberration. Since the spherical aberration of such special lenses was discussed for lens systems of equal focal length f, length should be normalized by f. Thus, a solution r_α of the path equation has to satisfy initial and boundary conditions of the form $r_\alpha(\zeta_0) = 0$, $r'_\alpha(\zeta_0) = 1$ at the object $z_0 = \zeta_0 f$, and $r_\alpha(\zeta_i) = 1$ and $r'_\alpha(\zeta_i) = 0$ at the image $z_i = \zeta_i f$.

In such a normalized system, the third-order spherical aberration coefficient C_s is given by the integral based on Glaser's equation (1933):

$$C_s/f = \frac{1}{2}\int_{\zeta_0}^{\zeta_i} r_\alpha^4\{[[fK(f\zeta)]^2 + (r'_\alpha/r_\alpha)^2]^2 - f^2K(f\zeta)K''(f\zeta)\}\,d\zeta \qquad (1.1)$$

where $K(f\zeta) = K(z)$ is defined by the paraxial electron path equation

$$r''(z) = -K^2(z)r \qquad (1.2)$$

With the assumption that the constant saturation field is of the form $fK = \frac{1}{2}K$ while for the decreasing field $fK = 6/13(\zeta - c)$, with $c = 0.0997$, as shown in Fig. 1.3, the spherical aberration coefficient C_s is given by

$$\frac{C_s}{f} = \frac{f^2}{2}(r_\alpha^4 KK')_{\zeta=a} + \frac{1}{2}a - \frac{1}{8K_m f}\sin 4K_m af$$

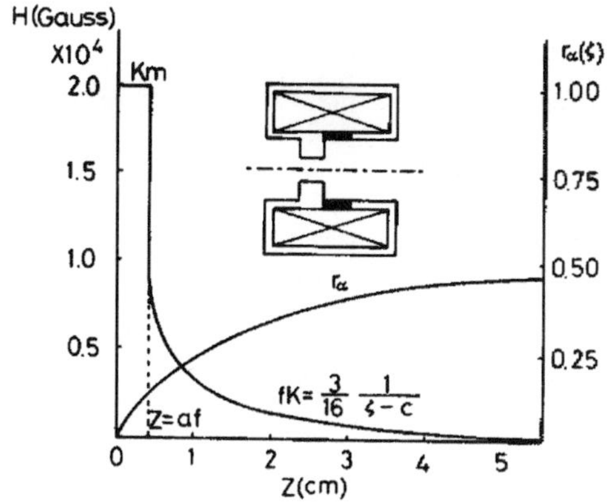

Figure 1.3 Magnetic field distribution and r_α curve along the axis of a unipole magnetic lens.

$$+\left(\frac{3}{8\alpha}\right)^3 \int_{r_\alpha(a)}^{1} r_\alpha^{-9/4}(-20r_\alpha^{-18/4} + 52r_\alpha^{-9/4} - 5)\, dr_\alpha \qquad (1.3)$$
$$= 0.00669 \quad (f = 0.5 \text{ cm})$$

where α is the angular aperture in the well-known equation $\Delta u = C_s\alpha^3$. According to the above equation, C_s is given approximately by the simple form

$$C_s = 1/30K_m \qquad (1.4)$$

where $K_m = (0.022/\Phi H_m)^{1/2}$ (Φ in volts, H_m in Gauss). Similarly, the axial chromatic aberration coefficient C_c is given by

$$\frac{C_c}{f} = \frac{1}{2}\int_{\zeta_0}^{\zeta_i} [r_\alpha'^2 + (fK)^2 r_\alpha^2]\, d\zeta$$
$$= \frac{1}{K_m} \qquad (1.5)$$

When $H_m = 2 \times 10^4$ G ($= 2$ T), $\Phi = 8 \times 10^4$ V, a resolving power $\delta = 2.4$ Å with $\alpha_{opt} = 3.6 \times 10^{-6}$ rad for $\Delta\Phi/\Phi = 10^{-5}$ was obtained.

The unipole magnetic lens was developed in practice by K. Kanaya as an asymmetrical lens with different bore radii (1949b; 1950) in which the spherical and axial chromatic aberration coefficients C_s and C_c take their minimum values when the front and rear half-width ratios of the magnetic field distribution are $a_1/a_2 = \frac{1}{4}$ and $a_1 = a_2 = \frac{1}{1.5}$, respectively.

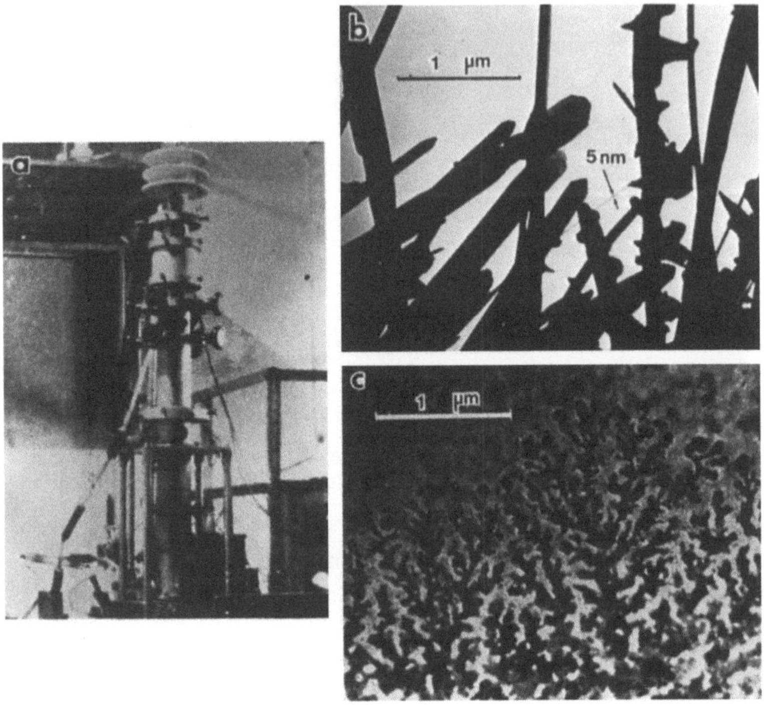

Figure 1.4 Electron microscope: (a) column, (b) litharge crystal grown after electron bombardment, and (c) zinc crystal grown (S. Suzuki and K. Kanaya, 1947). Courtesy of E. Sugata (1968).

Fig. 1.4a shows the magnetic electron microscope made by Shimadzu Seisakusho Co. Ltd. under the guidance of the Electrotechnical Laboratory. Figs. 1.4b and 1.4c show the crystal growth of PbO and zinc taken at 50 kV in 1948; the microscope was designed by S. Suzuki and K. Kanaya; the ripple in the applied voltage is 5×10^{-4}, the resolution 20 Å. The column is constructed of a three-lens system and is a modified version of a von Ardenne design. Early investigations of the temperature of specimens and their structural changes due to electron bombardment in the electron microscope were made by K. Kanaya (1948) using the microscope, as shown in Figs. 1.4b and 1.4c.

When using the microscope, it was found that the specimens changed very much due to electron bombardment. Kanaya (1949a) has reported a theoretical formula for calculating the temperature of specimens, considering the energy absorbed by the specimen and that radiated at its surface. The specimen temperature t is given theoretically by

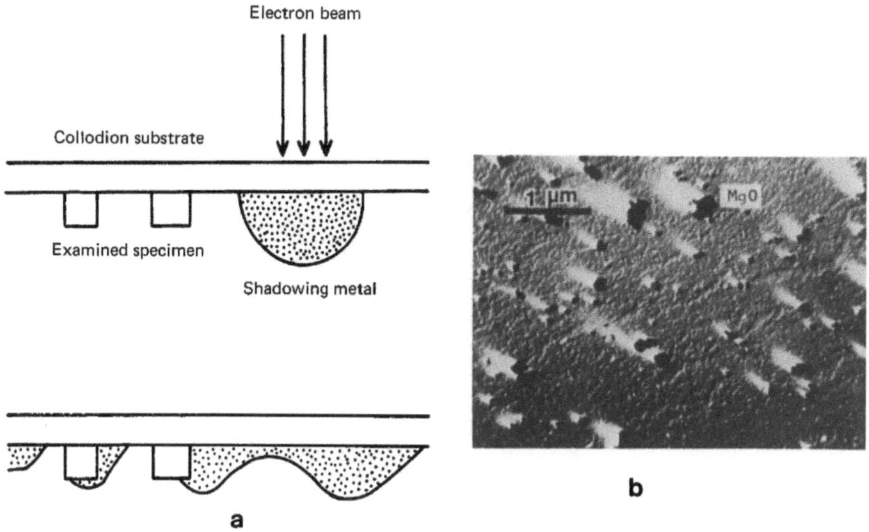

Figure 1.5 Direct shadowcasting principle (a) and micrograph (b) of an MgO crystal shadowed by evaporated Pb, where Pb was bombarded with electron beam $i_0 = 0.8$ A/cm^2 for 5 min at 50 kV.

$$t = \left[293^4 + \frac{0.112Fdi_0}{c\Phi A}\sum NZ\left(\log\frac{e\Phi}{IZ} + \frac{1}{2}\right)\right]^{1/4} - 273 \qquad (1.6)$$

The summation is taken over all species of atoms involved; it was assumed that the temperature of the instrument is 20 °C (corresponding to 293 K) in this calculation; F is the surface area surrounding the specimen, d the thickness, i_0 the electron beam current density, c the emissibility, Φ the accelerating voltage, A the cross section bombarded by the beam, N the number of atoms with atomic number Z, e the electronic charge, and I the ionization energy.

Fig. 1.4b shows one of the changed structures of red litharge crystals, where crystals have grown after heating to 500–890 °C in the dense electron beam (Kanaya & Oshina, 1949b, 1949c; Kanaya et al., 1949); Fig. 1.4c also shows the growth of zinc crystals, produced by heating the brass mesh in the electron microscope and suddenly introducing air into it (Kanaya & Oshina, 1949a).

From the remarkable facts that metals and metallic oxides melt and evaporate under dense electron bombardment with a current density $i \geqq 0.1$ A/cm^2 at 50 kV, the direct shadowcasting method *in situ* in the electron microscope has been successfully employed as shown in Fig. 1.5. The specimens supported on the collodion film are shadowed very clearly by

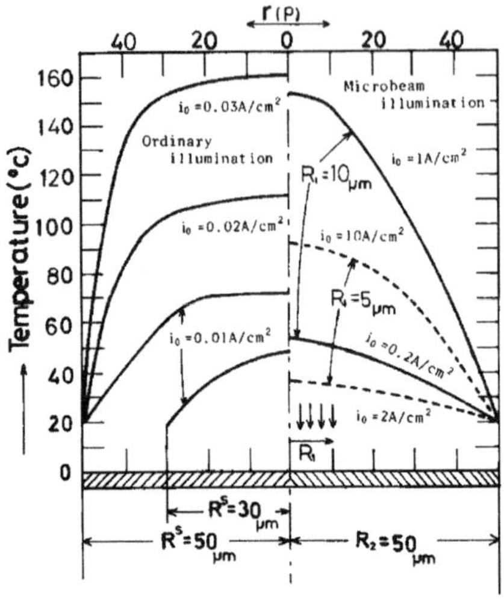

Figure 1.6 Distribution of temperature in a collodion substrate 10 nm thick in comparison with that of microbeam illumination, where $\phi = 50$ kV.

the evaporated metallic films of Pb and Bi (Kanaya, 1949a). Afterwards, following the work of B. von Borries and W. Glaser (1944), the distribution of temperature in thin films supported over a circular opening (Kanaya, 1954b), the temperature distribution of specimens on thin substrates (Kanaya, 1955b, 1955c), and the temperature distribution along a rod specimen (Kanaya, 1955d, 1956d) were theoretically discussed and formulated by K. Kanaya, thermal conduction as well as radiation being taken into account.

It is a remarkable fact that under microbeam illumination, part of the energy absorbed is used for the conduction of energy and then the temperature at the center of a substrate can be reduced to $(R_2/R_1)^{1/2}T$ in comparison with that of the ordinary illumination as shown in Fig. 1.6, where R_2 and R_1 are the radii of a microbeam and a circular hole, respectively.

This was confirmed by the fact that with microbeam illumination produced by a double-condenser system, the structure of Pb_3O_4 is not affected, but it melts under ordinary illumination with $R_1 = R_2 =\backsim 2$ μm at 50 kV.

Since the beginning of electron microscopy, wide contour lines following the outline of particles have often been observed. von Borries and Ruska (1938) attributed these contour lines to variation in the spherical aberration of the objective for different regions of the specimen, while J. Hillier (1939) explained such contour lines at the line of separation of two areas of the specimen, having widely different atomic mass thickness, as due to chromatic aberrations.

In Japan, straight-edge contours due to the spherical aberration and scattering were first discussed by Y. Sugiura and S. Suzuki (1943b), and in addition, the variation of the chromatic aberration with scattering for different aperture and focusing conditions was investigated and combined with the experimental results expressed by densitometer tracings (Kanaya & Ishikawa, 1951; Kanaya et al., 1951). Adopting the viewpoint of the wave optical theory given by Glaser (1943), the minimum resolving power δ of the high-resolution electron microscope was expressed as a function of the aperture angle α for the different modes of image formation as follows:

for coherent phase-contrast image:

$$\delta = 0.71 C_s^{1/4} \lambda^{3/4} \qquad \text{for } \alpha_{opt} = 0.895(\lambda/C_s)^{1/4} \tag{1.7}$$

for an incoherent amplitude-contrast image:

$$\delta = 0.785(\mu\lambda)^{1/2} \qquad \text{for } \alpha_{opt} = 0.71(\lambda/\mu)^{1/2}$$

for an image dominated by astigmatism:

$$\delta = 0.766(\lambda\xi)^{1/2} \qquad \text{for } \alpha_{opt} = 0.89(\lambda/\xi)^{1/2}$$

where λ is the electron beam wavelength; C_s, C_c, and ξ are the spherical, axial chromatic, and astigmatic aberration coefficients; $\Delta E/E$ and $\Delta\Phi/\Phi$ are the energy loss and fluctuation of the applied voltage, used here in the combined form $\mu = C_c(\Delta E/E \pm \Delta\Phi/\Phi)$ (see the review by Kanaya, 1955a).

Also, based on the electron scattering theory of L. Marton and L. I. Schiff (1941) and R. Uyeda (1953) and the phase-contrast theory taking account of aberrations and defocusing given by O. Scherzer (1949), the fundamental equation of image formation for a disk specimen (Kanaya et al., 1956) and for a crystalline specimen (Kanaya, 1956a, 1956b, 1956c) have been formulated and confirmed with the aid of experimental results.

The most important results are that, if the numerical aperture of the optical system is sufficiently reduced, Fresnel diffraction fringes can be observed in out-of-focus images (Boersch, 1943, 1944). T. Hibi and S. Takahashi (1958) made some beautiful observations of Fresnel fringes, which are discussed analytically by K. Kanaya (1957), and obtained highly resolved images of slip bands of MgO (111) twin or (110) sections (Hibi &

Yada, 1954) using a so-called pointed filament, the tip of which is polished mechanically and by chemical etching (Hibi, 1955, 1956).

According to the early books of Y. Tani (1959) and Y. Sugata (1968), miscellaneous devices for observing the structural changes induced in materials by heating (Hashimoto, 1960; Hashimoto et al., 1958; Takahashi & Ashinuma, 1957; Takahashi et al., 1956), cooling (Honjou et al., 1956), chemical reaction (Ito & Hiziya, 1958), and tensile deformation (Takahashi et al., 1960) were constructed and assembled with some commercial electron microscopes; these accessories were first derived in Japan.

Based on the work of Boersch (1936), the selected-area diffraction method has been developed to identify inclusions in crystals, with special reference to the determinations in MoO_3 crystals (Ito & Ito, 1953). G. Ruthemann (1941, 1942) observed characteristic energy losses due to plasma oscillation at X-ray levels when a beam is passing through thin layers of materials. In Japan, such plasmon losses were more clearly obtained by using the electrostatic lens as an analyzer (Watanabe, 1956, 1960).

High-resolution shadowing using a nozzle (Hibi, 1952), an improved replica technique (Fukami & Akiyama, 1953), and the production of a microgrid (Fukami & Adachi, 1964) were noteworthy contributions to specimen preparation techniques.

2. Lens aberrations

2.1 Fifth-order Seidel aberrations and third-order astigmatism

Early in the development of the electron microscope, it was recognized that the astigmatic aberration is one of the most serious limiting factors in the performance of high-resolution electron microscopes. From the beginning, efforts were directed toward the elimination of this defect.

One of the earliest attempts in Japan was that of S. Suzuki and Y. Inoue (1948), who demonstrated that the astigmatisms are due to an asymmetrical potential given by

$$\begin{aligned}\phi_n(r,z) &= \sum_{\nu=0}^{\infty} \frac{(-1)^\nu n!}{\nu!(n+\nu)!} \Phi_n(z)^{2\nu+n} \left(\frac{r}{2}\right)^{2\nu+n} \\ A_n(r,z) &= \sum_{\nu=0}^{\infty} \frac{(-1)^\nu n!}{\nu!(n+\nu)!} H_n(z)^{2\nu+n-1} \left(\frac{r}{2}\right)^{2\nu+n}\end{aligned} \tag{2.1}$$

where n is an integer indicating the rank of the asymmetry. Using the above electrostatic potential ϕ_n and magnetic vector potential A_n, the generalized aberration formula has been extended by K. Kanaya (1949) and K. Kanaya, H. Kawakatsu, and A. Kato (1959). A similar treatment had already been found by F. Bertein (1941).

The exact total aberration for any electron ray could, in principle, be found by calculation of its Hamiltonian characteristic, the point eiconal ϵ (Glaser, 1952):

$$\epsilon = \frac{1}{\Phi_0^{1/2}} \int_{z_a}^{z_b} F\,dz \tag{2.2}$$

with

$$F = [\Phi(1 + r'^2 + r^2\psi'^2)]^{1/2} - (e/2m)^{1/2}(A_r r' + A_z + A_\psi r\psi')$$

As a simple example, let a magnetic field with asymmetry be considered; F can be expanded as follows:

$$\begin{aligned}
F &= J_0 + \tfrac{1}{2}J_2 + \tfrac{1}{4}J_4 + \tfrac{1}{8}J_6 + \cdots + L_1 + \tfrac{1}{2}L_2 + \tfrac{1}{3}L_3 + \cdots \\
J_0 &= \Phi^{1/2} \\
J_2 &= \Phi^{1/2}(r'^2 + r^2\psi'^2) - (e/2m)^{1/2}Hr^2\psi' \\
J_4 &= -\tfrac{1}{2}\Phi^{1/2}(r'^2 + r^2\psi'^2) - \tfrac{1}{4}(e/2m)^{1/2}H''r^4\psi' \\
J_6 &= \tfrac{1}{2}\Phi^{1/2}(r'^2 + r^2\psi'^2) - \tfrac{1}{48}(e/2m)^{1/2}H''''r^6\psi' \\
&\ \vdots \\
L_1 &= -\tfrac{1}{2}(e/2m)^{1/2}H_1 r\sin(\psi - \lambda_1) \\
L_2 &= -\tfrac{1}{2}(e/2m)^{1/2}H_2'r^2\sin 2(\psi + \lambda_2) \\
L_3 &= \tfrac{3}{16}(e/2m)^{1/2}H_1''r^3\sin(\psi + \lambda_1) + \tfrac{3}{8}(e/2m)^{1/2}H_1'r^2r'\sin(\psi + \lambda_1) \\
&\quad - \tfrac{3}{8}(e/2m)^{1/2}H_1'r^3\psi'\cos(\psi + \lambda_1) \\
&\quad - \tfrac{3}{8}(e/2m)^{1/2}H_3''r^3\sin 3(\psi + \lambda_3) \pm \cdots
\end{aligned}$$

where Φ_0 is the accelerating potential and Φ the electrostatic potential; r, z, ψ are cylindrical coordinates; e is the electronic charge and m the electronic mass; A_r, A_z, A_ψ are the vector potential components in the r, z, and ψ coordinates; the J_n are the symmetrical parts of series containing terms of order n in the coordinates; the L_n are the asymmetrical parts of order n in the coordinates; and the H_n are the nth components of asymmetrical magnetic field with the phase angle λ_n. By using the well-known relationships for the electron ray

$$u = re^{i(\psi-\kappa)}, \quad \bar{u} = re^{-i(\psi-\kappa)}$$

$$\kappa = \frac{1}{2}\left(\frac{e}{2m\Phi_0}\right)^{1/2}\int_{z_a}^{z_b} H\,dz, \quad K = \left(\frac{e}{8m\Phi_0}\right)^{1/2} H$$

J_4 and J_6 can be obtained as

$$\begin{aligned} J_4 &= -\tfrac{1}{2}\Phi_0^{1/2}[u'\bar{u}' + K(u'\bar{u} - u\bar{u}')/i + K^2 u\bar{u}]^2 \\ &\quad + \tfrac{1}{2}\Phi_0^{1/2}K''[(u'\bar{u} - u\bar{u}')/2i + Ku\bar{u}]u\bar{u} \\ J_6 &= \tfrac{1}{2}\Phi_0^{1/2}[u'\bar{u}' + K(u'\bar{u} - \bar{u}'u)/i + K^2 u\bar{u}]^3 \\ &\quad - \tfrac{1}{24}\Phi_0^{1/2}K''''(u\bar{u})^2[(u'\bar{u} - \bar{u}'u)/2i + Ku\bar{u}] \end{aligned}$$

Also, L_2 and L_3 lead to

$$\begin{aligned} L_2 &= N_{10}u\bar{u} + N_{20}u^2 + \bar{N}_{20}\bar{u}^2 \\ L_3 &= -(M_{10}u^2\bar{u} + \bar{M}_{10}u\bar{u}^2 + M_{20}u'u\bar{u} + \bar{M}_{20}u\bar{u}\bar{u}' + M_{30}u^3 + \bar{M}_{30}\bar{u}^3) \end{aligned}$$

Now, we set

$$u = u_a r_\gamma + (u_B/r_{\alpha B})r_\alpha, \quad \bar{u} = \bar{u}_a r_\gamma + (\bar{u}_B/r_{\alpha B})r_\alpha$$

where

$$u_a = x_a + iy_a = r_a e^{i\psi_a}, \quad u_B = x_B + iy_B = r_B e^{i\psi_B}$$

The eiconal function of symmetrical aberration of higher order can be arranged as follows:

third-order aberration ($2\nu = 4$):

$$\begin{aligned} \epsilon_4 = \frac{1}{\Phi_0^{1/2}}\int_{z_a}^{z_b}\frac{J_4}{4}\,dz = -\Big(&\frac{AR^2}{4} + \frac{B\rho^2}{4} + Ck^2 + \frac{DR\rho}{2} \\ &+ ERk + F\rho k + ck\sigma + eR\sigma + f\rho\sigma\Big) \end{aligned} \tag{2.3}$$

fifth-order aberration ($2\nu = 6$):

$$\epsilon_6 = \frac{1}{\Phi_0^{1/2}} \int_{z_a}^{z_b} \frac{J_6}{8} dz = S_1 R^3 + S_2 R^2 \rho + S_3 R^2 k + S_4 R \rho^2$$
$$+ S_5 R\rho k + S_6 R k^2 + S_7 \rho^3 + S_8 \rho^2 k$$
$$+ S_9 \rho k^2 + S_{10} k^3 + S_{11} R^2 \sigma + S_{12} R \rho \sigma$$
$$+ S_{13} R k \sigma + S_{14} \rho^2 \sigma + S_{15} \rho k \sigma$$

where

$$R = x_z^2 + y_a^2, \quad \rho = x_B^2 + y_B^2, \quad k = x_a x_B + y_a y_B, \quad \sigma = x_a y_B - y_a y_B$$

The above fifth-order aberration formula was introduced in Japan by N. Kato and T. Inoue (1941).

On the other hand, the eiconal function of asymmetrical aberration of higher order leads to

second-order aberration ($n = 2$):

$$\epsilon_2 = \frac{1}{\Phi_0^{1/2}} \int_{z_a}^{z_b} \frac{L_2}{2} dz = N_1 u_a \bar{u}_a + N_2 u_a \bar{u}_B + \bar{N}_2 \bar{u}_a \bar{u}_B + N_3 u_B \bar{u}_B + \bar{N}_3 \bar{u}_B u_B$$
$$+ N_4 u_a^2 + N_5 u_a u_B + N_6 u_B^2 + \bar{N}_4 \bar{u}_a^2 + \bar{N}_5 \bar{u}_a \bar{u}_B + \bar{N}_6 \bar{u}_B^2 \tag{2.4}$$

third-order aberration ($n = 3$):

$$\epsilon_3 = \frac{1}{\Phi_0^{1/2}} \int_{z_a}^{z_b} \frac{L_3}{3} dz = M_1 u_a^2 \bar{u}_a + \bar{M}_1 u_a \bar{u}_a^2 + M_2 u_a \bar{u}_a u_B$$
$$+ \bar{M}_2 u_a \bar{u}_a \bar{u}_B + M_3 \bar{u}_a u_B^2 + \bar{M}_3 u_a \bar{u}_B^2 + M_4 u_a^2 \bar{u}_B$$
$$+ \bar{M}_4 \bar{u}_a^2 u_B + M_5 u_a u_B \bar{u}_B + \bar{M}_5 \bar{u}_a u_B \bar{u}_B + M_6 u_B^2 \bar{u}_B$$
$$+ \bar{M}_6 u_B \bar{u}_B^2 + M_7 u_a^3 + \bar{M}_7 \bar{u}_a^3 + M_8 u_a^2 u_B + \bar{M}_8 \bar{u}_a^2 \bar{u}_B$$
$$+ M_9 u_a u_B^2 + \bar{M}_9 \bar{u}_a \bar{u}_B^2 + M_{10} u_B^3 + \bar{M}_{10} \bar{u}_B^3$$

Accordingly, both functions due to the second- and third-order astigmatisms are given by the terms of C, c, S_6, S_{13} and M_3, $\bar{M}_3$, M_9, $\bar{M}_9$, N_6, $\bar{N}_6$, and S_{10} and M_{10}, $\bar{M}_{10}$ in Eqs. (2.3) and (2.4), respectively,

$$\epsilon_2 = -C r_a^2 r_B^2 \cos^2(\psi_B - \psi_a) - i c r_a^2 r_B^2 \sin(\psi_B - \psi_a)\cos(\psi_B - \psi_a)$$
$$+ S_6 r_a^4 r_B^2 \cos^2(\psi_B - \psi_a) + i S_{13} r_a^4 r_B^2 \sin(\psi_B - \psi_a)\cos(\psi_B - \psi_a)$$
$$- 2|M_3| r_a r_B^2 \cos(2\psi_B - \psi_a + \psi_{M3}) - 2|M_9| r_a r_B^2 \cos(2\psi_B + \psi_a + \psi_{M9})$$

$$
\begin{aligned}
&+ 2|N_6| r_B^2 \cos(2\psi_B + \psi_{N6}) \\
\epsilon_3 &= S_{10} r_a^3 r_B^3 \cos^3(\psi_B - \psi_a) - 2|M_{10}| r_B^3 \cos(3\psi_B + \psi_{M10})
\end{aligned}
$$

The second- and third-order astigmatic aberrations can be deduced by differentiating the above terms in the following:

$$
\begin{aligned}
\Delta u_2 &= N_0 \theta e^{i\delta_N} e^{-i\psi_B} \\
N_0 e^{i\delta_N} &= S' r_a^2 e^{i2\psi_a} + S'' r_a^4 e^{i2\psi_a} + M' r_a e^{-i\psi_{M3}} e^{i\psi_a} \\
&\quad + M'' r_a e^{-i\psi_{M9}} e^{-i\psi_a} + N e^{i\psi_{N6}} \qquad (2.5) \\
\Delta u_3 &= M_0 \theta^2 e^{i\delta_M} e^{-i2\psi_B} \\
M_0 e^{i\delta_M} &= S r_a^3 e^{i3\psi_a} + M e^{-i\psi_{M10}}
\end{aligned}
$$

where $S' = -r_{\alpha B}^2(C + ic)$, $S'' = r_{\alpha B}^2(S_6 + iS_{13})$, $M' = -4r_{\alpha B}^2|M_3|$, $M'' = -4r_{\alpha B}^2|M_9|$, $N = 4r_{\alpha B}^2|N_6|$, $S = \frac{3}{4}S_{10}r_{\alpha B}^3$, $M = -6M_{10}r_{\alpha B}^3$, $u_B/r_{\alpha B} = e^{i\psi_B}$.

2.2 Aberrations of a focused charged-particle beam caused by space charge

In electron beam work, one of the most important effects of space charge is to spread the beam. It is this effect that makes it necessary to apply various forms of focusing to long, dense beams. Several papers discussing the effect of mutual electron repulsion on the focus of electron beams have appeared in the past (Glaser, 1956; Sturrock, 1955; Thomson & Headrick, 1940; von Borries & Ruska, 1932; Watson, 1927). Even in electron microscopy, where the beam current is much lower, it has been shown by H. Grümm (1952) and W. Meyer (1958) that, under normal conditions, the space charge may limit the resolving power more than the spherical aberration, and, moreover, may compensate the Seidel aberrations.

For a beam-focusing system such as the high-magnification objective in the electron microscope or the reducing lens in an ion-beam device, it has been shown by K. Kanaya, H. Kawakatsu, and H. Yamazaki (1965) that the vacuum is so perfect that no appreciable number of positive ions is formed by collision and also that the deviation of the ray paths by the radial force giving rise to space-charge spreading is of the same order as that of the lens aberrations. The aberration caused by space charge, which opposes the normal aberration of the lens, can be expressed in terms of the space-charge potential distribution along the optic axis z, satisfying the Poisson equation;

this may be expressed in cylindrical coordinates as

$$\Phi(r, z) = \Phi - \frac{1}{4}\left(\Phi'' + \frac{1}{\epsilon_0}\rho(z)\right)r^2 + \frac{1}{64}\left(\Phi'''' - \frac{\rho''(z)}{\epsilon_0}\right)r^4 - \cdots \tag{2.6}$$

in which ϵ_0 is the permittivity of free space, Φ the potential of the beam, and $\rho(z)$ the space-charge density given by

$$\rho(z) = j/v = i/\pi R^2 v$$

where v is the velocity of the charged particles, which depends on the electronic charge e and the mass m; i is the total current, j is the current density; and R is the outer radius of the beam.

The aberration caused by space charge can be obtained by a method similar to Glaser's (1953). Kanaya et al. (1965) introduced the basic formula for aberration due to space charge with the aid of P. W. Hawkes's comment (1966). Using Eq. (2.2) and expressing the results in terms of the aperture angle $\alpha = |u_B/r_{\alpha B}|$, we find

$$\begin{aligned}(-1)\Delta u_s &= \frac{\partial \epsilon}{\partial (u_B/r_{\alpha B})} \\ &= -\Delta f_s \frac{u_B}{r_{\alpha B}} - \Delta r_s u_a + (B - B_s)\left(\frac{u_B}{r_{\alpha B}}\right)^2 \frac{\bar{u}_B}{r_{\alpha B}} \\ &\quad + [F - F_s - i(f - f_0)]\bar{u}_a\left(\frac{u_B}{r_{\alpha B}}\right)^2 \\ &\quad + 2[F - F_s + i(f - f_s)]u_a \frac{u_B}{r_{\alpha B}}\frac{\bar{u}_B}{r_{\alpha B}} \\ &\quad + (C - C_s + D - D_s)u_a \bar{u}_a \frac{u_B}{r_{\alpha B}} \\ &\quad + [C - C_s + i(c - c_s)]u_a^2 \frac{\bar{u}_B}{r_{\alpha B}} + [E - E_s + i(e - e_s)]u_a^2 \bar{u}_a\end{aligned} \tag{2.7}$$

Note that the sign of $\rho(z)$, denoting the repulsion force, is negative with respect to the ordinary aberration coefficients and so all aberration coefficients concerning $\rho(z)$ have negative signs; C_s is sometimes used to denote the spherical aberration coefficient—here it stands for a different aberration as explained below.

The first-order terms Δf_s and Δr_s correspond to defocusing and change of magnification, respectively, and other higher order aberration terms B_s, $F_s - if_s$, $C_s + ic_s$, $D_s + C_s$ and $E_s + ie_s$ are aberration coefficients acting in opposition to the ordinary third-order lens aberration coefficients corresponding to spherical aberration B, coma $F - if$, astigmatism $C + ic$, curvature of field $C + D$, and distortion $E + ie$.

The most important space-charge terms may be written

$$\begin{aligned} \Delta f_s &= \frac{1}{4\epsilon_0 \Phi_0^{1/2}} \int_{z_a}^{z_b} \frac{\rho(z) r_\alpha^2}{\Phi_0^{1/2}(z)} dz \\ \Delta r_s &= \frac{1}{4\epsilon_0 \Phi_0^{1/2}} \int_{z_a}^{z_b} \frac{\rho(z) r_\gamma r_\alpha}{\Phi_0^{1/2}(z)} dz \\ B_s &= \frac{1}{4\epsilon_0 \Phi_0^{1/2}} \int_{z_a}^{z_b} \frac{\rho(z)(r_\alpha^2 r_\alpha'^2 + K^2 r_\alpha^4)}{\Phi_0^{1/2}(z)} dz \end{aligned} \tag{2.8}$$

where r_α and r_γ are the two particular solutions of the paraxial-ray equation $u'' + K^2 u = 0$, which are defined by the initial conditions $r_\alpha(z_a) = 0$, $r_\alpha'(z_a) = 1$, $r_\gamma(z_a) = 1$, and $r_\gamma'(z_a) = 0$.

Consider the objective lens of a high-magnification electron microscope; defocusing change of magnification and spherical aberration are given by

$$\begin{aligned} -\Delta f_s \alpha_0 &= -\tfrac{1}{2}\sigma(3 - 2\log 2) f/\alpha_0 \\ -\Delta r_s |u_a| &= -(\sigma D/\alpha_0^2 f)|u_a| \\ -B_s \alpha_0^3 &= -\sigma(3 - 4\log 2)\alpha_0 f \end{aligned} \tag{2.9}$$

in which σ is the space-charge parameter given by

$$\begin{aligned} &\sigma = 3.04 \times 10^4 i\ (\mathrm{V}^{3/2})/\Phi^{3/2}\ (\mathrm{A}) \quad \text{(electron beam)} \\ &\sigma_i = \sigma(M_i 1840/n)^{1/2} \quad \text{(ion beam)} \end{aligned} \tag{2.10}$$

where M_i is the ratio of ion mass to proton mass and n the number of positive charges, r_a the radius of the beam, f the focal length, α_0 the limiting aperture angle, and $M = D/f$ (D is the image distance) may be assumed.

Figs. 2.1 and 2.2 show the calculated results for a high-magnification electron microscope; Fig. 2.3 shows the results for an ion beam instrument.

It should be possible in practice to ignore all space-charge aberrations in high-magnification electron microscopes except defocusing and change of magnification. However, in order to obtain high-resolution images at the angstrom level indicated by Δu, the total current of the beam should be restricted by the relationship

$$i \leqq 4.4 \times 10^{-4}(\alpha_0/f)\Phi^{3/2} \tag{2.11}$$

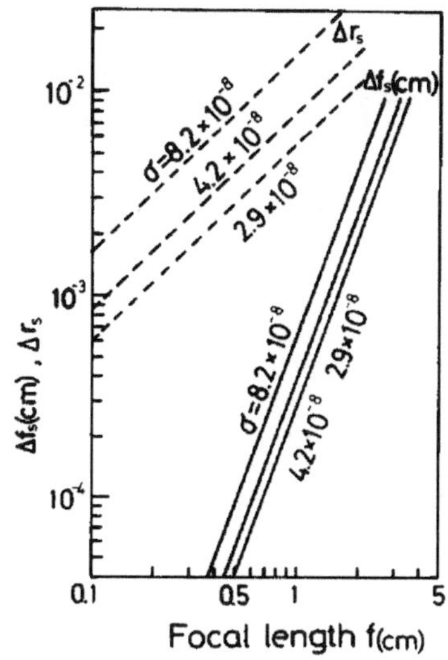

Figure 2.1 Defocusing Δf_s and magnification difference Δr_s due to space charge as a function of focal length f for various values of σ in the case of a high-magnification electron microscope, where $r_a = 100$ μm and $D = 20$ cm.

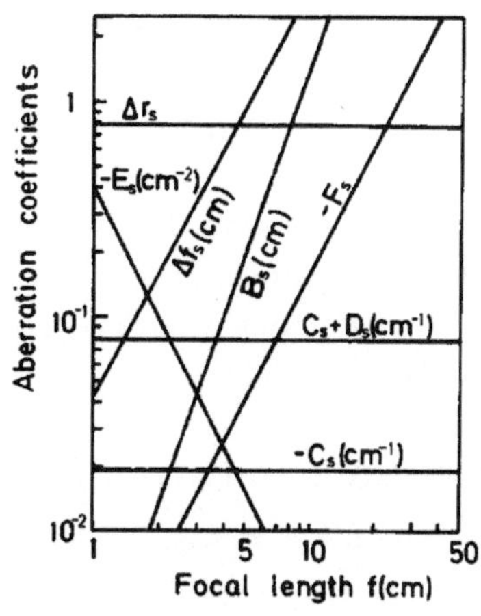

Figure 2.2 Space-charge aberration coefficients as a function of focal length in the case of a high-magnification electron microscope, where $\sigma = 8.2 \times 10^{-8}$.

Thus the permissible total current i at the specimen at 50 kV is less than 1 μA.

Consider an ion beam machine with an electrostatic reducing lens; defocusing, change of magnification, and spherical aberration are given by

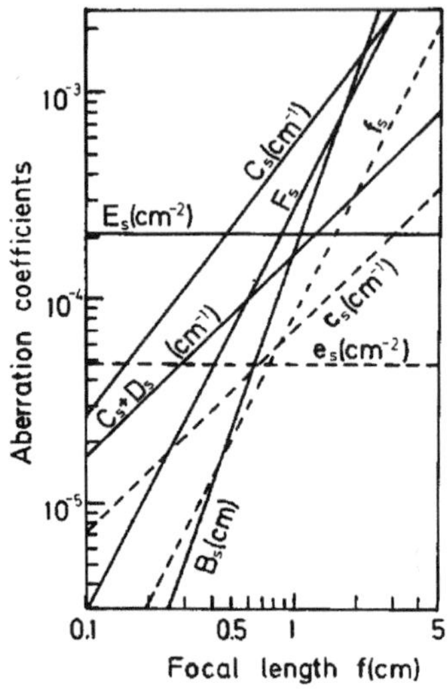

Figure 2.3 Space-charge aberration coefficients as a function of focal length for an ion beam reduced by an electrostatic lens, where $\sigma_i = 3.6 \times 10^{-4}$.

$$\begin{aligned} -\Delta f_s\alpha_0 &= -\tfrac{1}{6}\sigma_i D/\alpha_0 \\ -\Delta r_s|u_a| &= -\tfrac{1}{12}\sigma_i (D/\alpha_0 f)^2 |u_a| \\ -B_s\alpha_0^3 &= -\tfrac{1}{4}\sigma_i \alpha_0 f \end{aligned} \tag{2.12}$$

These are the most important factors in the realization of a fine spot with high power density. The caustic pattern of the focused ion beam is strongly affected by space charge, since the focal length is greatly increased, as shown in another paper (Kanaya et al., 1964).

2.3 Asymmetric aberrations and manufacture accuracy

If the objective polepieces contain an asymmetric hole or are inclined to each other, an asymmetric image contour frequently appears in micrographs taken both in focus and out of focus, in the latter case, in directions perpendicular to each other.

According to the basic Eq. (2.2) of asymmetric aberration, the image displacement Δu and the third-order astigmatism Δu_ξ (astigmatic spherical aberration) are given by

$$\Delta u = -i\left(\frac{e}{8m\Phi_0}\right)^{1/2} \int_{z_a}^{z_b} H_1(z) r_\alpha e^{-i\kappa}\, dz$$

$$
\begin{aligned}
\Delta u_\xi &= \xi \alpha \\
\xi &= -i\left(\frac{e}{8m\Phi_0}\right)^{1/2} \int_{z_a}^{z_b} H_2'(z) r_\alpha^2 e^{-i2\kappa}\, dz
\end{aligned} \tag{2.13}
$$

with

$$
\alpha = \frac{\bar{u}_B}{r_{\alpha B}}, \quad \kappa = \left(\frac{e}{8m\Phi_0}\right)^{1/2} \int_{z_a}^{z_b} H(z)\, dz
$$

in which ξ represents the astigmatic spherical aberration coefficient and the subscripts 1 and 2 show the order of axial field strength $H(z)$ (Kanaya & Kato, 1951a, 1951b).

2.3.1 Astigmatism due to polepiece disfigurement

As the result of a theoretical calculation based on Ollendorff's potential distribution, Δu and Δu_ξ are given by K. Kanaya and A. Kato (1951a; 1951b) and K. Kanaya (1952):

$$
\begin{aligned}
\Delta u &= K_{\mathrm{I}}^{\mathrm{m}} \int_0^{2\pi} \epsilon e^{-i\theta}\, d\theta = K_1^{\mathrm{m}} A(\epsilon) \\
\xi &= K_{\mathrm{II}}^{\mathrm{m}} \int_0^{2\pi} \epsilon e^{-i2\theta}\, d\theta = K_2^{\mathrm{m}} B(\epsilon)
\end{aligned}
$$

with

$$
\begin{aligned}
K_1^{\mathrm{m}} &= -i\frac{2\rho^3 K_0}{\pi^2} \int_{z_a}^{z_b} \frac{r_\alpha e^{-i\kappa}}{(\rho^2+z^2)^2}\, dz \\
K_2^{\mathrm{m}} &= -i\frac{\rho^4 K_0}{\pi^2} \int_{z_a}^{z_b} \frac{r_\alpha^2 e^{-i2\kappa}}{(\rho^2+z^2)^3}\, dz \\
K_0 &= \left(\frac{e}{8m\Phi}\right)^{1/2} H_0
\end{aligned} \tag{2.14}
$$

where $\epsilon = \epsilon(\theta)$ expresses the disfigurement of the polepiece and ρ is the radius of the polepieces. Then, using r_α for the bell-shaped field curve (Glaser, 1941) and for the case $2a = d$ (d is the gap between the polepieces) and the radius $\rho = a$, K_1^{m} and K_2^{m} are given by Table 2.1, where (I) and (II), express the disfigurements of the polepieces in the front and rear directions, respectively.

Table 2.1 Calculated values of K_1^{m} and K_2^{m}.

ω	1.1	1.2	1.3	1.4
$K_1^{\mathrm{m}}(\mathrm{I})$	$0.33e^{i1.3}$	$0.12e^{i1.2}$	$0.073e^{i1.07}$	$0.044e^{i0.89}$
$K_2^{\mathrm{m}}(\mathrm{I})$	$1.45e^{i1.05}$	$0.17e^{i0.62}$	$0.056e^{i0.53}$	$0.025e^{i0.4}$
$K_1^{\mathrm{m}}(\mathrm{II})$	$0.33e^{i0.75}$	$0.21e^{i0.51}$	$0.18e^{i0.39}$	$0.17e^{i0.29}$
$K_2^{\mathrm{m}}(\mathrm{II})$	$2.1e^{-i0.19}$	$0.75e^{-i0.73}$	$0.47e^{-i1.04}$	$0.34e^{-i1.24}$

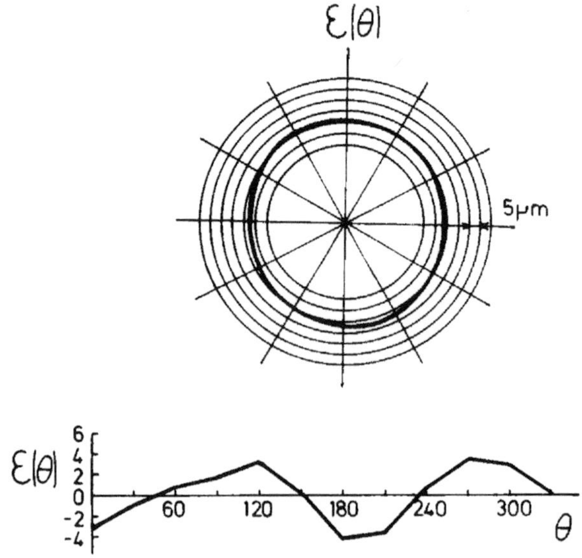

Figure 2.4 Hole contour $\epsilon(\theta)$ of polepiece measured by pneumatic micrometer.

Generally, when the disfigurement $\epsilon(\theta)$ can be expressed by a Fourier series, $A(\epsilon)$ and $B(\epsilon)$ are given by

$$A(\epsilon) = \pi(a_1^2 + b_1^2)^{1/2}, \quad B(\epsilon) = \pi(a_2^2 + b_2^2)^{1/2}$$
$$\epsilon(\theta) = \sum_1^n a_n \sin(n\theta) + \sum_0^n b_n \cos(n\theta) \tag{2.15}$$

with

$$a_n = \frac{1}{\pi}\int_0^{2n} \epsilon(\theta)\cos(n\theta)\, d\theta, \quad b_n = \frac{1}{\pi}\int_0^{2n} \epsilon(\theta)\sin(n\theta)\, d\theta$$

Fig. 2.4 shows the measured hole contour of a polepiece given by the pneumatic micrometer where $\epsilon(\theta)$ is calculated from the above equation

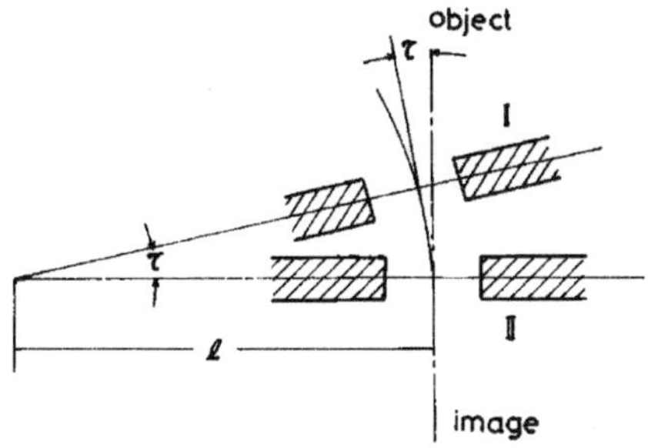

Figure 2.5 Representation of inclination of polepieces.

by K. Kanaya (1955). Accordingly, let ϵ_0 be the maximum disfigurement value; $A(\epsilon)$ and $B(\epsilon)$ are then obtained as follows:

$$A(\epsilon) = 0, \quad B(\epsilon) = \tfrac{1}{2}\pi\epsilon_0 \quad \text{(ellipticity)}$$

$$A(\epsilon) = \tfrac{3}{4}\pi\epsilon_0 e^{-i\pi/2}, \quad B(\epsilon) = \tfrac{1}{4}\pi\epsilon_0 \quad \text{(half-ellipticity)}$$

$$A(\epsilon) = \tfrac{1}{2}\pi\epsilon_0\ (n=1), \quad B(\epsilon) = \tfrac{1}{4}\pi\epsilon_0 \quad (n=2)$$

$$[\epsilon(\theta) = \tfrac{1}{2}\epsilon_0(\cos^n\theta + 1)]$$

$$A(\epsilon) = \begin{cases} \tfrac{1}{4}\pi\epsilon_0\ (n=1), \\ \tfrac{1}{8}\pi\epsilon_0\ (n=3), \end{cases} \quad B(\epsilon) = \tfrac{1}{8}\pi\epsilon_0 \quad (n=4)$$

$$[\epsilon(\theta) = \tfrac{1}{2}\epsilon_0 \sin^2\theta(\cos^n\theta + 1)]$$

$A(\epsilon) = 2\epsilon_0$, $B(\epsilon) = 2\epsilon_0$, for the case in which $A(\epsilon)$ and $B(\epsilon)$ are at their maximum.

2.3.2 Astigmatism due to polepiece inclination

When the polepieces are inclined to each other as shown in Fig. 2.5, the asymmetric image contour has the same appearance as those due to disfigurements of polepieces. In the following l is the radius of the envelope, indicating the inclination, and τ the angle of inclination of the polepieces. The asymmetric aberrations Δu and $\Delta u_\xi = \xi\alpha$ are calculated as follows:

$$\Delta u = (a^2/l)f(\omega) = \tfrac{1}{2}\tau a f(\omega), \quad \xi = (a^3/l^2)g(\omega) = \tfrac{1}{4}\tau^2 a g(\omega)$$

with

$$f(\omega) = \frac{i(\omega^2-1)^{1/2}}{\omega\sin(\pi/\omega)} \int_{\pi/\omega}^{0} \tfrac{3}{4}\cot\psi\cos^3\psi\sin\omega\psi e^{-i\kappa}\,d\psi$$

Table 2.2 Calculated values of $f(\omega)$ and $g(\omega)$.

ω	**1.1**	**1.2**	**1.3**	**1.4**
$f(\omega)$	$0.48e^{i0.82}$	$0.49e^{i0.38}$	$0.52e^{i0.21}$	$0.42e^{i0.12}$
$g(\omega)$	$0.79e^{-i1.43}$	$0.57e^{-i1.11}$	$0.32e^{-i0.76}$	$0.16e^{-i0.096}$

$$g(\omega) = \frac{-i(\omega^2 - 1)^{1/2}}{[\omega \sin(\pi/\omega)]^2} \int_{\pi/\omega}^{0} [\tfrac{1}{3} + \tfrac{15}{32} \cos^2 \psi (4 - 7 \sin^2 \psi)] \times \cot \psi \cos^3 \psi \sin^2 \omega\psi e^{-i2\kappa} \, d\psi \tag{2.16}$$

where

$$\kappa = (\omega^2 - 1)^{1/2} (\cos \psi - \cos \psi_a)$$

$$\omega^2 = 1 + (K_0 a)^2, \quad K_0 = \left(\frac{0.022}{\Phi \text{ (V)}}\right) h_0 \text{ (Gauss)}$$

Table 2.2 gives the calculated results $f(\omega)$ and $g(\omega)$ as a function of the objective-lens strength parameter ω, which depends on the maximum field strength H_0 (Gauss) and the half-width a (cm).

2.3.3 Manufacture accuracy

The resolution limit in conventional transmission electron microscopes with an accelerating voltage of 50 kV is considered to be 20 Å from the photographic characteristics and the chromatic aberration. The manufacture accuracy can then be calculated so as to make the value of asymmetric aberration 20 Å. The value of the polepiece disfigurement of the objective lens, $B(\epsilon)$ and their inclination τ can be determined as follows for $\Delta u_\xi \leqq 20$ Å and $\alpha = 5 \times 10^{-3}$ rad:

$$B(\omega) \leqq 4 \times 10^{-5}/K_2^{\mathrm{m}}, \quad \tau \leqq [1.6 \times 10^{-4} \text{ (cm)}]/[a \text{ (cm)} g(\omega)]^{1/2} \tag{2.17}$$

On the other hand, the case in which only the objective-lens image displacement exists is not important. Conversely, it is very important that the field aberrations of both intermediate and projective lenses be increased by it. Consequently, by considering the allowable displacement Δu for which the condition that field aberration $\Delta u_{\mathrm{f}} \leqq 20$ Å at $u_{\mathrm{a}} = 1.5 \times 10^{-3}$ cm is satisfied, it is found that

$$A(\epsilon) \leqq \Delta u/K_1^{\mathrm{m}}, \quad \tau \leqq 2\Delta u/af(\omega) \tag{2.18}$$

Table 2.3 Calculated values of manufacture accuracy.

	Objective	Projector
Two-stage lens system		
Ellipticity (μm)	2	∽10
Inclination (rad)	8/1000	∽1/100
Three-stage lens system		
Ellipticity (μm)	2	5
Inclination (rad)	2/1000	5/1000

Table 2.3 gives the calculated results of the manufacture accuracy that must be attained to contain these aberration defects due to asymmetric aberrations (Kanaya, 1955).

The results are as follows: for the aberration defects, the astigmatism due to polepiece disfigurement is larger than that due to polepiece inclination, and as to disfigurement, that of the rear polepiece is more serious than that of the front polepiece.

The asymmetric lens, having large diameters and $2\rho_1{:}2\rho_2 = 2{:}4$ (mm), where $2\rho_1$ and $2\rho_2$ are the diameters in the front and rear directions, respectively, is very attractive owing to the reduced value of the astigmatic spherical aberration. At low magnification, we must especially consider the increase of the field aberrations caused by image displacement due to the asymmetric aberration; as to the image displacement, the asymmetric aberration due to polepiece inclination is larger than that due to polepiece disfigurement.

Based on the above assessment of manufacture accuracy, a console-type microscope consisting of a two-lens system without condenser was made by Shimadzu Seisakusho Ltd. in 1950 as shown in Fig. 2.6. Its specifications are easy operation and a resolution of 50 Å; lens alignment is not necessary. Figs. 2.6c and 2.6d show the micrographs of the surface of $BaTiO_3$ and a phage-lysed typhoid bacillus taken at 50 kV. It has been shown by T. Taoka et al. (1967), using a magnetic torque meter, that the magnetic anisotropy of the material affects the astigmatism of the objective lens.

3. Practical magnetic lens design

3.1 Minimum aberration coefficients

The most important papers about strong magnetic lenses used for a practical lens design were published by W. Glaser (1940, 1941) and J. Dosse

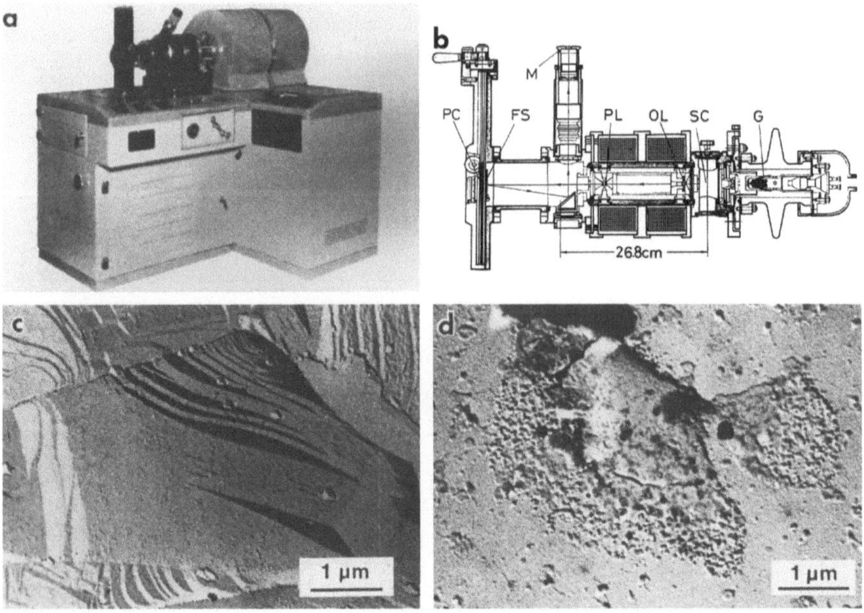

Figure 2.6 Console-type electron microscope made by Shimadzu Ltd. in 1950. (a) Column and (b) sectional view: G, electron gun; SC, specimen chamber; OL, objective lens; PL, projection lens; M, optical magnifier; FS, fluorescent screen; PC, plate chamber. (c) $BaTiO_3$ and (d) phage-lysed typhoid bacillus.

(1941a, 1941b). Glaser first assumed that the measured field distribution $H(z)$ on the optic axis z can be approximated closely by the expression

$$H(z) = H_0/[1 + (z/a)^2] \tag{3.1}$$

Here H_0 is the maximum field strength at $z = 0$ and a is the point at which $H(z)$ drops to $H_0/2$; $2a$ therefore may be called the half-width (full width at half-height) of the field curve. The paraxial-ray differential equation for a magnetic field of rotational symmetry is

$$r''(z) + (e/8m\Phi)H^2(z)r(z) = 0 \tag{3.2}$$

where $r = r(z)$ is the equation for the paraxial-ray path of an electron in the magnetic field whose axial field distribution is given by $H(z)$; e/m is the specific charge of the electron and Φ is the accelerating potential. By substituting Eq. (3.1) in Eq. (3.2) Glaser obtained the general solution

$$r(z) = a\left[1 + \left(\frac{z}{a}\right)^2\right]^{1/2}\left[c_1 \sin\left(\omega \cot^{-1}\frac{z}{a}\right) + c_2 \cos\left(\omega \cot^{-1}\frac{z}{a}\right)\right] \quad (3.3)$$

where c_1 and c_2 are two arbitrary constants of integration, $\omega = (1 + k^2)^{1/2}$, and $k^2 = eH_0^2a^2/8m\Phi$ is a parameter characterizing the lens strength.

Based on the bell-shaped distribution represented by Eq. (3.3), much elaborate work on electron lenses, devices, and electron lens systems has been carried out. It has been well recognized that the fundamental theory of electron optics for electron microscopes had already been accomplished in the early stages of electron microscopy by W. Glaser (1941, 1952), O. Scherzer (1937), and L. Marton (1946). However, electron-optical theory on the design of electron lenses may be not satisfactory for designing the electron microscope itself in practice, especially a high-resolution electron microscope free from aberrations.

The review paper on the electron optical theory underlying the design of magnetic lenses for electron microscopes by K. Kanaya (1955) was designed to give a consistent practical explanation of image confusions due to aberrations, and it was also intended to bring together the available data for the practical design purposes of high-resolution electron microscopes; part of this had been published in the *Journal of Electronmicroscopy (Japan)* or the *Bulletin of the Electrotechnical Laboratory* and included in the report of the Cooperative Research Committee on Electron Microscopy held in Japan during the first 10 years after World War II.

In this review, we discuss the principal topics which were of interest in Japan and were chiefly noted and discussed in miscellaneous works by K. Kanaya (1949a, 1949b, 1949c, 1949d, 1950, 1951a, 1951b, 1951c, 1951d, 1951e, 1952a, 1952b, 1952c). Some of these articles seem to represent some duplication of effort, and the same is true of the investigations on the applications of electron microscopy in Japan. This was due to the lack of familiarity with the papers to which reference is made and also to the absence of information about what was happening in other countries, especially in Europe, during the war and for some years after.

Based on the bell-shaped field of W. Glaser (1941) and the aberration theory by Scherzer (1937), expressions for the minimum focal length f and aberration coefficients were derived which are much easier to use than the theoretical expressions. If H_0 and Φ are kept constant, the aberration coefficients become functions of a alone. It is then possible to determine

Table 3.1 Minimum values of focal length f and aberration coefficients of spherical ζ, curvature of field β, astigmatic γ, coma δ and ϵ, and distortion α[a].

Minimum value	a_{opt}	ω_{opt}
Objective		
$f_{min} = 8.4\ \Phi^{1/2}/H_0$ (cm)	$6\ \Phi^{1/2}/H_0$	1.34
$\zeta_{min} = 3.24\ \Phi^{1/2}/H_0$ (cm)	$10.85\ \Phi^{1/2}/H_0$	1.9
$\beta_{min} = 0.284\ H_0/\Phi^{1/2}$ (cm^{-1})	$5.46\ \Phi^{1/2}/H_0$	1.32
$\lvert\gamma_{min}\rvert = 0.0725\ H_0/\Phi^{1/2}$ (cm^{-1})	$6\ \Phi^{1/2}/H_0$	1.34
$\lvert\delta_{min}\rvert = 0.62$	$6\ \Phi^{1/2}/H_0$	1.34
$\lvert\epsilon_{min}\rvert = 1.24$	$6\ \Phi^{1/2}/H_0$	1.34
Projector		
$f'_{min} = 8.93\ \Phi^{1/2}/H_0$ (cm)	$5.19\ \Phi^{1/2}/H_0$	1.262
$\lvert\alpha_{min}\rvert = 0.0165\ H_0^2/\Phi$ (cm^{-2})	$6.3\ \Phi^{1/2}/H_0$	1.36
$\beta_{min} = 0.26\ H_0/\Phi^{1/2}$ (cm^{-1})	$6.6\ \Phi^{1/2}/H_0$	1.4
$\lvert\gamma_{min}\rvert = 0.125\ H_0/\Phi^{1/2}$ (cm^{-1})	$5.19\ \Phi^{1/2}/H_0$	1.262
$\lvert\delta_{min}\rvert = 1$	$6.6\ \Phi^{1/2}/H_0$	1.4
$\lvert\epsilon_{min}\rvert = 2$	$6.6\ \Phi^{1/2}/H_0$	1.4

[a] Aberration coefficients follow the notation of O. Scherzer (1937).

the optimum value $a = a_{opt}$ for which each of these quantities reaches its minimum value, as shown in Table 3.1.

After some years, following the condenser–objective lens of W. D. Riecke (1962a, 1962b), the second-zone objective lens was constructed by S. Suzuki, K. Akashi, and H. Tochigi (1968) to obtain a high-resolution image.

3.2 Aberrations of three-stage electron microscopes

It has been demonstrated theoretically by L. Marton and K. Bol (1947) that the reduction of spherical aberration can be achieved by forming a virtual image with a strong lens and transforming this into a real one with a weak lens. For practical design purposes of three-stage electron microscopes, and extending the Marton's theory to other aberrations, numerical calculations (Kanaya, 1952a, 1952b, 1952c) were carried out for the bell-shaped field.

The variations of axial aberrations such as the spherical and chromatic aberrations are shown in Figs. 3.1 and 3.2, where f_{com} indicates the focal length of a compound lens.

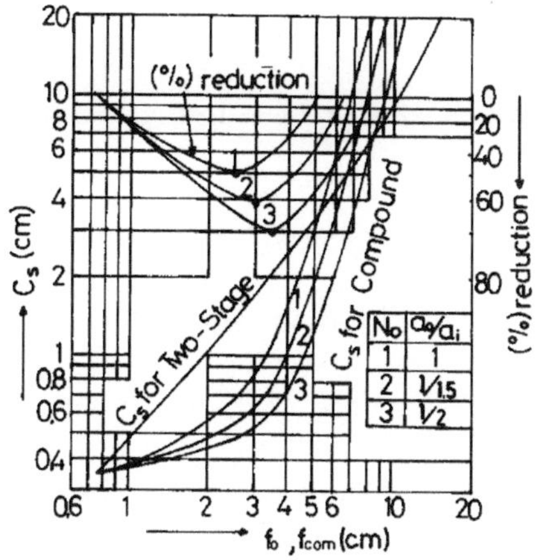

Figure 3.1 Spherical aberration coefficient of a compound lens versus the focal length when forming a virtual image with a strong objective and transforming this into a real one with a weak intermediate lens; a_o and a_i refer to the half-widths of the field of the objective and intermediate lenses.

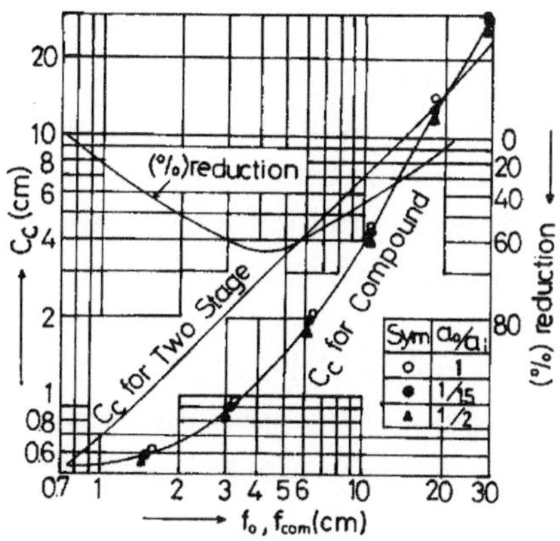

Figure 3.2 Chromatic aberration coefficient of a compound lens as a function of focal length.

The results are compared with those for a two-stage electron microscope, and the percentage reduction P is defined as

$$P = (\Delta u_2 - \Delta u_3)/\Delta u_2 \times 100\%$$

where Δu_2 and Δu_3 are the axial aberrations for two-stage and three-stage microscopes, respectively.

Maximum reduction of the spherical aberration

$$P_{max} = 51\% \quad (a_o/a_i = 1,\ f_{com} = 2.5\ \text{cm})$$
$$= 61\% \quad (a_o/a_i = \tfrac{1}{1.5},\ f_{com} = 3\ \text{cm})$$
$$= 70\% \quad (a_o/a_i = \tfrac{1}{2},\ f_{com} = 3.5\ \text{cm})$$

Maximum reduction of the chromatic aberration

$$P_{max} = 65\% \quad (f_{com} = 2.5\ \text{cm})$$

Here, $2a$ is the so-called half-width given by Glaser's bell-shaped field, and the subscripts o and i indicate the objective and intermediate lenses, respectively.

Furthermore, from the results of the field aberration calculations the following conclusions may be drawn:

1. *Distortion.* When the intermediate lens is always producing a real image of a virtual object, its spherical aberration results in a barrel distortion. The projection lens, on the other hand, is always producing a real image of a real object and hence, as is well known, produces a pincushion distortion. Thus, the distortion in the intermediate image is always opposite to that produced by the projector and tends to compensate it. Fig. 3.3 shows the measured distortion of a three-stage lens system in comparison with an ordinary two-stage system.

2. As the curvature of field and the astigmatism (third-order aberration) contain contributions from all the lenses, each lens should be designed so that the half-width of the field curve is as large as possible. The aberration due to the astigmatism is found to be less than a third of that due to the curvature of the field; its real part is always positive, but its imaginary part may be positive or negative. Thus, according to the lens strength and the polarity of each lens system, it is possible to remove this aberration by appropriately designing the lenses.

3. The magnitude of coma aberration is negligibly small in comparison with the other field aberrations. It has been found by K. Kanaya (1954a) that the chromatic difference in magnification, when the objective and the intermediate lenses are used as a compound lens, varies through negative values and vanishes for its optimum lens strength. On the other hand, when the intermediate lens is used similarly as the projector lens, it ranges from positive to negative infinity, as shown in Fig. 3.4. Of course, the achromatic

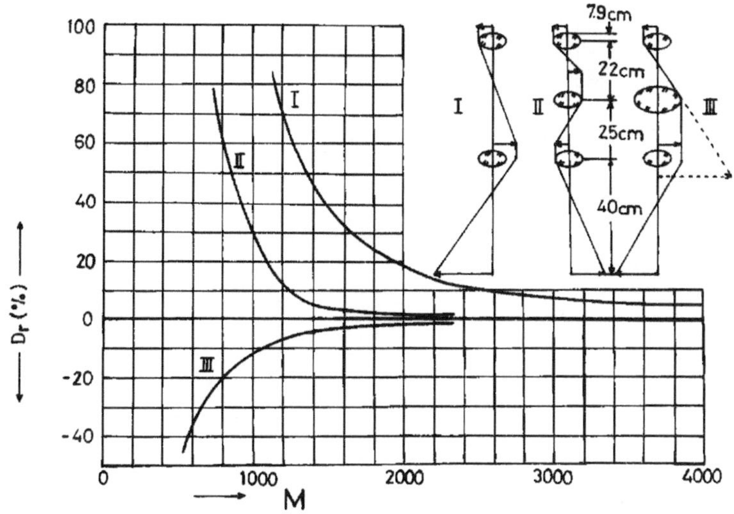

Figure 3.3 Variation of the radial distortion coefficient D_r measured with the aid of a replica of an optical grating with magnification *M*, where D_r was measured as the distance 2.5 cm from the center of the field, with $\phi = 40$ kV.

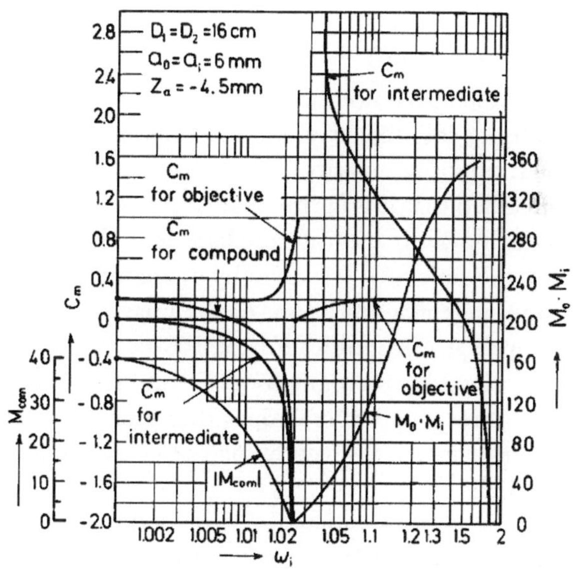

Figure 3.4 Variation of chromatic aberration of magnification coefficient C_m for two-stage objective and compound objective.

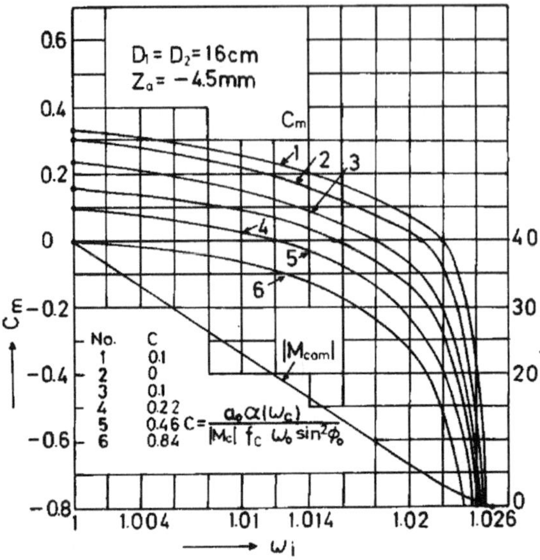

Figure 3.5 Variations of chromatic aberration of magnification coefficient C_m in compound objective for various illuminating conditions with the angular aperture $\alpha(\omega_c)$ and the convergent action of $1/|M_c|f_c$, where ω_o and ω_i are the lens strength parameters of objective and intermediate lenses.

condition $C_m = 0$ must depend on the illuminating beam, as represented in Fig. 3.5.

Another chromatic aberration, the chromatic difference in rotation C_r, can be canceled using the objective and intermediate lenses at the proper lens strengths with opposite polarity.

4. It was concluded that it is far more advantageous to obtain low magnification by reducing with the intermediate lens the size of the virtual image formed by the objective lens than by changing the positions of real images as for three-stage magnification. As the image quality of three-stage electron micrographs at a magnification of 1000× is equal to that of the existing two-stage electron micrographs at a magnification of 3000×, it is possible to design three-stage microscopes that give at a magnification of 500× images superior to those given by two-stage electron microscopes at 1500× magnification.

5. The experimental measurement of field distribution (Ishikawa & Kanaya, 1956) was performed by means of a long exploring solenoid, the winding direction of which is reversed at the center; it is positioned with its axis coincident with that of the lens and then connected to a source, and the static force exerted upon an exploring coil causes the pickup (a sapphire

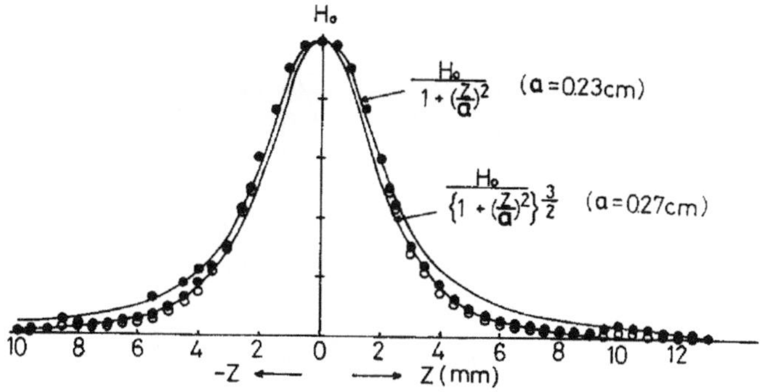

Figure 3.6 Axial field distribution for an experimental objective, $d = 3$ mm, $2\rho = 5$ mm ϕ. (•) $IN = 4400$, (○) $IN = 1600$.

needle) to vibrate; this vibration is amplified, and finally the magnetic field distribution can be directly read on an ammeter.

According to the results as shown in Fig. 3.6, the field of saturated polepieces approximates relatively closely to the bell-shaped one, but that of unsaturated polepieces comes closer to the field represented by

$$H(z) = H_0/[1 + (z/a)^2]^{3/2}$$

Hence the relation between the lens strength and ampere-turns IN must be corrected as follows:

$$\omega^2 = 1 + [6 \times 10^{-3}/\Phi \text{ (V)}](IN)^2, \quad IN = 1.91H_0a \qquad (3.4)$$

The half-width value of a, referring to results introduced by K. Ito and T. Ito (1950), K. Ito, T. Ito, K. Kanaya, and N. Morito (1953), and F. Lenz (1950), can be obtained from Fig. 3.7, which has been used for the magnetic lens design of early electron microscopes in Japan.

3.2.1 Effects of the front field of the objective on the illuminating system

It is known that the highest resolving power can be achieved only when both the inclination of the paraxial ray and illuminating aperture at the object plane have the proper values. This aperture and inclination can be adjusted to the correct values by varying the condenser lens strength. But, because the object is situated well within the objective lens, in order to reduce the spherical and chromatic aberrations, these values are not determined by the focusing action of the condenser lens alone, but are also

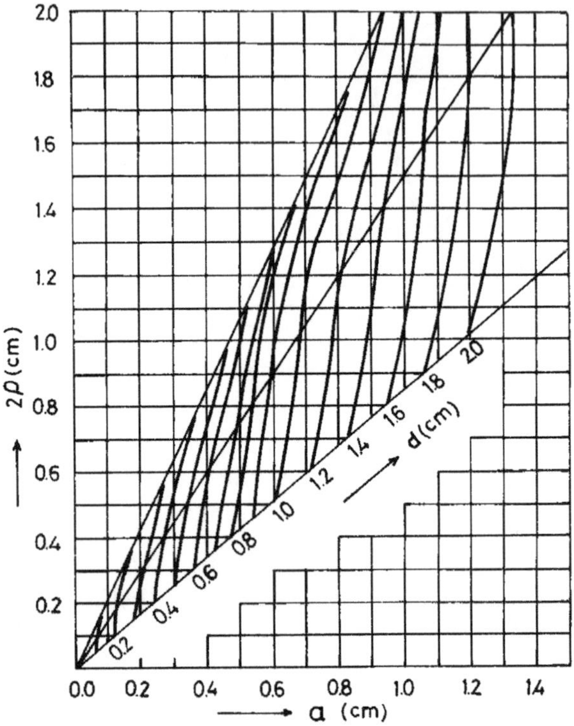

Figure 3.7 Variation of the diameter 2ρ of polepieces with half-width a of the field for various values of the gap d in a symmetrical magnetic lens.

affected by the front field of the objective as follows:

$$\begin{aligned} \alpha_o &= \alpha_c \alpha(\omega_o) \\ \tau_o &= \tau^p - \alpha(\omega_c)\alpha(\omega_o)/M_c f_c \quad \text{(convergent beam)} \\ &= \tau^p + \alpha(\omega_o)/M_c f_c \quad \text{(divergent beam)} \end{aligned} \tag{3.5}$$

with

$$\alpha(\omega_o) = \sin(\omega_o \pi - \pi)\omega_o \sin(\pi/\omega_o)$$
$$\tau^p = \frac{\sin(\pi/\omega_o)}{a_o}\left(\omega_0 \cot(\omega_0 \pi) \sin\frac{\pi}{\omega_0} + \cos\frac{\pi}{\omega_0}\right)$$

Here the subscripts o and c refer to the objective and condenser lenses, respectively, and M represents the magnification. The function $\alpha(\omega)$ increases in proportion to the lens strength ω and the inclination of the principal ray τ^p as shown by Fig. 3.8. In order to reduce the axial aberration, the aper-

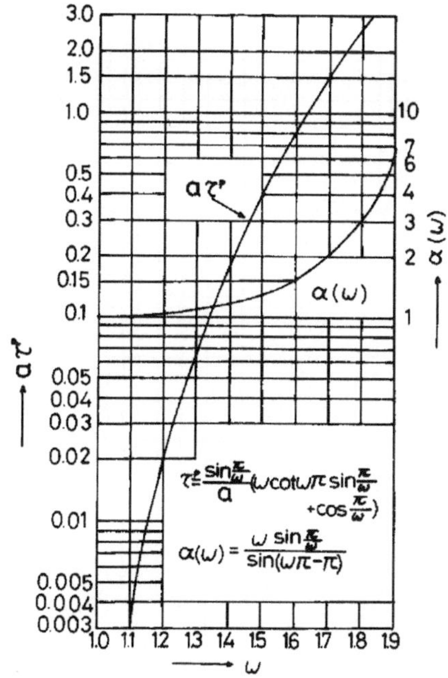

Figure 3.8 Inclination of the principal ray τ^{p} (cm^{-1}) and the function $\alpha(\omega)$ versus lens strength ω.

ture angle $\alpha(\omega_{\mathrm{o}})$ is kept as small as possible. On the other hand, to eliminate the field aberrations, τ should be negative and as large as possible.

It is impossible to satisfy these conditions with a single-condenser lens, but the compound double-condenser lens system introduced by K. Kanaya (1953) is very useful to increase the convergence effect so as to eliminate the field aberrations, and it is also possible to change the aperture over a wide range so as to obtain high resolving power.

Fig. 3.9 represents the illuminating aperture produced by the two-stage double-condenser lens system, where a, b_1, and b_2 are the distances between the first condenser c_1, the second condenser c_2 and the object plane, and between each other; r_{e} is the radius of crossover at the electron source. When the compound condenser lenses are used to focus an illuminating beam, the inclination at the object plane τ_{o} is given by

$$\begin{aligned}\tau_{\mathrm{o}} &= \left(-\frac{\alpha(\omega_{\mathrm{c1}})\alpha(\omega_{\mathrm{c2}})}{M_{\mathrm{c1}}f_{\mathrm{c1}}} - \frac{\alpha(\omega_{\mathrm{c2}})}{M_{\mathrm{c2}}f_{\mathrm{c2}}}\right)\alpha(\omega_{\mathrm{o}}) + \tau^{\mathrm{p}} \quad \text{(convergent beam)} \\ &= \left(\frac{1}{M_{\mathrm{c1}}f_{\mathrm{c1}}} - \frac{1}{M_{\mathrm{c2}}f_{\mathrm{c2}}}\right)\alpha(\omega_{\mathrm{o}}) + \tau^{\mathrm{p}} \quad \text{(divergent beam)} \qquad (3.6)\end{aligned}$$

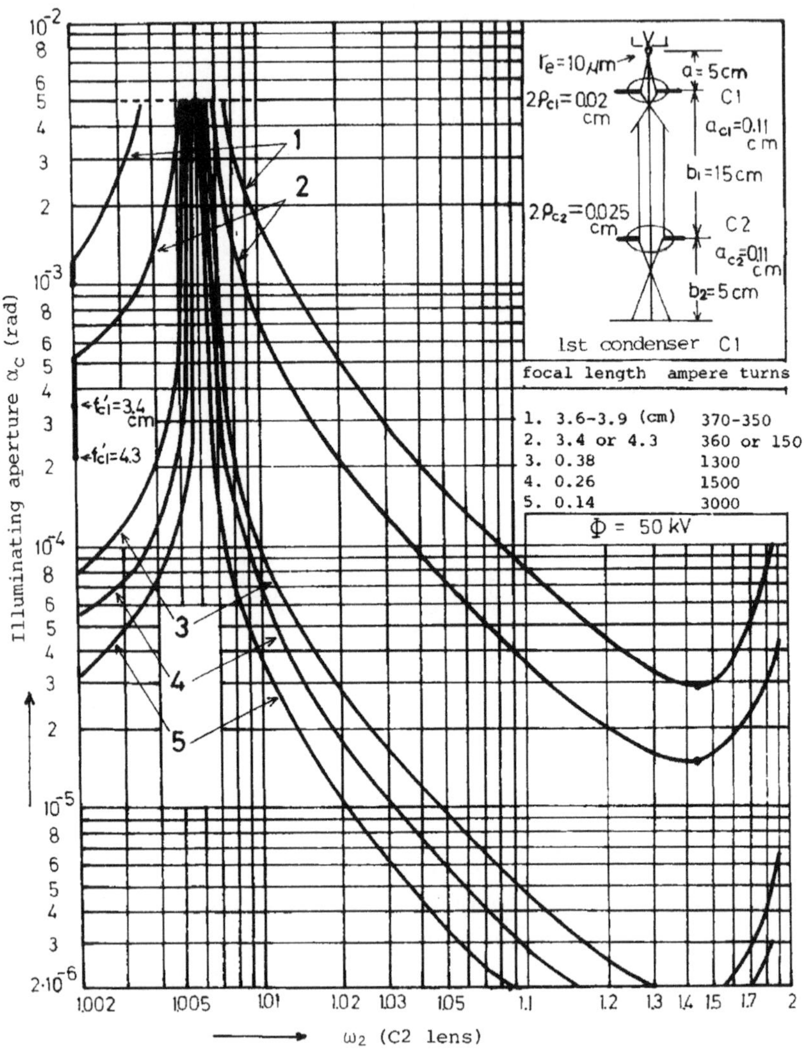

Figure 3.9 Angular aperture α_C of a double condenser as a function of second lens strength ω_2(C2) for various conditions of first lens (C1), where a_{C1} and a_{C2} are the half-widths of first and second lenses, respectively.

which varies as a function of the objective-lens strength ω_o, as shown in Fig. 3.10. Since τ_o may be negative or positive, the field aberrations can be removed by using the two-condenser lenses as a compound lens, provided that $M_{c2}f_{c2} \leqq 10$ cm.

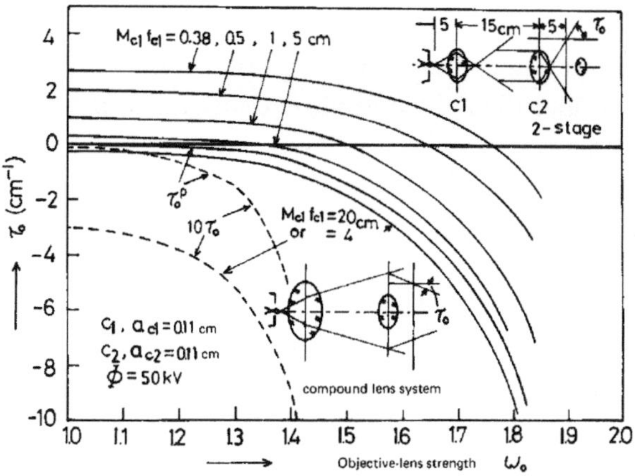

Figure 3.10 Variation of the inclination of the illuminating beam with the objective-lens strength.

3.3 Minimum aberrations and achromatization

The axial aberrations such as the spherical and chromatic aberrations, considering the effect of the objective field in front of the object, can be obtained as a function of the objective-lens strength ω_o

$$\Delta u = C_s \alpha_c^3 \alpha^3(\omega_o) + C_c(\Delta\Phi/\Phi)\alpha_c \alpha(\omega_o) \tag{3.7}$$

From numerical calculation of Eq. (3.7) using the bell-shaped field, the following expressions (Kanaya, 1951a) of minimum aberration result.

Spherical aberration:

$$\Delta u = 1.1\alpha_c^3 a \quad [(IN/\Phi^{1/2})_{opt} = 13.2 \;\; (\omega_o = 1.43)] \tag{3.8}$$

Axial chromatic aberration:

$$\Delta u = \alpha_c(\Delta\Phi/\Phi)a \quad [(IN/\Phi^{1/2})_{opt} = 13.9 \;\; (\omega_o = 1.27)]$$

In a strong magnetic lens, the specimen must lie in the position at which the axial field strength is maximum. The spherical aberration of such a field distribution, which decays from a maximum, is about 10 times smaller and the resulting resolution limit nearly two times better than in weaker magnetic objectives, where the specimen is generally situated in front of the whole field. These conclusions agree with the results on the optimum objective position which have been already pointed out by Kanaya (1949a).

K. Kanaya (1954a) has introduced the idea that the effect of the spherical aberration ζ on the field aberrations is theoretically reduced.

We now consider the field aberration constants α, β, γ, and δ and ϵ, the so-called distortion, field curvature, astigmatism, and coma, respectively. These field aberration coefficients are calculated on the assumption that the general solution of the paraxial ray r_γ can be replaced by $r_\gamma^{\mathrm{p}} + \tau r_{\alpha'}$

$$\begin{aligned} \alpha &= \alpha^{\mathrm{p}} + \tau(\beta^{\mathrm{p}} + \gamma^{\mathrm{p}}) + \tau^2(\delta^{\mathrm{p}} + \epsilon^{\mathrm{p}}) + \tau^3\zeta \\ \beta &= \beta^{\mathrm{p}} + 4\tau\delta^{\mathrm{p}} \text{ (real part)} + 2\tau^2\zeta \\ \gamma &= \gamma^{\mathrm{p}} + 2\tau\delta^{\mathrm{p}} + \tau^2\zeta \\ \delta &= \delta^{\mathrm{p}} + \tau\zeta, \quad \epsilon = 2\bar{\delta} \end{aligned} \tag{3.9}$$

Here r_γ^{p} is the principal ray satisfying the boundary conditions $r_\gamma^{\mathrm{p}} = 1$, $r_\gamma^{\mathrm{p}} = 0$, at $z_{\mathrm{a}} = -\infty$, and τ is the inclination; all aberration coefficients follow the notation given by the Bahnmethode (Scherzer, 1937).

Similarly, the chromatic aberration of magnification is rewritten as

$$\Delta u_{\mathrm{c}} = C_{\mathrm{m}}(\Delta\Phi/\Phi)u_{\mathrm{a}}, \quad C_{\mathrm{m}} = C_{\mathrm{m}}^{\mathrm{p}} + \tau C_{\mathrm{m}} \tag{3.10}$$

where $\Delta\Phi/\Phi$ is the fluctuation of the accelerating potential and u_{a} the coordinate of an object point. When an illuminating beam can be adjusted to the proper values by varying the condenser lens strength, the condition of "achromatization" due to the chromatic aberration changes as a function of the lens strength (Kanaya, 1954b) as shown in Fig. 3.11.

For practical design purposes, it is difficult to get the large number of ampere-turns ($IN > 3000$ for $\Phi = 50$ kV) needed to keep the objective at the optimum position. Accordingly, to keep the axial aberration at its minimum value and simultaneously make the chromatic field aberration zero, it is most advantageous to use convergent beam illumination.

The perfect "achromatization" must eliminate another chromatic field aberration, that of rotation, so that $C_{\mathrm{r}} = 0$. It is well-known that the sign of the chromatic difference in rotation is reversed if the polarity of the excitation is changed, and "achromatization" of such an aberration may be easily obtained in the three-stage electron microscope with another intermediate lens; however, it is difficult to set all the chromatic field aberrations zero over a wide magnification range.

The double-gap lens with two exciting coils, which has been introduced by Kanaya (1949a, 1953), is very useful for satisfying the achromatic condition, because it can be used to act in various modes, as shown in Fig. 3.12.

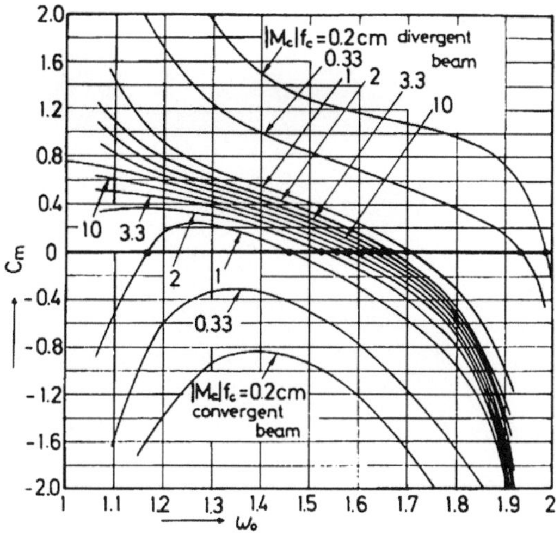

Figure 3.11 Representation of the effects of condenser lens system on chromatic aberration of magnification coefficient.

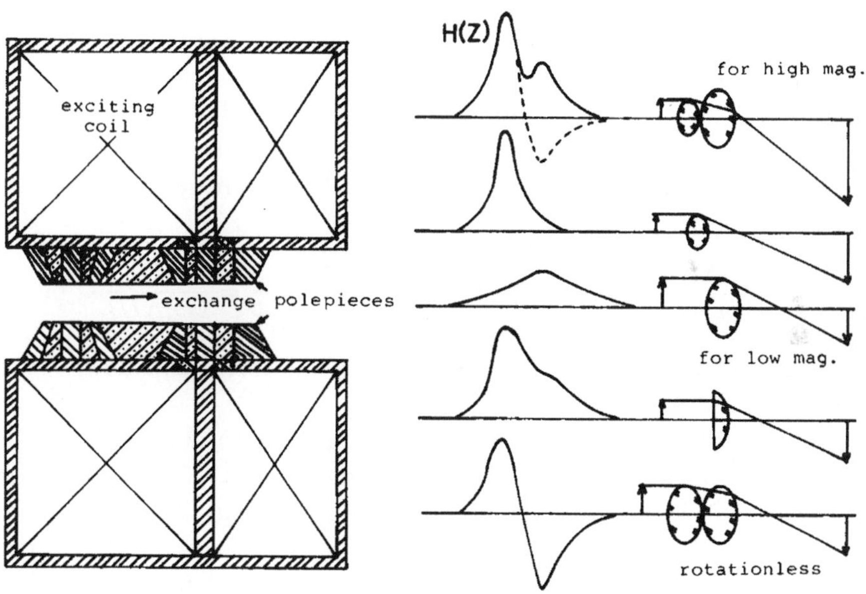

Figure 3.12 Double-gap lens having two exciting coils and its lens action.

The image confusions due to the chromatic field aberrations have been pointed out by B. von Borries (1949) and experimentally demonstrated by S. Katagiri (1953). M. Iwanaga and K. Kanaya (1953) introduced a simple

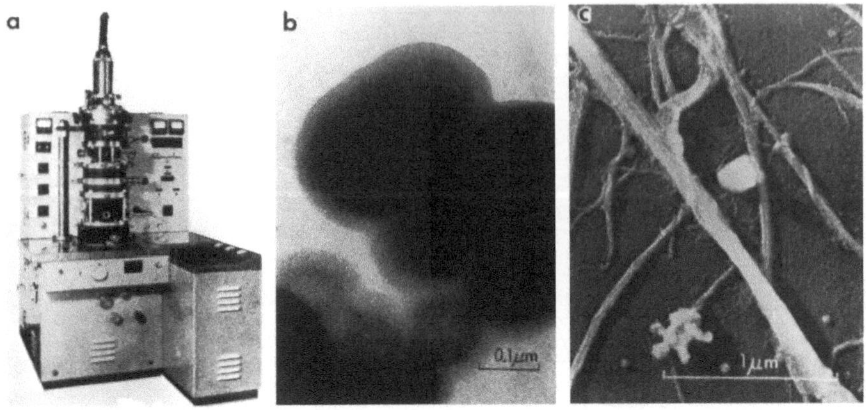

Figure 3.13 Three-stage electron microscope with a double-condenser system made by Shimadzu Ltd. in 1955: (a) column, (b) carbon black, (c) rayon fiber.

method of measuring chromatic field aberrations, in which the chromatic field aberration coefficients can be readily obtained from double-focused images, with variation of exciting coil current but without fluctuation of the accelerating potential applied.

Evidently, the deviation in the circumferential direction is $\Delta D_c = M\Delta u_m$. That in the radial direction is given by $\Delta D_r = M\Delta u_r$. Since $\Delta\Phi/\Phi = -2\Delta I/I$, the chromatic field aberration coefficients can be obtained as follows:

$$C_m = \frac{\Delta D_c}{Mu_a\Delta\Phi/\Phi} = \frac{\Delta D_c}{2D_a\Delta I/I}, \quad C_r = \frac{\Delta D_r}{Mu_a\Delta\Phi/\Phi} = \frac{\Delta D_r}{2D_a\Delta I/I} \tag{3.11}$$

where D_a is the distance from the center of the image and $\Delta I/I$ is the variation of exciting coil current.

A magnetic three-stage electron microscope with a double-condenser system and double-gap intermediate lens designed by K. Kanaya was made by Shimadzu Co. in 1955, as shown in Fig. 3.13a. Figs. 3.13b and 3.13c show micrographs of carbon black and rayon fiber taken on this instrument 100 kV. Its functional and operational features are as follows: resolution 20 Å at 40–100 kV, magnification 500–20,000×; electron-microscopic images and selected-area diffraction patterns did not rotate when the magnification was changed. Using the microscope, achromatic compensation of chromatic aberration in magnification by the projector at lens strength $\omega_p = 1.5$ has been obtained very clearly. The disagreement of voltage and current centers due to misalignment of the electron lens system, as well as the theoretical basis of the achromatic lens, has been discussed at length by K. Kanaya (1958).

3.4 Anastigmatic lens

As has been theoretically discussed by K. Kanaya and A. Ishikawa (1959), an obvious effect of misalignment of the electron lens system is to spoil the axial symmetry of images and make it difficult to interpret the results of observation. The coma-free point was calculated for various misalignments and tilts of a lens, a lateral shift, and an inclination of the incident beam. The anastigmatic condition for a magnetic lens was deduced.

For an objective lens:

$$\left[\left(C_o - \frac{F_o^2 + f_o^2}{B_o}\right)^2 + \left(c_o - \frac{2F_o f_o}{B_o}\right)\right]^{1/2} = 0, \quad (C_o^2 + c_o^2)^{1/2} = 0 \qquad (3.12)$$

For a projective lens:

$$(C_p^2 + c_p^2)^{1/2} = 0$$

where B, F and f and C and c are the respective spherical, coma, and astigmatic aberration coefficients following Glaser's results, and o and p refer to the objective and projector lenses, respectively.

From Eq. (3.12), the anastigmatic conditions relative to each coefficient were obtained, namely, $B_o C_o = F_o^2$, and $f_o = C_o = c_o = 0$ for an objective lens, $C_p = c_p = 0$ for a projective lens. To realize such an anastigmatic lens, it should be constructed of three cylindrical polepieces with double gaps independently excited by reverse polarities. Such a lens has the great merit that all the imaginary terms of the aberrations $f_o = c_o = c_p = 0$ can be canceled by the effect of reverse polarities. Fig. 3.14 shows an example of the construction of an anastigmatic lens (Kanaya, 1958).

The condition of $B_o C_o = F_o^2$ or $C_o = 0$ can be satisfied by using the strong excitation $\omega_o = 1.24$. Furthermore, when we consider the theoretical limit of allowable misalignment of the existing magnetic lens to obtain high-resolution images, the tolerances of lateral shift, objective tilt, and incident beam inclination τ are given by

$$d \leqq [10^{-5}\ (\text{cm})/(C_p + D_p)]^{1/2}, \quad \delta \leqq [10^{-5}\ (\text{cm})/(C_p + D_p)]^{1/2}/f_p \qquad (3.13)$$

and

$$\tau \leqq \{[10^{-5}\ (\text{cm})(F_o^2 + f_o^2)]/[B_o^2(C_o + D_o) - 2B_o(F_o^2 + f_o^2)]\}^{1/2}$$

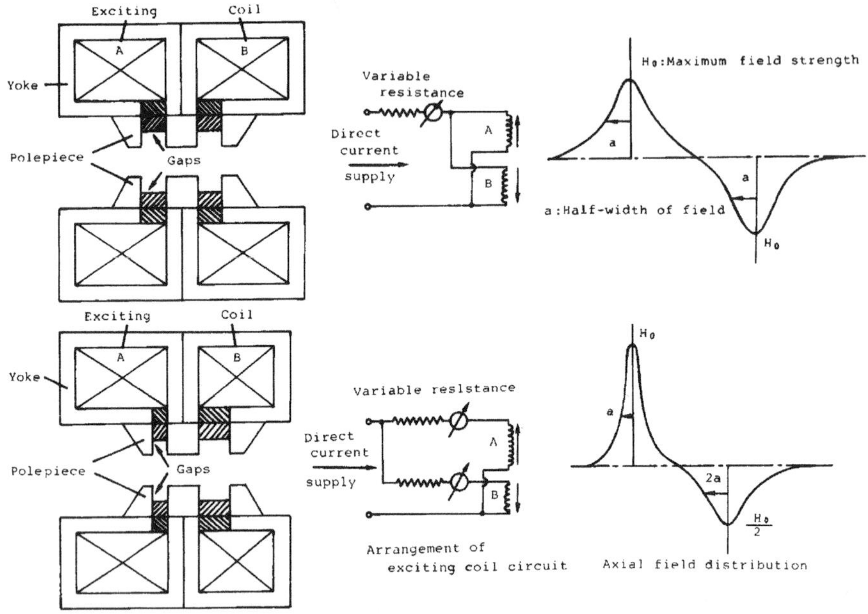

Figure 3.14 Constructions of an anastigmatic lens and its field distributions; the lens consists of three cylindrical polepieces and double separated gaps independently excited with reverse polarities; the gap length and coil turns are the same in (A) and different in (B).

where f_p is the projector lens focal length. Such tolerances expressed numerically lead to $d \leqq 10$ μm, $\delta \leqq 0.005$ rad, and $\tau \leqq 0.006$ rad.

4. Formation of caustic and shadow images

4.1 Third-order caustic

It is well known that the pattern of concentration of the electron beam near the back focal point of a objective lens is a more complicated pattern, which is called the "caustic," than that obtained with light optics, since, on account of the anisotropic aberrations of magnetic lenses, its basic shape is rotated bodily through a certain angle. W. Glaser (1941, 1953), W. Scheffels, B. von Borries and F. Lenz (1954), and F. Lenz (1956a, 1956b) have shown that the geometrical locus of the points of intersection of neighboring trajectories in the beam is a kind of surface with folds. It is also known that, in the case of any constructional inaccuracies in the polepieces, a second-order caustic appears, second-order astigmatism combining with spherical aberration and any defocusing. It has been confirmed by K. Kanaya, H. Kawakatsu, and

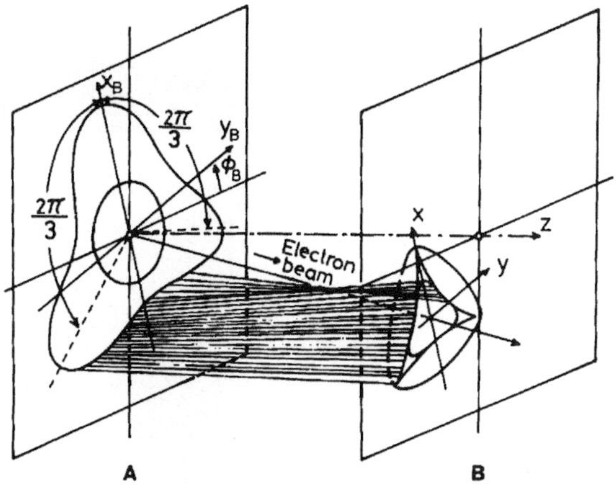

Figure 4.1 Third-order caustic formation. (A) Aperture plane; (B) caustic plane.

A. Kato (1959a) that, in the case of oblique illumination, a third-order caustic appears due to the third-order astigmatism, which is increased by misalignment, together with the spherical aberration and defocusing, which are experimentally known.

The aberration formula giving the third-order caustic has been found to be as follows:

$$\Delta u = \theta(\Delta f + B_0\theta^2)e^{i(\psi_B+\alpha)} + M_0\theta^2 e^{-i(2\psi_B-\delta)} \tag{4.1}$$

where θ and ψ_B are the coordinates of the beam, B_0 is the spherical aberration coefficient, and M_0 and δ represent the third-order astigmatism as given in Eq. (2.5). Following W. Glaser, we write the dimensionless formula for the third-order caustic as follows:

$$\begin{aligned} \xi &= (\zeta + \rho_B^2)\rho_B \cos\psi + \rho_B^2 \cos 2\psi, \\ \eta &= (\zeta + \rho_B^2)\rho_B \sin\psi - \rho_B^2 \sin 2\psi \end{aligned} \tag{4.2}$$

where $r^2 = x^2 + y^2$, $x = \xi M_0^3/B_0^2$, $y = \eta M_0^3/B_0^2$, $\theta \cos\psi = \xi_B M_0/B_0$, $\theta \sin\psi = \eta_B M_0/B_0$, $\xi_B^2 + \eta_B^2 = \rho_B^2$, $\Delta f = (M_0^2/B_0)\zeta$, and $\psi_B + \frac{1}{3}\alpha - \frac{1}{3}\delta = \psi$. From the basic curve of the caustic (Fig. 4.1) given by

$$3\rho_B^4 - 4\rho_B^2 \cos 3\psi + 4(\zeta - 1)\rho_B^2 + \zeta^2 = 0$$

the caustic lines in different caustic planes are shown in Fig. 4.2.

From a series of micrographs of the caustic, the spherical aberration coefficient B_0 and the third-order astigmatism M_0 can be determined:

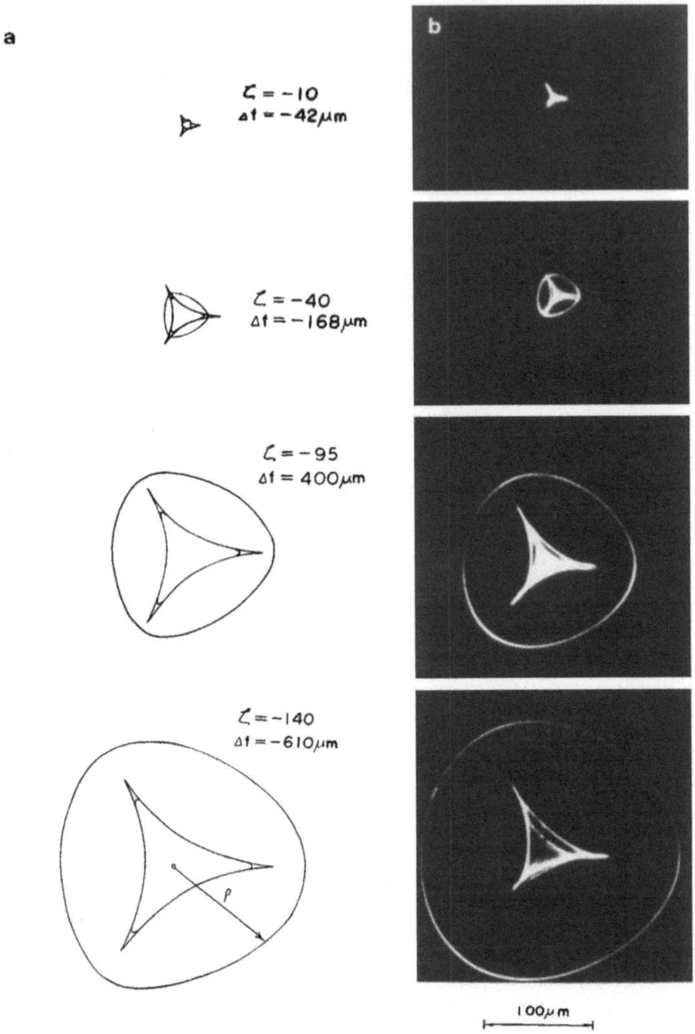

Figure 4.2 Comparison of theoretical results (a) with experimental results (b) of caustic lines in different caustic planes.

$$B_0 = (\rho/r)^2(\Delta f/\zeta)^3, \quad M_0 = (\rho/r)(\Delta f/\zeta)^2 \tag{4.3}$$

where $\rho = (\xi^2 + \eta^2)^{1/2}$. In a typical experiment, these are given by

$$B_0 = 0.23 \text{ cm}, \quad M_0 = 90 \ \mu\text{m}$$

where $r = 85$ μm, $\rho = 558$, $\Delta f = 300$ μm, and $\zeta = 80$.

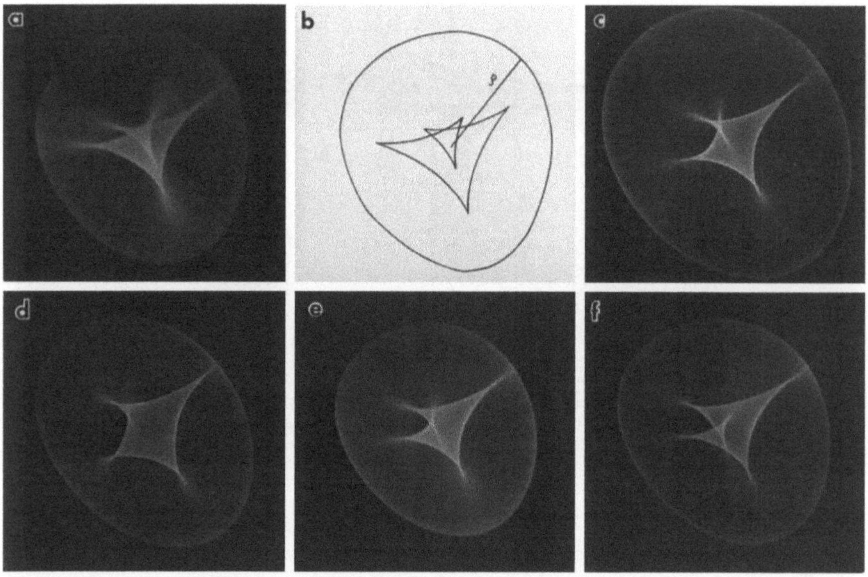

Figure 4.3 Caustic figures (a) agree with theoretical ones (b) for $\gamma = 0.25$, $\zeta = -20$, $\psi_\delta = -90°$; (c) for $\gamma = 0.15$, $\zeta = -18$, $\psi_\delta = -65°$; (d) for $\gamma = 0.12$, $\zeta = 12$, $\psi_\delta = -60°$; (e) for $\gamma = 0.14$, $\zeta = -16$, $\psi_\delta = -50°$; (f) for $\gamma = 0.22$, $\zeta = -19$, $\psi_\delta = -40°$.

4.2 Caustic properties combined with second- and third-order astigmatism

It has been shown by K. Kanaya and H. Kawakatsu (1960b; 1960c), using geometrical optics and taking account of the higher order aberration, that the combined caustic pattern including the second- and third-order astigmatism due to any misalignment such as oblique illumination appears as shown in Fig. 4.3 and its direction depends very sharply on the phase angle ψ_{NM} between the second- and third-order astigmatisms N and M and the ratio $\gamma = M/(BN)^{1/2}$. By comparing a micrograph of the combined caustic pattern experimentally obtained with the charts calculated for γ with the phase angle ψ_{NM}, the generalized defocus ζ and the generalized distance ρ can be determined. On the other hand, the value of the defocusing Δf and the actual value of the caustic locus Δu measured from the caustic center are easily measured. Accordingly, B, N, and M can be estimated from

$$N = \Delta f/\zeta, \quad B = (\rho/\Delta u)^2 N^3, \quad M = \gamma(BN)^{1/2} = \gamma N^2(\rho/\Delta u) \tag{4.4}$$

It has been confirmed that the combined caustic appears with an oblique illumination $|u_a| = 5$–20 μm deviating from the optic center and the third-

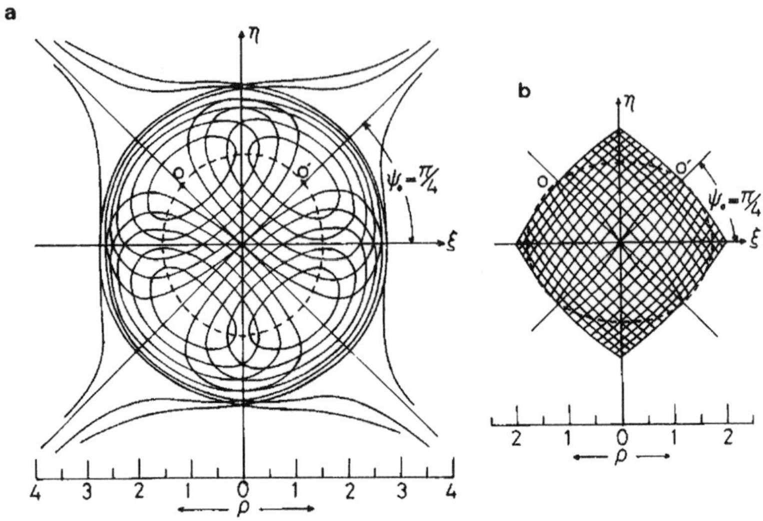

Figure 4.4 Results of calculations of shadow images for a square mesh as a function of the defocusing; (a) $\Delta f < 0$, $\zeta = -8$ and (b) $\Delta f > 0$, $\zeta = 8$ and $\psi_0 = \pi/4$; the spacing of the mesh $\rho_{n+1} - \rho_n = 2$.

order caustic appears at a deviation of $|u_a| > 50$ μm. Then, as for the best compensation, it is most important to make the alignment perfect until the second-order caustic alone appears in place of the combined caustic patterns.

4.3 Shadow image

It is well known that a shadow image of a cross diffraction grating is often used to measure the spherical aberration in the light optics. In the electron microscope, Y. Sakaki and S. Maruse (1954) introduced a similar method of measuring the spherical and astigmatic aberrations from the shadow image of a cross diffraction grating. It has been confirmed by K. Kanaya, H. Kawakatsu, and A. Kato (Kanaya & Kawakatsu, 1960a; Kanaya et al., 1959b) that the locus of a square mesh can be obtained by solving the third-order algebraic equation of the caustic coordinate. The patterns seen vary with the defocusing; they may be of simple barrel type, pincushion type, or a combination of the two in which a pincushion is seen in the inside region and a barrel in the outside region of the image field, or conversely, as shown schematically in Fig. 4.4.

According to the theory of caustic formation (Glaser, 1952; Kanaya et al., 1959b), the shadow image of a square mesh has been found to be the solution of the third-order equation of the generalized aperture coordinate $\rho_B = \theta(B/N)^{1/2}$ with the phase angle ψ_B and the coordinate of the square mesh $\rho = r(B/N^3)^{1/2}$ with the phase angle ψ_0

$$\rho_B^3 + \rho_B\left(\zeta - \frac{\sin(\psi_B + \psi_0)}{\sin(\psi_B - \psi_0)} - \frac{\rho}{\sin(\psi_B - \psi_0)}\right) = 0 \tag{4.5}$$

where one utilizes $\psi_0' = \psi_0 - \pi/2$ for another mesh in the perpendicular direction, ζ is the generalized defocus $\zeta = \Delta f/N$, and N is the astigmatism coefficient. The image of a square mesh can be solved as a function of defocus as follows:

1. $\Delta f > 0$:

$$\rho_B = (\tfrac{4}{3}p)^{1/2}\sinh\theta$$

where

$$\sinh 3\theta = \left(\frac{27q^2}{4p^3}\right)^{1/2}; \quad q = \frac{\rho}{\sin(\psi_B - \psi_0)}, \quad p = \zeta - \frac{\sin(\psi_B + \psi_0)}{\sin(\psi_B - \psi_0)}$$

2. $\Delta f < 0$:

a. $27q^2 > 4(-p)^3$,

$$\rho_B = [\tfrac{4}{3}(-p)]^{1/2}\cosh\theta \tag{4.6}$$

where

$$\cosh 3\theta = [27q^2/4(-p)^3]^{1/2}$$

b. $27q^2 \leqq 4(-p)^3$, three solutions of ρ_B are given by

$$\begin{aligned}\rho_B = &-[\tfrac{4}{3}(-p)]^{1/2}\cos\theta\\ &-[\tfrac{4}{3}(-p)]^{1/2}\cos(\theta + \tfrac{2}{3}\pi)\\ &-[\tfrac{4}{3}(-p)]^{1/2}\cos(\theta + \tfrac{4}{3}\pi)\end{aligned}$$

where

$$3\theta = [27q^2/4(-p)^3]^{1/2}$$

Although the solution of Eq. (4.6) may be considered to be threefold in this last case, one has a negative sign and then the solution for ρ_B, in general, has two or one value. Thus the shadow image of a square mesh appears as a barrel-distorted pattern in $\Delta f > 0$, and it appears as a pincushion pattern near the center of the image field [Eq. (4.6b)] and as a barrel pattern far

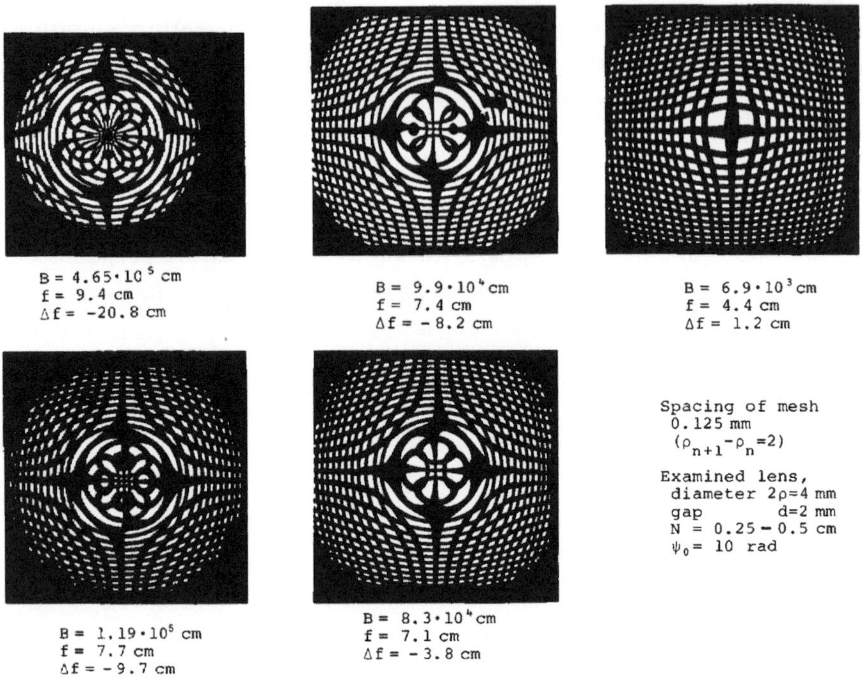

Figure 4.5 Through-focus series of the shadow image for a square mesh.

from the center [Eq. (4.6a)] in $\Delta f < 0$, as shown by the experimental results in Fig. 4.5.

The critical point at which the image of a straight line collapses to a point can be calculated for $d\rho/d\rho_{\mathrm{B}} = 0$, $3\rho_{\mathrm{B}}^2 = -\zeta + \sin(\psi_{\mathrm{B}} + \psi_0)/\sin(\psi_{\mathrm{B}} - \psi_0)$. Fig. 4.6 shows the ellipsoid fylfot locus of the shadow image for a square mesh.

The spherical aberration coefficient B and the astigmatic aberration coefficient N are obtained from the mesh widths in perpendicular directions, r_0 and r_0':

$$B = \tfrac{32}{27}(-\Delta f)^3/(r_0^{2/3} + r_0'^{2/3})^3, \quad N = \Delta f/\zeta \tag{4.7}$$

where the generalized defocus is obtained from the equation

$$-\zeta = [(\rho_{\mathrm{B0}}^2 + \rho_{\mathrm{B0}}'^2)/(\rho_{\mathrm{B0}}^2 - \rho_{\mathrm{B0}}'^2)]\cos 2\psi_0$$

in which ρ_{B0} and ρ_{B0}' are the distances of the critical points from the image center.

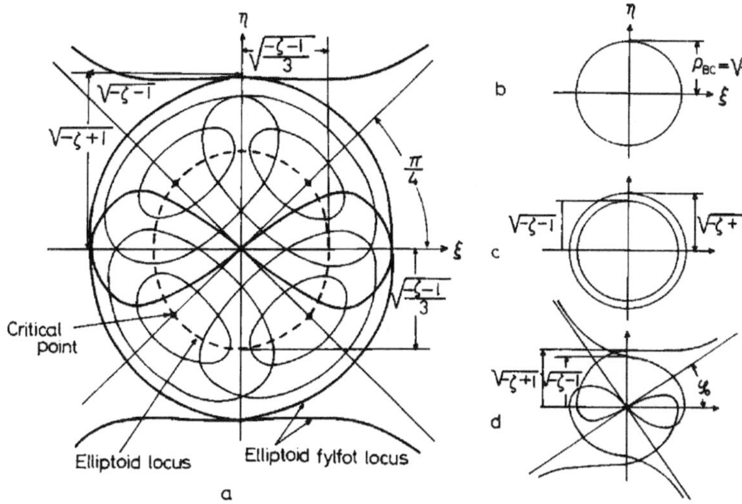

Figure 4.6 Variation of elliptoid fylfot locus of shadow image with ψ_0 in overfocus: (a) $\psi_0 = \pi/4$, (b) $N = 0$, (c) $\psi_0 = 0$, and (d) general case.

The theory of caustic pattern and shadow image formation with a convergent beam has been developed to measure the current density distribution of electron beams at an electron gun as well as a reducing lens (Kanaya et al., 1962, 1967, 1968; Kawakatsu et al., 1965). Furthermore, the shadow image technique was applied to measure the current density distribution, power density, and minimum spot size at the focal point of the electron beam. One of the results of this work is that a microfocus X-ray unit with a telefocal electron gun has been made by JEOL Co.; a beam spot 10 μm in diameter for an electron beam current of 145 μA on the target at 50 kV is used to observe the "Pendellösung fringe" appearing in a section of (111)-reflection silicon (Kanaya et al., 1967).

Fig. 4.7 shows the constructional diagram of this microfocus X-ray unit.

5. Stigmators

5.1 Electrostatic stigmator used to correct second- and third-order astigmatisms

Early in the development of the electron microscope, it was recognized that the astigmatic aberration is one of the most serious factors limiting the performance of the high-resolution electron microscope. From the beginning, efforts were made to eliminate this defect. One of the earliest

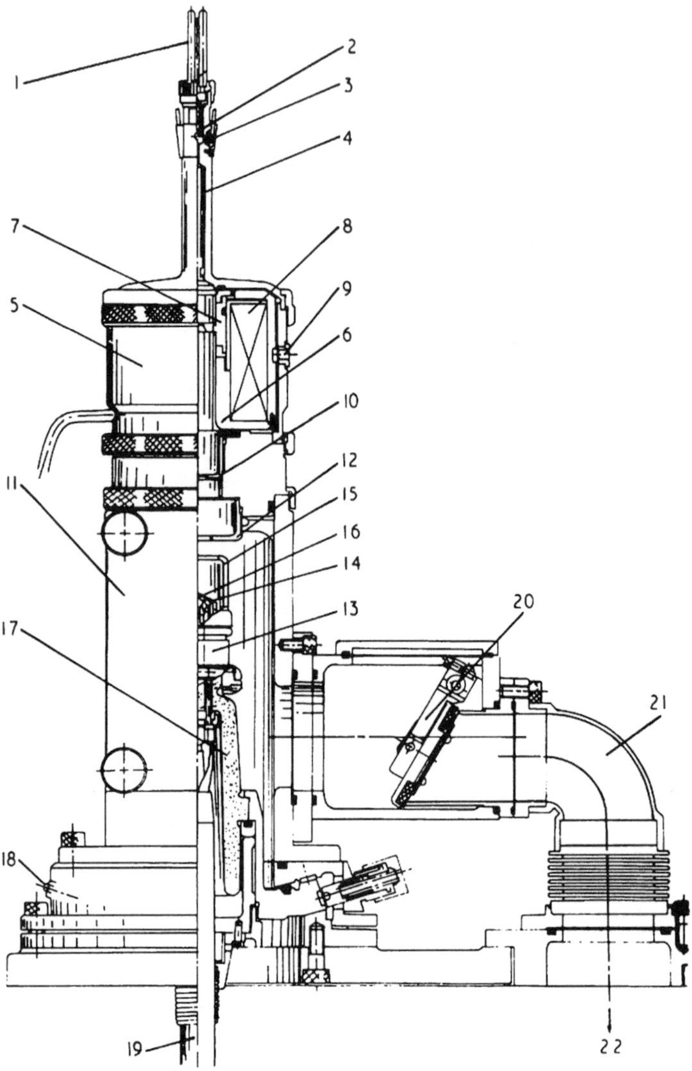

Figure 4.7 Cross-sectional diagram of a microfocus X-ray unit: (1) water-cooled pipe, (2) target, (3) X-ray outlet, (4) target tower, (5) magnetic lens, (6) (magnetic) lens core, (7) magnetic lens polepiece, (8) magnetic lens coil, (9) magnetic lens adjustment, (10) diaphragm, (11) electron gun chamber, (12) anode, (13) grid assembly, (14) electrode 1, (15) electrode 2, (16) filament, (17) ceramic insulator, (18) electron gun tilting adjustment, (19) high-tension cable (to oil tank), (20) main vacuum valve, (21) exhaust pipe, (22) to diffusion pump.

attempts was that of Hillier and Ramberg[1] (1947), who demonstrated that the second-order astigmatism could be corrected by special objective pole-pieces fitted with compensating screws consisting of eight soft-iron sections. Similar work on the corrector known as the stigmator was introduced by O. Scherzer (1947, 1949). Later, O. Rang (1949), N. Morito and K. Koizumi (1951), S. Leisegang (1953), and R. Seeliger (1953) contributed to the same subject. Theoretical aspects of the astigmatism and its compensation are discussed by F. Lenz and M. Hahn (1953), G. D. Archard (1954), and K. Kanaya, H. Kawakatsu, and I. Tanaka (1961).

Throughout these publications it is asserted that it may be possible to attain, in practice, the theoretical limit of existing lenses if such asymmetries could be completely removed. A consistent theory of compensation by an electrostatic stigmator with n electrodes (Kanaya & Kawakatsu, 1961c) was introduced as follows. Consider the electrostatic stigmator consisting of n insulated metal cylinders arranged round the optic axis. In the first approximation, we examine the two-dimensional case, in which the axial potential along the optic axis is assumed to be constant in the region of its length l; the potential distribution normal to the optic axis indicating nth-order astigmatism can be expanded as a Fourier series in the form

$$\Phi_n = A_n(r/r_0)^n \cos n(\lambda - \lambda_n) \tag{5.1}$$

where A is the potential function applied to each electrode, r_0 the stigmator radius, and λ_n the phase angle of order n.

According to the calculation of Φ_n, a stigmator with 2 or 4 electrodes cannot produce rotatable second-order astigmatism; however, a more complicated arrangement—a 6-, 8-, 9-, 10-, or 12-electrode system—can produce rotatable second- or third-order astigmatism. As for the rotatable second-order stigmator used to correct second-order astigmatism in any direction, 12-, 10-, 8-, and 6-electrode systems are suitable for the elimination of this astigmatism.

Electrode systems with 12, 9, and 8 electrodes, in that order, are useful for the rotatable third-order stigmator. Consider the case in which the stigmator is located at the focus of the objective lens; the astigmatic aberration of order n can be expressed in the form

$$\Delta u_n = -(2nA_n/4)(f/r_0)^n l\alpha^{n-1} e^{-i[(n-1)\chi - n\lambda_n]} \tag{5.2}$$

[1] See also the article by P. Grivet in volume 220 of these Advances.

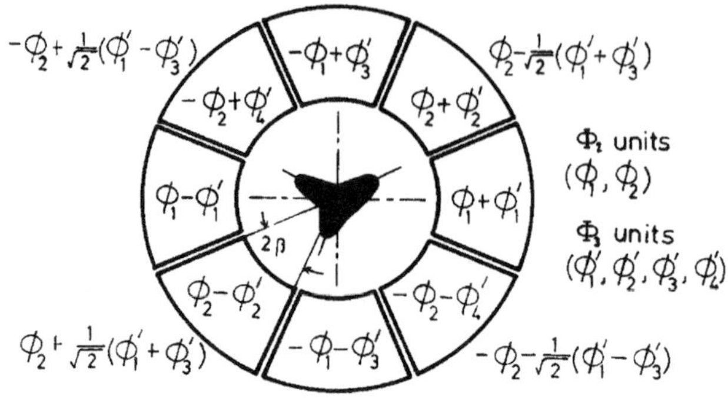

Figure 5.1 Potential arrangement of eight-electrode stigmator producing both second- and third-order astigmatisms which can independently rotate toward any values of the directions $\lambda_2 = \frac{1}{2}\tan^{-1}(\phi_2/\phi_1)$ and $\lambda_3 = \frac{1}{3}\tan^{-1}(\phi_3'/\phi_1')$.

where f is the focal length of the objective lens, l the length of the stigmator, and α the half-angle of the aperture with the phase angle χ.

It is clear that, for the accurate compensation of astigmatism needed to attain highly resolved images, the third-order astigmatism as well as the second-order one should be eliminated. For practical design purposes, the stigmator with eight electrodes is most useful for correcting both astigmatisms.

The compensation condition in such a system becomes

$$2\lambda_n = \tan^{-1}\frac{\psi_2}{\psi_1} = -2\lambda_N, \quad N = \frac{2\sqrt{2}}{\pi\Phi}(\psi_1^2+\psi_2^2)^{1/2}\left(\frac{f}{r_0}\right)^2 l \tag{5.3}$$

and

$$3\lambda_n = \tan^{-1}\frac{\psi_3'}{\psi_1'} = -3\lambda_M, \quad M = \frac{2(2+\sqrt{2})^{1/2}}{\pi\Phi}(\psi_1'^2+\psi_3'^2)^{1/2}\left(\frac{f}{r_0}\right)^3 l$$

where $Ne^{-i2\lambda_N}$ and $Me^{-i3\lambda_M}$ are the second- and third-order objective-lens astigmatism coefficients; Φ is the accelerating potential; and ψ_1, ψ_2, and ψ_1', ψ_3' are the potentials applied to the electrodes, as shown in Fig. 5.1.

5.2 Magnetic stigmator used to correct second- and third-order astigmatisms

For each magnetic pole of the magnetic stigmator, the familiar relation connects the ampere-turns and the magnetic potential; the potential distribution of a magnetic stigmator can then be obtained from the expansion of a Fourier series as a function of ampere-turns.

Consider a magnetic stigmator with n poles arranged around the optic axis (Kawakatsu & Kanaya, 1961); its scalar potential distribution can be expressed by

$$\psi(r,\chi) = \sum_{n=0}^{\infty} C_n \left(\frac{r}{r_0}\right)^n \cos n(\chi - \lambda_n) \tag{5.4}$$

where C_n is the potential function at the surface of each pole.

Consider the case in which the stigmator is located at the focus of an objective lens; the astigmatic aberration of order n produced by the stigmator can be expressed in the form

$$\Delta u_n = i\eta \sum_{n=1}^{\infty} \frac{nC_n}{r_0^n} f^n l\alpha^{n-1} e^{-i[(n-1)\chi - n\lambda_n]} \tag{5.5}$$

where $\eta = (e/2m\Phi)^{1/2}$, e is the electron charge, m the electron mass, and Φ the accelerating potential. The major axis of the second-order astigmatism produced by the stigmator lies in the direction $\lambda_2 - \pi/4$ and that of third-order $\lambda_3 - \pi/6$. It was found that the magnetic stigmator is theoretically a more accurate corrector than the electrostatic stigmator because, in the magnetic case, the combination and squaring terms of higher order astigmatism are negligible.

It was shown that the magnetic stigmator consisting of 12 or 8 poles is capable of eliminating astigmatisms of both orders. For example, the compensation conditions in an 8-pole stigmator become

$$\begin{aligned} N &= (8\eta f^2 lk/\pi r_0^2)(I_1^2 + I_2^2)^{1/2} \sin 2\beta, \quad \lambda_N = -(\lambda_2 - \tfrac{1}{4}\pi) \\ M &= (8\eta f^3 lk/\pi r_0^3)(I_1'^2 + I_2'^2)^{1/2} \sin 3\beta, \quad \lambda_M = -(\lambda_3 - \tfrac{1}{6}\pi) \end{aligned} \tag{5.6}$$

where β is the half-angle subtended by a pole at the center of the stigmator; I_1, I_2, and I_1', I_3' are the exciting coil currents corresponding to the scalar potentials ψ_1, ψ_2 and ψ_1', ψ_2' as in the electrostatic case, where k is the form factor of the polepiece depending on the ampere-turns IN given by $\psi = 0.4\pi IN/k$.

The electron microscope made by JEOL first used eight-screw compensation of Hillier's type (1951). Afterward, a magnetic eight-pole stigmator was used for the high-resolution electron microscopes made in Japan (Watanabe & Someya, 1963). Unfortunately, there are no electron microscopes using a stigmator compensating both second- and third-order astigmatisms.

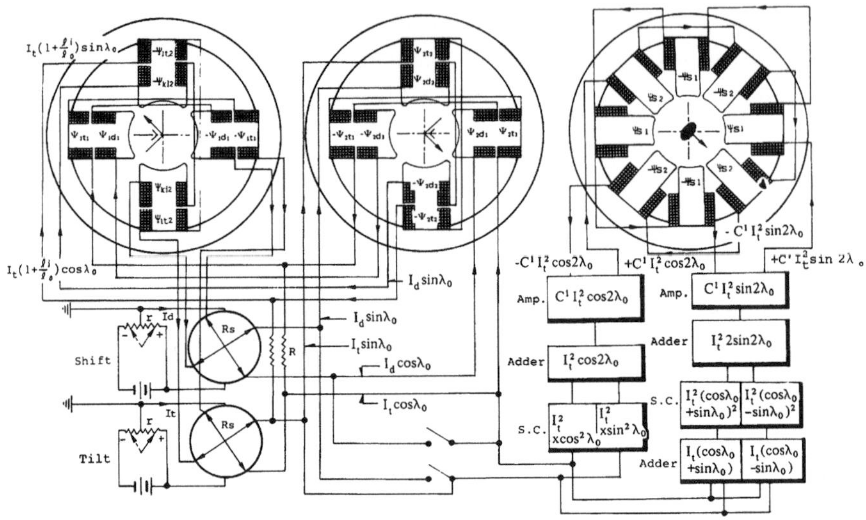

Figure 5.2 Simplified diagram of control circuit for an anastigmatic alignment system using double magnetic deflectors with stigmator.

5.3 Anastigmatic alignment

In electron beam instruments of conventional design, it is a most serious problem that the electron gun is aligned with the other lenses by means of mechanical tilting and displacement control.

An electrostatic-magnetic alignment section introduced by M. E. Haine, A. W. Agar, and T. Mulvey (1958) and a magnetic double-deflection system described by G. D. Archard and T. Mulvey (1958) contributed to the alignment of the optical column in the electron microscope, but these are limited in that there is only one beam deflection direction and, moreover, the production of the polepieces seems to be very difficult in practice because of the complicated shapes of the polepieces. An improved alignment system free from astigmatism has been introduced by K. Kanaya (1962), as an electronic alignment system in place of mechanical alignment as shown in Fig. 5.2.

The alignment principle is as follows: with the magnetic deflector with four poles arranged around the optic axis, it is possible to rotate in any direction by varying the currents of two different exciting coils. Next, it is observed that the alignment action needed to produce a lateral shift of the beam without a tilt for the displaced gun relative to the optic axis can be obtained by combining two deflectors with the same dimensions and deflecting powers. It is also found that the alignment action needed to

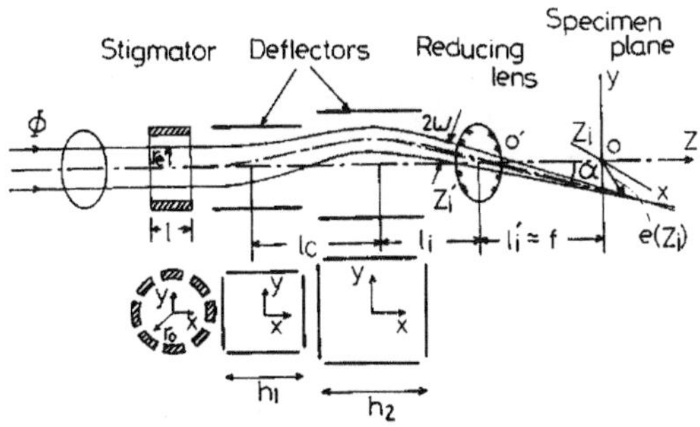

Figure 5.3 Double-deflection system with astigmatic corrector and reducing lens for scanning system.

produce a tilt for an inclined gun relative to the optic axis can be achieved compensating the deflection angles α as

$$\alpha_{1t} = \alpha_0(l_0 + l_i)/l_i, \quad \alpha_{2t} = \alpha_0(l_0/l_i) \tag{5.7}$$

where α_{1t} and α_{2t} are the deflection angles generated by the first and second deflectors, l_0 and l_i, are the distances between the electron gun and the first deflector and between the first deflector and the second deflector, respectively.

Also, if a purely magnetic deflection system is employed in the double-deflection system, the astigmatic aberration it causes is a most serious defect; since the phase angle of the astigmatism caused by the stigmator becomes twice that due to the deflectors, the electron beam passing forward through the alignment system coinciding with the optical angle can be corrected by an electronic control circuit. T o realize this, a pair of currents of magnetizing coils satisfying the conditions for the compensation of astigmatism caused by deflection,

$$\pm N_s I_s = c(NI)^2 \sin 2\lambda_0, \quad \mp N_s I_s = c(NI)^2 \cos 2\lambda_0 \tag{5.8}$$

can be produced by an electronic control circuit consisting of an adder, squaring circuit, and current amplifier; $N_s I_s$ and NI are the ampere-turns of stigmator and deflector, respectively; c is the functional constant; and λ_0 is the phase of incident beam angle α_0.

An electronic double-alignment system together with the double-condenser lens system has been used for most high-resolution electron microscopes made in Japan since 1960. Similar work on the design of

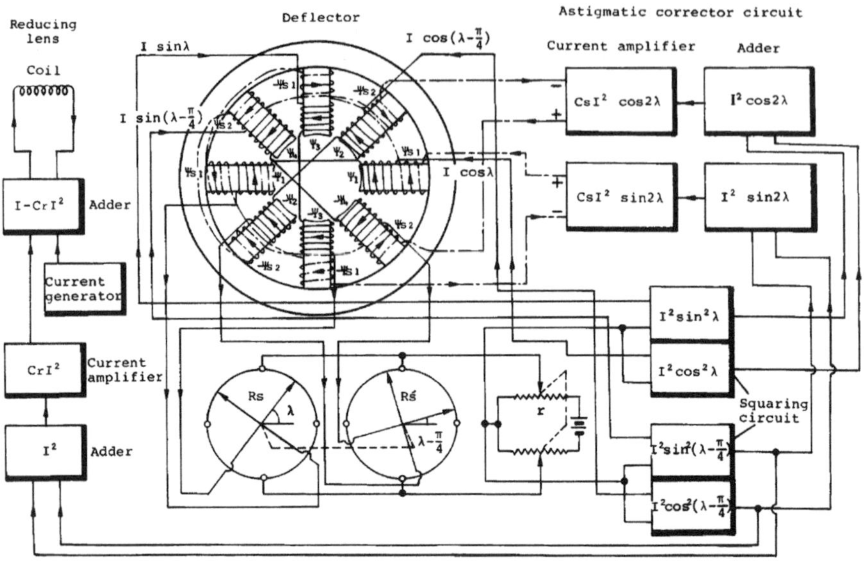

Figure 5.4 An anastigmatic deflection system correcting astigmatism and its defocusing.

an improved deflection system free from aberration consisting of a stigmator, double electronic deflection system, and magnetic reducing lens without anisotropic astigmatism was based on the compensation theory of astigmatism combined with the defocusing defect (Kanaya & Kawakatsu, 1961a, 1961b, 1961c; Kanaya et al., 1961, 1963, 1964). Such a double-deflection system with stigmator and reducing lens for a scanning system (Fig. 5.3) has been used for the main optical column of an electron beam machine first made by JEOL in Japan (1961).

5.4 Anastigmatic deflection system for correcting astigmatism and its defocusing action

An anastigmatic deflection system consisting of eight magnetic poles around the optic axis for use with an electron beam deflection or scanning system has been designed by K. Kanaya and H. Kawakatsu (1962); this provides the deflection and also acts as a stigmator without the necessity of including an additional astigmatic corrector. This enables the electron probe to be deflected and scanned across the specimen and at the same time permits the astigmatism of the deflected beam to be corrected electrically by means of an electronic control circuit as shown in Fig. 5.4.

From the theoretical analysis of the deflection system consisting of eight magnetic poles around the optic axis, in which the magnetic scalar potential applied to each pole is such that $\psi_2 = (1/\sqrt{2})(\psi_1 + \psi_3)$, $\psi_4 = (1/\sqrt{2})(-\psi_1 + \psi_3)$, $\psi_5 = -\psi_1$, $\psi_6 = -\psi_2$, $\psi_7 = -\psi_3$, and $\psi_8 = -\psi_4$, its deflection angle can be given as

$$\alpha = (8\eta/\pi)(\psi_1^2 + \psi_3^2)^{1/2}(l/r_0)\sin\beta e^{i\lambda} \tag{5.9}$$

with

$$\lambda = \lambda_1 - \tfrac{1}{2}\pi, \quad \lambda_1 = \tan^{-1}(\psi_3/\psi_1)$$

The deflection angle α rotates in any desired direction depending on the phase angle of the deflector λ, like the deflector with four magnetic poles. Furthermore, such a deflector is capable of correcting astigmatism if two magnetic quadrupoles, displaced $\pi/4$ from each other, are magnetically excited at two different ampere-turn levels, $\pm\psi_{s1}$ and $\mp\psi_{s2}$, so that their polarities alternate from one pole to the other. It can also operate as an astigmatic corrector, since the astigmatic aberration produced by it is given by

$$\Delta u_s = (8\eta/\pi)(\psi_{s1}^2 + \psi_{s2}^2)^{1/2}\sin 2\beta_s(L/r_{s0})^2 l_s\omega e^{-i(\chi - 2\lambda_s + \pi/2)} \tag{5.10}$$

with $\lambda_s = \frac{1}{2}\tan^{-1}(\psi_{s2}/\psi_{s1})$, where L is the working distance, l and r are the length and radius of the polepieces, and ω is the angle of the convergent beam. If one section of the deflector consisting of eight magnetic poles around the optic axis is independently magnetized by two kinds of coils for use with the deflector and the stigmator, electrically gauged using an electronic circuit, an anastigmatic deflection system can be created by operating it as a deflector as well as a stigmator.

Consider an application of such an anastigmatic deflection system combined with a reducing lens in the field of electron beam micromachining and X-ray microanalysis. The astigmatic aberration caused by deflection, which distorts the cross section of a beam into a long ellipse and leads to a loss of intensity, can be corrected by applying the feedback potentials ψ_{s1} and ψ_{s2}, generated by the potentials applied to poles for use with deflection ψ_1 and ψ_3, to poles for use with a stigmator:

$$\pm\psi_{s1} = c(2\psi_1\psi_3), \quad \mp\psi_{s2} = c(\psi_1^2 - \psi_3^2) \tag{5.11}$$

with $c = (1.8/\pi)\eta\tan\beta(l/L)$.

Another serious defocusing defect of the beam deviated from the focus point can be corrected by varying the magnetizing current of the reducing lens

$$\frac{\Delta i}{i} = \frac{2.8L\omega_0^3\alpha^2}{f\pi(\omega_0^2 - 1)\cot(\pi - \pi/\omega_0)} \tag{5.12}$$

where ω_0 is the lens strength of a reducing lens with focal length f.

In the practical design of such an anastigmatic deflection system the magnetizing coils for use with the deflector are wound so that poles facing each other have opposite polarity and the following phase relations between the ampere-turns IN are also satisfied:

$$\begin{aligned} N_1I_1 &= NI\cos\lambda, & -N_5I_5 &= NI\cos\lambda \\ N_3I_3 &= NI\sin\lambda, & -N_7I_7 &= NI\sin\lambda \\ N_2I_2 &= NI\cos(\lambda - \tfrac{1}{4}\pi), & -N_6I_6 &= NI\cos(\lambda - \tfrac{1}{4}\pi) \\ N_4I_4 &= NI\sin(\lambda - \tfrac{1}{4}\pi), & -N_8I_8 &= NI\sin(\lambda - \tfrac{1}{4}\pi) \end{aligned} \tag{5.13}$$

Another set of magnetizing coils in the same section of each pole for use with an astigmatic corrector are wound so that each pole is magnetized by two different currents alternating in sign from one pole to the next; the following relationships between the coils must be satisfied to correct the astigmatism:

$$\pm I_{s1} = C_sI^2\sin 2\lambda, \quad \mp I_{s2} = C_sI^2\cos 2\lambda \tag{5.14}$$

where

$$C_s = \frac{0.72\eta}{k}\tan\beta\frac{l}{L}\frac{N}{N_s}$$

and k is the form factor between the scalar potential ψ and the number of ampere-turns IN of the exciting coil given by $\psi = 0.4\pi IN/k$. The feedback current from the deflector to the reducing lens,

$$-C_rI^2 = \frac{2.8}{f}\frac{L\omega_0^3}{\pi(\omega_0^2 - 1)\cot(\pi - \pi/\omega_0)}\left(\frac{3.2\eta Nl\sin\beta}{kr_0}\right)^2 \tag{5.15}$$

is necessary to correct the astigmatic defocusing.

The electronic control circuit of such system can be realized by using a current amplifier, adder, squaring circuit, and resistor, producing a sinsoidal current. The deflection angle under consideration is then given by

$$\alpha = (3.2/k)\eta NI \sin\beta(l/r_0)e^{i(\lambda-\pi/2)} \tag{5.16}$$

A 150-kV electron beam machine was made by JEOL (1964), designed by K. Kanaya at the Electrotechnical Laboratory (Kanaya et al., 1967), in which a telefocus tetrode electron gun (Kawakatsu et al., 1965), electronic alignment system, and octupole deflection system were operated with a deflector and an astigmatic corrector, using a flying spot control system and a tape-controlled programming system. In a machining example, a microboat, a spinning nozzle, diamond dice, and microwelds were produced effectively at a maximum beam current of 2 mA; the power density ranged from 10^6 to 5×10^8 W/cm^2 under continuous operation with pulses up to 10^9 W/cm^2.

6. Electron phase microscope

Since the beginnings of electron microscopy, it has been well known that the phase contrast due to interference of coherent waves plays a very important role in image formation (see Kanaya, 1954, 1956a, 1956b, 1956c; Kanaya et al., 1954; Scherzer, 1949). Noteworthy examples include Fresnel diffraction fringes as demonstrated by H. Boersch (1940), T. Hibi (1955), and K. Kanaya (1957); interference fringes in crystals by R. D. Heidenreich (1949) and J. W. Menter (1955); biprism interference fringes obtained by G. Möllenstedt and H. Düker (1956) and J. Faget and C. Fert (1957); and a possible interferometer introduced by L. Marton (1953). In view of Zernike's theory (1942) of the phase-difference microscope, if an appropriate phase shift is inserted into the back focal plane of the objective and if, in addition, it is possible to cause a phase difference in the transmitted wave or the diffracted one, it is reasonable to expect that phase contrast will be observed.

According to E. G. Ramberg's calculation (1949), a collodion film 100 Å thick should result in a phase delay of the order of $\pi/2$ provided that the accelerating potential is 60 kV. For this reason the phase plate must be very thin. In this respect it is different from interchromatic phase contrast as tested by M. Locquin (1956), in which a metallic cone diagram is used as phase plate. Later, K. Kanaya, H. Kawakatsu, and A. Ishikawa (1957), K. Kanaya, H. Kawakatsu, and H. Yotsumoto (1958), and K. Kanaya, Kawakatsu, Ito, et al. (1958) tried to use a collodion or carbon film instead of a metallic film. For the phase plate, it has been shown that a hole

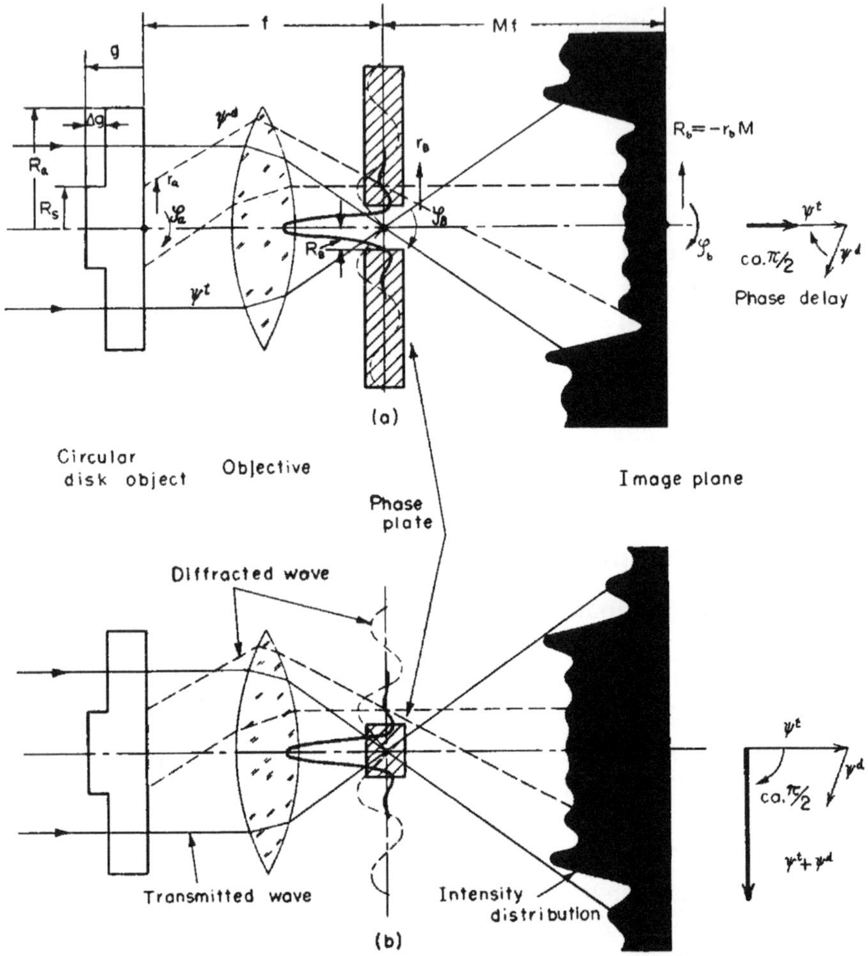

Figure 6.1 Representation of image formation by electron phase microscopy, in the case of dark contrast image using a phase plate with a hole (a) and of bright contrast image using a phase plate consisting of a disk (b).

in a collodion membrane or disk on a thin collodion membrane may be suitable for dark or bright contrast observation, as shown in Fig. 6.1.

However, when the electron beam is exactly focused at the back focal point, its intensity may be 10 times as high as that at the specimen, and the resulting destruction, contamination, and charging up of the phase plate have frequently been observed. Accordingly, it is concluded that a phase plate with holes may be considered to be superior to one without holes. Furthermore, an appropriate potential to create various phase differ-

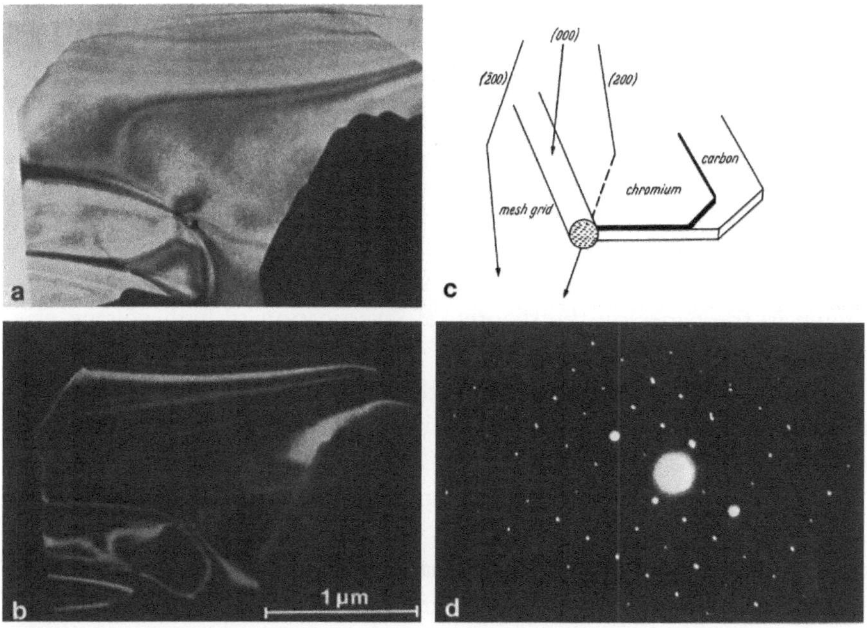

Figure 6.2 Images of the same MoO_3 crystal: (a) ordinary observation with 50-μm aperture; (b) interference bright contrast image with the phase plate; (c) as (b) but with a potential of 10 V applied; (d) selected-area diffraction pattern ($\lambda = 0.053$ Å).

ences is applied to the carbon phase plate, which is shadowed as thinly as possible with chromium. Successful results using a phase plate with many holes ∽20–50 μm in diameter in a carbon membrane, together with the application of a small potential ∽10 V, were reported by K. Kanaya and H. Kawakatsu at the Fourth International Congress for Electron Microscopy at Berlin in 1958.

The images of thin crystals of MoO_3 show the influence of the position of the hole in the phase plate and of the potential applied to it. Fig. 6.2a was taken under ordinary observation conditions without any limiting aperture. Fig. 6.2b shows dark-field images of the same crystal. Such an image can be obtained by arranging that the primary wave be stopped by the mesh grid while the ($\bar{2}$00)-diffracted wave passes directly through the holes and the (200)-diffracted wave passes through the thin phase plate, changing its phase provided that a potential of 11 V is applied, as shown in Fig. 6.2c; Fig. 6.2d is a selected-area diffraction pattern. Phase microscopic observations have been reported subsequently by many authors, one of whom observed a single mercury atom in a triacetoxy-quicksilver-aurin (TAM A) molecule (Willasch, 1973).

7. High-voltage electron microscopes

High-voltage 300-kV electron microscopes constructed by RCA (Zworykin et al., 1941) and Philips (Dorsten et al., 1947) strongly stimulated the Japanese Cooperative Committee on Electron Microscopy, as shown in the paper on the theoretical investigation of high-voltage electron microscopes (Kanaya, 1950).

In order to observe viable biological specimens enclosed in environmental cells, three groups at Hitachi (B. Tadano, Y. Sakaki), Shimadzu (S. Shimadzu, K. Kobayashi, K. Kanaya), and JEOL (K. Ito, S. Seto, H. Yamashita, Y. Tani) began the development of high-voltage electron microscopes in 1950, sponsored by various research funds (Ministry of International Trade and Commerce, and Ashahi-shinbun, a newspaper company).

The 300-kV microscope, with high voltage supplied by a Van de Graaff generator (Tadano et al., 1958), contributed greatly to metallurgical studies. Carbide reactions in steel have been examined with respect to their selected diffraction patterns by using a carbon extraction replica method (Akutagawa et al., 1958).

Fig. 7.1a shows the 200-kV microscope made by Shimadzu Ltd. in 1950, in which the high-voltage supply of the two-stage gun was based on cascade transformers, the design being based on a model studied with an electrolytic plotting tank at the Electrotechnical Laboratory. Later, the high-voltage supply was raised to 350 kV by using a three-stage electron gun. Figs. 7.1b and 7.1c show electron micrographs taken at 300 kV, where the former represents a stacking fault in a 0.8-μm thickness of stainless steel and the latter grain boundaries in a 0.75-μm thickness of the same stainless steel. K. Kobayashi et al. (1958) have reported the irradiation changes induced in organic polymers by its use. K. Kobayashi and K. Sakaoku (1965) later discussed such radiation damage more quantitatively.

These prototype high-voltage electron microscopes contributed considerably to complete high-voltage electron microscopes such as the Toulouse 1500-kV instrument (Dupouy & Perrier, 1964a, 1964b) and the 1000-kV electron microscopes with a Cockcroft-Walton (1932)-type high-frequency voltage supply made by Hitachi and JEOL in 1965.

Fig. 7.2 shows a 500-kV high-voltage electron microscope made by Shimadzu Ltd. in 1963; its 6.9-Å resolution was confirmed by a half-lattice

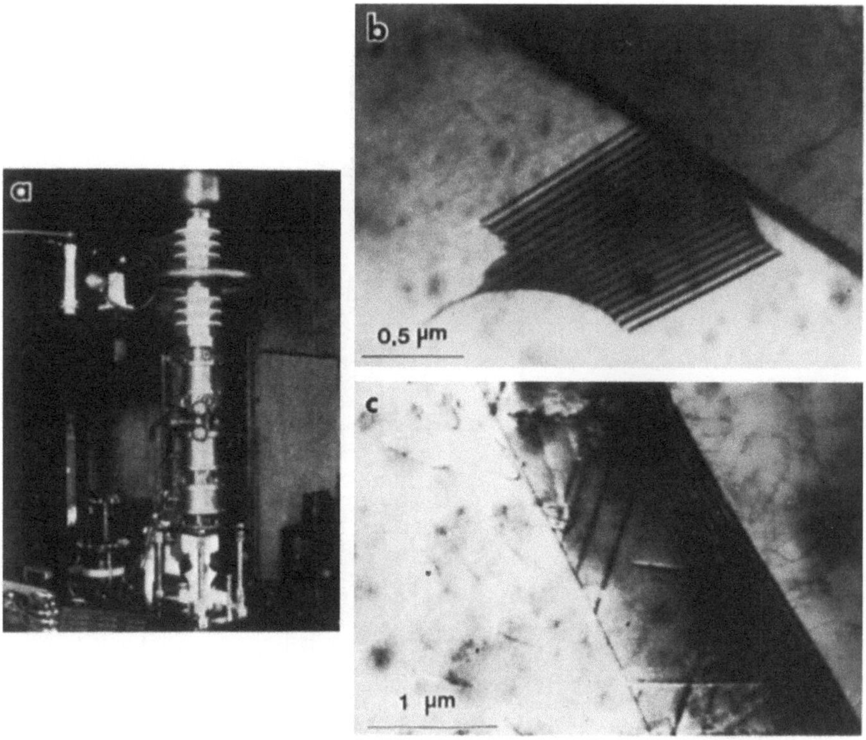

Figure 7.1 Preliminary 200-kV electron microscope: (a) column, (b and c) stacking fault and grain boundaries in stainless steel taken at 300 kV.

image of copper phthalocyanine. The design of the 10-stage and later 20-stage accelerator tube providing the accelerating voltage was based on the focusing properties of a bipotential electrostatic lens (Kanaya et al., 1965; Taoka et al., 1967).

8. Afterword by Peter Hawkes

Koichi Kanaya was born in 1919 and died on 22 May 1998 after a lifetime of research in the fields of electron optics and related teaching. After graduating from the Tokyo Institute of Technology in 1942, he joined the Electrotechnical Laboratory, where he spent nearly 30 years. His work there is recorded in very numerous publications in the *Bulletin of the Electrotechnical Laboratory*. He then moved to the Kogakuin University, where he spent a further 19 years on research, teaching and supervising doctoral theses. He was editor-in-chief of the *Journal of Electron Microscopy* and active

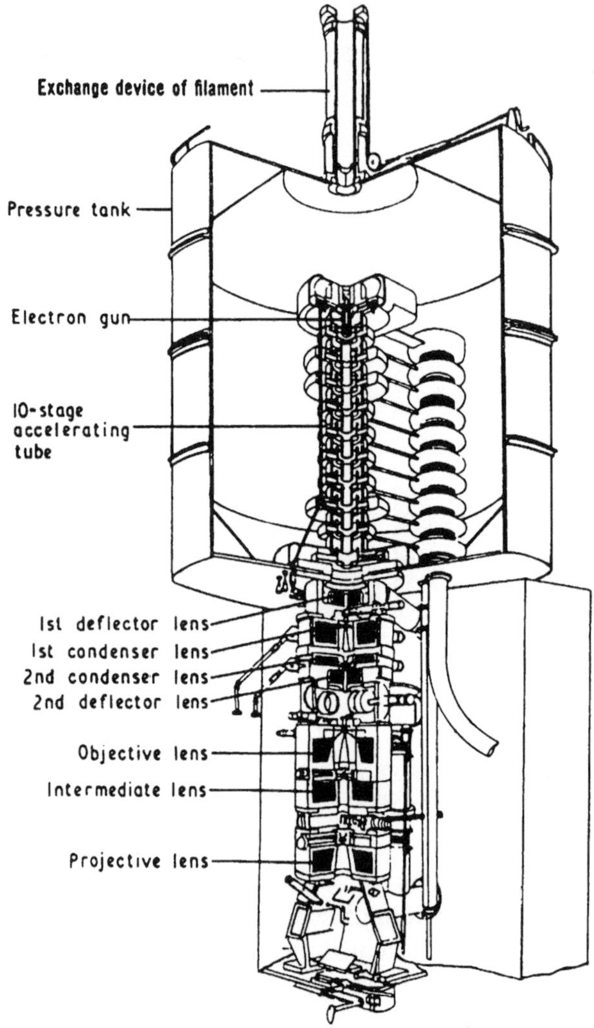

Figure 7.2 High-voltage 500-kV electron microscope (Shimadzu Ltd., 1963).

in the Japanese Society of Electron Microscopy (JSEM), of which he was President in 1970. His work was recognized by the award of several prizes, notably the Asahi Scientific Promotion Prize for 'Experimental research on high-voltage electron microscopes' and the Seto Prize of the JSEM for 'Electron optical theory in the design of magnetic electron lenses for electron microscopes'. An obituary appeared in the *Journal of Electron Microscopy* **48** (1999) 181–182

References

Section 1

Boersch, H. (1936). *Annalen der Physik (Leipzig) [5], 27*, 75.
Boersch, H. (1943). *Physikalische Zeitschrift, 44*, 32.
Boersch, H. (1944). *Kolloid-Zeitschrift, 106*, 169.
Fukami, A., & Adachi, K. (1964). *Journal of Electron Microscopy, 14*, 112.
Fukami, A., & Akiyama, T. (1953). *Journal of Electron Microscopy, 1*, 28.
Glaser, W. (1933). *Annalen der Physik (Leipzig) [5], 18*, 557.
Glaser, W. (1943). *Zeitschrift für Physik, 121*, 647.
Hashimoto, H. (1960). *Journal of Electron Microscopy, 9*, 130.
Hashimoto, H., Tanaka, K., & Yoda, E. (1958). *Journal of Electron Microscopy, 8*, 6.
Hibi, T. (1952). *Journal of Applied Physics, 23*, 957.
Hibi, T. (1955). *Journal of Electron Microscopy, 3*, 15.
Hibi, T. (1956). *Journal of Electron Microscopy, 4*, 24.
Hibi, T., & Takahashi, S. (1958). *Journal of Electron Microscopy, 7*, 15.
Hibi, T., & Yada, K. (1954). *Journal of Electron Microscopy, 2*, 45.
Hillier, J. (1939). *Canadian Journal of Research, 17*, 64.
Honjou, G., Kitamura, N., Shimoda, K., & Mihama, K. (1956). *Journal of the Physical Society of Japan, 11*, 527.
Ito, K., & Ito, T. (1953). *Journal of Electron Microscopy, 1*, 18.
Ito, T., & Hiziya, K. (1958). *Journal of Electron Microscopy, 6*, 4.
Kanaya, K. (1948). *On the temperature-rise of specimens in electron microscopes*. Mem. Res. Rep. Electrotech. Lab. (p. 33) (in Japanese).
Kanaya, K. (1949a). *Investigation on the temperature of specimens and their structure changes due to electron bombardment in electron microscope*. Res. Rep. Electrotech. Lab., No. 509 (in Japanese).
Kanaya, K. (1949b). *Bulletin of the Electrotechnical Laboratory (Tokyo 1937–1970), 13*, 338.
Kanaya, K. (1950). *Journal of Electron Microscopy, 1*, 35 (in Japanese).
Kanaya, K. (1954a). *Electron microscope-theory and practice*. Kyoto: Denkishoin Book Co. (in Japanese).
Kanaya, K. (1954b). *Bulletin of the Electrotechnical Laboratory (Tokyo 1937–1970), 18*, 911.
Kanaya, K. (1955a). *Electron optic theory on the design of magnetic lenses for electron microscopes*. Res. Rep. Electrotech. Lab., No. 548.
Kanaya, K. (1955b). *Journal of Electron Microscopy, 3*, 1.
Kanaya, K. (1955c). *Bulletin of the Electrotechnical Laboratory (Tokyo 1937–1970), 19*, 217.
Kanaya, K. (1955d). *Bulletin of the Electrotechnical Laboratory (Tokyo 1937–1970), 19*, 680.
Kanaya, K. (1956a). *Bulletin of the Electrotechnical Laboratory (Tokyo 1937–1970), 20*, 610.
Kanaya, K. (1956b). *Bulletin of the Electrotechnical Laboratory (Tokyo 1937–1970), 20*, 801.
Kanaya, K. (1956c). *Tokyo, 1956*. (p. 114).
Kanaya, K. (1956d). *Journal of Electron Microscopy, 4*, 1.
Kanaya, K. (1957). *Bulletin of the Electrotechnical Laboratory (Tokyo 1937–1970), 21*, 455.
Kanaya, K., & Ishikawa, A. (1951). *Bulletin of the Electrotechnical Laboratory (Tokyo 1937–1970), 15*, 264 (in Japanese).
Kanaya, K., & Oshina, M. (1949a). *Bulletin of the Electrotechnical Laboratory (Tokyo 1937–1970), 13*, 104 (in Japanese).
Kanaya, K., & Oshina, M. (1949b). *Bulletin of the Electrotechnical Laboratory (Tokyo 1937–1970), 13*, 112 (in Japanese).
Kanaya, K., & Oshina, M. (1949c). *Bulletin of the Electrotechnical Laboratory (Tokyo 1937–1970), 13*, 284 (in Japanese).
Kanaya, K., Oshina, M., & Shibuya, S. (1949). *Bulletin of the Electrotechnical Laboratory (Tokyo 1937–1970), 13*, 152 (in Japanese).

Kanaya, K., Kato, A., & Ishikawa, A. (1951). *Bulletin of the Electrotechnical Laboratory (Tokyo 1937–1970), 15*, 257 (in Japanese).
Kanaya, K., Inoue, Y., & Ishikawa, A. (1956). *London, 1954*. (p. 46).
Marton, L. (1968). *Early history of the electron microscope*. San Francisco, California: San Francisco Press.
Marton, L., & Schilf, L. I. (1941). *Journal of Applied Physics, 12*, 759.
Ruthemann, G. (1941). *Naturwissenschaften, 20*, 648.
Ruthemann, G. (1942). *Naturwissenschaften, 30*, 145.
Scherzer, O. (1949). *Journal of Applied Physics, 20*, 20.
Sugata, E. (1956). *Electron microscopes*. Tokyo: Ohm Co. Ltd. (in Japanese).
Sugata, E. (Ed.). (1968). *A historical survey of electron microscopes of Japan 1936-1965*. Tokyo: Maruzen Co. Ltd.
Sugiura, Y., & Suzuki, S. (1943a). *Proceedings of the Imperial Academy (Tokyo), 19*, 293.
Sugiura, Y., & Suzuki, S. (1943b). *Proceedings of the Imperial Academy (Tokyo), 19*, 546.
Takahashi, N., & Ashinuma, K. (1957). *Journal of Electron Microscopy, 6*, 29.
Takahashi, N., Ashinuma, K., & Watanabe, M. (1956). *Journal of Electron Microscopy, 5*, 22.
Takahashi, N., Ashinuma, K., Watanabe, M., Okazaki, I., & Nagahama, Y. (1960). *Journal of Electron Microscopy, 9*, 104.
Tani, Y. (1956). *Tokyo, 1956*. (p. 50).
Tani, Y. (1959). *Theory and applications in electron microscopes: Vols. I–III*. Tokyo: Committee of Japan Society of Electron-Microscopy, Maruzen Co. Ltd. (in Japanese).
Uyeda, R. (1953). *Journal of the Physical Society of Japan, 12*, 267.
von Ardenne, M. (1940). *Elektronen-Übermikroskopie*. Berlin and New York: Springer-Verlag.
von Borries, B., & Glaser, W. (1944). *Kolloid-Zeitschrift, 106*, 123.
von Borries, B., & Ruska, E. (1938). *Wissenschaftliche Veröffentlichungen aus den Siemens-Werken, 17*, 99.
Watanabe, H. (1956). *Journal of Electron Microscopy, 4*, 24.
Watanabe, H. (1960). *Journal of Electron Microscopy, 9*, 123.
Zworykin, V. K., Morton, G. A., Ramberg, E. G., Hillier, J., & Vance, A. W. (1945). *Electron optics and the electron microscope*. New York: Wiley.

Section 2

Bertein, F. (1941). *Comptes Rendus Hebdomadaires des Séances de L'Académie des Sciences, 224*, 560.
Glaser, W. (1941). *Zeitschrift für Physik, 117*, 285.
Glaser, W. (1952). *Grundlagen der Elektronenoptik*. Berlin and New York: Springer-Verlag.
Glaser, W. (1956). *Handbuch der Physik, 33*, 123.
Grümm, H. (1952). *Annalen der Physik (Leipzig) [6], 6*, 131.
Hawkes, P. W. (1966). *British Journal of Applied Physics, 17*, 143.
Kanaya, K. (1949). *Theory of aberrations for an axisymmetrical magnetic type electron microscope*. Res. Rep. Electrotech. Lab., No. 495 (in Japanese).
Kanaya, K. (1952). *Journal of Electron Microscopy, 1*, 1 (in Japanese).
Kanaya, K. (1955). *Electron optic theory on the design of magnetic lenses for electron microscopes*. Res. Rep. Electrotech. Lab, No. 548.
Kanaya, K., & Kato, A. (1951a). *Bulletin of the Electrotechnical Laboratory (Tokyo 1937–1970), 15*, 203 (in Japanese).
Kanaya, K., & Kato, A. (1951b). *Bulletin of the Electrotechnical Laboratory (Tokyo 1937–1970), 15*, 646 (in Japanese).
Kanaya, K., Kawakatsu, H., & Kato, A. (1959). *Bulletin of the Electrotechnical Laboratory (Tokyo 1937–1970), 23*, 431.
Kanaya, K., Kawakatsu, H., Yamazaki, H., Okazaki, I., & Tanaka, K. (1964). *Optik, 8*, 399.

Kanaya, K., Kawakatsu, H., & Yamazaki, H. (1965). *British Journal of Applied Physics, 16*, 991.

Kato, N., & Inoue, T. (1941). *Journal of the Institute of Electrical Engineers of Japan, 61*, 122 (in Japanese).

Meyer, W. (1958). *Optik, 15*, 398.

Sturrock, P. A. (1955). *Static and dynamic electron optics*. London and New York: Cambridge Univ. Press.

Suzuki, S., & Inoue, Y. (1948). *Rep. Coop. Res. Comm. Electron Microsc.* No. 46-2 (in Japanese).

Taoka, T., Fujita, H., Kanaya, K., Iwanaga, M., & Iwasa, N. (1967). *Scientific Instruments, 44*, 748.

Thomson, B., & Headrick, L. (1940). *Proceedings of the IRE, 28*, 318.

von Borries, B., & Ruska, E. (1932). *Zeitschrift für Physik, 76*, 649.

Watson, E. (1927). *Philosophical Magazine [7], 3*, 849.

Section 3

Dosse, J. (1941a). *Zeitschrift für Physik, 117*, 437.

Dosse, J. (1941b). *Zeitschrift für Physik, 117*, 722.

Glaser, W. (1940). *Zeitschrift für Physik, 116*, 19.

Glaser, W. (1941). *Zeitschrift für Physik, 117*, 285.

Glaser, W. (1952). *Grundlagen der Elektronenoptik*. Berlin and New York: Springer-Verlag.

Ishikawa, A., & Kanaya, K. (1956). *Bulletin of the Electrotechnical Laboratory (Tokyo 1937–1970), 20*, 292 (in Japanese).

Ito, K., & Ito, T. (1950). *Journal of Electron Microscopy, 1*, 47 (in Japanese).

Ito, K., Ito, T., Kanaya, K., & Morito, N. (1953). *Paris, 1950*. (p. 207).

Iwanaga, M., & Kanaya, K. (1953). In *Proc. 9th annu. meet. soc. electron microsc.*

Kanaya, K. (1949a). *Theory of aberrations for an axisymmetrical magnetic type electron microscope*. Res. Rep. Electrotech. Lab., No. 495 (in Japanese).

Kanaya, K. (1949b). *Bulletin of the Electrotechnical Laboratory (Tokyo 1937–1970), 13*, 1 (in Japanese).

Kanaya, K. (1949c). *Bulletin of the Electrotechnical Laboratory (Tokyo 1937–1970), 13*, 338 (in Japanese).

Kanaya, K. (1949d). *Bulletin of the Electrotechnical Laboratory (Tokyo 1937–1970), 13*, 453 (in Japanese).

Kanaya, K. (1950). *Journal of Electron Microscopy, 1*, 53 (in Japanese).

Kanaya, K. (1951a). *Bulletin of the Electrotechnical Laboratory (Tokyo 1937–1970), 15*, 86 (in Japanese).

Kanaya, K. (1951b). *Bulletin of the Electrotechnical Laboratory (Tokyo 1937–1970), 15*, 91 (in Japanese).

Kanaya, K. (1951c). *Bulletin of the Electrotechnical Laboratory (Tokyo 1937–1970), 15*, 193 (in Japanese).

Kanaya, K. (1951d). *Bulletin of the Electrotechnical Laboratory (Tokyo 1937–1970), 15*, 199 (in Japanese).

Kanaya, K. (1951e). *Bulletin of the Electrotechnical Laboratory (Tokyo 1937–1970), 15*, 203 (in Japanese).

Kanaya, K. (1952a). *Bulletin of the Electrotechnical Laboratory (Tokyo 1937–1970), 16*, 24 (in Japanese).

Kanaya, K. (1952b). *Bulletin of the Electrotechnical Laboratory (Tokyo 1937–1970), 16*, 134 (in Japanese).

Kanaya, K. (1952c). *Bulletin of the Electrotechnical Laboratory (Tokyo 1937–1970), 16*, 184 (in Japanese).

Kanaya, K. (1953). *Bulletin of the Electrotechnical Laboratory (Tokyo 1937–1970), 17*, 528 (in Japanese).
Kanaya, K. (1954a). *Journal of Electron Microscopy, 3*, 166 (in Japanese).
Kanaya, K. (1954b). *Bulletin of the Electrotechnical Laboratory (Tokyo 1937–1970), 18*, 13 (in Japanese).
Kanaya, K. (1955). *Electron optic theory on the design of magnetic lenses for electron microscopes.* Res. Rep. Electrotech. Lab., No. 548.
Kanaya, K. (1958). *Bulletin of the Electrotechnical Laboratory (Tokyo 1937–1970), 22*, 615.
Kanaya, K., & Ishikawa, A. (1959). *Journal of Electron Microscopy, 7*, 13.
Katagiri, S. (1953). *Journal of Electron Microscopy, 1*, 13 (in Japanese).
Lenz, F. (1950). *Optik, 7*, 243.
Marton, L. (1946). *Reports on Progress in Physics, 19*, 204.
Marton, L., & Bol, K. (1947). *Journal of Applied Physics, 18*, 522.
Riecke, W. D. (1962a). *Optik, 19*, 81.
Riecke, W. D. (1962b). *Optik, 19*, 169.
Scherzer, O. (1937). In *Beiträge zur Elektronenoptik* (p. 33). Leipzig: Barth.
Suzuki, S., Akashi, K., & Tochigi, H. (1968). *Proceedings — Annual Meeting, Electron Microscopy Society of America, 26*, 1.
von Borries, B. (1949). *Die Übermikroskopie*. Aulendorf/Württ: Cantor-Verlag.

Section 4

Glaser, W. (1941). *Zeitschrift für Physik, 117*, 284.
Glaser, W. (1953). *Paris, 1950*. (p. 171).
Glaser, W. (1952). *Grundlagen der elektronenoptik*. Berlin and New York: Springer-Verlag.
Kanaya, K., & Kawakatsu, E. (1960a). *Journal of Electron Microscopy, 8*, 1.
Kanaya, K., & Kawakatsu, E. (1960b). *Journal of Electron Microscopy, 9*, 71.
Kanaya, K., & Kawakatsu, H. (1960c). *Bulletin of the Electrotechnical Laboratory (Tokyo 1937–1970), 24*, 721.
Kanaya, K., Kawakatsu, E., & Kato, A. (1959a). *Bulletin of the Electrotechnical Laboratory (Tokyo 1937–1970), 23*, 431.
Kanaya, K., Kawakatsu, E., & Kato, A. (1959b). *Bulletin of the Electrotechnical Laboratory (Tokyo 1937–1970), 23*, 801.
Kanaya, K., Kawakatsu, E., & Matsui, S. (1962). *Bulletin of the Electrotechnical Laboratory (Tokyo 1937–1970), 26*, 881.
Kanaya, K., Shimizu, K., Yamazaki, H., Okazaki, I., & Okamoto, T. (1967). *Bulletin of the Electrotechnical Laboratory (Tokyo 1937–1970), 31*(Spec. Issue), 185.
Kanaya, K., Yamazaki, E., Takano, Y., & Kobayashi, M. (1968). *Journal of Physics E: Scientific Instruments, 1*, 289.
Kawakatsu, H., Matsui, S., & Kanaya, K. (1965). *Optik, 22*, 476.
Lenz, F. (1956a). *London, 1954*. (p. 83).
Lenz, F. (1956b). *Tokyo, 1956*. (p. 105).
Sakaki, Y., & Maruse, S. (1954). *Journal of Electron Microscopy, 2*, 8.
Scheffels, W., von Borries, B., & Lenz, F. (1954). *Naturwissenschaften, 4*, 83.

Section 5

Archard, G. D. (1954). *British Journal of Applied Physics, 5*, 294.
Archard, G. D., & Mulvey, T. (1958). *Journal of Scientific Instruments, 35*, 278.
Haine, M. E., Agar, A. W., & Mulvey, T. (1958). *Journal of Scientific Instruments, 35*, 357.
Hillier, J., & Ramberg, E. G. (1947). *Journal of Applied Physics, 18*, 48.
Kanaya, K. (1962). *Bulletin of the Electrotechnical Laboratory (Tokyo 1937–1970), 26*, 161.
Kanaya, K., & Kawakatsu, H. (1961a). *Journal of Electron Microscopy, 10*, 218.

Kanaya, K., & Kawakatsu, H. (1961b). *Bulletin of the Electrotechnical Laboratory (Tokyo 1937–1970), 25*, 241.
Kanaya, K., & Kawakatsu, H. (1961c). *Bulletin of the Electrotechnical Laboratory (Tokyo 1937–1970), 25*, 641.
Kanaya, K., & Kawakatsu, H. (1962). *Bulletin of the Electrotechnical Laboratory (Tokyo 1937–1970), 26*, 241.
Kanaya, K., Kawakatsu, H., & Tanaka, I. (1961). *Bulletin of the Electrotechnical Laboratory (Tokyo 1937–1970), 25*, 482.
Kanaya, K., Kawakatsu, H., Okazaki, I., & Takizawa, T. (1963). *Bulletin of the Electrotechnical Laboratory (Tokyo 1937–1970), 27*, 401.
Kanaya, K., Kawakatsu, H., Okazaki, I., & Takizawa, T. (1964). *Journal of Electron Microscopy, 13*, 80.
Kanaya, K., Matsui, S., Shimizu, K., Okazaki, I., & Tanaka, I. (1967). *Bulletin of the Electrotechnical Laboratory (Tokyo 1937–1970), 31*(Spec. Issue), 329.
Kawakatsu, H., & Kanaya, K. (1961). *Bulletin of the Electrotechnical Laboratory (Tokyo 1937–1970), 25*, 801.
Kawakatsu, H., Matsui, S., & Kanaya, K. (1965). *Optik, 22*, 476.
Leisegang, S. (1953). *Optik, 10*, 5.
Lenz, F., & Hahn, M. (1953). *Optik, 10*, 15.
Morito, N., & Koizumi, K. (1951). *Journal of Electron Microscopy, 2*, 84 (in Japanese).
Rang, O. (1949). *Optik, 5*, 518.
Scherzer, O. (1947). *Optik, 2*, 114.
Scherzer, O. (1949). *Optik, 5*, 497.
Seeliger, R. (1953). *Optik, 10*, 29.
Watanabe, M., & Someya, T. (1963). *Optik, 20*, 99.

Section 6

Boersch, H. (1940). *Naturwissenschaften, 28*, 709.
Faget, J., & Fert, C. (1957). *Cahiers de Physique, 83*, 285.
Heidenreich, R. D. (1949). *Journal of Applied Physics, 20*, 993.
Hibi, T. (1955). *Journal of Electron Microscopy, 3*, 15.
Kanaya, K. (1954). *Journal of Electron Microscopy, 2*, 1.
Kanaya, K. (1956a). *Bulletin of the Electrotechnical Laboratory (Tokyo 1937–1970), 20*, 610.
Kanaya, K. (1956b). *Bulletin of the Electrotechnical Laboratory (Tokyo 1937–1970), 20*, 802.
Kanaya, K. (1956c). *Tokyo, 1956*. (p. 114).
Kanaya, K. (1957). *Bulletin of the Electrotechnical Laboratory (Tokyo 1937–1970), 21*, 455.
Kanaya, K., & Kawakatsu, H. (1958). *Berlin, 1958*. (p. 308).
Kanaya, K., Inoue, Y., & Ishikawa, A. (1954). *Bulletin of the Electrotechnical Laboratory (Tokyo 1937–1970), 18*, 517.
Kanaya, K., Kawakatsu, H., & Ishikawa, A. (1957). *Bulletin of the Electrotechnical Laboratory (Tokyo 1937–1970), 21*, 825.
Kanaya, K., Kawakatsu, H., Ito, K., & Yotsumoto, H. (1958). *Journal of Applied Physics, 29*, 1045.
Kanaya, K., Kawakatsu, H., & Yotsumoto, H. (1958). *Journal of Electron Microscopy, 6*, 1.
Locquin, M. (1956). *London, 1954*. (p. 285).
Marton, L. (1953). *Science, 118*, 470.
Menter, J. W. (1955). *Proceedings of the Royal Society of London, 236*, 119.
Möllenstedt, G., & Düker, H. (1956). *Zeitschrift für Physik, 145*, 377.
Ramberg, E. G. (1949). *Journal of Applied Physics, 20*, 441.
Scherzer, O. (1949). *Journal of Applied Physics, 20*, 20.
Willasch, D. (1973). *Versuche zur Kontrastverbesserung in der Elektronenmikroskopie durch Hellfeldabbildung mittels Phasenplatten und Dunkelfeldabbildung bei hohlkegelförmiger Beleuchtung* (Ph.D. Dissertation), Universität zu Tübingen.
Zernike, F. (1942). *Physica (Amsterdam), 1*, 686.

Section 7

Akutagawa, T., Tanino, M., Uchiyama, I., & Katagiri, S. (1958). *Berlin, 1958*. (p. 673).
Cockcroft, J. D., & Walton, E. T. S. (1932). *Proceedings of the Royal Society of London*, *136*, 619.
Dorsten, A. C., Oosterkamp, W. J., & LePoole, J. B. (1947). *Philips Technical Review*, *9*, 193.
Dupouy, G., & Perrier, F. (1964a). *Comptes Rendus Hebdomadaires des Séances de L'Académie des Sciences*, *258*, 4213.
Dupouy, G., & Perrier, F. (1964b). *Journal of Microscopy (Paris)*, *3*, 233.
Kanaya, K. (1950). *Bulletin of the Electrotechnical Laboratory (Tokyo 1937–1970)*, *14*, 599.
Kanaya, K., Yamazaki, H., Taoka, T., Fujita, H., Iwanaga, H., & Shibata, S. (1965). *Journal of Electron Microscopy*, *14*, 196.
Kobayashi, K., & Sakaoku, K. (1965). In G. F. Bahr, & E. H. Zeitler (Eds.), *Quantitive electron microscopy: Vol. 14* (p. 359).
Kobayashi, K., Suito, E., & Shimadzu, S. (1958). *Berlin, 1958*. (p. 165).
Tadano, B., Katagiri, S., Ichige, K., Sakaki, Y., & Maruse, S. (1958). *Berlin, 1958*. (p. 166).
Taoka, T., Fujita, H., Kanaya, K., Iwanaga, M., & Iwasa, N. (1967). *Scientific Instruments*, *44*, 748.
Zworykin, V. K., Hillier, J., & Vance, A. W. (1941). *Journal of Applied Physics*, *12*, 738.

CHAPTER FIVE

Early electron microscopy in the Netherlands

J.B. Le Poole[a,c] **and Pieter Kruit (Afterword)**[b,*]
[a]Formerly Technische Natuurkunde, Technische Hogeschool, Delft, The Netherlands
[b]Department of Imaging Physics, Faculty of Applied Sciences, Delft University of Technology, Delft, The Netherlands
*Corresponding author. e-mail address: p.kruit@tudelft.nl

Contents

1. Afterword by Pieter Kruit: about Jan Bart Le Poole (1917–1993) 168
References 171

My account of the early development of electron microscopy in Holland is a very personal one. I have never been great at keeping notes, therefore everything is written from memory only. Yet, by relating the succession of events to experiences in my private life I was able to do reasonably accurate dating. Some experiments and theoretical work described in this contribution have been published before. I apologize for seeming to claim any priority on ideas. Any scientist not publishing his views in time loses these rights automatically. It was impossible, however, to give a coherent history without mentioning gradual gains in insight. The rather ample description of the living conditions has been included to give the reader some feeling for the "romantic" atmosphere in which the work of a young scientist was performed in those days.

I started my studies at Delft University in 1937. Before that I was at the MTS (Polytechnic), now HTS—higher technical school, where I learned mechanical engineering. It was not a very highbrow education, but since out of sheer boredom I had completely ruined my reputation in the HBS (high school, Realschule) and since, halfway through the school term, it was easy to predict that I was heading for the disaster of failing another year, I had asked my father to put an end to my misery and allow me to take this technical education. I was fascinated by the system. It was a purely concentric education. Anything was taught exactly at the time it

[c] Present address: 1822 South Camino Seco, Rolling Hills, Tucson, Arizona 85710.

Advances in Imaging and Electron Physics, Volume 221
ISSN 1076-5670
https://doi.org/10.1016/bs.aiep.2022.03.005

was needed. For mathematics we had excellent teachers and for the first time I got into trouble with a teacher in a field where I had very little to fear: physics. This contributed very much to my joy as well as to the misery of my teacher. To do him justice, though, it must be added that not only did he treat me extremely fairly but he also taught me many things I did not know the first thing about, in a most penetrating way. I am very glad I had the opportunity some 30 years later to tell him how much I learned from him, at which time tears came to his eyes.

In this paradise, which the school was to me, I was even prepared to learn German, be it only technical German, given by a very inspiring teacher. Before this, I had lavishly abused the fact that my father would never frown upon anybody who lacked interest in that language. The "Totale Krieg" (Total War) started in 1917, my year of birth, and had caused him to chose sides very decidedly. Still, in my memory, I can hear him imitate the U-boat captains' cry, "Fünf Minuten," referring to the "five minutes" before a defenseless ocean liner was torpedoed.

The MTS gave me some self-confidence because for the first time in my life I enjoyed going to school, since it was not keeping me from more interesting things. As a result I was an interested student and found out that I was not all that stupid.

In December, 1935, my father died. This meant that the family income was reduced to one-third, and with eight children the finances were never abundant. However, seeing my results at school, friends and relations got together and put up money for a "tight" start at the university. But before that, I had either to finish high school or to pass the state exams "extraneous" to qualify. So after graduation from MTS with a certificate, not a diploma, because first I had to finish a year of "practical work," I had to demonstrate competence equivalent to an HBS diploma. For this I badly needed some extra tutoring in history, French, German, economics, bookkeeping, geography, and biology. Never shall I forget my history lessons: "Now here is where some Dutch history is described. You have to do that yourself. I flatly refuse to tell you in detail how these grocers were behaving. After that, there's a chapter on civilized nations, led by people like Bismarck, Gladstone, Garibaldi, etc. So come back in a week." Did my tutor turn me into an arch-conservative with strong social feelings at that time?

To my great surprise, I made the state exam and the road to a university education was open! Now I had to make my choice: was it to be mathematics and physics with chemistry or astronomy as a minor, or chemistry and physics with mathematics as a minor? The answer came from a little note in our local newspaper, stating that Delft, some years ago, had started an education in Technical Physics (Engineering Physics) "to investigate technological processes and to find means to improve these." That was exactly what I was after. No mathematics, no theoretical physics could ever beat that! Moreover, 15–20 people a year enlisted for that course. It proved to be, in practice, nearly individual courses, and when I say nearly individual I mean I made sure to get the absolute maximum out of it.

My previous technical education at last allowed me to make up for my "year lost to boredom" at high school. My "paradise" at the MTS was only the beginning of the real thing. I heard top people such as C. Zwikker explaining how delicate a Weston cell was, then banging it on the counter and showing it was just a wooden box. Or "I take this one-inch iron bar and bend it" (at which he just bent it to a 30-cm radius). It was a rubber rod carefully painted to resemble steel! Or Dorgelo, connecting two soap bubbles, the greater one growing at the expense of the smaller one: "This is a demonstration of a recent political happening" (annexation of Austria in 1937). Or else "an infinite time lapse, say one microsecond." Or else E. C. Wiersma with "the thermodynamic temperature scale." Later P. von de Leeden: "There exist equipotential planes, intersecting, perpendicularly" (the quadrupole), or A. C. S. van Heel: "Parson's father wanted his son to be a blacksmith. The son however wanted to do optics. So he invented the Parson's turbine and asked permission to go into optics now." Or the physics laboratory teacher telling his students: "Apparently you think glass is as strong as steel" after which, to my glee, he forgot to apply the very rule for which his statement was designed and pushed the objective lens of a microscope through the object slide. Or H. A. Kramers: "Electrons to me are cold, hard, blue spheres," which later on left us with the problem of explaining why the images in an electron microscope were green. Of course H. J. de Heer had the answer: "Yes madam, you could have a blue image, but blue electrons are much more expensive, and we are only poor students."

Somehow, in my first year, van Heel, with all his anecdotes, had not instilled one iota of understanding in me. So I was absolutely scared stiff by the exam I had to pass in optics. I really did my best, but as I went up I prayed to heaven that the examiners would not ask me to explain the Gaussian theory of imaging. Heaven, that day, appeared to have time off. The examiners did ask that one question before I ever sat down! But then, the same good heaven obviously realized the awful mistake and in an absolute flash I realized that the whole thing was just a matter of linear relations. Armed with this sudden revelation I answered with more enthusiasm than ever before, and from then on, no question could scare me. Afterward van Heel asked me "Did you ever specialize in optics?" I told him it scared me more than anything in the world, because notwithstanding his exciting lectures I never made heads or tails of it, so I just worked a little (no: *much*) harder at it. By which I gained my first stage in real optics with him. It turned into a disaster. But van Heel was a humane observer. One happy day after 3 months' precision toiling he said "I seem to notice the subject is not quite in your line. Why don't I introduce you to Dorgelo's group?"

My first very uncertain steps on the road to electron optics were to check Busch's equation:

$$\frac{1}{f} = \frac{0.022}{V} \int B^2\, dx$$

A nickel disk with a T-shaped hole served as the object. Electrons were provided by a BaO–SrO cathode. An external magnetic lens, copied from W. G. Burgers (Burgers & Ploos van Amstel, 1937), was to form an image of the T on a fluorescent screen. The image was very weak and could only be seen in a darkened room, although the magnification was only 3–5×. This didn't seem too odd, though; it was the first electron image I had ever seen. The glass tube was pumped via a rubber vacuum hose, with an inner diameter of 8 mm and a length of some 70 cm. We now know that this reduced the pumping speed to barely 70 cm^3/sec and the glass pumping rack was hardly any better. The pressure was measured by a McLeod manometer, connected far away from the experiment, so it can be safely assumed that no relation existed between the pressure in the electron tube and the McLeod reading. No wonder I had religiously to carry out "formation" of the cathode every day.

More than a year passed before we discovered that the calcium tungstate used as fluorescent material had an efficiency equal to that of the glass in the poor vacuum. Yet there was just enough light to measure f as a function of the lens current. The next step was to measure the magnetic field and its distribution along the axis of the lens. In those days this was done with a ballistic galvanometer. Knowing next to nothing about magnetic fields this gave me the unattractive prospect of measuring some 30–100 points at some 10 different currents. Being very lazy I looked for ways to determine $\int B^2\, dx$ in one measurement, but alas, there was no usable effect proportional to B^2. The measurements were finally carried out by what could best be described as an inverse moving-coil ammeter. A small coil (diameter, 3 mm) was suspended between torsion ribbons. For damping the coil was immersed in oil. This allowed very fast measuring, i.e., 60 points in 20 min.

When I try to reconstruct my way of thinking in those days it becomes increasingly clear that I was very ignorant. But what I fail to understand is that there is no indication that older and more experienced people were less ignorant as far as the magnetic field distribution in a lens configuration was concerned. By "no indication" I mean that nobody was helping me by pointing out a more direct approach, such as "Look here, you know that the field must be bell shaped and clearly the line integral of the field along the axis has to equal *NI*. So you do know quite a bit about the field, enough to make an educated estimate."

Also I do remember quite distinctly that I was puzzled by the hole in the polepieces, and had no inkling how to deal with it. In my desperation I finally tried the following approach: I could easily calculate the field between polepieces without a hole, assuming the gap to be small compared to the radius of the polefaces. In the assumption that the field was caused by a magnetic charge on the faces, I could then calculate the field on the axis simply assuming no charge on the face where the hole was. Clearly there were also field lines ending on the inside of the bore. These I took into account by assuming a charge there, which was inversely proportional to the distance from the center, knowing that this contribution would be small anyway. In so doing I found the same distribution that, to my amazement, Glaser later used in his famous work, and which we know now to be rather different from actual lens fields.

At the time I was doing the magnetic measurements Dorgelo was also interested to know whether the new permanent magnetic material Ticonal could be used for electron lenses. Knowing so little we believed that a simple disk of Ticonal with a hole could be a useful electron lens. So I measured the field of six or eight pieces of Ticonal arranged around the axis, and was very surprised that the field changed sign at some distance from the magnets. But everybody else seemed surprised too! I did find afterward that this was what could be expected from dipoles, but it took me nearly 3 years to discover that, without this change of sign, the line integral on the axis could never be zero. It still puzzles me how I could ever have graduated without anybody noticing how little I knew. Because it seems so devastating, I hesitate to draw the apparently inevitable conclusion that my surroundings were equally ignorant(?).

Meanwhile I had become very fascinated by electron optics. Therefore, in the summer of 1939 I asked my professor, H. B. Dorgelo, if I could build an electron microscope as a project for my Ingenieur degree. To my amazement he immediately agreed. This was rather unlike him and it was not until 12 years later that the explanation came up. Professor Dorgelo, on this fateful day, had just returned from Berlin, where he had visited Siemens with Professor A. J. Kluyver, the famous microbiologist, and Ir. F. G. Waller, director of the yeast factory in Delft. They were very impressed by the work of Ernst and Helmut Ruska and Bodo von Borries. Waller thought that the electron microscope offered great possibilities in biology, but that the price was prohibitive. Dorgelo then allegedly answered that it should be possible to build one in his laboratory, if an interested and sufficiently qualified student came along. I am sure that at that time it was my interest more than my qualifications that got me the job, although my work for Professor Dorgelo in preparing experiments for his lectures may have pleased him.

As I started the design of the microscope I had some very somber premonitions about the intensity of two-stage images. As I mentioned before the images obtained so far with a 3–5× magnification were of very low intensity and everything I tried to obtain better illumination had failed. But since Ruska did not seem to have encountered such problems I hoped and prayed that some answer would come in due time. As it happened, Professor Dorgelo was asked to give a lecture about luminescence to an interested laymen's society, "Diligentia," in The Hague in November, 1939. For the demonstrations he ordered a variety of phosphors from Levy West in London. As I witnessed the experiments it occurred to me that I might try some in the electron beam. Lo and behold—there was the answer to my prayers! We never found out why what was called fluorescent powder in our lab was so absolutely dead. Nor did I ever discover the creator of this "tradition." It is amusing to note that it took me quite some convincing to have the useless calcium tungstate dumped where it belonged. Since then I have learned that a very reliable method for selecting phosphors is to expose them to X rays.

A few weeks later the first stage of the electron microscope was ready and produced bright images of a broken collodion foil on a small (diameter, 3 cm) fluorescent screen and I knew that my worst worries were over!

The high-voltage supply was rebuilt from the parts of an earlier 40-kV electron diffraction experiment of A. Nawijn (Internal report). It consisted of a 500-Hz transformer fed by a frequency converter of rather fascinating design in that the armature field was excited by three-phase ac so that the resulting magnetic field rotated at double the rpm of the armature itself. The ac high voltage was rectified by two valves, thus doubling the high voltage and the ripple frequency. At that point I started to think about high-voltage stabilization. In my opinion this could be done electronically, but when I consulted electrical engineering experts they thought I had gone out of my mind! "Everybody knows that electronics are only good for 1% accuracy." So we made the standard bet for one cigar. These 8-cent cigars were the leftovers from birthday parties, and according to tradition became the property of the concierge of our laboratory, who sold them, mostly for betting purposes, for 10 cents. It was agreed that I should show the principle only, not the HT application. Amazing how good such a cigar tastes and how well one can show the pride in one's victory with the proper gestures!

Soon after that, the curtain came down, i.e., the Nazis invaded Holland. Clearly it took a while after our capitulation to get things going again. After the first shock was over I was surprised by the politeness of the German army, apart from announcing the death penalty for so many "crimes" that one soon became immune to the threatening effect it was supposed to have, only to be shocked now and then by finding out that the Germans meant what they said.

Much more serious was the fact that Holland was to have a "civilian" administration, at the top of which Hitler appointed a man who had served him so well by his treachery in Austria: Seyss Inquart. He proved to be in a hurry to make his presence felt, and in late November, 1940, the first measures against our Jewish compatriots were published. This caused a chain of demonstrations at our universities. Leiden and Delft went on strike and were closed. The other universities were very severely reprimanded, but could resume soon afterward, having been less pronounced in their outcries. A curious rule was that everybody on the payroll of the universities was to go to work normally. Being an assistant to Professor Dorgelo, I was on the payroll too. To this day I have been in doubt whether it was fair for me to accept the condition, but many of us felt at the time that a refusal would have endangered the university even more. Even so, I myself had been quite active in advocating the strike.

Fair or not, I now had less work grading student reports and consequently more time to finish the microscope, and in view of the very unknown but certainly threatening future I decided to take advantage of the situation. Another stroke of good luck of a totally different nature was

Figure 1 The Delft electron microscope, April, 1941.

that the crystallography group in our department had newly acquired electron diffraction equipment from Seemann which was provided with a very simple camera. The second stage of the microscope was nearing completion. Meanwhile there were signs that our university might be allowed to function again, so I felt I should try to complete my degree (Ir.) as soon as possible, before further mishaps could prevent me from doing so altogether.

On April 8, 1941, the first pictures were taken with the two-stage microscope (see Fig. 1). The camera was borrowed from the Seemann diffraction equipment. I soon decided to copy it and complete the microscope. Early in May the university was opened again and as a result I got my degree on July 26.

My job then became to upgrade the microscope and promote research with it, being paid by the yeast factory. By the end of the year 1941 a bright physics student, H. J. de Heer, joined me. He was an excellent amateur photographer, which helped us very much in improving the quality of the pictures. I had been intrigued by the influence of diffraction on the spherical aberration, and tried to work out the proper theory. This resulted in the now well-known integral, which I had no way of solving. So model experiments were made with an optical lens having a known spherical aberration. The scaling rules of course followed directly from the theory. For the rest the experiment was much too simple. The sample had 100% contrast, and not knowing anything yet about the importance of coherent illumination and phase contrast, I used a wide condenser aperture. In short, it was all wrong—but it did give me more insight. In particular I found that the size of the objective aperture was hardly relevant.

When I reported my findings to my optics teacher, A. C. S. van Heel, he asked me his standard question: "What does Lord Rayleigh say on this?" So I went to the library and found the complete theory, applied to star observation with spherical mirrors. This obviously stopped me from publishing. Later W. Glaser published the same story in a much more complete form (Glaser, 1943).

Working with the electron microscope we soon found that 40-kV accelerating voltage made life extra difficult for us. So it was decided that we were to buy a 150-kV supply with electronic stabilization to 1:10,000. Philips had recently come out with a 300-kV supply with much greater stability. Lindenhovius (Lindenhovius & Rinia, 1941) and A. C. van Dorsten (1948), who was an expert in high voltage, believed the same system could be used for 150 kV. The pricetag was 5000 Dutch guilders; needless to say that the Philips company was making a very friendly gesture. The system, I am proud to say, was the same as the one I had earlier earned my cigar with! To help us out until this supply was finished we got an HT rectifier tube on loan, plus an 80-kV condenser. So we could work at higher voltage, albeit without stabilization. In order to see what better stability would do for us, H. J. de Heer and I decided to take pictures at night. This involved working a full 7 hours because Holland had a curfew from 9 PM to 4 AM, which made it a rather memorable occasion.

Here I should like to include an amusing little incident: one day, as I was enlarging some photographs in the common dark room, my then and now colleague P. M. de Wolff came in with some X-ray diffraction patterns he had just taken and wanted to develop. While he was waiting for me to finish what I was doing and looking over my shoulder he was intrigued by the very interesting structure slowly appearing in the developer. I was too! I had borrowed the Leica camera from the lab to take pictures of a new biological object: my few-weeks-old son lying in his cradle. The curtains showed a regular design, which I assume drew the crystallographer's attention. While visiting Philips Natuurkundig Laboratorium ("Nat Lab") a little later to see how the new HT supply was progressing I got an opportunity to meet Dr. F. A. Heijn, who was operating a 1-MV accelerator for nuclear physics

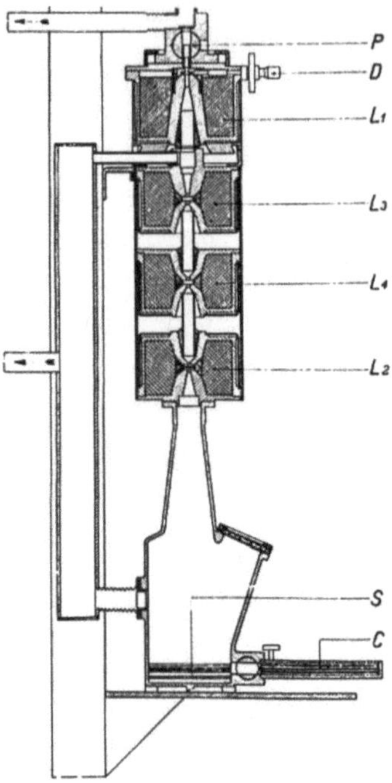

Figure 2 Cross section of the 450-kV electron microscope column. Design started early 1943. (Courtesy *Philips Techn. Rev.*)

research. After von Ardenne's work at 220 kV I had become interested in HVEM. Furthermore, von Borries published his beautiful work on photographic plates. It looked to me that by optimizing everything one should be able to get good quality images of thick specimens at very high voltages. Therefore I suggested to Professor G. Holst that we build a 1-MeV microscope, and to my pleasant surprise he was willing to consider such an undertaking. It was decided that 1 MV would be too large a step to take for a start and that 450 kV would be a better choice.

The high-voltage generator would be van Dorsten's responsibility. W. J. Oosterkamp, who had played a major role in the development of a 1-MeV X-ray source for therapeutic purposes, would take care of the accelerator tube and I would design the column (van Dorsten et al., 1947) (see Fig. 2). In scaling up the lenses I soon found that a more intensive study of the magnetic circuit would be needed. As it happened, quite a bit of research was carried out in the Philips labs on the design of magnets, and the description of the big Bellevue magnet in Paris (Dreyfus, 1931) also contained a wealth of information, as was pointed out to me by Professor E. C. Wiersma. I think this was the first time I started to know what I was doing. Also, I soon found that the conventional two-stage column would be too long for comfort, so I decided on four stages. As a by-product, the widely variable magnification was born! The two-stage microscope allows for only minor variations in magnification, because of the distortion of the projector. For this reason Ruska and von Borries had introduced a multiple-polepiece system, which, by rotation, could be inserted into the projector circuit. An intermediate lens, however, would have to produce a distortion-free image only the size of the projector aperture. I had not realized before that things could be so simple. So the connecting tube between objective and projector was sawed in two, a magnetic spacer inserted, and inclusion of a coil and magnetic shield completed the three-stage microscope in less than a week!

At the end of 1942 von Borries visited our laboratory. I had already installed the third lens on my Mark I microscope and found it to be extremely useful. For this reason I was reluctant to give this information to a representative of our invaders, even less so when he was wearing a swastika. So I regretted that the microscope was "dismantled for necessary upgrading" and not available for demonstration. I did show some pictures we had taken and von Borries took a much friendlier attitude to me than I to him in freely discussing what he was interested in. In particular, he told me that my pictures showed very little astigmatism. So now I knew: the rotational symmetry of the lenses was really important! I had sometimes wondered about that. (See Fig. 3.)

In the afternoon von Borries gave a public lecture on the Siemens microscope. During lunch Dorgelo must have talked the swastika off his lapel. Wisely, von Borries accepted; if not his audience would have been very small. I vividly remember that, of course, he showed a slide with the ray paths, in which the rays leaving the object ran through the center of the objective. Since I knew that it was only with some care that one could align the shadow image of the objective aperture around the projector bore, I was very well aware that the rays in his slide did not actually exist. Or could they perhaps sometimes? It then struck me that diffracted rays would show this sudden change of direction, and, with that, selected-area diffraction was reinvented. It was a much more elegant way to obtain diffraction patterns on the final screen than the removal of the projector polepieces. Armed with this information I went to see Professor Holst,[1] and suggested that the

[1] The reader should keep in mind that Holst carried the tremendous responsibility of keeping things going in the laboratories and trying to protect some of his co-workers against

Figure 3 Group photo at the symposium in Toulouse. Front row: Cosslett, Marton, Ruska, Dupouy, Le Poole, Fert. Second row (above Marton): von Borries. Third row (second from right): Bernhard. Top row (second from right): Castaing.

electron microscope could be an interesting commercial item for Philips, now that we could include some features that other microscopes did not have. To show me big-industry philosophy he asked me how many I could hope to sell in Holland. Maybe three? I suggested that Philips could sell outside Holland. "OK, take Europe: shall we say ten? That is one million guilders. Philips is only interested in many-million-guilder projects."

Somewhat earlier I had made an experiment with a projector lens corrected for distortion. Calculation had shown me that if the crossover of the projector was in the center of a second lens of equal focal length a corrected system would result. Of course with this I reinvented the Ramsden ocular of the optical microscope. The experiment went very well. By reversing the second lens current the rotational distortion was reduced to invisible proportions. Also, the lens proved to be achromatic. Now this may seem very nice. In reality the complication of an extra lens is not worth the trouble. This was my conclusion after an emission microscope was built

deportation. The allied bombs dropped on the Philips factories also, causing damage to his famous creation, the Nat Lab, and may well have impaired his visionary and enterprising leadership and made him hesitant to take on new and rather ambitious projects.

for Rathenau at the Nat Lab, into which it was incorporated. It did make me easily understand, however, how W. Glaser's calculations (Glaser, 1941) could show a change of sign of the distortion at a certain value of *k*. So a single lens could do the same trick, except for the rotational distortion!

I had now some elements that would enable me to design a Mark II microscope with a number of attractive features. The projector lens would be at constant power, thus giving low distortion, even at a short projection distance, say 250 mm, giving some 100× magnification. The intermediate lens would enable me to have a very short column. Together with the objective lens an overall magnification of 100,000×, much too great for the image quality of those days, would be attainable in a column of barely 50 cm between specimen and final screen. The intermediate lens would not be very suitable for low magnification and diffraction, however, since its small bore would cause far too much distortion. For this reason a separate lens was inserted immediately under the objective. This lens would only be used for reducing the magnification of the two-stage microscope and to provide a sharp image of the back focal plane of the objective on the object plane of the projector. (In the newest microscopes its range has been extended for the very high-magnification mode.)

The objective lens was to be provided with an aperture, alignable during operation. In order to allow diffracted beams to pass, the aperture could be pushed up close to the specimen. The design was completed by the end of the summer of 1943, with the exception of the specimen air lock. Not finding any original solution we finally decided to copy the Siemens air lock. The design was carried out by H. J. de Heer during my absence. Since I am ambidextrous I had envisaged the rotary knob to be operated by my left hand. de Heer, being right-handed, designed the knob on the other side. I never got used to this, and always started to put my left hand up before I noticed, very much to everybody's amusement.

During the manufacture of the microscope a number of experiments were carried out to produce ultra-fine-grain fluorescent screens. von Ardenne's suggestion to use single-crystal ZnS screens was not very suited for our laboratory since I had no access to such crystals. I did explore other fluorescent materials. The most promising one seemed barium platinocyanide. However, I found out that the crystal water was an essential ingredient; therefore it could not be used *in vacuo*.

The gain to be obtained by partial optical magnification had impressed me enough to push on. The fine-grain screens were obtained by grinding willemite and centrifuging it onto a substrate. This way a resolution of

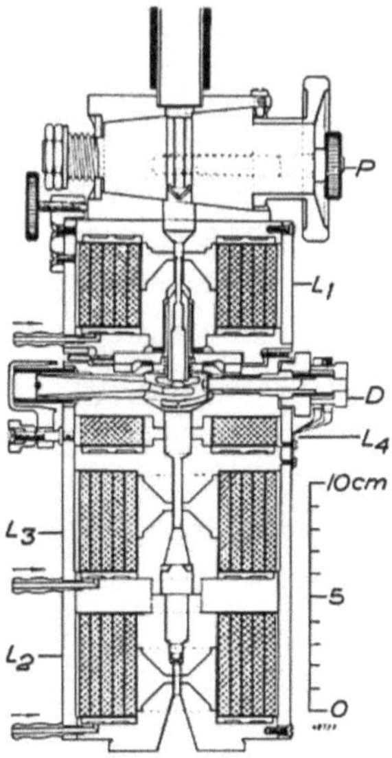

Figure 4 Cross section of Delft microscope, Mark II, August, 1944.

less than 3 μm was obtained. We also tried to thin-section a suspension of willemite in polymerized styrene of nearly the same refractive index as willemite, but we never succeeded in obtaining good sections.

When the Mark II microscope was completed and installed, it worked when we turned the switch! And with our fine-grain screen we could see immediately that the image quality was much better than that of the first microscope. (See Fig. 4.)

And here I made the mistake of my lifetime! With the fine-grain screen I did see Fresnel fringes, which I had never seen before. Then when I did some calculations I invariably got the result that the width of the fringes was proportional to the square root of the defocusing. This, I believed, could not be true! So I decided I did not understand Fresnel fringes and stopped further research. Two years later, as I was contemplating the apparent barrier preventing microscopes all over the world from obtaining resolution substantially under 3 nm, the square-root relation came back to my mind. It would mean that for better resolution the required accuracy of the HT stabilization would increase with the inverse square of the resolution. A chromatic aberration constant of 4 mm would require an accuracy $\Delta V/V$ close to 10^{-4}, which was the best most microscopes, except RCA's, did. But since astigmatism, being also a first-order error, behaves in the same way as chromatic aberration, it also followed that the distance between focal lines should be less than 1 μm! This, I believed, was probably the bottleneck. And my talk with von Borries in late 1942 came back to me.

So I prepared a paper for the London conference in the spring of 1947, organized by the British EM group, and earlier gave a colloquium in Philips Nat Lab. A week or so before the conference, Hillier's paper came out, showing that he had beaten me by many, many lengths. Although I should have withdrawn my paper, I did give it with apologies.

As I said, the Mark II microscope worked successfully but only for a few weeks. I had gone vacationing and soon afterward the Germans discovered a secret transmitter sending classified information to England in the physics lab. H. J. de Heer made the wise decision to dismantle the microscope and hide it in the attic of a very old small house in the center of the old part of Delft. My vacation lasted quite a bit longer than originally intended. With my wife and my son of 20 months I was to stay at a farm just north of Eindhoven. It was in a beautiful neighborhood and there was plenty of food, an important item at that stage of the war. Moreover, there were plenty of plums, which we were free to pick.

Being so close to Eindhoven I went there one day to check the progress on the high-voltage microscope. It was at that time that the Allies had broken through the German lines in France; Paris had been liberated and tanks were making rapid progress in our direction. Allied planes were busy all day paralyzing all traffic. This included shooting up civilian trains. Though the Dutch were always very good at accepting the necessary actions and offering friendly explanations, some actions could not be explained so easily. Shooting up trains with fairly heavy artillery was definitely considered doubtful. Well, on my half-hour trip back our train suddenly stopped and everybody was ordered out to seek protection in ditches and behind trees. Fortunately no planes showed up, but for me it marked the beginning of a many-weeks-long period of constantly having to watch the skies. As the Allies were making very fast progress and we were to leave the farm within a few days to spend an additional week with relatives near Zutphen (north of Arnhem), we discussed whether to wait for the liberation where we were. In the end we agreed that 1 or 2 days would not make that much difference after 4 years under the German heel, so we decided to go to Zutphen. The 1 or 2 days turned out to last until May 4, 1945!

After two unscheduled stops for nearby planes we finally arrived. The first thing I was told by my brother-in-law was the escape route to take if unwanted black- or green-uniformed visitors came to the house.[2] Mind you, my rather long vacation was meant for recovering from a high degree of nervousness resulting from the fact that up to then I had been unreasonably lucky in never having any direct confrontation with Holland's uninvited visitors. Next morning at breakfast we made our first acquaintance with large-scale actions of the R.A.F. Planes raced at very low level over the rooftop, firing what seemed to be all their guns. From then on this was to be the regular (or rather *irregular*) program for the day.

The Allies had warned the Dutch to keep away from railway lines, waterways, main roads, and barracks. Where we stayed there were within less than half a mile two connecting main roads, a canal with a much-used lock, three railway lines, and barracks used by our oppressors. Soon the electricity supply stopped, so we had to go to bed early. My brother owned a stationery factory and happened to have large quantities of methylated spirit. After an eerie evening with some light obtained by adding salt to the spirit I constructed a two-stage lamp that would boil spirit in an old copper polish can. The vapor emerging from a small orifice in the screw-on lid would be mixed with air in a bunsen-like burner and heat the old-fashioned gaslight gauze. That really made the evenings very cosy!

But our worries were not quite over yet. After the shocking disappointment of the failure at Arnhem, bombs started dropping around the house. It was time to leave and go home to Delft. So at the end of October we mounted our bikes. First we had to cross the river IJssel. The bridge had had some near misses when the Allies tried to bomb it a week before and the bombs had unfortunately destroyed some houses, killing several people. Since some weeks before a German munition train had exploded for 6 hours, the people in and around Zutphen were easily frightened. Just as we were approaching the bridge, the air-raid warning sounded! My wife and son and I ended up in two different shelters. When it was over we braced ourselves for the crossing and I can assure you we never biked faster!

Halfway to Delft we spent a night at my sister's, arriving there unwittingly after curfew. She advised me to sleep under the floor with her oldest son because our "friends" were rounding up people under 40 for forced labor. Finally, after about a week, we arrived in Delft, having had to pump our leaky tires every half hour or so for the last 80 miles.

In our house, we already knew, we would find my sister and her husband, who is half Jewish. They had had to leave their home in Wassenaar,

[2] German and Dutch SS (even worse) were dressed in black; even more feared and hated were the German Security Police, wearing green uniforms.

where V2 sites were set up. Now that we were free of the scary planes we could use our time for procuring some food and fuel. I was never any good at this. My wife did pretty well. My brother-in-law of course could not leave the house. Work in the lab was no longer possible; there was no electricity in Delft. However, our workshop had changed into a small factory making miniature stoves and all the scientists were happily doing platework under supervision and with much help from the mechanics. I say *miniature* stoves, because soon we would have to cook our meals on just a spoonful of fuel. We agreed with our neighbors to share the wooden fence.

This was a good time to do some theoretical work, especially if it was time consuming. So I started to do numerical integration of the paraxial equation for strong excitations, assuming a Gaussian distribution. Then I tried to develop the results in a power series as a function of I^2/V. It was only then that I discovered that each term could be described as some integral of the previous one, the first one being Busch's well-known

$$\frac{0.022}{V}\int B^2\,dx$$

This work was rudely interrupted by the Germans, who rang the doorbell at 7 AM one morning in January to tell us that all males between 18 and 40 were to stand at the front door at 8 AM with food for 2 or 3 days. We knew it would come: Rotterdam and The Hague had been the victims of similar raids.[3] So we had prepared a shelter under the floor. A few weeks before a neighbor had asked me "How do you like your hiding place?" When I asked him how he knew about it he said "There's only one possibility, we all have the same kind!"

So at 7:15 AM my brother-in-law and I took the chessboard and went under the floor. At noon our wives offered some warm watery liquid called soup, and we sat on the edge of the trap door to savor it. As it turned out later, this was exactly the time that the house behind ours was raided. From there we were in full view! The Germans then apparently went for lunch and somehow lost the count. Anyway the small street where we lived was never raided! Too bad they didn't tell us before: it was bitterly cold under the floor. The big difference from The Hague was that Delft was raided after the Battle of the Bulge so our "friends'" confidence in the final victory was not as healthy as it had been.

[3] In Rotterdam some 4000 people were picked up; after this warning far fewer people got caught in The Hague.

I am forever grateful for the powerful BBC transmitter, allowing me to receive the news with a one-tube receiver fed by a dilapidated 4-V accumulator, probably having a capacity of 1 or 2 A-hr, which I could recharge in half an hour at the yeast factory while I gulped down my daily quart of porridge. After this I walked home so hungry that I could not resist eating a half-inch tip of a breadroll we got with the porridge that I wanted to save for the family. The receiver was dismantled immediately after the nine o'clock news, for safety, and rebuilt when needed. I was amused to see my neighbor across the street always go to the same place in his house when I started rebuilding. He later confirmed that his purpose was as I had suspected.

It was in this way that one Sunday morning late in April, 1945, I heard that food planes were standing ready to fly to Holland as soon as the weather improved. Now didn't that sound very exciting! I had heard much earlier how Churchill was asked in the House of Commons if he had an explanation for the sand storms in the desert always hampering British tanks and apparently leaving the sunny weather for the Germans. At the time there was nothing wrong with the weather in Holland, nothing exceptionally, that is, so I took it that maybe in a month or so the planes would come. That morning I started a new experiment: I was preparing a crocus-bulb cake. It was said to be edible but I must confess it looked much more poisonous than tulip bulbs. I had just started to fire up our spoonful of coke dust when I heard the familiar sound of Lancaster bombers, only this time flying much lower than usual. I ran outside and through their open bomb doors I saw neat rows of bags. The food had arrived! I threw the crocus bulbs in the fire (they were cut up so finely that planting them was out of the question) and climbed on the roof. How I did it will forever be unknown, because I needed the rest of the family to help me off. We celebrated with bacon that we had kept in reserve!

Clearly the end was in view now. Two weeks later Holland was free! It took many weeks before things started to run again. We didn't even bother to take the microscope out of its hiding place. We still had no electricity and no gas to heat the diffusion pumps.

However, one day of this beautiful, beautiful summer some visitors from the United States were to come to the Shell Labs in Amsterdam (see Fig. 5). For this occasion we were asked to give a demonstration with the Mark II microscope, on 2 days' notice. de Heer and I wondered if we could make provisions. For the occasion we were allowed to draw the lens currents from the batteries, which, having been well looked after, were still in good

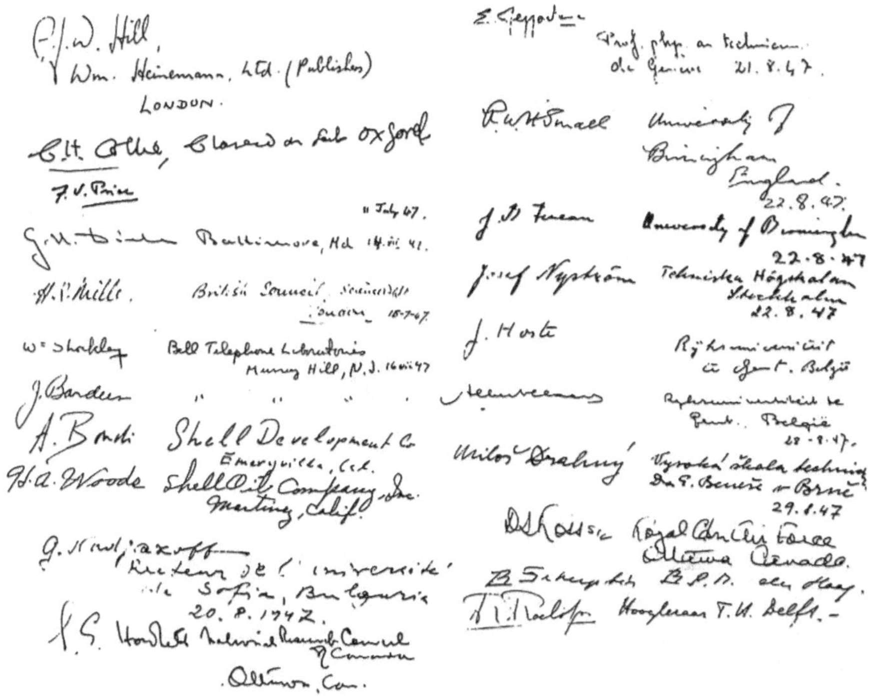

Figure 5 A page, not completely randomly selected, from our visitors' book.

shape. For the ac we could run a dc to ac converter, meant for generating variable frequency. There remained the heating of the pumps. de Heer had a solution: an acquaintance of his owned a housepainting firm. Surely he would lend us two petrol burners! So we bought some pink petrol from the Canadians, who were around everywhere.

The microscope was recovered and assembled and in the nick of time we got it in operation. For better demonstration value we took an old 8-in. oscilloscope tube, removed the electrostatic deflectors, covered the stubs with glass thimbles sealed with Picein, redid the fluorescent screen, and then we were in business. The demonstration was a success. We had molybdenum oxide as a specimen, which, apart from being very interesting to look at, also gave an easy and very clear spot diffraction pattern; this also made quite clear how the dark bands in the platelets came into being, because the diffraction pattern could be defocused at will. At very high illumination the platelets disintegrated into a large number of small crystals, giving diffraction rings.

By the way, at that time I could not get any crystallographers sufficiently interested to do some work on the mosaic structure and other lattice defects. Not being a crystallographer myself, I decided that it was obviously not important.

Well, the Americans had not seen a versatile microscope like ours and were quite enthusiastic. There was, however, one thing they wanted to know: what was the deeper thought behind the four black stubs on the side of this beautifully wide viewing tube? We were not prepared for this question, but although we have subsequently been asked the same thing by practically everyone, we have never been able to dream up any intelligent mock reason, so we were forced to tell the truth. (Thinking back, there was one very intelligent man who did not ask the question, so I asked him: "Don't you want to know what these are for?" His answer: "Sure, I was just going to ask you!")

We liked the big screen so much that we kept it for exploring the specimens. For recording we then had to replace it by our three-exposure 6 × 6-cm camera with a 6-cm peephole.

The enthusiasm of the Americans made me go again to G. Holst, director of Philips Lab, hoping he would reconsider going into the commercial production of electron microscopes, but he repeated what he had told me before: "Maybe we could sell 20 now that the world is open again, instead of 10. That still isn't a commercial proposition." "But," I said, "they told me a thing like that doesn't exist in the United States or anywhere they had been up till then." His answer: "You don't know the Americans, they are so polite." The first part of that last sentence was obviously correct, but it took me another 3 years to find the same for the second part. However, I met very few people in the United States who put politeness before truth. My collaborator, Miss W. van Iterson, who took care of the microbiological applications, had access to the wartime literature in the Allied countries. The only thing I'd ever heard was that electron microscopes were included in the lend–lease agreement. She made a careful abstract of all she could find, which pretty much assured us that nobody in the Allied countries had been ahead of us. So she talked to her father, F. K. Th. van Iterson, managing director of the State Mines. Apart from being a family acquaintance he had at one time expressed his confidence in my creativity. Now father van Iterson was on good terms with Dr. A. F. Philips, and as it happened he had a lunch appointment with him shortly after his daughter told him the sad story. "Well," he said, "Why don't you come with me?" Apparently Dr. Philips was too polite to show his surprise at meeting father and daughter.

After the gentlemen had talked business for a while he asked "And what, Miss van Iterson, are you occupying yourself with?" "Tell your story," her father said. I'm sure he must have added something to his daughter's story. I shall come back to this later, mainly because, although I heard rumors about what happened, I confirmed them only in November, 1982.

Although we were a free nation once again, there was still very little communication with the outside world and literature from the Allied countries was not available apart from what Miss van Iterson had found out for us. All transportation was needed for rebuilding and feeding a devastated Europe. Therefore, the Delft Hogeschool Fonds (Delft University Fund) took a very laudable initiative: a number of young and promising graduates from Delft were enabled to spend some weeks in England and study what progress had been made in their respective fields during the war. Selection was done mainly on the basis of their knowledge of English. This is where my 4 years of listening to the BBC news paid off, although English had always been the only foreign language I did well in at school. There had been, until then, very few indications that I was "promising." In fact, I think the marks I scored during my studies were all but impressive. I had badly failed a test in partial differential equations about 5 weeks before getting my degree.

The day before I went up, I was telling my colleagues that I saw no way I could pass, but they encouraged me, saying "No one ever failed; that is to say, one girl did, but she was plain stupid." Well, I was the second one that did so, and very convincingly too. The mere fact that announcements for my wedding 3 days after graduation had been sent out and that I already had secured a job saved me from postponing the graduation and I was given another chance 5 weeks later instead of 3 months or so. It upset my schedule but with extreme effort I just made it.

The trip to England was made under the supervision of J. M. Burgers. How well qualified he was, not only as a most prominent scientist, but also as a human being. By his clear introduction to the map of London I could take pride in hardly ever needing a map there. On foot I could even take short cuts by just remembering his lessons and arrive at places I had never been before!

The literature studies confirmed what Miss van Iterson's abstracts had indicated before. The Delft microscope was really an innovation. It occurred to me that it had been a blessing to be a (forced) solitary worker who was, as somebody expressed in a very friendly way, "not biased by knowledge." It so happened that during my visit to England (January, 1946) there

Figure 6 Our group in 1947. Front row: de Heer, van Ments, Miss Schaap, Dr. van Iterson, Dr. Le Rütte. Second row: Groenheide, Fraase Storm. Back row: Le Poole, Miss Le Poole, Miss van der Wees, Miss van de Heuvel, Kramer.

was a meeting of the British Electron Microscope Group in Manchester, of which I was informed by M. E. Haine, who was showing the Metropolitan Vickers microscope at the annual exhibition of the Physical Society. The train ride to Manchester was very long and very cold. The length of the trip made me break the sacred rule that fellow passengers are not to be talked to in Britain, and I got into a lively conversation with Mr. Smiles, who turned out to be going to the same meeting and did electron microscopy at the National Institute of Medical Research in Hampstead. He told me he had established that the quality of his pictures could be noticeably improved by choosing the correct height of the specimen. As far as my theoretical knowledge went, I told him that the stronger the objective, the better the results should be. But he insisted there was an optimum height. So I said: "I'm sure to learn a lot at the meeting." Years later I realized that his objective lens must have had rather large astigmatism, and that he had minimized that by the proper image rotation. (See Figs. 6–9.)

At the Midlands Hotel in Manchester I met van Dorsten, who was over for a business visit. In the hall I met many electron microscopists who were exchanging results obtained. van Dorsten, seeing the quality of their

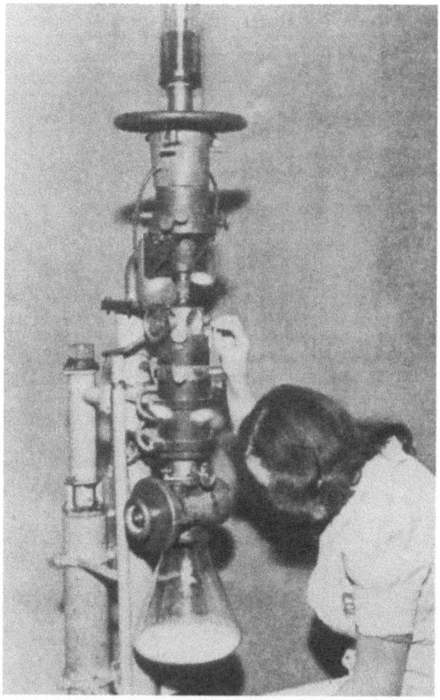

Figure 7 Miss Schaap at our Mark II microscope, 1947.

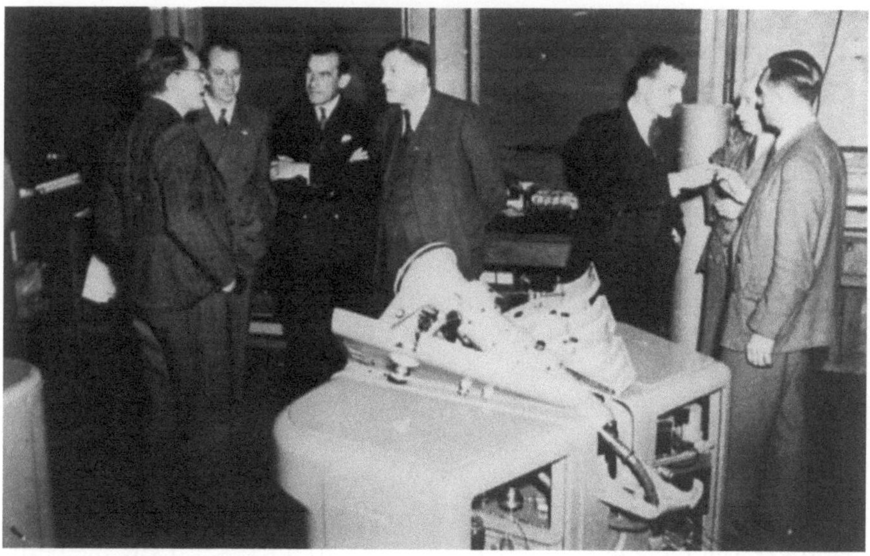

Figure 8 Inauguration of the commercial EM 100. Left to right: H. B. G. Casimir, W. J. Oosterkamp, H. Alting, H. B. Dorgelo, J. B. Le Poole, A. Verhoeff, A. C. van Dorsten.

Figure 9 The 1000th electron micrograph taken by Professor H. B. Dorgelo, 1946.

pictures, talked me into getting ours, which I must admit did look better, not the least because of de Heer's superior photographic techniques. "Did you say you would learn much from us?" Mr. Smiles asked. Well, although I didn't answer that one, I can assure the reader that I did learn a lot, but also that the little incident provided a very useful introduction, as, in view of my age, I would otherwise have been very hesitant to enter into discussions.

There was Dennis Gabor, who proposed an off-axis system to correct spherical aberration. I asked him whether he was not afraid of severe astigmatism (I well remembered von Borries' remark in 1942). I saw the beautiful work done with "shadowcasting" by the group in Leeds, headed by T. Astbury. Shadowcasting was a technique used by Wyckoff and Williams to bring out three-dimensional information. It had been used earlier by H. O. Müller in Berlin, who had, however, missed the point that negative prints made the pictures look like objects illuminated at a low angle. In 1937 Wyckoff made headlines by showing molecules in the regular arrays of crystals. In a discussion with V. E. Cosslett on lens calculation, he told me he had worked with the same method on electrostatic lenses. All in all a very memorable meeting!

But there was more to come. In discussions with van Dorsten plans were made as to what a commercial Philips microscope could look like, and when at the end of January I returned to Holland electron microscopes had gained first priority at Philips. van Dorsten's report had fallen on ground prepared by Miss van Iterson's plea.

Having accustomed ourselves to the large fluorescent screen so easily viewed by several observers, it occurred to me that a camera using 35-mm film could be built above the screen. This would have to be taken out of the beam for visual observation. No change of focus would be required, the focal depth at the projector side being some 100 m. A simple mechanism made it possible to rotate the film out of the way and take care of film transport at the same time. Moreover, calculating the useable area of the film we decided not to use perforated film. The slightly increasing distance between successive exposures resulting from the increasing winding radius is far less wasteful than the perforation. The picture size was therefore set at 30×30 mm^2.

Clearly the photographic emulsion had to be chosen with some care. It had to be very fine grained, which in itself was no real problem, because its speed could be chosen 10 times slower on account of the 10 times higher current density at the reduced magnification. Second, the emulsion had to be thin to avoid deterioration due to electron scattering.

Talking this over with L. V. Chilton of the Ilford Co., he recommended a fine-grain positive release film, containing a maximum of silver halide in a minimum of gelatin. Advantages of the 35-mm procedure were small storage space, developing in sodium light, fast operation, and low cost. After this system became standard in the Philips microscope, Ilford changed the code number of the film, and sold it as electron recording film at 10 times the price. We were unable to spot any difference in behavior.

Now we were to perform a stunt. In September there was to be a similar meeting in Oxford and we would be showing a prototype, based primarily on the Delft microscope. Earlier I had received permission to attract a qualified technician, so I approached the director of my old school (MTS)

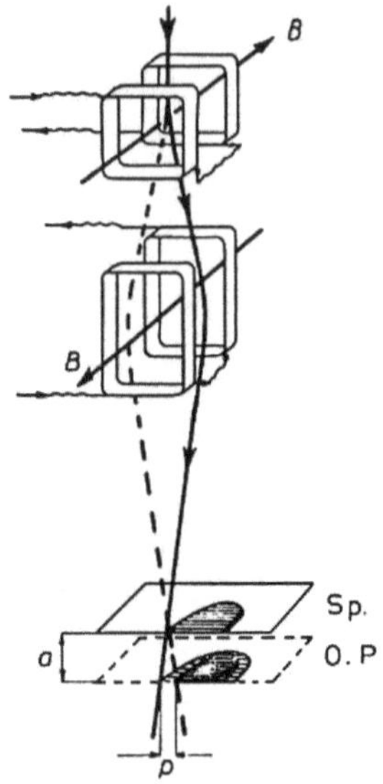

Figure 10 The "wobbler" focusing aid, 1946.

for somebody who was interested in something more than the regular curriculum. Thus I got J. Kramer. His first commitment was to design and build the "wobbler," the focusing aid that is still incorporated in all Philips microscopes (see Fig. 10). Soon Kramer showed himself to be a wizard in electron microscope alignment. In the late summer of 1945, M. van Ments joined us. He and his father, mother, and sister had been hiding for over 3 years and all were spared the hell of the concentration camps. Max felt he was given an opportunity to live a second life. His positive attitude and his tremendous sense of humor (his ethnic jokes were better and more numerous than ours) were an inspiration to us all. He agreed to measure a number of magnetic field distributions. The conception of the Mark II microscope had made it clear to me that there is a decided advantage in designing the complete magnetic circuit in such a way that saturation would play no role. Only in that way would it be possible to calibrate magnifications once for all voltages. The method was very simple: the current at which the diffraction pattern was focused was used as a standard. A variable shunt on the meter measuring this current was adjusted so that this current would always be indicated by the same point on the scale. Using the same meter for the whole range of magnifications, the calibration was then automatically correct. The projector was simply set at maximum magnification. The consequence of avoiding the influence of saturation was that measurements

Figure 11 Some sophisticated electron optics on the beach at St. Andrews, 1951. Right to left: M. E. Haine, G. Liebmann, V. E. Cosslett, J. B. Le Poole.

of field distributions in real lenses were not necessary and that scaled-up models could be used. Furthermore, we agreed that the field distribution would only depend on the ratio of gap and bore. So it was decided to use a model magnet with a variable gap.

After the measurements were made, Max carried out the multiple integrations. We had hoped to derive clear design rules. Instead we found that all lenses are equal (van Ments & Le Poole, 1947)! Even when subsequently we also measured asymmetric distributions, the same equality resulted! This was later (1950) confirmed and formulated more precisely by M. E. Haine (see Fig. 11), on the basis of more accurate measurements on G. Liebmann's resistor network: reducing all focal lengths and spherical aberration constants to dimensionless quantities by dividing them by the minimal focal length when used as a projector and similarly reducing the lens currents, dividing by the current for minimum focal length, the curves overlapped within the linewidth!

The polepiece configuration can therefore be chosen on other than electron-optical grounds, a truly simple design rule!

The stunt of building a prototype commercial microscope was not quite as simple as I had thought. The experience with the Mark II microscope had made me rather optimistic! It had taken me less than a year to design it and have it made. No doubt, with the more efficient methods of industry, 6 or 7 months would be ample. But I did not know the industrial methods. In my ignorance I had always carried out my experiments by designing part for part, having the part made and tested, then proceeding to the next item. Of course industry has not one component made before all the drawings are ready, even in a laboratory. The drawings were completed in early June. By early August the components were pouring out. Now the results of speed were showing up. Only a few components fitted exactly as intended. So van Dorsten, Le Rütte, Verhoeff, and many others were happily filing, drilling, sawing, and sanding away. In our frustration we also changed parts that were properly sized after all! The beautifully cream-white-painted instrument was getting darker and darker with all this handling. One of the mechanics said with tears in his eyes "They painted this white; it would have been more convenient if they had tarred it." Le Rütte remarked "One could say this microscope is very special because it is completely handmade." van Dorsten's immediate reply: "Yes, all handwork, no brainwork at all."

The microscope was flown over to Oxford to be there in time. But despite 65 hours of work in the next 3 days by everyone involved, we could not give a demonstration with a working microscope. Amazingly, however, this did not doom the enterprise to failure. Very friendly onlookers were still willing to admire the sound basis of the design, and everybody liked the big screen.

No doubt the large impact of van Dorsten's and my visit to the Manchester meeting had made me feel somewhat indebted to the British Electron Microscope Group, and we became regular attendants of its spring and autumn meetings, establishing many lasting friendships. I believe it was at the spring meeting of 1948 that G. Dupouy expressed his admiration for these meetings and invited them over to the continent. When no follow-up was given at the autumn meeting, I asked permission from Professor Dorgelo to invite the meeting over to Holland, which resulted in the First International Conference on Electron Microscopy in Delft, July, 1949.

It was there that Castaing gave his famous paper on X-ray microanalysis, which was decidedly the most important contribution in my view. Other important contributions: a joint paper by Agar and Revell on preshadowed replicas, which caused their identities to be confused for many years. I am no judge on whether Bretschneider's paper on thin sections has inspired many others in this important direction. (See Fig. 12.) His pictures certainly belonged among Delft's showpieces for quite some time. The interesting thing was Bretschneider's early recognition of the importance of

Figure 12 Celebration of the 10,000th micrograph. L. H. Bretschneider, top left.

the embedding material. He used carnauba wax mixed with beeswax. Since at room temperature this was not quite hard enough, Bretschneider made his sections in a cold room, at some 4 °C, so doing it he was dressed in a heavy winter coat!

After D. Gabor had first published his ideas on holography, M. E. Haine and his co-workers started to try and produce the necessary electron probe. Although this proved too difficult at the time, Haine did some brilliant work on electron sources, which I consider the undisputed foundation of our knowledge of electron gun behavior. His discovery that space charge played hardly any role was a revelation, which after 30 years is still far too little recognized. When Gabor and Haine later showed some results I remarked how much the holograms looked like out-of-focus micrographs. Gabor insisted that holography was something very different. A little later it occurred to me that Fresnel fringes have the same structure as a zone lens, and that therefore a sharp edge could be reconstructed from a defocused one. Holography or not, I tried the experiment and it worked.

A year later, in my absence, Bragg visited the lab and got the usual EM demonstration. My co-workers proudly showed my efforts on image reconstruction. As a demonstration there was an in-focus picture, a fairly good fringe picture, and the reconstruction. Bragg had been very impressed, my

Figure 13 Old soldiers never die ... (photo: B. J. G. M. Roelofs).

co-workers told me later. Not thinking very much of the reconstruction myself, I found out that they had shown the in-focus picture as the reconstruction! I am sure Gabor himself would have been thoroughly impressed too. I never met Bragg to apologize for the mistake.

We have now arrived at the early 1950s. The electron microscope, although still far from being fully developed, had reached a stage of maturity that made it a generally recognized and powerful tool in science. Very soon publications became so numerous that it was no longer possible for any one person to follow the entire field of applications. Taking part in laying the foundation for this new phase has been a privilege. (See Fig. 13.)

1. Afterword by Pieter Kruit: about Jan Bart Le Poole (1917–1993)

An excellent and very complete memoir of the life and career of Jan Le Poole was published in 2001 by his old friend Tom Mulvey and his long-term secretary and friend Thea van de Laak-Tijssen. What follows is a summary from that document.

Le Poole's own story about "Early microscopy in The Netherlands" ends in the early 1950's, so we'll take it from there. Philips had started production of commercial microscopes and Le Poole was heading the electron microscope institute in the applied physics lab (TPD) of the Netherlands Organization for Applied Research (TNO). At the same time, he had an appointment at the Department of Applied Physics of Delft University which gave him the opportunity to involve master's students in his projects. His developments and inventions in those years were only presented in conference proceedings and published in patent applications. His former teacher, Prof. Dorgelo, realized that this lack of a paper trail would hamper his further career and pushed him to write a Ph.D thesis. This 30,000 word thesis was, in fact, his only extensive publication. It covered the period 1941–1954, a very creative and inventive time of his life, and enabled his 1957 appointment as a professor. This did not end his relation with the TPD; he was part of its scientific counsel until his retirement. I remember that in the seventies he went to the TPD regularly, which was conveniently located in the same building, to have lunch and play a rubber of bridge with his old friends. Those were the days.

In his research group he hired Joop Kramer to work with him and later Jim Barth and Erico Koets. His interest widened and the innovative instruments that he designed in the following 25 years ranged from X-ray projection microscopes and time-of-flight mass spectrometers to electron and ion lithography machines. In the field of electron microscopy it is worth mentioning his design of the mini-lens, his work on a Wien-filter for energy-filtered imaging and energy loss spectroscopy, his design of a scanning mirror microscope, and his very compact 1 MV electron microscope. He was a pioneer of electron beam lithography and experimented very early on with shaped beams and multi-beam systems. The first multiple-beam lithography concept was tested in 1974, followed by an electron mirror as a multibeam modulator in 1981, and a 10 Gpixel/s concept in 1982. His group produced Fresnel-type X-ray lenses to be used in satellites. He was well aware of the brightness limitations of existing electron and ion sources and worked on improvements. He initiated the study on stochastic electron-electron interactions that led to Jansen's seminal work on the subject. There were several experiments on the correction of spherical aberration, using grid lenses, superconducting grids, or space charge. An attempt was made to build a demonstration unit for the Kapitza–Dirac effect.

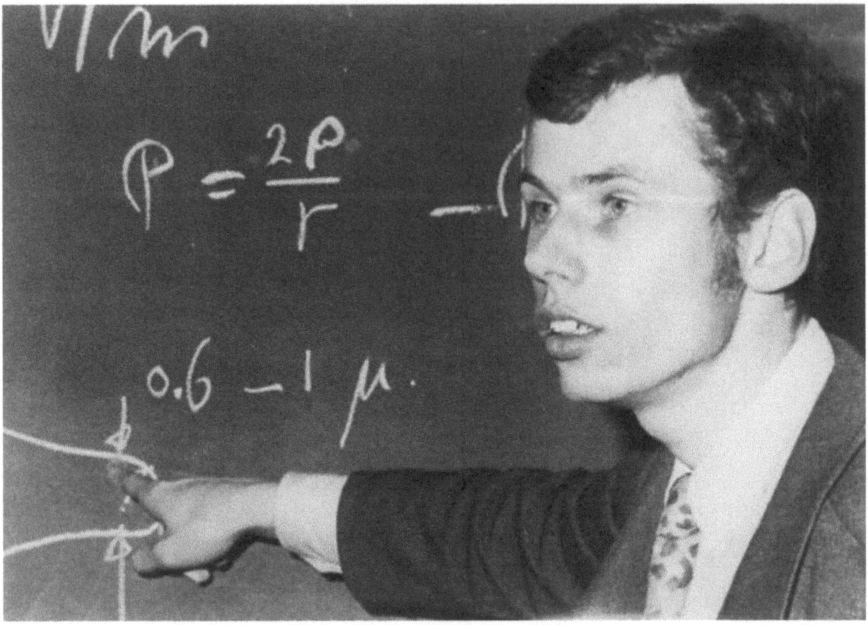

Figure 14 The author's successor, during a visit to Ernst Ruska's lab, explaining the laser-heated gun, 1973 (photographer: H. J. Elema).

Jan Bart Le Poole was the organizer of the first international conference on electron microscopy in 1949. He was the founder of the Dutch Society for Electron Microscopy and a co-founder of the International Federation of Societies for Electron Microscopy, where he later served as General Secretary and as President from 1978–1982.

In 1982, Le Poole turned 65 and retired from his position as research professor. K. D. (Karel) van der Mast, who had obtained both his master's degree and Ph.D. under Le Poole, was selected as his successor (Fig. 14). Jan moved to Tucson, Arizona where he continued as a consultant to the Philips Company and took on an extra consultancy at Gatan in California. He passed away in 1993.

Van der Mast returned to Philips Electron Optics in 1989 and was succeeded by the present author, Pieter Kruit, also a graduate of Le Poole with a PhD in atomic physics. I am still a professor of charged particle optics at the time of writing this short appreciation (2021).

Those of us who knew Le Poole personally recall his creativity during discussions and his penetrating questions at presentations. His way of interacting was as uplifting as his way of teaching: he tried to convince students that they knew enough to solve almost any technical problem; it was just a question of applying their knowledge correctly. That was the way he himself had approached the research challenges in his career so successfully.

References

Burgers, W. G., & Ploos van Amstel, J. J. A. (1937). *Physica (Amsterdam)*, *4*, 5.
Dreyfus, H. (1931). *Archiv für Elektrotechnik (Berlin)*, *25*, 392.
Glaser, W. (1941). *Zeitschrift für Physik*, *117*, 137.
Glaser, W. (1943). *Zeitschrift für Physik*, *121*, 647.
Lindenhovius, H. J., & Rinia, H. (1941). *Philips Technical Review*, *6*, 54.
van Dorsten, A. C. (1948). *Philips Technical Review*, *10*, 137.
van Dorsten, A. C., Oosterkamp, W. J., & Le Poole, J. B. (1947). *Philips Technical Review*, *9*, 193.
van Ments, M., & Le Poole, J. B. (1947). *Applied Scientific Research. Section B*, *1*, 3.

Afterword reference

Mulvey, T., & van de Laak-Tijssen, D. J. J. (2001). Jan Bart Le Poole (1917-1993), pioneer of the electron microscope and particle optics. *Advances in Imaging and Electron Physics*, *115*, 287–354.

CHAPTER SIX

L. L. Marton, 1901–1979

Charles Süsskind[a] and Peter Hawkes (Afterword)[b,*]
[a]Formerly College of Engineering, University of California, Berkeley, Berkeley, CA, United States
[b]CEMES-CNRS, Toulouse, France
*Corresponding author. e-mail address: hawkes@cemes.fr

Contents

1. Bibliographical appendix: publications of L. L. Marton 192
2. Afterword by Peter Hawkes 198
References 198

In 1961–1962, during a sabbatical year spent at Imperial College in London, I had several conversations with Dennis Gabor about the history of electron optics in general and of the electron microscope in particular. In 1957 Gabor, who had received his doctorate in Berlin, gave a paper on the topic as part of a celebration of the seventy-fifth birthday of his old professor, Adolf Matthias, which was subsequently published in a German engineering journal (Gabor, 1957). (Gabor had written it in English and it had been translated into German; I have never seen the English original.) And in 1967, at a special session during the IEEE Ninth Annual Symposium on Electron, Ion, and Laser Beam Technology in Berkeley, L. L. Marton gave a special lecture on the "Early History of the Electron Microscope," which was subsequently published by the San Francisco Press in its "History of Technology Monographs" series (Marton, 1968a). We all met once more (after Gabor had received the Nobel Prize in physics for his contributions to holography) during the 1972 IEEE International Convention in New York, at which Marton had organized a session on the history and philosophy of technology, and Gabor and I were both speakers. Over coffee afterward, the talk naturally turned to the history of the electron microscope, especially since H. D. Rüdenberg (the son of Reinhold Rüdenberg, its "legal" inventor) made the fourth of our little group. I have drawn heavily on these several encounters in what follows, as well as on published and unpublished written materials and recollections (Mulvey, 1962; Ruska, 1979).

L. L. Marton (Fig. 1) was born in Budapest on August 15, 1901; his given names, Ladislaus Lászlo, were too difficult for America, where he

Advances in Imaging and Electron Physics, Volume 221
ISSN 1076-5670
https://doi.org/10.1016/bs.aiep.2022.03.006

Figure 1 L. L. Marton (1975).

became just plain Bill. He was of Hungarian and Swiss extraction, which meant that he early became fluent in several languages. Following high school in Hungary he went to the University of Zurich to study physical chemistry and remained there for a year as an assistant spectroscopist after he obtained the doctorate in 1924. He then returned to Hungary and worked for 3 years at the famed Tungsram Lamp Co. (in full, the United Incandescent Lamp & Electric Co.; the better-known trademark name Tungsram was a contraction of Tungsten and its German equivalent, Wolfram, the commercial use of which in filaments the firm helped pioneer). Although the company was part of a five-member international lightbulb cartel and had a research laboratory of world caliber, Marton left after 3 years on the first leg of what ultimately proved to be a fruitless quest for a professorial appointment. He became a research associate at the Free (that is, private) University of Brussels. He remained there for 10 years, even after he married another young physicist, Claire, who was to share his professional life to the end. They had no children.

His thesis at Zurich had been on infrared spectroscopy, and that is what he continued to do at Brussels. For a time he worked in photoelectricity and to some extent on X rays, but soon came back to infrared. With his varied interests and linguistic abilities, he assiduously kept up with the physics and applied physics literature of the day. Electron optics scarcely existed as a subject, even after Hans Busch at the University of Jena had shown in 1926 that an analogy between light and electron lenses existed, for that result remained of interest mainly to the relatively small band of physicists and engineers who were trying to improve the cathode-ray oscilloscope, which Ferdinand Braun had invented in 1897 (Kurylo & Susskind, 1981). Even after 1931, the year the electron microscope was born, few physicists paid much attention to the reports that were coming out of Berlin, its birthplace. Interest in electron optics would ultimately spread to other countries, beginning with America (Davisson at the Bell Laboratories), Britain (Martin at Imperial College), and Canada (Hillier at the University of Toronto), but, with one exception, all the work of the early 1930s was being done in Berlin. The exception was Marton's tiny laboratory in Brussels.

By the summer of 1932 Marton had read most of the relevant literature on electron optics. He found the implications tremendously exciting. Harking back to his interest in photoelectricity, he convinced himself that a connection could be established between that field and electron optics. A phenomenon that immediately came to mind was the so-called "patch effect" in photoelectricity. Photoelectric emission tended to be uneven; the reason, it was hypothesized, was that the work function was not uniform over the cathode surface. If that were so, thought Marton, one should be able to see the nonuniformities by means of this splendid new tool, the electron microscope.

Such an experiment would stretch the resources of a small laboratory to the utmost, but Marton was fortunate in having a chief, Emile Henriot (1855–1961), who was sufficiently impressed by his young associate's enthusiasm to give him his head. The laboratory's resources were rather meager, but Henriot encouraged Marton and provided the means for a pilot study. It became Marton's first primitive electron microscope, which was in operation before the end of 1932.

The configuration was of the simplest. The cathode was a small coil of tungsten wire pressed into a flat spiral of square shape. The emitted electrons were made to travel on the inside of a brass tube to a fluorescent screen at the other end. There was a single lens, a coil of wire wound loosely on the outside of the brass tube, so that the coil could be slid along it. A specimen holder that could also be moved along the axis was placed inside the tube. Of course the whole contraption had to be airtight so that it could be evacuated, and it had to be returned to atmospheric pressure and evacuated again each time the specimen had to be changed or even moved. Primitive as it all was, the very first pictures showed reasonably good imaging quality—good enough to encourage him to undertake the construction of a better instrument. Marton never did achieve his original purpose of using it for research on photoelectrity, but even that aspect of the work was not a total loss. In viewing images of the tungsten spiral cathode he was able to confirm that emission along the various parts of the spiral was not uniform. Since he was familiar with the crystalline appearance of tungsten wire from his earlier work at Tungsram, he was led to conclude that the several crystalline faces at the surface must have different work functions, which would account for the differences in emission. He never published this observation, partly because research along the same lines had been started in Germany at AEG, a major electrical equipment firm to which emission from tungsten was of considerable commercial importance. He need not have been quite so diffident—his results were at least as convincing as those from Berlin—but he decided that it made no sense for him to compete against a huge industrial laboratory, with its vastly superior means for supporting and advancing research of this sort. For example, when Brüche and Johannson at AEG undertook to use their electrostatically focused electron microscope to investigate the process by which oxide-coated cathodes were activated, they were able to mount a rather elaborate setup. The still camera in front of the fluorescent screen was replaced by a movie camera and individual frames were exposed at intervals, so that a speeded-up movie of the 2-hr activation process was compressed into 2 min. The investigation was later extended to the observation of pure-tungsten cathodes: even a slight improvement in cathode emission would have a large payoff for a manufacturer of electron tubes. At the end of the paper that described this later investigation, Marton saw a short note by Johannson, a calculation of the focal length of an "einzel" lens (a three-electrode configuration in which the two outer electrodes are at the same potential), which the author had done jointly with Otto Scherzer—the first time this name appeared in the literature of electron optics (Johannson & Scherzer, 1933).

In addition to experimental work, Marton took an early interest in the theoretical underpinnings of electron optics, in the development of which Scherzer and his contemporary Walter Glaser were to play such an important part. In his very first paper Glaser showed that the formalism derived from the Fermat–Hamilton analogy could be used to describe the imaging properties of electron-optical systems (Glaser, 1933a), work that led him to calculate image aberrations in an electron microscope, to show that spherical aberration would loom more important than in light optics, and to predict aberrations that were not even known to light optics (Glaser, 1933b). The mathematical elegance of these results made a deep impression on Marton, as did the way Scherzer's calculations could be readily adapted to representation by physical models. He thought that the two summaries of their respective approaches contributed by Glaser and Scherzer to the volume "Beiträge zur Elektronenoptik" that Busch and Brüche edited in

Figure 2 Marton's No. 2 electron microscope (Brussels, 1934).

1937 were classics that would remain the most concise presentations of the two approaches to electron-optical theory for all time (Glaser, 1937; Scherzer, 1937).

Marton used to say in this connection that wave mechanics had to be invented before the development of the electron microscope was possible. Although Glaser's approach started with the assumption of the wave nature of the electron, his entire treatment (after the initial assumption had been made) turned out to be completely classical—just as classical as Scherzer's trajectory method. Later, Glaser had to modify his treatment so as to carry the wave-mechanical approach through his whole treatment of electron-optical theory.

Marton had got off to a slow start, as might have been expected in view of the limited facilities and resources available to him. During 1933 (the year he married Claire) he explored the limitations of his first instrument, started to construct a second one, and published one short review paper on electron optics, in Flemish, jointly with Maurice Nuyens.[1] His new instrument had a cold-cathode electron gun, condenser, objective and projector lens, and a translucent metallized fluorescent screen (Fig. 2). The instrument was mounted horizontally, partly to facilitate the taking of pho-

[1] For papers by Marton, see the Bibliographical Appendix (p. 192).

tographs, but also because much of the early work in electron microscopy was done by people whose consciousness was governed by the conventional optical bench. The coil that made up the projector lens was mounted with a sliding fit on the outside of the brass housing of the microscope, with exchangeable polepieces inside the vacuum. (In an instrument that Ruska was then developing and which he described in print at a time when Marton's instrument was already being tested, separate polepieces were used in a similar way.) A rubber sleeve fitted tightly over the brass tubing that connected the electron gun with the miroscope proper and allowed a relatively easy exchange of specimens. The primary objective of the pioneers was to increase the resolving power rather than the magnification, which in the earliest instruments scarcely approached that of ordinary light microscopes. Marton set himself two additional goals: first, to explore some of the possible applications of the instrument and to make it most useful for such purposes; and second, to gain an understanding of the mechanism of image formation.

With regard to the first goal, he very quickly concluded that biological objects would offer the richest field of application. That was a courageous decision, since it was universally believed that any specimen exposed to the electron beam would, as the saying went, "burn to a cinder." That was what Gabor believed, too. In a conversation held in Berlin in 1928 Leo Szilard had suggested to Gabor, who had just spent 3 years researching and designing a high-voltage oscilloscope as part of his doctoral dissertation, that he should make a microscope based on electrons; after all, who knew more about electron beams at that time than Gabor? Once Busch had shown how one could make an electron lens, had not Gabor demonstrated that the action of such a lens could be considerably improved if the coil were surrounded by an iron mantle? True, said Gabor, but he had been around electron beams too long and had become temporarily allergic to electrons; time to go on to something else. Such a project might take several years, and what would be the use? No living matter could survive in a vacuum.

This viewpoint (which was generally held) was based on the assumption that contrast in an electron microscope would be obtained, as it was in the light microscope, by differential absorption of the beam in the several parts of the specimen. If not everything, *almost* everything would burn to a cinder under the electron beam. The first specimens used by Knoll and Ruska were things that would not easily "burn," such as tungsten wires. (Of course nothing would really burn—that is, be consumed by fire—in a vacuum, in which the conditions for combustion were absent; but

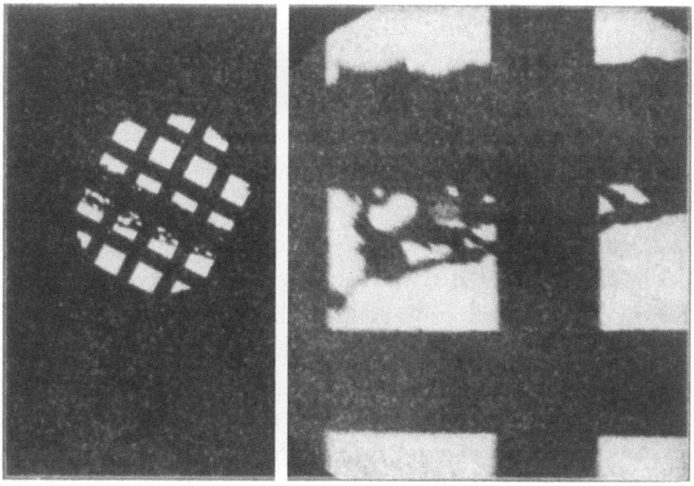

Figure 3 First electron micrographs of biological sample (15-μm-thick slice of *Drosera intermedia* leaf on copper net), April 4, 1934.

the energy absorbed from electron bombardment would surely cause the specimen to disintegrate or melt, which is why specimens of high melting points were favored at first.)

Marton believed that too, but thought that he could make biological objects heat resistant and also less subject to disintegration by impregnating them with substances such as osmium, which had a high melting point (3045 °C). Another idea that occurred to him was to mount the specimens on a support with good heat conductivity, which would carry away some of the absorbed energy. He went around to consult with some of his university colleagues in the life sciences: the zoologist Paul Brien and the botanist Marcel Homès (who later became president of the Royal Academy of Belgium). They were able to provide some microtome slices impregnated with osmium tetroxide (OsO_4), but these were quite thick cuts by modern standards. Ultramicrotomes, instruments that could cut slices a fraction of a micrometer in thickness, did not become available until the 1950s; the best one could to in the 1930s was a specimen that was several micrometers thick. In a relatively thick cut like that, one would indeed expect that the specimen would heat up as it absorbed energy from the impinging electrons, and that it would be ultimately destroyed. Nevertheless, the technique was sufficiently successful to yield the first histological photographs ever made with an electron microscope, on April 4, 1934 (Fig. 3). They were presented at the May 8, 1934, meeting of the Belgian Academy and appeared in its *Bulletin* of that date, which came out somewhat later; but Marton managed to get a short note into the June 16, 1934, issue of *Nature*, which was actually the first appearance of these results in print.

A little later, Marton and his wife traveled to Berlin and met Knoll, Ruska, and Brüche for the first time. Marton handed around his first micrographs and was gratified to see their initial complete disbelief—"This is not possible!"—gradually give way to requests for prints. It was a memorable visit. At one point the Martons went to see a nonscientist friend, who drove them several kilometers out of Berlin to a spot from which one could see a goodly distance in all directions. He stopped the car, closed all the windows, and then said, *"Now* we can talk." Small wonder: it was June 30, 1934, the infamous Night of the Long Knives, when Heinrich Himmler's SS storm troopers massacred Ernst Röhm and the other leaders of the rival Nazi gang, the SA.

Marton's important advance in pioneering the electron microscopy of biological specimens was accompanied by an even more fundamental piece of work that did a great deal to clear up the underlying principles and that banished the belief that such specimens would "burn to a cinder" once and for all. The priority of that particular insight is usually credited to Ruska and von Borries, but Gabor thought that Marton deserved the credit.

> *The great discovery dawned on [the pioneers] gradually, in the course of several years [wrote Gabor]. The electron microscope is similar to the light microscope in its basic design, but entirely different in its action, that is to say in the way it forms contrast in the image. Absorption [in sufficiently thin specimens] played almost no part. The main source of contrast was* scattering, *and the first to recognize this fact clearly was Bill Marton. He was also the first to open thereby the way to the biological application of the electron microscope. But even that was not the end of surprises, because in very thin objects, where even scattering was too weak, phase contrast took over. As Otto Scherzer put it:* "Erst war es Vielfachstreuung, dann war es Einmalstreuung, jetzt ist es Keinmalstreuung" *[First it was multiple ("many-fold") scattering, then it was single ("one-time") scattering, now it is no ("not-even-one-time") scattering] (Gabor, 1968).*

That is not to say that the progression was quite so orderly. The basic predicament facing Marton in 1933–1934 was that ultrathin cuts were hard to get and harder to mount in the electron microscope, and the more easily mounted thicker slices did not last long under the electron bombardment. Marton tried to cool them by supporting them on very thin aluminum foil, about half a micron thick. The electrons would get through, but the foil would somewhat disperse them and spoil the contrast. Still, it was an improvement: one could see the internal details of the image better. In his first experiments, Marton had been able to discern only the cell walls before the specimen crumbled away. After he mounted it on the aluminum foil, he could actually pick out cell nuclei—but with poorer contrast. He next tried gold foil, which could be beaten down to a very small thickness, but that was no better, since the foil proved to be too uneven.

Evaporation techniques—the solution that comes so readily to mind today—were still in their infancy in the 1930s. Then Marton remembered a technique that had been developed for "windows" that would transmit infrared rays: the casting of collodion foils, of the sort used in photographic emulsion, onto a water surface. That method yielded reasonably good films within the desired thickness range; moreover, they turned out to stand up very well in the electron beam. That was an important advance, since it proved that heavy-metal impregnation of specimens was not absolutely necessary, provided the specimens were thin enough.

As to which investigator had first established the importance of scattering in electron microscopy, that was one discovery that Marton, with uncharacteristic insistence on his priority, claimed for himself.

> *Simultaneously with these results I slowly attacked the problem of interpretation of the image [wrote Marton]. I felt that there was little advantage in higher resolution if we did not know what the image represented. Available in the literature were two statements. In 1933 von Borries and Ruska had written that "metallic foils … as a result of their small density and thickness variations (varying magnitude of absorption) allow a 'structure' to be discerned." Eight months later Ruska wrote: "The photographs that follow are meant to demonstrate the considerably improved optical performance capability of the new apparatus, without our entering into particulars as to the question of image interpretation." In this same paper, however, he says earlier: "In the limiting case one might speak of absorption images and of diffusion images, yet for the most part the effect of the diffusion phenomenon predominates." If he had written in French, where* diffusion *means both diffusion and scattering, this would be a relatively modern statement and it has been interpreted as such by some writers. In German, however, the word for scattering is* Streuung. *As he used the German word* Diffusion *I am inclined to believe that the first clear statement ascribing the contrast in the image to differences in the scattering properties of the object was contained in my paper of 8 May 1934, where I calculated and tabulated mean scattering angles as a function of thickness and of electron velocity for two substances, beryllium and aluminum. The calculation assumed multiple scattering and showed the need for very thin objects or for electrons of higher energy. It was based on Bothe's formulation of multiple Rutherford scattering for small angles (Marton, 1968b).*

However, Marton's interpretation is challenged by Ruska himself, who (in a private communication) points out that the sentence about the diffusion phenomenon cited by Marton is preceded by another in which scattering angles (*Streuwinkel*) are mentioned in defining *Diffusion:*

> *On the other hand, in objects that are very thin in relation to beam potential the reason the image intensity varies considerably despite the transmitted beam-current intensity being kept constant is that individual points of the object give rise to rays of nonuniform [optical] aperture (intensity distribution to scattering angles) corresponding to their mass density, so that if only the lens apertures are sufficiently small, currents of differing intensities appear to originate from various points of the object and show up in the image.*

But Ruska does acknowledge that immediately following his brief reference to these questions, Marton pursued them more fully.

The discovery that biological specimens could be safely examined under a microscope that was potentially much more powerful than any instrument then available did not exactly revolutionize biology overnight. In late 1934 Marton got a chance to report on the results achieved up to that time, and on future prospects, to an audience of physicians and life scientists, the Royal Society of Medical and Biological Scientists of Brussels. He thought he had made a good case for the new tool, but when he finished, the eminent bacteriologist Jules Bordet, Nobel prizewinner for his discovery of immunity factors in blood serum, got up and said, "Oh no—not an electron microscope! We have enough trouble trying to interpret the images we get with a light microscope!"

Marton's enthusiasm could scarcely have been expected to carry the day at a time when all he could point to was an electron microscope that came nowhere near the light microscope in magnification, although even some of the earliest instruments could compete in resolution. But the besetting problem of early electron microscopy was the ultimate destruction of all but the thinnest specimens by electron bombardment. To deal with the problem Marton hit upon a new solution: the "electron shutter." Instead of using the specimen as an object during the lengthy and laborious adjustments needed for a sharp image, he put a special highly resistant phantom target in its place. When all the adjustments had been made, he moved the beam aside by switching on a deflecting voltage, mechanically swung the phantom target out of the way and the specimen into the exact spot were the phantom had been, and then briefly switched off the beam-deflecting voltage, so that the beam was sent through the specimen for a fraction of a second, long enough to expose the recording film but not so long as to damage the specimen. The length of the exposure was controlled by an electronically operated timing circuit.

Another, equally vexing problem was the interminable wait each time a specimen was changed. The vacuum had to be let down to atmospheric pressure and then laboriously restored after the specimen had been replaced. Not only that: it became evident before long that the recording film would have to be somehow introduced into the vacuum as well, to be exposed by the electron beam itself rather than by the light coming off the screen. Not much improvement was possible as long as one was confined to external photography of an image on a fluorescent screen. Marton tried to improve the quality of the phosphors that made up the screen, to get a better match between the color sensitivity of the photographic material and the light coming off the phosphors, and to use more efficient photographic optics. All he could gain by these combined methods was something like a 50-fold improvement, which was certainly impressive but still not sufficient. Internal recording of electron micrographs was the answer. To avoid excessive "down time" during which the instrument was inoperable, air locks were introduced for the first time for both specimens and photographic materials, so that only the portions of the vacuum envelope containing them had to be evacuated each time, which could be done more quickly. These experiments were performed long before reliable rubber gaskets had been introduced in vacuum technology. The air locks of Marton's new instrument had one gasket, in the specimen air lock, but the rest used greased-paper joints. Moreover, the specimen air lock was constructed so as to allow motion of the specimen in two directions at right angles—a crude mechanical "stage." Next, the air lock was modified to provide for insertion of six specimens at a time, interchangeable without opening of the air lock, and there were external scales to monitor the motion of any stage position within the limited accuracy of the device. The photographic air lock could accommodate a single plate at a time, but the plate could be moved within the exposure chamber to yield several partial exposures.

Marton also sought to simplify lens settings by compiling a nomogram. That idea did not work out too well. The grease and other insulating surfaces in and around the beam path tended to charge up from stray electrons and thus cause beam deflections and image instabilities. Only by leaving the beam on long enough could he establish equilibrium conditions for a sufficiently long time to record a reasonably stable image. If the beam was switched on and off, it took several minutes each time to stabilize the image within acceptable limits.

Marton started the construction of a third microscope in the autumn of 1934 and could soon report some improvements, although each bit of progress cost a tremendous amount of effort. University funds were practically nonexistent. His family helped a little. With their slight resources, the Martons became frequent visitors at the flea market. Some German equipment that had been abandoned after World War I was still to be found and several components of the new microscope benefited from this availability at very low prices. The only other source of funds was a modest grant from the Institut International de Physique Solvay. Marton's chief, Professor Henriot, was the permanent scientific secretary of the Solvay physics congresses, and thanks to his intervention Marton had a Solvay fellowship for several years.

Ernest Solvay (1838–1922) had made a fortune after he patented a cheap way of making soda ash, which was used in huge amounts in the manufacture of soap and glass. He endowed institutes of physics, of chemistry, and of sociology that held periodic international conferences attended by the leaders in each of these fields. These institutes reflected Solvay's own interests, which did not include the life sciences. In fact, throughout the early years of electron microscopy, the biomedical professions played a negligible part in its development. Professor Bordet's reaction was typical of this wait-and-see attitude. One consequence was that the physicists had to do everything for themselves, learning the techniques of biological research as they went along, and suffering countless delays and false starts. Marton acknowledged that some of the failures in his first attempt with bacteria were due to insufficient washing of the specimens and to crystals of the inorganic salts that he had used as fixatives masking the essential features. "My lack of biological training was only too apparent," he said.

In the 1930s biophysics, which combines the skills of both sciences in one practitioner, had not yet established itself as a separate academic discipline. Apart from his note in *Nature*, and the abortive talk to the biomedical audience that had included Jules Bordet, Marton's presentations were mainly to his fellow physicists. Even his debut outside Belgium was before physicists, at a meeting of the French Physical Society, which also led to his first monograph. As a consequence of this talk, he was asked to prepare a review of the subject. Professor Henriot added a review of the theory of electron optics and the two contributions appeared together in a booklet published by Editions de la Revue d'Optique in 1935.

The talk before the French Physical Society led to an amusing incident after the meeting. The Martons were staying with friends in a Paris suburb—a Solvay fellow's stipend did not run to hotel accommodation—and Frédéric Joliot and Pierre Auger lived in the same area. The four young

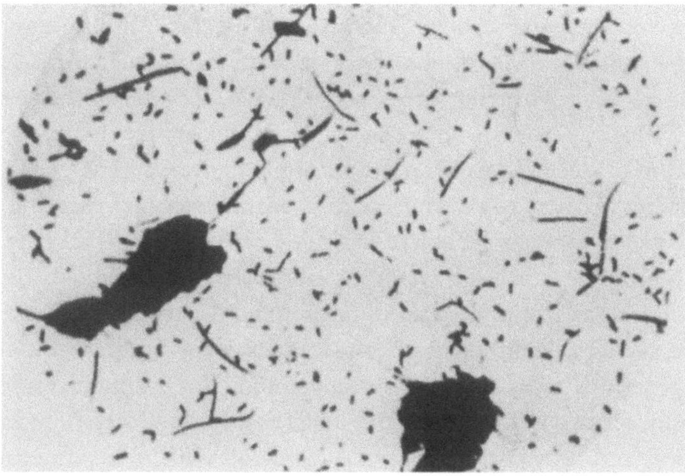

Figure 4 First electron micrograph of bacteriological specimen (*Chromobacterium prodigiosum* fixed in chromic acid), 1937.

people started walking toward the subway but on the way got sidetracked to a bistro, where they began to play some of the "fruit machines," as the one-armed bandits of Nevada were called in Europe. They ended by losing every sou they had on them and had to walk all the way back to the suburb, arriving well after midnight.

Consorting with the likes of Auger, the discoverer of the effect named after him, and Joliot, who had discovered artificial radioactivity with his wife Irène (the daughter of Pierre and Marie Curie) meant that Marton was well launched in the world of science, especially after the Joliots duplicated the feat of Irène's parents of sharing a Nobel Prize later that year. But biomedical applications of the electron microscope would get going only if rugged and foolproof instruments could be fabricated, and no manufacturer would undertake fabrication unless there was a prospective market of some size—the usual impasse. Marton got one Belgian firm interested, the Société Belge d'Optique in Ghent, whose founder, André Callier, had been trained as a scientist and was fascinated by the new tool's possibilities. His untimely death put an end to that company's efforts.

Marton continued to improve his instruments and to obtain results of increasing interest with them, among them the first electron micrograph of a bacteriological specimen (Fig. 4). But he also went on to make contributions to the understanding of contrast phenomena in the image. In 1936 he wrote a paper for *Physica* on "Some considerations concerning resolving power in electron microscopy" that was still based on multiple scattering, but went beyond the mere tabulation of scattering angles for a couple of substances, as his pioneering paper of 1934 had done. Now he spelled out the implications. Given an object of varying thickness or density, an objective aperture of a limited solid angle, a scattering distribution that depended on the object's constitution, and contrast production by scattering, he calculated the scattering angles as a function of beam energy, thickness, density (atomic number and atomic weight), and aperture solid angle. By assuming a photometric minimum in the photographic record, he also defined a "depth resolving power" analogous to optical resolving power. Moreover, he recommended the "staining" of specimens with substances of higher atomic number in cases in which the contrast afforded by the specimen was insufficient, and showed that contrast decreased with increasing numerical aperture, so that there was an upper bound on attempts to increase resolving power by reduction of spherical aberration.

Marton's contributions played a part in the employment of Ruska's electron microscope for biomedical research in Germany, especially after resolving powers in excess of those of the light microscope were achieved (Driest & Müller, 1935). Among the contributions of others during this period, Marton thought the most fundamental were three German papers, by Boersch (1936), Scherzer (1936), and Henneberg and Recknagel (1935).

Marton remained at the Free University of Brussels until the summer of 1938, the year before World War II broke out. Conscious of the gathering storm, he emigrated to America in the autumn of 1938, to continue his work in electron microscopy in a different setting altogether.

Meanwhile, contributions began to be made in Britain, France, and the United States. The British work was done at Imperial College and Metropolitan Vickers and resulted in an instrument whose magnetic lenses were entirely inside the vacuum envelope and which had a rotatable disk that supported the specimen, so that it could be positioned alternately before a light and before an electron microscope (Martin et al., 1937). The device was not a success and when Metrovick took up the manufacture of electron microscopes in earnest after the war, its instruments omitted these two special features.

In France the theoretician M. Cotte, who had published a paper on electron-lens aberrations in 1938 (Cotte, 1938), made a serious attempt, together with a collaborator named Cardan, to make a start in electron optics, an effort that was interrupted by the start of World War II, in which Cardan was killed. In the United States, J. H. McMillen and C. H. Scott built an electron microscope in which biological specimens were deliberately destroyed in the hope that something like a "skeleton" would be left to be examined (McMillen & Scott, 1937).

In Germany, Ruska and von Borries were also finding it difficult to persuade physicists (let alone physicians and other scientists) that the electron microscope could yield useful new knowledge. Not until the great firm of Siemens in Berlin became interested in the project did any progress occur. The disruptions caused by World War II and its aftermath slowed down these efforts considerably, but the company ultimately resumed production of electron microscopes and achieved a position of considerable importance in the world market.

Marton's first job in America was with RCA Manufacturing Co. in Camden, New Jersey, where he stayed from 1938 to 1941. (During 1938–1939, he was also a lecturer at the University of Pennsylvania.) RCA had been long active in electron optics because of its interest in electron multipliers and in television cameras and receivers; the monograph written at the RCA Laboratories by Zworykin, Morton, Ramberg, Hillier, and Vance, "Electron Optics and the Electron Microscope" (affectionately known to several generations of students as "We, the People" because of its multiple authorship) became the leading textbook in English on the subject (Zworykin et al., 1945). Marton was able to bring his last (1935) Belgian instrument—his No. 3, as he called it—with him to America, and continued to work with it at RCA while a properly engineered instrument, the RCA Type A, was being developed there—"his" No. 4 (Fig. 5). Marton published a description of it in the *Physical Review* in 1940 and remarked afterward how several of its features later reappeared elsewhere as "new" inventions, including a radically different objective lens that he analyzed in detail in the same journal in 1944 with R. G. E. Hutter. But the RCA Type A instrument was a team effort, and Marton singled out one engineering contribution, the use of a highly regulated power supply designed by Arthur Vance, as particularly significant (Vance, 1941). Another RCA researcher, James Hillier, had done some very original work on an electron microscope with A. Prebus at the University of Toronto before he had come to RCA in 1940 (Prebus & Hillier, 1939). At RCA, Hillier took a lead in the design of the next version, the Type B microscope, which became the first instrument marketed by the company and became a great commercial success; its excellent alignment system and stable power supply (again designed by Vance) made it very useful, since it could be successfully operated by researchers unfamiliar with electronics (Fig. 6).

Figure 5 Marton at RCA Type A electron microscope (1940), with his Brussels No. 3 instrument (1935) in background.

Marton's name is absent from publications originating at RCA after 1940, for in 1941 he had found another venue: the Department of Physics at Stanford University. He stayed there for 5 years, under the terms of a Rockefeller grant that assured him a laboratory somewhat grandiosely named Division of Electron Optics and the courtesy title of associate professor—without tenure. At the University of Pennsylvania, he had worked in part with Leonard Schiff, an expert on quantum mechanics who later wrote the standard text on the subject (Schiff, 1949) and with whom he collaborated on a first analysis of the scattering phenomena in the electron microscope. At Stanford, Marton was more or less on his own; although Schiff later came there and ultimately became executive head of the physics department, that happened after Marton had left. Before

Figure 6 Marton (right) demonstrating RCA Type B microscope.

the war the Stanford physics department, not yet the star-studded group it would later become, was host to all sorts of experimental endeavors. The brothers Russell and Sigurd Varian had come there in 1937 and had been given a basement room in which to develop Russell Varian's invention, the klystron—the microwave tube that was to become the basis for the present-day multinational company Varian Associates (and that was to yield Stanford millions of dollars in royalties). Now Bill and Claire Marton had arrived to advance electron microscopy. The work proceeded slowly, especially since Bill was also expected to supervise graduate and postdoctoral students, mostly foreign-born ones since native young physicists were in short supply during the war years. The Dutchman Kees Bol, the Indian N. N. Das Gupta, and the aforementioned Rudolf Hutter (who had left a good job at Telefunken and emigrated from Germany after he married his Jewish girlfriend) all worked with Marton during his Stanford years. Marton pioneered the design of electron microscopes with intermediate lenses—a three-stage magnification and a double condenser lens—a scheme that later became very popular. In his introduction to the 1944 paper in *Science* describing this instrument (Fig. 7), which ranged from 30 to 100 kV in

Figure 7 Marton's Stanford electron microscope (1944).

voltage, Marton made some characteristic remarks about industrial versus academic research.

In the first few years of electron microscopy, development and research was entirely carried out by university laboratories. Soon, however, the commercial value of this new instrument was discovered, and the industrial development started, first abroad and then in this country too. Very valuable work was done by those industries and for a while it looked as if very little was left over for academic research. Also it seemed as if that little could not be efficiently tackled, if the means, put at the disposal of industrial and academic organizations, were compared.

However, it soon became apparent in the field of electron microscopy, as in other branches of science which were industrially developed, that continuation of academic research is not only valuable, but also necessary. One of the first institutions, or perhaps the first, to recognize this fact and to start a research program of its own in this field was Stanford University. As part of this research program an electron microscope was designed and constructed which shows in several respects departures from known instruments...

However, Stanford University was not destined to loom large in the history of electron microscopy. With the end of World War II the physics department's interest in this field flagged. In 1946 Marton moved to Washington as principal physicist at the National Bureau of Standards (in the Department of Commerce). Two years later, he was appointed the first chief of its newly founded Electron Physics Section, a post he held for 16 years, in the course of which NBS became a prime center of research on the design and use of electron microscopes and of their descendants. Despite his administrative duties, he remained active in research. His development of the electron interferometer and of the electron-optical technique of mapping electrostatic and magnetic fields of extremely small dimensions earned him the Commerce Department's Gold Medal for Exceptional Service in 1955. He organized symposia and conferences; he helped found the Electron Microscopy Society of America (which today has some 3000 members, divided approximately equally between the physical and the life sciences); a related organization, the Microbeam Analysis Society founded in 1964, elected him Honorary Member; and he achieved the highest grade of membership, Fellow, in the American Physical Society, the American Association for the Advancement of Science, the Institute of Electrical and Electronics Engineers, and the Physical Society (London).

Marton remained at NBS for 24 years, the last 8 of them as head of its Office of International Relations. He retired in 1970, just before his sixty-ninth birthday—and then proceeded to have yet another career! He became an unsalaried research associate at the Smithsonian Institution, where he was able to indulge his interests in the history and philosophy of science and technology and make several professional contributions to what had until then been an avocation. And throughout his U.S. government career, from 1948 until his death, he also served as editor for Academic Press of the first 55 volumes of *Advances in Electronics* (renamed *Advances in Electronics and Electron Physics* in 1955) and as editor-in-chief of *Methods of Experimental Physics*, jointly with Claire Marton after 1958. He died in Washington on January 22, 1979, of a heart attack, at the age of 77. Claire followed him 2 years later. His honors were many: the medals of the Union Internationale Contre le Cancer (1938) and of the University of Brussels (1947), election to the Royal Academy of Belgium (1955), a Fulbright Award to spend a term at the University of Paris (1962), and visiting professorships at several other institutions. But he will be best remembered for the two contributions he made before he was 35—the first use of the electron microscope for biological specimens, and his recognition that it was scattering, not absorption, that produced contrast in the instrument.

1. Bibliographical appendix: publications of L. L. Marton

Ultra-rote Absorptionsspektra. *Z. phys. Chemie* **117**, 97 (1925).

Photocellak elektrolytikus eloallitasa es azok gyakorlati alkalmazasa. *Chem. Rundschau* **4**, 156 (1927). With E. Rostas.

Über die elektrolytische Herstellung von Photozellen und deren Verwendung. *Z. tech. Phys.* **10**, 52 (1929). With E. Rostas.

L'effet photoélectrique. *Bull. Tech. Assoc. Ingénieurs Ecole Polytech. de Bruxelles* **27**, Nos. 4–5 (1931).

Mercury traps. *Nature* **130**, 739 (1932).

Meetkundige optic der electronen. *Wis- en natuurkundig tijdschrift* **6**, 159 (1933). With M. Nuyens.

Note sur la photométrie des phares d'auto. *Bull. Soc. Belge des Electriciens*, April–May (1933). With G. A. Homès and G. Marchal.

Note sur la photométrie des phares d'auto. *L'électricité pour tous*, October (1933). With G. A. Homès and G. Marchal.

Electron microscopy of biological objects. *Nature* **133**, 911 (1934).

La microscopie électronique des objets biologiques: I. *Bull. Acad. Roy. Belgique* **20**, 439 (1934).

Electron microscopy of biological objects. *Phys. Rev.* **46**, 527 (1934).

Le microscope électronique: Premiers essais d'application à la biologie. *Ann. et Bull. Soc. Roy. Sci. Med. Nat. de Bruxelles*, 92 (1934).

La microscopie électronique des objets biologiques: II. *Bull. Acad. Roy. Belgique* **21**, 553 (1935).

La microscopie électronique des objets biologiques: III. *Bull. Acad. Roy. Belgique* **21**, 606 (1935).

Le microscope électronique et ses applications, en particulier en biologie. *2me Congrès National des Sciences*, 928 (1935).

Quelques observations sur les propriétés des cellules photoélectriques à couche d'arrêt au sélénium. *2me Congrès National des Sciences*, 408 (1935). With G. Marchal.

Le microscope électronique et ses applications. *Rev. d'optique* **14**, 129 (1935).

"Le microscope électronique et ses applications." Editions Rev. d'Optique, Paris, 1935. Followed by E. Henriot, "Optique électronique des systèmes centrés."

Quelques observations sur les propriétés des cellules photoélectriques à couche d'arrêt au sélénium. *Rev. d'optique* **15**, 1 (1936). With G. Marchal.

Microscope électronique. *Rev. microbiologie appliquée*, 117 (1936).

Quelques considerations concernant le pouvoir séparateur en microscopie électronique. *Physica* **9**, 959 (1936).

La microscopie électronique des objets biologiques: IV. *Bull. Acad. Roy. Belgique* **22**, 1336 (1936).

La microscopie électronique des objets biologiques: V. *Bull. Acad. Roy. Belgique* **23**, 672 (1937).

"Les photomètres photoélectriques portatifs." Bruxelles, 1937. With G. Marchal.

Bemerkungen zu der Arbeit: 'Einige praktische Anwendungen der Elektronenoptik, von Dr. Fritz Luft.' *Röntgenpraxis* **10**, 352 (1938).

Le microscope électronique en biologie. *Compte rendus de la semaine internationale contre le cancer, Paris*, 221 (1939).

On the sensitivity of photographic emulsions for electrons between 50 and 100 keV. *Phys. Rev.* **56**, 290 (1939).

A new electron microscope. *Phys. Rev.* **58**, 57 (1940).

Electron microscope. *The Frontier*, September (1940).

A new electron microscope. *RCA Rev.* **5**, 323 (1940). With M. C. Banca and J. F. Bender.

The electron microscope. *J. Bacteriol.* **41**, 397 (1941).

An electron microscope study of curd fibers of sodium laurate. *J. Am. Chem. Soc.* **63**, 1990 (1941). With J. W. McBain and R. D. Void.

Determination of object thickness in electron microscopy. *J. Appl. Phys.* **12**, 759 (1941). With L. I. Schiff.

The electron microscope and its development. *Chem. Products*, July–August (1942).

Applications of the electron microscope in colloid chemistry. *J. Phys. Chem.* **46**, 1023 (1942).

Alice in electronland. *Am. Scientist* **31**, 247 (1943).

The electron microscope in biology. *Ann. Rev. Biochem.* **12**, 587 (1943).

The transmission type of electron microscope and its optics. *Proc. IRE* **32**, 3 (1944). With R. G. E. Hutter.

On apertures of transmission type electron microscopes using magnetic lenses. *Phys. Rev.* **65**, 161 (1944).

A 100 kV electron microscope. *Science* **100**, 318 (1944).
Optical constants of a magnetic type electron microscope. *Proc. IRE* **32**, 542 (1944). With R. G. E. Hutter.
Stereoscopy with the electron microscope. *J. Appl. Phys.* **15**, 726 (1944).
A 100 kV electron microscope. *J. Appl. Phys.* **16**, 131 (1945).
The electron microscope and its applications. *Frontiers in Chemistry* **4**, 17 (1945).
Adaptation of special specimen holders to commercial electron microscopes. *J. Appl. Phys.* **16**, 387 (1945).
Modifications of specimens in electron microscopy. *Science* **104**, 35 (1946). With N. N. Das Gupta and C. Marton.
Electron microscopy. *Reports of Progress in Physics* **10**, 204 (1944–45).
Electron microscopy of purified Lansing virus. *Proc. Soc. Exp. Biol. Med.* **62**, 291 (1946). With H. S. Loring and C. E. Schwerdt.
Spherical aberration of compound magnetic lenses. *J. Appl. Phys.* **18**, 522 (1947). With K. Bol.
Tracer micrography. *Science* **107**, 70 (1947). With P. H. Abelson.
Ferromagnetic domain observation. *Phys. Rev.* **73**, 1475 (1948).
Electron optical "schlieren" effect. *J. Appl. Phys.* **19**, 687 (1948).
Electron optical observation of magnetic fields. *J. Appl. Phys.* **19**, 863 (1948).
Electron microscopic observations on *Pseudomona aeruginosa* bacteriophage. *Proc. Soc. Exp. Biol. Med.* **68**, 451 (1948).
Aperçus sur les récénts progrès scientifiques aux Etats-Unis. *Chimie et industrie* **63**, 49 (1948).
Electron optical observation of magnetic fields. *J. Res. NBS* **43**, 409 (1949). With S. H. Lachenbruch.
Electron optical mapping of electromagnetic fields. *J. Appl. Phys.* **20**, 1171 (1949). With S. H. Lachenbruch.
Fringe fields of ferromagnetic domains. *J. Appl. Phys.* **20**, 1258 (1949). With S. H. Lachenbruch, J. A. Simpson, and A. Van Bronkhorst.
Electrons vs photons: A comparison of microscopes. *J. Opt. Soc. Am.* **40**, 269 (1950).
Stroboscopic mapping of time-variable fields. *J. Appl. Phys.* **21**, 617 (1950). With D. L. Reverdin.
Exploration of electrostatic and magnetic fields. *Scientific Monthly* **71** (No. 1), 3 (July 1950).
Modification of electron microscope for electron optical shadow method. *Rev. Sci. Instr.* **21**, 669 (1950). With J. A. Simpson and A. Van Bronkhorst.

Bibliography of Electron Microscopy. *NBS Circ.* **502**, 1 (1950). Compiled by C. Marton, S. Sass, M. Swerdlow, A. van Bronkhorst, and H. Meryman.

Electron optical properties of space-charge clouds. *J. Appl. Phys.* **21**, 842 (1950).

Electron optical disk. *Rev. Sci. Instr.* **21**, 1027 (1950).

Electron-optical bench. *J. Res. NBS* **47**, 461 (1951). With M. M. Morgan, D. C. Schubert, J. R. Shah, and J. A. Simpson.

Electron interferometer. *Phys. Rev.* **85**, 1057 (1952).

Electron beam interferometer. *Phys. Rev.* **90**, 490 (1953).

Electron interferometry. *Science* **118**, 470 (1953).

Electron-optical shadow method of magnetic-field mapping. *J. Res. NBS* **52**, 97 (1954). With J. A. Simpson and S. H. Lachenbruch.

X-ray shadow microscope. *NBS Circ.* **527**, 265 (1954).

Energy loss of electrons in passage through thin films. *Phys. Rev.* **94**, 203 (1954). With L. B. Leder.

Effect of chemical combination on the characteristic energy loss of electrons. *Phys. Rev.* **95**, 1345 (1954). With L. B. Leder.

An electron interferometer. *Rev. Sci. Instr.* **25**, 1099 (1954). With J. A. Simpson and J. A. Suddeth.

Distribution of electrons scattered by gold. *Phys. Rev.* **99**, 495 (1955). With J. A. Simpson and T. F. McGraw.

Automatic instrument for electron scattering measurements. *Rev. Sci. Instr.* **26**, 855 (1955). With J. A. Simpson and T. F. McGraw.

Characteristic energy losses of electrons in solids. *Adv. Electronics Electron Physics* **7**, 183 (1955). With L. B. Leder and H. Mendlowitz.

Electron physics tables. *NBS Cir.* **571**, 1 (1956). With C. Marton and W. G. Hall.

Comparison of the characteristic energy losses of electrons with the fine structure of the x-ray absorption spectra. *Phys. Rev.* **101**, 1460 (1956).

Les pertes d'énergie des électrons et leur rôle en microscopie électronique. *Colloque CNRS Toulouse*, 175 (1955). With C. Marton, L. B. Leder, J. A. Simpson, and H. Mendlowitz.

Electron interference phenomena. *London, 1954* p. 272.

Electron optical image of an atomic beam. *J. Appl. Phys.* **27**, 419 (1956). With D. C. Schubert and S. R. Mielczarek.

Experiments on low-energy electron scattering and energy losses. *Rev. Mod. Phys.* **28**, 172 (1956).

Conference on quantum interactions of the free electron. *Rev. Mod. Phys.* **28**, 171 (1956).

Distribution of electrons scattered by solids. *Phys. Rev.* **104**, 64 (1956). With J. A. Simpson and T. F. McGraw.

Electron scattering phenomena. *J. Sci. Ind. Res.* **16A**, 221 (1957).

Measuring with the electron. *J. Sci. Ind. Res.* **16A** (1957).

A field emitter point projection x-ray microscope. *X-ray Microscopy Microradiography*, 287 (1957). With R. A. Schrack and R. P. Placious.

Limiting resolution of electron spectrometers. *Rev. Sci. Instr.* **29**, 438 (1958).

Sur les limites du pouvoir de résolution des spectromètres corpusculaires. *Bull. Acad. Roy. Belg.*, 159 (1958).

Men and electrons. *J. Wash. Acad. Sci.* **49** (No. 4), 97 (1959).

Electron absorption spectrometer using an improved velocity analyzer. *Rev. Sci. Instr.* **29**, 567 (1958).

Effect of degree in orientation and crystal size on the scattering of 20-keV electrons by aluminum. *Phys. Rev.* **110**, 1057 (1958). With J. A. Simpson, J. A. Wagner, and H. Watanabe.

High-vacuum temperature-gradient furnace. *Rev. Sci. Instr.* **29**, 440 (1958).

Characteristic energy losses of electrons. *Berlin, 1958* Vol. 1, p. 281 (1960). With L. Leder, C. Marton, H. Mendlowitz, H. Simpson, J. A. Sudderth, and J. A. Wagner.

Note historique sur les premières années de la microscopie électronique. *Bull. Acad. Roy. Belg.* **5**, 119 (1959).

Cryostat for electron bombardment and electron diffraction work. *Rev. Sci. Instr.* **29**, 859 (1958). With E. M. Horl.

Temperature dependence of the characteristic energy loss of electrons in aluminum. *Phys. Rev.* **112**, 341 (1958). With L. B. Leder.

Electron beam magnetometer. *J. Res. NBS* **63C**, 69 (1959). With L. B. Leder, J. W. Coleman, and D. C. Schubert.

Apparatus for electron optical study of low density gas flow. *Trans. 6th National Symp. Vacuum Technology*, 206 (1960).

Electron diffraction studies on solid α-nitrogen. *Acta Crystallographica* **14**, 11 (1961). With E. M. Horl.

Prof. E. Henriot. *Nature* **190**, 861 (1961).

Electron interferometer studies of iron whiskers. *J. Appl. Phys.* **32**, 1153 (1961). With H. A. Fowler, J. A. Simpson, and J. A. Suddeth.

Improved electron filter lens. *Rev. Sci. Instr.* **32**, 802 (1961).

A method for the study of vector velocity distribution of low density molecular beams. *Rarefied Gas Dynamics*, 61 (1961). With S. R. Mielczarek and D. C. Schubert.

Image formation mechanism; and History of electron optics. *In* "Encyclopedia of Microscopy" (1961).

Electron microscope; and Electron microscopy. *In* "Physics Encyclopedia," Pergamon, 1961.

Time-resolved electron optical image of a pulsed atomic beam in flight. *J. Appl. Phys.* **33**, 1613 (1962). With S. R. Mielcsarek and D. C. Schubert.

Plural scattering of 20-keV electrons in aluminum. *Phys. Rev.* **126**, 182 (1962). With H. A. Fowler, J. A. Simpson, and N. Swanson.

Ultra high vacuum ultraviolet monochromater. *Trans. 8th Vacuum Symp.*, 633 (1962). With J. A. Simpson, J. A. Suddeth, and L. B. Leder.

Electron energy losses in solids and their influence on the electron diffraction diagram. *J. Phys. Soc. Japan (Suppl. B/2)*, 68 (1962).

Optical constants in the vacuum ultraviolet and electron energy losses. *J. Quant. Spectrosc. Radiation Transfer* **2**, 1 (1962).

Electron optical studies of low-pressure gases. *NBS Monograph* **66**, 37 (1963). With D. C. Schubert and S. R. Mielczarek.

Les pertes caractéristiques d'énergie des électrons dans les solides. *Actions chimiques et biologiques des radiations* **7**, 258 (1964). With C. Marton.

APS Division of Electron Physics: The first 20 years. *Physics Today*, 44 (October 1964).

Historical background of image formation. *Laboratory Investigation* **14**, 739 (1965).

Experiments in electron scattering: II. *Laboratory Investigation* **14**, 754 (1965).

Pokroky v elektronové fyzice v posledních dvaceti letech. *Pokroky matematiky, fyziky a astronomie* **11**, 14 (1965).

X-ray fiber optics. *Appl. Phys. Letters* **9**, 194 (1966).

A view of Sir Isaac Newton's philosophy. *Physics Teacher* **6**, 1966.

Width of transmission Kikuchi lines in silicon and diamond. *J. Appl. Phys.* **36**, 1986 (1965). With H. A. Fowler.

Correction of spherical aberration of axially symmetrical magnetic lenses. *Rev. Sci. Instr.* **38**, 130 (1967).

Progress in electron physics during the last 20 years. *Vacuum Physics*, 17 (1967).

Optical properties of germanium in the far ultraviolet. *Phys. Rev.* **160**, 602 (1967). With J. Toots.

Reflectance and $1/\epsilon$ resonance of beryllium in the far ultraviolet. *Phys. Rev.* **172**, 670 (1968). With H. A. Fowler and J. Toots.

"Early History of the Electron Microscope." San Francisco Press, San Francisco, 1968.

Photoelectricity. *In* "The Encyclopedia of Physics" (R. M. Besançon, ed.). Van Nostrand Reinhold, New York, 1974.

Photoelectric effect. *In* "Encyclopaedia Britannica," Vol. 14, 296 (1974).
Also, over 20 U.S. and foreign patents, and editorship of "Advances in Electronics and Electron Physics" and of "Methods of Experimental Physics," Academic Press, New York.

2. Afterword by Peter Hawkes

Charles Süsskind, the author of this life of Marton, was born on 19 August 1921 in Prague and escaped to England just before Germany invaded the country. He served as a radar specialist in the US Air Force (in England). In 1945, he married Teresa Gabriel (a "Wren") and they moved to Pasadena where Charles studied for his B.S at the California Institute of Technology. He moved to Yale, where he obtained his PhD and returned to Stanford where, in 1955, he accepted a post at the University of California in Berkeley; the history of technology was one of his interests. In later years he spent a year at Imperial College on a Fulbright Scholarship and a year in the University of Geneva on a NSF award. He also founded the San Francisco Press, which published Marton's short autobiography.

References

Boersch, H. (1936). *Annalen der Physik (Leipzig) [5], 26*, 631;
Annalen der Physik (Leipzig) [5], 27 (1936), 75.
Cotte, M. (1938). *Annales de Physique (Paris) [11], 10*, 333.
Driest, E., & Müller, H. O. (1935). *Zeitschrift für Wissenschaftliche Mikroskopie, 42*, 53.
Gabor, D. (1957). *ETZ. Elektrotechnische Zeitschrift. Ausgabe A, 78*, 522.
Gabor, D. (1968). *Preface to L. Marton "Early history of the electron microscope"*. San Francisco, California: San Francisco Press (p. vi).
Glaser, W. (1933a). *Zeitschrift für Physik, 80*, 451.
Glaser, W. (1933b). *Zeitschrift für Physik, 81*, 647;
Zeitschrift für Physik, 83 (1933), 104.
Glaser, W. (1937). In H. Busch, & E. Brüche (Eds.), *Beiträge zur Elektronenoptik* (p. 24). Leipzig: Barth.
Henneberg, W., & Recknagel, A. (1935). *Zeitschrift für Physik, 16*, 621.
Johannson, H., & Scherzer, O. (1933). *Zeitschrift für Physik, 80*, 183.
Kurylo, F., & Susskind, C. (1981). *Ferdinand Braun*. Cambridge, Massachusetts: MIT Press.
Martin, L. C., Whelpton, R. V., & Parnum, D. H. (1937). *Journal of Scientific Instruments, 14*, 14.
Marton, L. (1968a). *Early history of the electron microscope*. San Francisco, California: San Francisco Press.
Marton, L. (1968b). *Early history of the electron microscope*. San Francisco, California: San Francisco Press (p. 18). Translations of German citations revised by C. Süsskind;
They are from Ruska, E. *Zeitschrift für Physik, 87* (1934), 580.
McMillen, J. H., & Scott, C. H. (1937). *Review of Scientific Instruments, 8*, 288.
See, for instance Mulvey, T. (1962). *British Journal of Applied Physics, 13*, 197;
Proceedings of the Royal Microscopical Society, 2(Pt. 1) (1967), 201.

Prebus, A., & Hillier, J. (1939). *Canadian Journal of Research. Section A, 17*, 49 and 64.
Ruska, E. (1979). Die frühe Entwicklung der Elektronenlinsen und der Elektronenmikroskopie. *Acta Historica Leopoldina, 13*.
Scherzer, O. (1936). *Zeitschrift für Physik, 101*, 593.
Scherzer, O. (1937). In H. Busch, & E. Brüche (Eds.), *Beiträge zur Elektronenoptik* (p. 33). Leipzig: Barth.
Schiff, L. (1949). *Quantum mechanics*. New York: McGraw-Hill.
Vance, A. W. (1941). *RCA Review, 5*, 293.
Zworykin, V. K., Morton, G. A., Ramberg, E. G., Hillier, J., & Vance, A. W. (1945). *Electron optics and the electron microscope*. New York: Wiley.

CHAPTER SEVEN

The invention of the electron Fresnel interference biprism☆,☆☆

G. Möllenstedt[a] and Peter Hawkes (Afterword)[b,*]

[a]Formerly Institut für Angewandte Physik, Universität Tübingen, Tübingen, Germany
[b]CEMES-CNRS, Toulouse, France
*Corresponding author. e-mail address: hawkes@cemes.fr

Contents

1. Childhood years in Bielefeld 201
2. Gaining technical know-how as a student apprentice 202
3. A broad education in physics with Professor Walter Kossel 202
4. Influence of the work of Kikuchi and convergent beam diffraction 203
5. Electron-optical experiments with Brüche, Scherzer, and Mahl 205
6. Quantitative testing of the operation of the biprism 211
7. Measurement of the inner potential of solids 218
8. Electron interference microscope in the transmission mode 219
9. The intensity problem in electron interferometers 219
10. Atomic resolution electron holography 220
11. Editorial note 222
12. Afterword by Peter Hawkes 222
References 223

1. Childhood years in Bielefeld

In order to describe how I happened on the idea of realizing interference with electron waves by means of an electron optical biprism, I must, by way of introduction, say a few words about my early physics education. I should begin with the remark that I had the good fortune to matriculate in the Oberrealschule Helmholtz at Bielefeld. As the name Helmholtz may already suggest, the physical sciences held a prominent place there. In particular, Dr Lippert, my physics teacher, gave me a splendid grounding in the fundamentals of physics. Although he was suffering from cancer, he instructed us with unbelievable care and attention. He seemed, in particular,

☆ Translated by T. Mulvey.
☆☆ Reprinted from *Advances in Optical and Electron Microscopy* **12** (1991) 1–23.

Advances in Imaging and Electron Physics, Volume 221
ISSN 1076-5670
https://doi.org/10.1016/bs.aiep.2022.03.007

to have a very soft spot for me. On Saturday afternoons, I would go to his house, where he immediately brought me into the family circle; afterwards we went through the details of the current lessons. You can well understand the deep sadness we students felt at the death of our physics teacher. I have to thank him for my special liking for physics. On 2 March 1932 I passed the Matriculation examination; the commendation states "Gottfried Möllenstedt wants to be an engineer".

2. Gaining technical know-how as a student apprentice

At first I wanted to study aircraft construction. That meant that practical apprenticeship of several months was necessary. I worked in the Anker Works and in the Adler Works in Bielefeld from April to December in 1932, in the section locksmith's shop, the planing shop, the milling and grinding shops, the carpentry shop, hardening shop, the smithy and iron foundry. In the summer semester of 1933 I began the study of marine and aircraft technology in Danzig. Applying myself to constructional methods, materials science, machine drawing and hull-shape drawing gave me much pleasure as well as constituting a sound preparation for the technology of experimental physics. However, I soon came to the conclusion that it was the field of experimental physics that would give the best chance of carrying out actual experiments. So, on the advice of a member of the staff who was an *Assistent*, i.e. one who assisted the Professor, I transferred in May 1934 to the Faculty of Mathematics and Physics in the Kossel Institute.

3. A broad education in physics with Professor Walter Kossel

Here I learned what experimental physics is all about and how one carries out research. I had the good fortune soon to be appointed as an *Assistent* myself and to experience the full sweep of physics. To be sure, the Institute was not large, but work was going on in the most varied of research fields. I was fascinated by the X-ray interference patterns from lattice sources. These Kossel lines were to become well known throughout the world. The Borrmann anomalous transparency for X-rays in single crystals arose out of these X-ray investigations. Single crystal pulling and the growth of solid bodies, i.e. solid state physics, formed the principal research area of the Institute. With total love and devotion, Kossel (Fig. 1) carried out electron diffraction experiments at voltages of 30–800 kV with,

Figure 1 Walter Kossel, 1888–1956.

at that time, home-made high-voltage generators. The physics of thin films and electron microscopy was held in high esteem and vigorously promoted by Kossel and his research student, Georg Hass. The best education, however, for me was my role as technician in the great experimental lectures given by Kossel. These held sway over all other events in the Institute at that time. I emphasize these aspects of my multi-faceted learning phase as being particularly important for my ability to do independent research. If one wants to invent and develop something new, one must have observed and experienced how a productive and inspired researcher draws out new concepts. Walter Kossel is to be numbered among the great physicists of his time.

4. Influence of the work of Kikuchi and convergent beam diffraction

As a preparatory phase to my later discovery of electron biprism interference, I must include my diploma work in convergent beam electron diffraction. The wonderful electron diffraction patterns of Seichi Kikuchi are, of course, now well known. A parallel monochromatic electron beam

Figure 2 Seichi Kikuchi and Gottfried Möllenstedt in Tokyo in 1961.

that penetrates a single crystal several hundred nanometers thick or is reflected at the surface of a single crystal produces on a photographic plate bright and dark lines characteristic of the single crystal and frequently bounded by bright and dark bands. The explanation of these phenomena occupied a central position in the theoretical activity of the Danzig Institute. It would be no exaggeration to say that no name was uttered so frequently in the Danzig Institute as that of the Japanese physicist Kikuchi (Fig. 2). Now it is well known that the Kikuchi patterns are a statistical result of the interaction of electrons with the single crystal. One day, Kossel commissioned me to take up this field as a diploma investigation, in order to clarify the fundamental physical processes involved. I used an extremely thin single crystal, which I could irradiate with a convergent electron beam formed by electron optical means (Fig. 3). The technical details of how my diploma investigation was carried out cannot be discussed here. However, I must emphasize that I learnt as much from this experiment as I did from the theory! As a researcher, I learnt how Kossel went about things. Peer intently, and learn the existing state of knowledge. And then ask oneself "How can we go on further to new results?" These were to be provided later by the convergent beam method (Fig. 4).

But now, to come to the discovery of electron biprism interference. Already in the Kossel Institute there had been much discussion about the interaction of waves with matter. The concept of "coherence length"; for example, was already playing a role in the years around 1938. The book

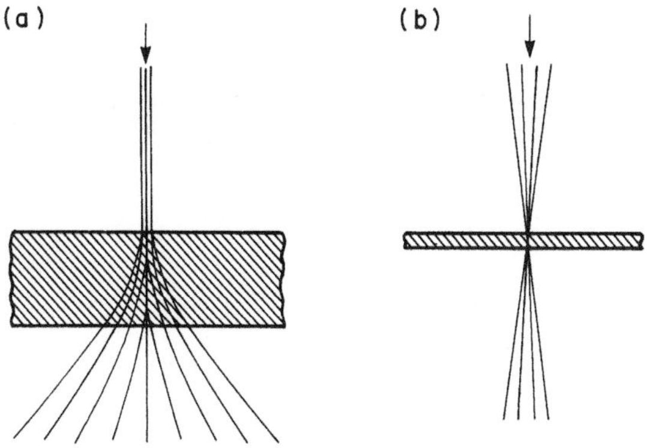

Figure 3 Electron diffraction. (a) Kikuchi method. (b) Convergent beam method of Kossel and Möllenstedt.

"Theory and Practice of Electron Diffraction" by the English physicists G. P. Thomson and W. Cochrane had been discussed in minute detail. From light optics I knew all about interferometry from the Danzig experiments. However, no one dared think about interferometry with electron waves at this time.

At about the time (1952) that I took up the Chair of Applied Physics at the University of Tübingen, there came the news of Marton's idea of arranging several crystals in series so as to split and then reunite a monochromatic electron beam, as shown in Fig. 5. The overlapping wave trains are coherent. The experiments were carried out by Arol Simpson at the National Bureau of Standards. They were fiendishly tedious and, as it turned out later, lacked reproducibility. We realized in Tübingen, on the basis of my experience with electron beams and single crystals, that this kind of interferometry could never provide a solution for electron beam interferometry.

5. Electron-optical experiments with Brüche, Scherzer, and Mahl

Now I come to the moment of inspiration for the idea of using an electron biprism for electron beam interferometry. From 1949 until 1952, I worked in the Süddeutsche Laboratorien in Mosbach-Baden where I had been appointed by Professor Brüche as a Section Leader after I had given up working as an independent manufacturer of simplified electron microscopes. At Mosbach, I worked with electrostatic microscopes of the type

Figure 4 Convergent beam electron diffraction pattern of Muscovite taken at 50 kV. Target diameter 120 nm.

AEG/Zeiss/EM7 shown in Fig. 6. I was able to produce dark-field images in the Zeiss EM7 by fixing a fine tungsten wire over the hole in the objective aperture mounted in the back focal plane of the objective lens. In this way, one could record, in a single micrograph, as shown in Fig. 7, both the bright-field and the dark-field image. After a prolonged period of operation, however, this novel aperture arrangement gave rise to imperfections in the final image. As a consequence of the electron bombardment of the aperture, a now well-known "contamination" layer was formed on its surface. This led to electrical charging-up effects at the wire. The most remarkable effect was that one frequently got "double" images, as may be seen in Fig. 8. Fig. 9 shows a light micrograph, taken in a reflection op-

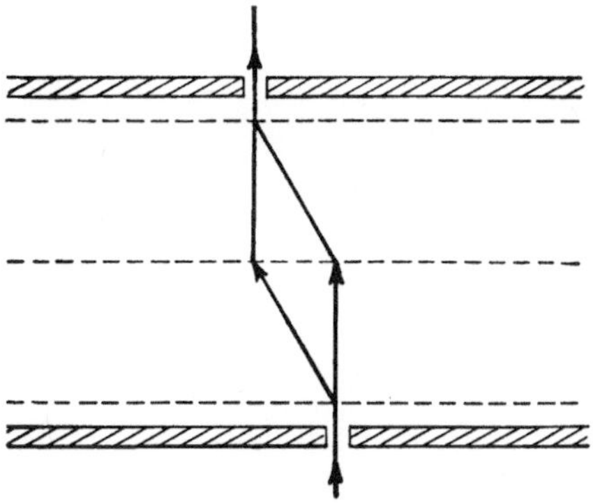

Figure 5 Marton and Simpson's arrangement of their electron interferometer. Dashed lines: thin crystals. Hatched area: aperture discs with fine bores.

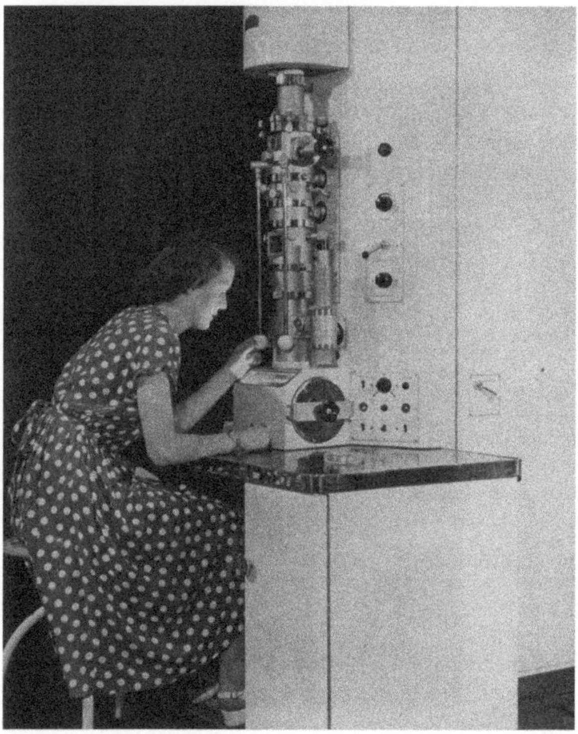

Figure 6 Zeiss/AEG electrostatic electron microscope type EM7/8.

tical microscope, of the 5 μm diameter tungsten wire across the aperture. This micrograph, accompanied by my handwritten marginal comments, is reproduced from my 1953 research record book. There, the (rhetorical)

Figure 7 Composite bright- and dark-field micrograph of zinc oxide crystals taken in the EM7 by means of a 5 μm diameter tungsten wire stop placed across the objective aperture. Field of view 270 μm.

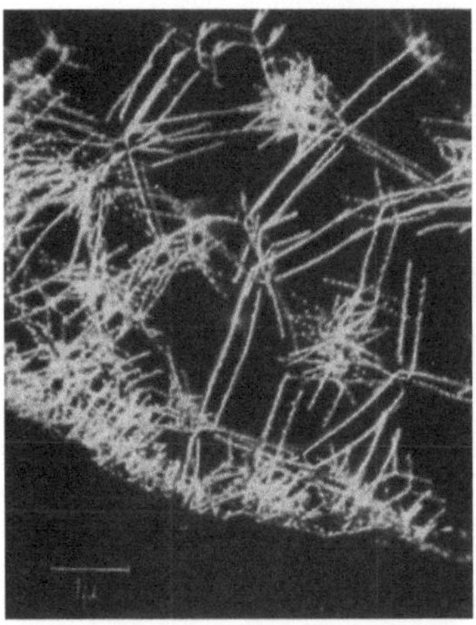

Figure 8 Double images of zinc oxide crystals, in dark field, caused by electrical charging-up of the contaminated tungsten wire stop across the objective aperture.

question is recorded: "What happens if we overlay the two images? Do we have coherence?" Just then, a spark flashed in my consciousness: "Double images are good for nothing, unless they are coherent!" How can one su-

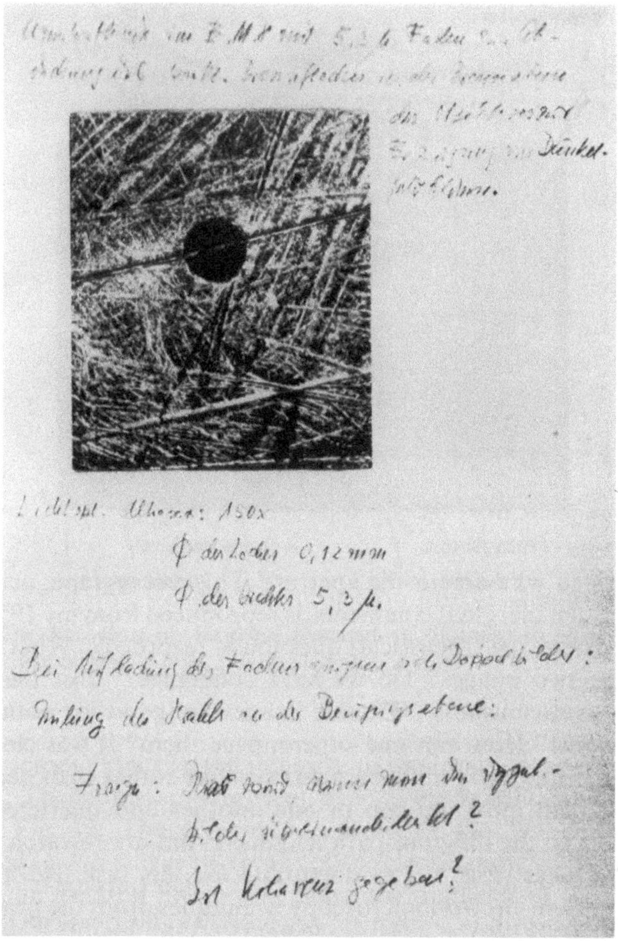

Figure 9 Photograph of a page from G. Möllenstedt's research notebook of 1953, showing reflection light optical micrograph of the objective aperture of a Zeiss EM8 microscope with a fine tungsten wire stretched across the 75 μm diameter hole. The handwritten notes may be translated as follows:
Aperture in EM8 with 5.3 μm wire to blank out demagnified focal spot in back focal plane of objective to produce dark-field images.
Optical enlargement. Ultrapak 150×; dia. of hole 0.12 mm; dia. of wire 5.3 μm.
Charging of wire shows double images: splitting of beam in diffraction plane.
Question: what happens if we overlay the two images? Do we have coherence?

perimpose them? It was clear that the electric charge on the wire must be made positive rather than negative. The images would then move on top of one another and interference fringes would arise due to the different path lengths. From my research students, I chose Heiner Düker (Fig. 10) as doctorand for this task. He had already gained experience in electron

Figure 10 Heiner Düker, 1923–1985.

microscopy techniques, from the practicals that we had introduced at the University of Tübingen. In addition, he possessed great technological experience, having been a Senior Electrical Engineer in U-boats during the War. The first apparatus that we built for the creation of biprism interference can be briefly described as follows. First, the electrical biprism itself: a quartz filament 1 μm in diameter, coated with evaporated gold, was electrically insulated from two earthed cheek plates which had a separation of about 5 mm (see Fig. 11). On electron-optical grounds, it was necessary to have a lateral shift adjustment, as well as a longitudinal shift. Rotation of the complete fiber and cheek plate assembly was also built in. In order to screen out stray magnetic fields, magnetic shielding was provided. A voltage lead to the biprism allowed a continuously variable d.c. voltage of up to several hundred volts to be applied via a potentiometer. After initial experiments, it became clear that in our 30 kV instrument with rotationally symmetric lenses (Fig. 12), the intensity of the interference fringes would be very small. Therefore we soon went over to electrostatic cylinder lenses as shown in Fig. 13. The crossover in front of the Wehnelt cylinder was demagnified in two stages by means of the electrostatic cylinder lenses I and II to form an electron line probe. Since the diameter of the gun crossover was around 50 μm, the line had a width, at 1000× electron-optical demagnification, of about 50 nm and a length of several millimeters. The image in the observation plane was, for intensity reasons, likewise imaged on the photographic plate with a cylinder lens at approx 200× on the fluorescent

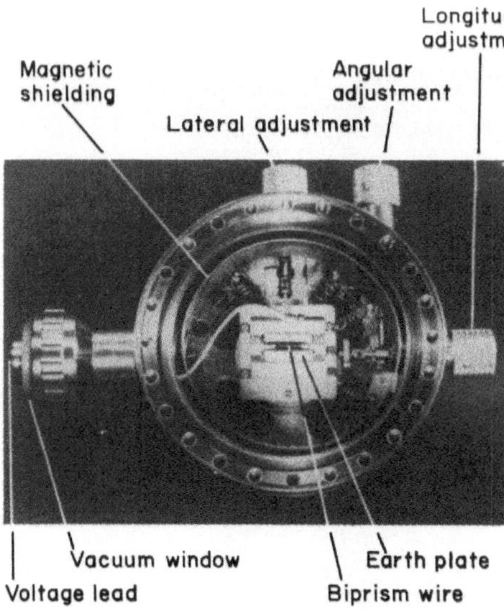

Figure 11 The first electron biprism. Distance of filament to earthed cheek plate, 2 mm. Width of cheeks, 5 mm. Filament length, 6 mm.

screen and/or photographic plate. Since, at the start of the first experiments, the metalized quartz fiber was insufficiently homogeneous over its entire length, one had to magnify a further five to ten times in the direction of the filament; this was done by the magnifying lens II. The interference pattern was observed in a powerful light-gathering telescope made by the firm Carl Zeiss Oberkochen and recorded photographically on repro plates supplied by the firm Krantz. The illumination time was typically several seconds. For historical interest, Fig. 14 shows the first ever electron biprism image, taken in the year 1954. The fringes are clearly recognizable in spite of the graininess of the plate.

6. Quantitative testing of the operation of the biprism

After we had obtained the first biprism interferences, we turned to systematic quantitative investigation of the phenomena. First of all the diffraction fringes of the electron projection images of the 2 μm filament were measured up. This showed good agreement of the maxima and minima with those of the well-known formula of light optics for diffraction at a half-plane, namely:

$$x = [((c - b)c/b)\lambda(n - \tfrac{1}{4})]^{1/2} \tag{1}$$

Figure 12 General construction of the first electron interferometer with rotationally symmetrical electrostatic lenses. The equipment was assembled from standard parts of electron microscopes.

which gives maxima for odd integer values of n and mimima for even values. In this formula x is the distance of the maximum or minimum respectively from the geometrical shadow edge in the observing plane, b is the distance between source and biprism, c is the distance between source and observation plane and λ is the wavelength of the electrons.

The distances of the maxima and minima (n up to 16) agree exactly with the relevant formula of light optics, as H. Boersch had already demonstrated previously. Moreover, even in the shadow region of the filament, the fringes could be made visible by long exposure. The fringe distance agreed with the expression:

$$\Delta = \lambda(c - b)/2r \tag{2}$$

where $2r$ is the diameter of the filament.

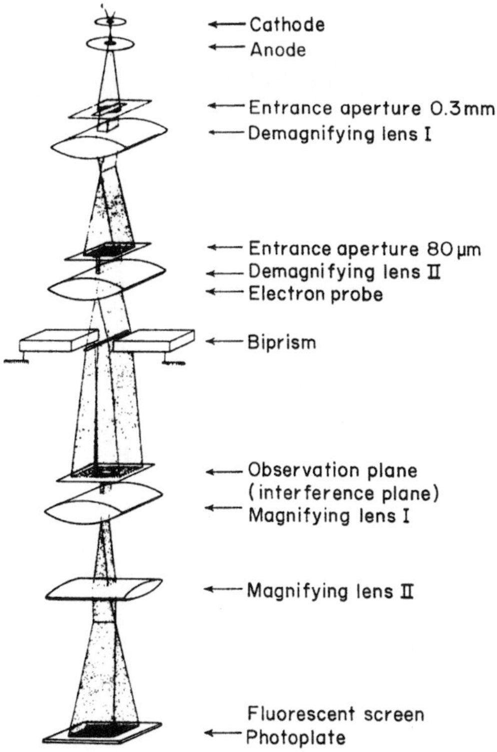

Figure 13 Ray paths for observing bright electron biprism interference patterns.

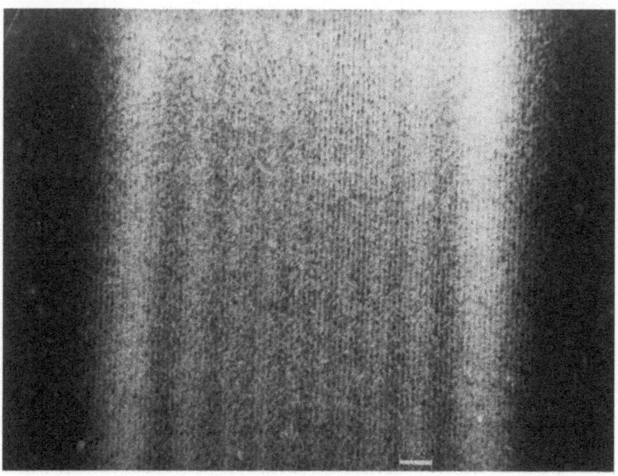

Figure 14 Interference pattern obtained with the electron biprism. The image on the photographic plate has been optically enlarged 50×.

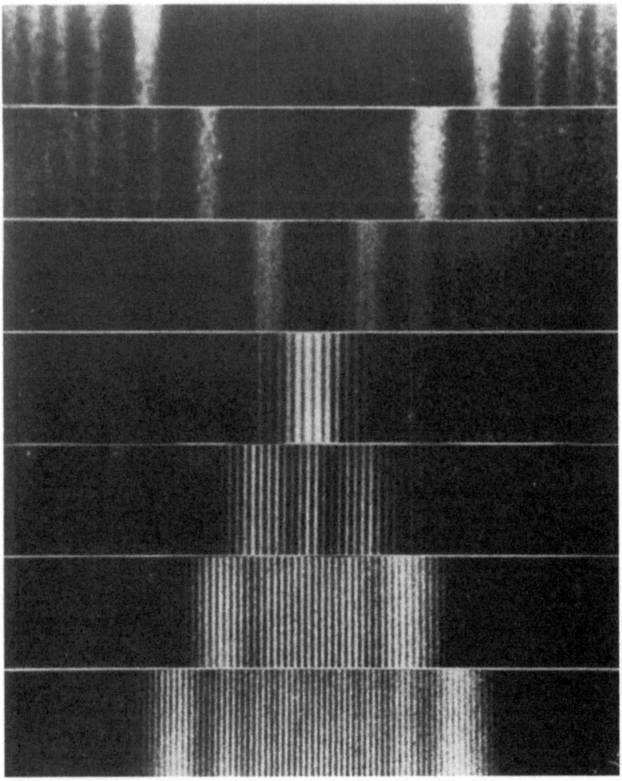

Figure 15 Illustrating the emergence and variation of electron biprism interference with increasing voltage applied to the metalized filament. The pattern in the interference plane has been magnified 2600×, electron-optically 167× and light-optically 15.5×.

The manner of the progressive overlapping of the fringes was studied by applying an increasing voltage to the filament. Fig. 15 demonstrates the reduction of the interference fringe width with increasing filament voltage. The filament voltage was increased in small steps from 0 to 7 V. The accelerating voltage for these experiments was about 20 kV. It counts as one of my greatest experiences as a physicist that I could, so to say, manually adjust the spacing of the interference fringes by means of a potentiometer. Fig. 15 also demonstrates this variation of fringe spacing with increasing filament voltage. The intensity of the interference pattern can be calculated as follows. If we replace the two plane-coherent electron waves by sinusoidal oscillations of amplitude A_1 and A_2 and phase difference δ, the resulting intensity I_g by superimposing these waves is given by:

$$I_g = A_1^2 + A_2^2 + 2A_1A_2\cos(2\pi\delta/\lambda) \tag{3}$$

Since in our case

$$A_1^2 = A_2^2 = I_0 \tag{4}$$

then

$$I_g = 2I_0[1 + \cos(2\pi\delta/\lambda)] = 4I_0\cos^2(\pi\delta/\lambda) \tag{5}$$

Theoretically, therefore, in the interference image an intensity variation from zero at the minimum to four times the primary intensity at the maximum is to be expected. In practice the intensity variations are not so marked.

We now proceed to the deflection of the wavefronts at the biprism filament. First, we calculated the potential distribution in the neighborhood of the biprism filament. It showed rotational symmetry. The form of the potential distribution corresponds to that of a cylindrical condensor whose inner radius is that of the biprism filament of radius r and whose outer radius is R_a. One finds that:

$$\phi_{(\hat{R})} = U_f \ln(R/R_a)/\ln(r/R_a) \tag{6}$$

From the known potential distribution, one can express the electric field distribution as a function of R. It is:

$$E(R) = -\operatorname{grad}\phi(R) = U_f\mathbf{R}_0([\ln(R_a/r)]^{-1}/R) \tag{7}$$

This shows that in the field close to the filament all rays have the same deflection angle. We have constant deflection so the deflected wavefronts stay planar. Experimentally, the fringes were always equidistant over the whole field. The constancy of the deviation angle of the electron rays in the electron biprism is demonstrated in Fig. 16. The biprism was placed in the ray path of an electron microscope between the objective and the projector plane. The earthed filament then appears in the electron microscope image, as shown in the upper image in Fig. 16. As a positive voltage is applied to the filament, the image halves originally on the left- and right-hand sides of the filament shadow move towards each other, as shown in Fig. 16 (lower image). The movement of the individual image points was measured as a function of their axial distance and plotted graphically. As the

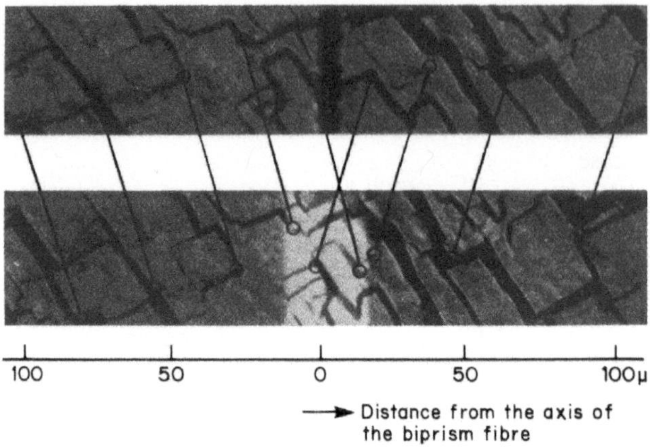

Figure 16 Experimental proof of the constancy of the deviation angle of the electron beam in the biprism.

parallelity of the lines joining the points shows, the deflection is constant over a significantly wide region, as might be used for biprism interference purposes.

De Broglie's relation $\lambda = h/mv$ had been confirmed by electron diffraction from crystals, using values of the lattice spacing known from X-ray diffraction and measuring the accelerating voltage as precisely as possible. The accuracy of this method was limited to "one or two per cent". The biprism interference method furnished another way of verifying de Broglie's relation. From Fig. 17, a relation can be derived between the wavelength λ, the fringe spacing Δ, the displacement δx of the shadow limit and the distances b and c.

$$\lambda = 2b\Delta\delta x/[(c-b)c]$$

All the distances occurring on the right-hand side of the expression can be measured. The experimental results, shown in Table 1, show that the difference between the measured and the calculated values does not exceed 0.5%. When I informed L. de Broglie (Fig. 18) of these results, he sent me a letter, the English translation of which reads as follows:

Paris, 19 June 1956.

Monsieur and dear Colleague,

I was extremely pleased to receive your kind letter and to learn of your beautiful experiments in which you have obtained electron interference by a method analogous to Fresnel's biprism. It was, of course, a great pleasure to see that you have obtained a new and particularly brilliant proof of the formula $\lambda = h/mv$, and I shall not fail to make known your experiments to my students.

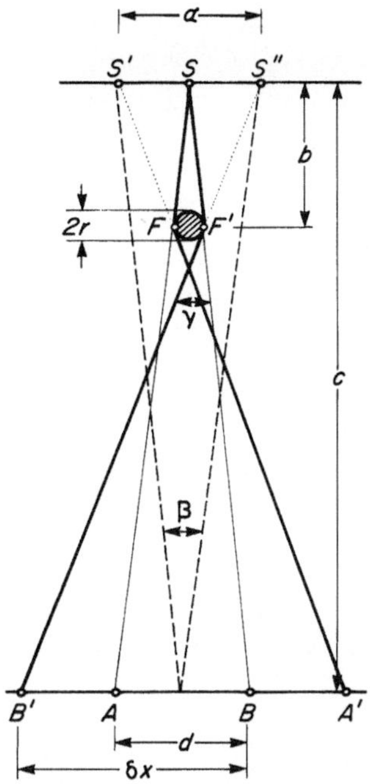

Figure 17 Geometry of the ray paths in the biprism experiment for determining the de Broglie wavelength.

Table 1 Comparison of the measured and calculated de Broglie wavelengths.

Plate No.	Measurement				Calculated from λ A.U.	de Broglie formula λ A.U.	Error (%)
	Accelerating voltage (kV)	U_f (V)	δx (µm)	Δ (µm)			
106	19.4	4.5	13.5	0.61	0.0873	0.0870	0.4
122	19.4	5.0	15.0	0.545	0.0866	0.0870	0.5
123	19.4	5.5	16.6	0.495	0.0870	0.0870	0.0
118	19.4	5.8	17.5	0.47	0.0872	0.0870	0.25
110	19.4	6.5	19.5	0.42	0.0868	0.0870	0.25
109	19.4	7.0	20.5	0.40	0.0869	0.0870	0.12
56	26.0	8.0	11.3	0.50	0.0747	0.0750	0.4
58	26.0	10.0	17.2	0.35	0.0752	0.0750	0.27
59	26.0	18.0	30.7	0.185	0.0748	0.0750	0.27

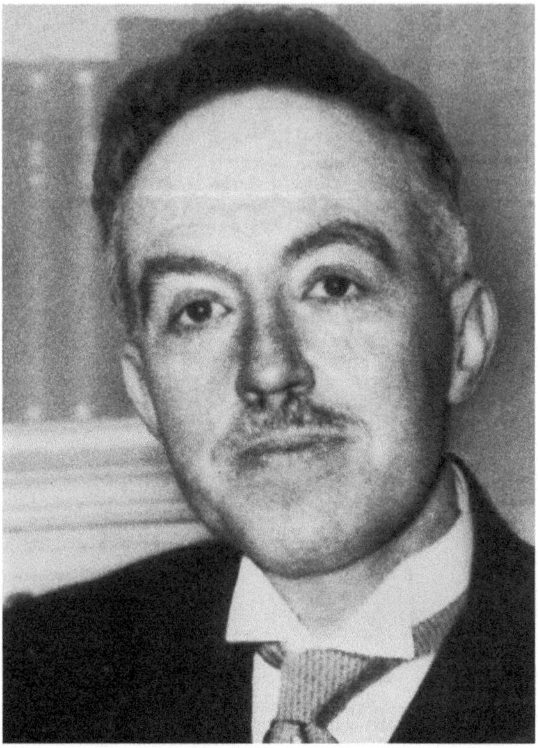

Figure 18 Louis Prince Victor de Broglie (born 1892).

Thanking you most gratefully for your communication,
I beg you to accept, Monsieur and dear colleague, the expression of my devoted sentiments,
Louis de Broglie.

7. Measurement of the inner potential of solids

Keller constructed a spectrometer with the optical paths shown in Fig. 19. Two partial waves transmitted through an object with thicknesses D_1 and D_2, in different regions are superimposed on each other. The fringes are shifted with respect to each other by an amount Δs due to the phase shift of one wave with respect to the other. The mean inner potential ϕ can be calculated using:

$$\phi = h\nu\Delta/\{e\, s(D_1 - D_2)\}$$

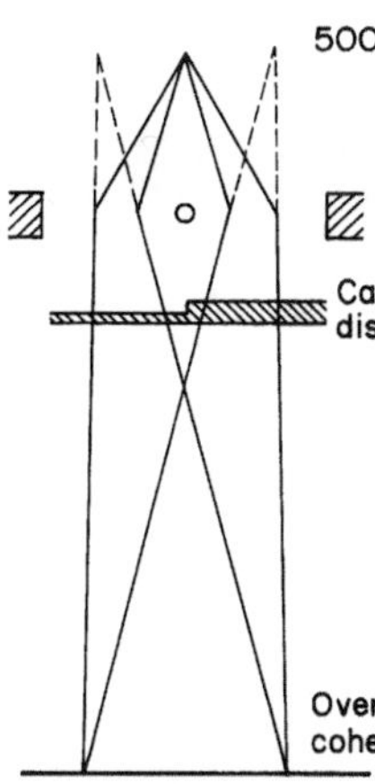

Figure 19 Ray paths in the electron interferometer for measuring the mean inner potential (M. Keller).

where v is the velocity of the electron and s is the spacing of the interference fringe.

An interference pattern taken in this way allows one to calculate the mean inner potential. The method has been shown to work for different materials and for accelerating voltages up to 300 kV. The results are in good agreement with the theoretical results of Siota.

8. Electron interference microscope in the transmission mode

Fig. 20 shows the ray paths in the interference microscope as devised by Möllenstedt and Buhl, using an electron biprism in conjunction with an axially symmetrical electrostatic lens. This microscope was used for measuring the mean inner potential of small regions of the specimen, as shown in Fig. 21.

9. The intensity problem in electron interferometers

The application of cylinder lenses has mainly been used for research investigations in laboratories. Commercial electron microscopes with built-in electron biprisms for electron waves, of course, use axisymmetrical lenses. So it is necessary when one wants to get intense fringes to turn to point sources with higher electron-optical brightness. An important aid, for us,

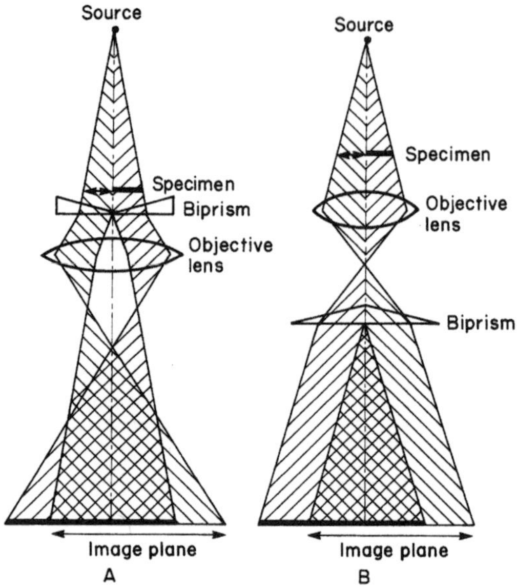

Figure 20 Electron biprism interference microscope with specimen and biprism filament at negative potential and rotationally symmetrical electrostatic lens (H. Buhl).

Figure 21 MoO_3 single crystal image taken in the interference microscope. Accelerating voltage 45 kV. Horizontal field of view 40 μm.

in the solution of these problems was the work carried out on pointed filaments (thermionic cathodes) by Professor Sakaki (Fig. 22) in Tübingen. A selection of such pointed filaments is shown in Fig. 23.

10. Atomic resolution electron holography

In recent times, electron interferometry in the transmission microscope with the use of field emission cathodes and computer methods of

Figure 22 Prof. Y. Sakaki, Electrotechnical Institute of Nagoya University, Japan, and Visiting Professor, Institute of Applied Physics, Universität Tübingen, 1956.

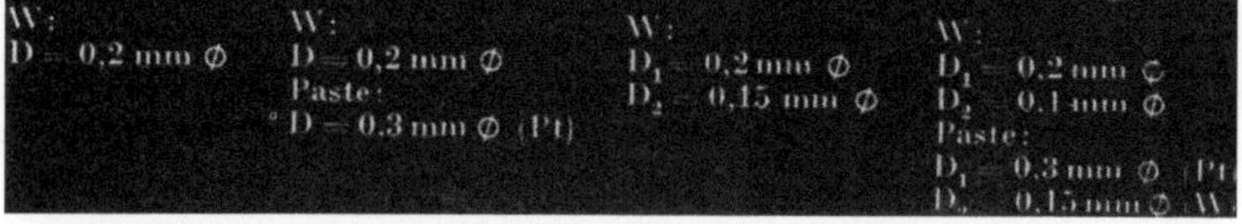

Figure 23 Various pointed filaments. (a) Point filament of E. W. Muller but with tip ground by the method of T. Hibi. (b) Normal tungsten filament after grinding. (c) Tungsten filament held by means of a wire sling and holder D. The current passes in the direction of the arrow. (d) Filament held from both sides so that at operating temperature the tip remains in the prescribed position.

image processing has led to a decisive step forward so that now, with the help of the electron biprism, two-beam electron holography, in accordance with the original concept of the Nobel prizewinner D. Gabor, has become possible, leading to the correction of spherical aberration and hence a genuine atomic resolution, as demonstrated by H. Lichte in *Advances in Optical and Electron Microscopy 12* (1991) 25–91, reprinted in vol. 218 (2021) 195-260 of these Advances.

11. Editorial note

The above article is a translation, with minor editing, by Tom Mulvey of Möllenstedt's original manuscript in German of a lecture given in Hakone, Japan, in 1986 at the Möllenstedt Symposium. It is published here as a tribute to the seminal part played by Gottlieb Möllenstedt's invention of the electron Fresnel biprism in enabling Gabor's ideas in electron beam holography to be fully realized in practice. This includes the correction of spherical aberration and the realization of true atomic resolution in which the phase and amplitude information in the wave scattered by the object is preserved intact in the image.

12. Afterword by Peter Hawkes

Gottfried Möllenstedt was born on 14 September 1912 in Versmold and died on 11 September 1997 in Tübingen. In 1939, he passed the Dipl.-Ing examination at the Technische Hochschule in Danzig and was awarded the title Dr-Ing.in 1940 for a thesis on 'Messungen an den Interferenzerscheinungen im konvergenten Elektronenbündel'. He won the title of 'fastest Danzig sprinter' having run 100 m in 10.6 s! During the wartime years, he was Kossel's scientific 'Assistant'. His Habilitationsschrift 'Elektronen-Optik bei Interferenzproblemen' was accepted in 1945. He escaped from Danzig aboard the *Deutschland* but was interned by the American authorities from June to October 1945. He spent the next years at the Suddeutschen Laboratorien of AEG-Zeiss in Mosbach, where he invented the "Möllenstedt energy analyser" and in 1952, he joined the University of Tübingen, where he was appointed full professor in 1960, moving into the newly established Institut fur angewandte Physik in the Morgenstelle in 1972. During his years in Tübingen, he and Heinrich Düker developed the electron biprism, with which many major observations were made, concerning the measurement of inner potential and the build-up of interference fringe, for example. With H. Wahl and later, H. Lichte, holography became a reality.

For more information, see Möllenstedt (1996), Lenz and Lichte (1998), Lichte (1998), Tonomura (1998), and the very detailed essay by Schulze (2010).

References

Lenz, F., & Lichte, H. (1998). In memory of Gottfried Möllenstedt. *Ultramicroscopy*, *74*, 95–103.

Lichte, H. (1998). Gottfried Möllenstedt and his electron biprism: four decades of challenging and exciting electron physics. *Journal of Electron Microscopy*, *47*, 387–394.

Möllenstedt, G. (1996). My early work on convergent-beam electron diffraction. *Advances in Imaging and Electron Physics*, *96*, 585–595.

Schulze, D. (2010). Gottfried Möllenstedts elektronenoptisches Biprisma, Schlüssel zur Wellenoptik von der Elektroneninterferometrie zur Holographie. *Elektronenmikroskopie*, *30*, 10–15. See too *ibid.* *16*, 42–44.

Tonomura, A. (1998). Prof Dr Gottfried Möllenstedt (1913–1997). *Journal of Electron Microscopy*, *47*, 363–364.

CHAPTER EIGHT

The industrial development of the electron microscope by the Metropolitan Vickers Electrical Company and AEI Limited

T. Mulvey[a] and Peter Hawkes (Afterword)[b,*]
[a]Formerly Department of Physics, University of Aston, Gosta Green, Birmingham, United Kingdom
[b]CEMES-CNRS, Toulouse, France
*Corresponding author. e-mail address: hawkes@cemes.fr

Contents

1. Introduction 225
2. Cathode rays and electron images—Gabor and the ironclad concentration coil 226
3. The first commercial electron microscope—EM1 228
4. Wartime developments 232
5. The EM2 electron microscope 233
6. A three-stage electron microscope—EM3 234
7. A simple electron microscope—EM4 236
8. High-voltage electron microscopy—the EM5 237
9. Improving the resolving power—EM3A and the emergence of electron holography 238
10. High-resolution microscopes—EM6 243
11. The image intensifier 246
12. The electron microscope in biology 247
13. Very high-resolution microscopy 248
14. Afterword by Peter Hawkes 248
References 252

1. Introduction

The rapid development of the electron microscope in the face of apparently insoluble technical and physical problems came about by the initial faith and enthusiasm of a few electrical engineers and physicists engaged in industrial-type research, backed eventually by the resources of a number of great electrical engineering firms. Here we are primarily concerned with the part played by the Metropolitan Vickers Electrical Company and AEI

Advances in Imaging and Electron Physics, Volume 221
ISSN 1076-5670
https://doi.org/10.1016/bs.aiep.2022.03.008

Ltd. in the development of the instrument itself; the contribution of these companies to specimen preparation techniques was also considerable but is beyond the scope of the present review.

2. Cathode rays and electron images—Gabor and the ironclad concentration coil

From about 1858 German physicists had been extremely interested in the so-called cathode rays that caused emission when they struck a glass plate. Although physicists had not made a great deal of headway in explaining the nature of these rays, German electrical engineers such as Ferdinand Braun had put them to commercial use in the form of the cathode-ray oscillograph. Of particular significance in the industrial history of the electron microscope, although it was not realized at the time, was the high-voltage cathode-ray oscillograph, shown in Fig. 1, which was designed and built by Dennis Gabor (1927) during the years 1924–1926 at the Institut fur Höchstspannungstechnik in Berlin under the enlightened directorship of Professor Matthias, who also held a chair at the neighboring Technische Hochschule Berlin.

The instrumental problem Gabor was trying to solve was that of recording, on a cathode-ray oscillograph, the fast surges or traveling waves (*Wanderwellen*) then afflicting the newly installed high-voltage grid supply network in Germany. Lightning strokes on the overhead lines and even switching surges were causing extensive damage to line equipment, especially to power transformers.

Gabor's oscillograph, of advanced design for its time, was operated at the high accelerating voltage of 50 kV in order to produce an electron beam of high intrinsic brightness. It was also provided with a four-plate camera which was located directly in the vacuum, thereby eliminating the inefficiency of the conventional method of indirect photography of the fluorescent screen. Gabor's most felicitous inspiration, however, was to dispense with the conventional long-focusing solenoid of Wiechert (1899) and replace it with a short coil encased in an iron shroud, except over the axial region. His chief reason for doing this was to contain the magnetic field within the confines of the coil itself so as to prevent the stray magnetic field from adversely affecting the operation of the cold-cathode source and also of the deflecting plates used to scan the electron beam.

Unwittingly, as he was always the first to admit, he had stumbled across the way to make an efficient focusing system that worked in an entirely

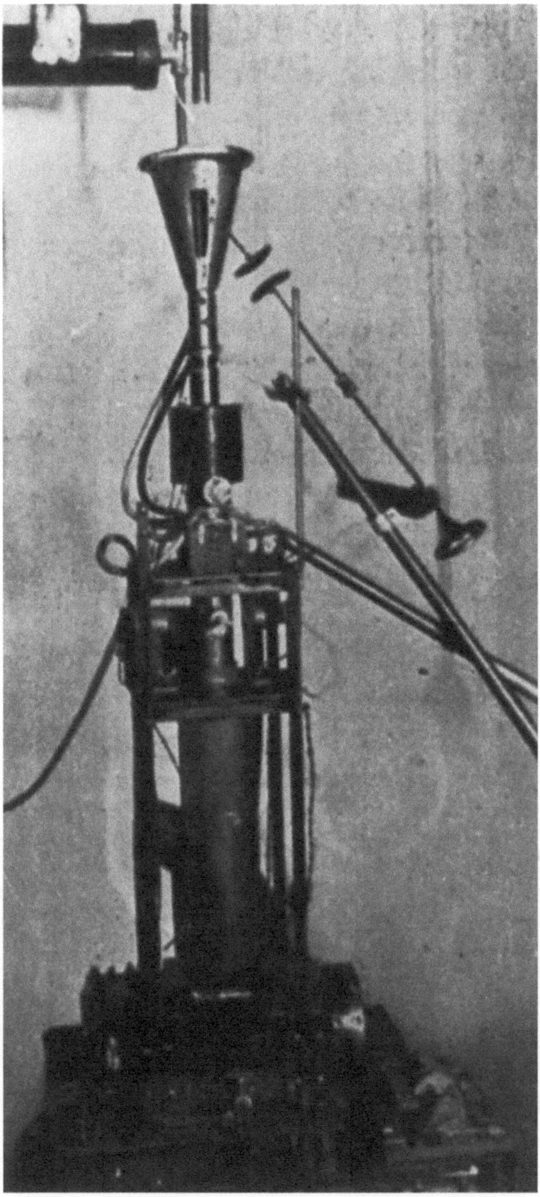

Figure 1 Gabor's high-voltage cathode-ray oscillograph equipped with the first ironclad magnetic lens (Gabor, 1927).

different way from that of the long-focusing solenoid. However, unusual for Gabor, he was unable to give an adequate explanation of the focusing action of his short solenoid. This is clearly revealed by a careful reading of his Ph.D. dissertation (Gabor, 1927). The difficulty was resolved with the appearance in the same year of the classic paper on the subject by Hans Busch (1927).

Busch therefore, and not Gabor, must be regarded as the founding father of electron optics. Nonetheless, Gabor's iron-shrouded lens may be regarded as the distant ancestor of the modern magnetic electron lens. Once Gabor had understood how such a lens worked and its possibilities for electron-optical instruments, he never let the subject depart very far from his mind and remained, on and off for the rest of his life, at the absolute frontier of ideas for improving the electron microscope. Later he was to come to Great Britain and be involved with the Metropolitan Vickers Electrical Company in electron microscopy, and later with the AEI Research Laboratory, where he introduced the idea of electron holography and persuaded the laboratory to undertake the first exploratory tests of this revolutionary idea. Knoll and Ruska (1932) were, in fact, the first to demonstrate experimentally the correctness of Busch's ideas. However, the step from the iron-shrouded coil to the crucial invention by von Borries and Ruska of the iron polepiece lens for high-resolution microscopes was a long and difficult one, as has recently been recorded by Ruska (1980). Present-day high-resolution electron microscopes stem directly from this invention and its subsequent refinement and industrial development.

3. The first commercial electron microscope—EM1

The Metropolitan Vickers Company was the first to become involved commercially in electron microscopy. By 1933 Ruska (1934) had demonstrated conclusively that a resolution appreciably better than that of the best optical microscope could be realized in an electron microscope. Moreover, Marton (1934, 1935) in Belgium and Driest and Müller (1935) in Berlin had demonstrated that such an instrument could produce high-resolution images of biological specimens, contrary to the prevailing climate of expert opinion. This was particularly the case in Germany, where expert opinion was, in general, skeptical, if not distinctly hostile, toward the scientific value of electron microscopy. To be fair, there were many problems in the early days with specimen preparation techniques, and the instruments themselves needed careful nursing and devoted attention. There seemed no possibility therefore of any commercial manufacturer becoming interested in producing an electron microscope. However, in the United Kingdom, things were more hopeful. At the Imperial College of London University there was already a strong tradition of electron diffraction instrumentation stemming from G. P. Thomson's classical experimental verification of the existence of de Broglie waves. In addition, there was a strong department of technical optics led by Professor L. C. Martin. It therefore seemed appropriate that an impartial investigation into the claims and counterclaims of electron microscopy vis à vis optical microscopy should be carried out at Imperial College. Thus, in 1935, Professor Martin approached the High Voltage Laboratory of the Metropolitan Vickers Research Department in Manchester to see if they would accept an order to manufacture an electron microscope with optical microscope viewing facilities. Money was scarce at the time, but financial support for the project was to be arranged through the good offices of the Royal Society. The company had already had several years experience with the high-voltage cathode-ray oscillograph and was familiar with Gabor's work as well as with the later developments by Knoll and Ruska. In July, 1928, Dr. T. E. Allibone, then a member of the Met-Vick Research Department, went to see Gabor in Berlin for discussions on the high-voltage oscillograph (Allibone, 1980). In 1929 experimental work had started in the Manchester laboratory on a high-voltage oscillograph (see Burch & Whelpton, 1932). The design had been influenced by the work both of the Rogowski school at Aachen and of the Matthias school at Berlin. This oscillograph was completed at the beginning of 1930 and was put on display at the official opening of the High Voltage Laboratory at Trafford Park, Manchester, in April, 1930. The complete instrument is shown in Fig. 2, in which the optical viewing port and cold-cathode source construction reveal it as indeed an ancestor of an electron microscope. It was therefore a natural step to use it as a starting point for the design of EM1, the first of a long series of Metropolitan Vickers electron microscopes. Fig. 3 shows the EM1 as installed at Imperial College, London, in 1936. The electron-optical system, as specified by Professor Martin (Martin et al., 1937), consisted of a condenser lens, a specimen stage, an objective lens, and a projector lens. All lenses were of the magnetic type. Two cameras were provided, both working under vacuum. The first could be placed in the image plane of the objective lens and was used for taking photographs at low magnification. The second camera was situated under the projector lens and was available for taking photographs at high magnification. As the instrument was intended for diagnostic purposes, it was considered essential to be able to move the same specimen rapidly between the electron-optical column and the built-in optical microscope. In order to achieve this, the specimen was mounted on a turntable and could be swung rapidly into the

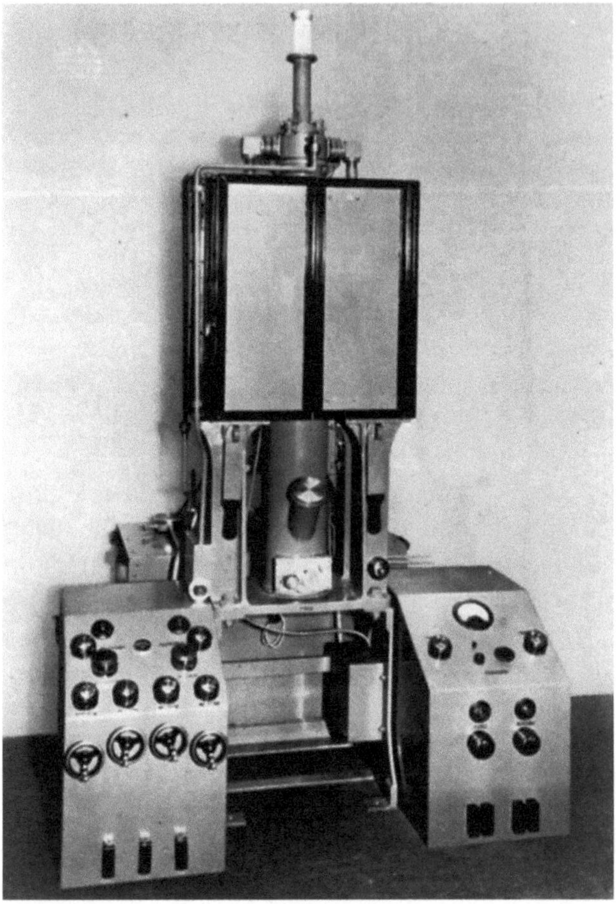

Figure 2 The Met-Vick high-voltage cathode-ray oscillograph (Burch & Whelpton, 1932).

field of view of either instrument. Little or nothing was known about lens aberrations at the time and so it was not realized that the resulting increase in working distance, and hence focal length of the magnetic objective lens, would seriously compromise the resolving power of the electron microscope. Consequently this electron microscope did not exceed the optical microscope in resolving power. However, it provided the company with an insight into challenging problems of design and manufacture of electron microscopes, problems several orders of magnitude greater than had been met previously by the engineering world.

At the time, Metropolitan Vickers was one of the few companies who were capable of producing scientific instrumentation requiring demount-

Figure 3 The Met-Vick EM1, the first commercially produced electron microscope, installed at Imperial College, London, in 1936 (Martin et al., 1937).

able vacuum apparatus. This was due largely to the invention by C. R. Burch, a member of the laboratory, of an oil diffusion pump using the so-called Apiezon oils. The whole electron microscope project was made possible by the generous and far-seeing attitude of the director of the laboratory, A. P. M. Fleming, a man of the same turn of mind as Professor Matthias in Berlin, who encouraged his gifted staff to undertake innovative work that could have long-term benefits for industry. The total cost of the EM1 was £1200. The DSIR (Department of Science and Industrial Research) and the Royal Society supplied £800 toward the cost of the project, while A. P. M. Fleming contributed £400 from his research budget. There was no suggestion at this exploratory stage, of course, that the company should undertake full-scale commercial production of electron microscopes. Moreover, in 1937, the company was becoming heavily committed to the development of "radio location" (later known as "radar"), a new idea for detecting the approach of aircraft. This demanded the design and development of high-power rf transmitting tubes operating with a demountable vacuum system. Thus engineering effort could simply not be spared to pursue the development of the electron microscope. In the meantime, in Berlin, after strenuous efforts, von Borries and Ruska had succeeded in persuading Siemens and Halske A.G. to allow them to join them in Berlin and undertake the development of an electron microscope for serial production. This work proceeded quickly and the first production model (von Borries & Ruska, 1939a) was installed in an industrial laboratory in January, 1939. It was the first serially produced instrument to surpass the optical microscope in resolving power (von Borries & Ruska, 1939b).

4. Wartime developments

Just before the EM1 microscope was shipped to Imperial College, M. E. Haine, then a young graduate from Bristol University, joined the Metropolitan Vickers Company and was immediately put to work on the radar development program in the Radio Department. This was fortunate since the design and manufacture of demountable vacuum systems was central to both radar and electron microscopy. On successful completion of this project he was transferred to the High Voltage Laboratory to work on highspeed oscillographs. He took readily to electron optics and had soon replaced the cold-cathode gun on his oscilloscope with a thermionic gun, thereby improving the electron-optical performance by an order of magnitude. His interest in the electron microscope was stimulated by a visit in 1944 to the Shirley Institute in Manchester, where Dr. Drummond was working on cotton fiber research with an RCA model B microscope supplied under the lend–lease agreement with the United States. This instrument, designed by Hillier and his team (Zworykin et al., 1940) in the United States, was a very advanced commercial design for its time, especially in regard to its electronically stabilized lens current and high-voltage supplies. However, it was not an easy machine to keep in good working order and RCA users in Britain soon founded a "self-help" club which proved invaluable for the sharing of technical information and "know-how" and the rapid development of electron microscopy in the United Kingdom.

Haine found out that the National Physical Laboratory (NPL) also had an RCA microscope and in fact were not using the Met-Vick EM1 that had been transferred from Imperial College to the NPL. At this time, Haine, as part of his war service, was a part-time fireman in the AFS (Auxiliary

Fire Service) stationed at the factory. At this stage in the war, however, air raids were practically nonexistent, and so the evenings spent on these "fire-watching" duties were extremely boring. Haine got the brilliant idea of "borrowing" the EM1 microscope from NPL in order to work on it on such evenings. Professor Preston at NPL and Dr. Allibone, the director of the Met-Vick High Voltage Laboratory, readily agreed to this unusual arrangement for combating boredom in the fire brigade!

The EM1 was duly rescued from a basement at NPL. Haine found that its design was ingenious but its construction was crude, especially the vacuum system. All joints were made by lapping mating faces together and liberally smearing them with grease. The microscope was, in fact, aligned by sliding the relevant parts over the grease seal. The first thing that had to be removed was the turntable arrangement for transferring the sample into the optical microscope. It leaked incurably!

A thermionic cathode gun was fitted in place of the cold-cathode gun and eventually a promising image was obtained on the fluorescent screen! It was not long before the Ministry of Supply somewhat unexpectedly placed an order with the company for six electron microscopes. A small ad hoc team of draftsmen and engineers was hurriedly formed and the design of the EM2 began. There were many technical and scientific problems to be solved in the years ahead.

5. The EM2 electron microscope

The design of the EM2 that quickly emerged is shown in Fig. 4. It was a two-stage instrument (Haine, 1947) retaining several of the features of the EM1, but was engineered on more solid lines; for example, vacuum seals made with grease were replaced by specially developed rubber cord seals. The mechanical stability was improved and a thermionic cathode was fitted as standard. Electronic stabilization of lens supplies and accelerating voltage was also incorporated. The maximum magnification was about 30,000× and the guaranteed resolving power was 100 Å or better. In the loving hands of the expert and by careful but tedious adjustment of the objective polepieces to minimize astigmatism, a resolution of 25 Å could occasionally be coaxed out of the instrument. The provision of robust specimen and camera air locks made routine operation particularly easy on this instrument. Development proceeded rapidly and the first instrument was installed at the Dunlop Rubber Company in Birmingham in December, 1946.

Figure 4 General view of the Met-Vick two-stage EM2 electron microscope.

6. A three-stage electron microscope—EM3

In the interests of producing a dependable electron microscope as quickly as possible, many useful features such as a large range of magnification, dark-field illumination, and stereo electron microscopy had to be omitted from the EM2. These were now included in the EM3, shown in Fig. 5. Thinking at Metropolitan Vickers had now passed from the initial concepts of the high-speed oscillograph to that of realizing with electrons the full analogy of the optical microscope (Haine et al., 1950). Haine therefore completely changed the design concept of the new microscope. The instrument should be as compact as possible in order to reduce to a minimum the volume of air to be pumped so that air locks could be omitted. It was also desirable to have as large a range of magnification as possible. These requirements could best be satisfied by the insertion of an intermediate projector lens in addition to the normal final projector lens. Moreover, such a lens could also be used for focusing on the electron diffraction pattern

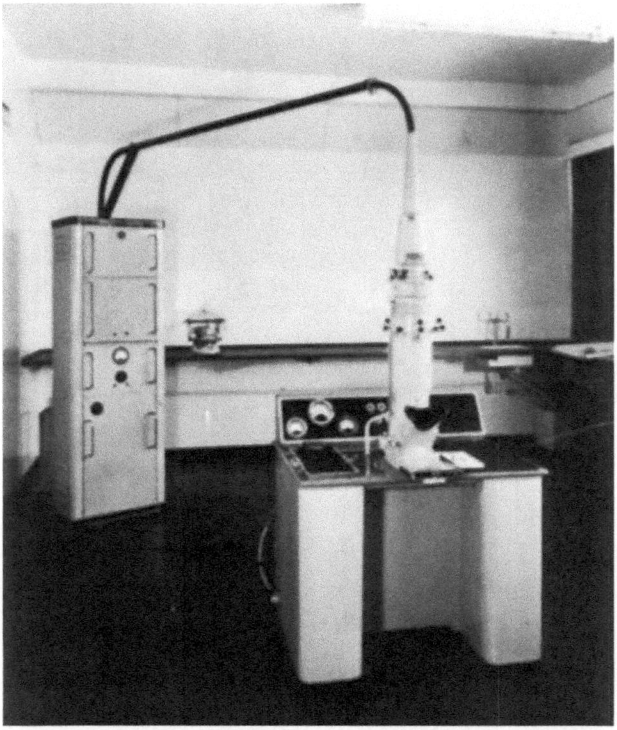

Figure 5 The Met-Vick EM3 electron microscope, a three-stage instrument with selected-area diffraction, stereo-imaging, and a wide range of magnification (Haine et al., 1950).

that would be formed in the back focal plane of the objective lens. These ideas were new in Britain but would not have seemed new to Boersch, who in 1936 had previously demonstrated in a crude but telling experiment (Boersch, 1936) how to form such selected-area diffraction patterns. In The Netherlands, during the German occupation, Le Poole (1947) had also independently conceived and realized the idea of a diffraction lens in an experimental electron microscope, but was obliged to delay publication until the end of hostilities. In the United States, Marton, during the wartime period, had also realized the advantages of multilens imaging, though not of selected-area diffraction, and incorporated intermediate lenses in his experimental microscope (Marton, 1945) at Stanford University. These examples perhaps illustrate the fact that when instrument designers are concentrating intensively on the solution of the same problem, the same basic solution often occurs almost simultaneously over a wide geographical area.

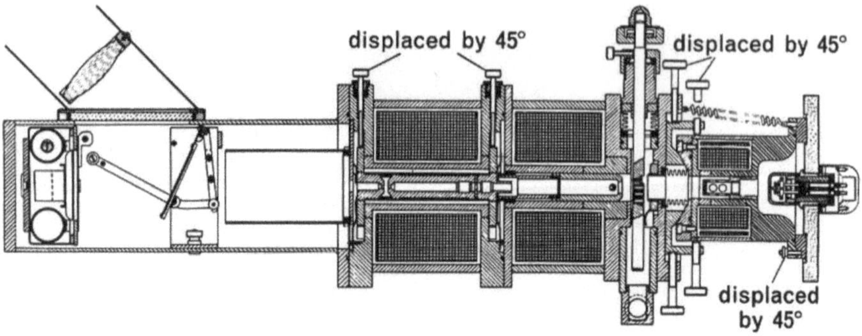

Figure 6 Electron-optical column of the EM4 showing the compact layout and the double-polepiece projector lens (Page, 1954).

The EM3 had a range of magnification of about 1000–100,000×. The guaranteed resolving power was 50 Å, completely adequate for the specimen preparation techniques of the day. The first production model was installed in May, 1949, in the Department of Biomolecular Studies at Leeds University.

7. A simple electron microscope—EM4

A constantly recurring need in electron microscopy is for a small, relatively inexpensive electron microscope of moderate resolving power and purchase price suitable for routine operation by relatively unskilled operators. The EM4 was designed by R. S. Page to meet these requirements (Page, 1954). Page had previously worked with Haine on the high-voltage oscillograph and after returning from war service took charge of the commercial production of the company's electron microscopes, Haine having left the Research Department at Manchester to lead an electron physics team at the newly formed AEI Research Laboratory, located at Aldermaston, Berkshire. In order to meet the above specification the EM4 departed considerably from the design principles of the EM2 and EM3 microscopes. Fig. 6 shows the general arrangement of the horizontal electron-optical column, which had a top magnification of 10,000×. The horizontal column made viewing through a transmission screen comparatively easy. A simple but effective rod-type specimen holder similar to that previously devised by Liebmann (1948) was fitted. This holder enabled three specimens to be examined in quick succession without breaking the vacuum. At the time Liebmann was not employed by Metropolitan Vickers, but he

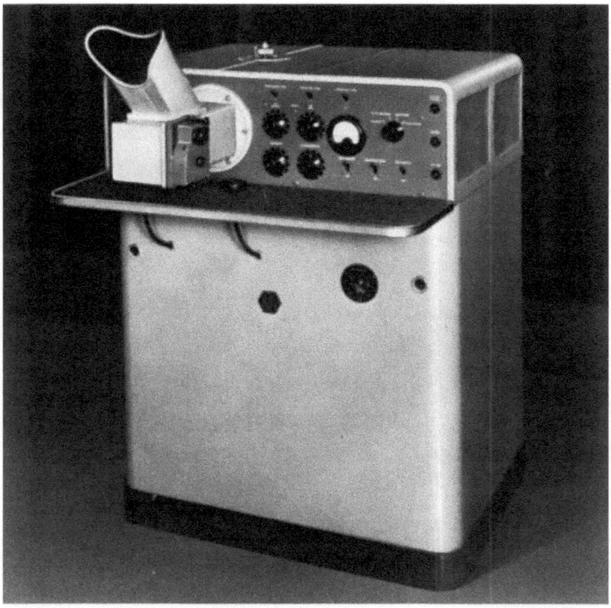

Figure 7 General view of the EM4 console microscope for routine electron microscopy.

was soon to join the AEI Research Laboratory and make a big impact on lens calculations. The electron-optical column of the EM4 consisted of a condenser lens, an objective lens, and a projector lens provided with two magnetic polepiece gaps in series. This enabled many of the advantages of a three-stage imaging system to be incorporated into a two-lens unit design. The high-tension set, lens supplies, and vacuum system were mounted in one compact console as shown in Fig. 7. The prototype instrument was demonstrated at the Physical Society Exhibition of 1950 at Imperial College, London, and the first production instrument was delivered in June, 1953, to the French military Attaché. The EM4 had a guaranteed resolving power of 100 Å.

8. High-voltage electron microscopy—the EM5

The problem of examining thick specimens has existed from the beginning of electron microscopy, and although ways have usually been found of thinning the sample to an acceptable thickness, in the region of 1000 Å for a 100-kV microscope, there has always been a desire to increase the accelerating voltage in order to be able to penetrate thick specimens without loss of resolution. In 1950 exploratory work was commenced by Page, Beadle, and Garfitt in the Research Department at Manchester to investigate the feasibility of designing and constructing a 500-kV electron microscope intended primarily for the examination of thick biological sections. Basically the equipment consisted of the imaging system of an EM3 electron microscope, pushed, however, well beyond its design limits, together with an experimental illuminating system operating at 500 kV. In spite of the limitations thus imposed on the system, it was nevertheless possible to achieve a resolution of about 100 Å. However, as often happens in industrial development, by 1953, when the instrument was ready for service, considerable improvements had been made in the performance of ultramicrotomes and so the interest of biologists in high-voltage microscopy waned and the EM5 never reached the production stage. However, as a mark of respect the name EM5 was not transferred to the next production instrument! The effort on the EM5 was not wasted, for soon afterward metallurgists began to have misgivings about the scientific value of their observations of dislocations in thin foils in 100-kV instruments. There was evidence that the behavior of dislocations in such foils was strongly influenced by surface effects and therefore not representative of their behavior in bulk material. Specimen thicknesses of 1 μm or more appeared to be necessary in order to reproduce bulk material effects. Electron penetration was here more relevant than high resolution and several high-voltage microscopes working at 300 kV or more were built in the Soviet Union (Popov, 1959), Japan (Kobayashi et al., 1963; Tadano et al., 1956), and France (Dupouy et al., 1961), the latter being capable of operating at 1.5 MeV. In the United Kingdom an experimental 500-kV microscope was built at the Cavendish laboratory (Cosslett, 1961), and the AEI Research Laboratory carried an initial design study for a 1-MeV electron microscope in collaboration with the National Physical Laboratory (Franks & Halliday, 1961). This led to the production of the very successful series of EM7 1000-kV microscopes.

9. Improving the resolving power—EM3A and the emergence of electron holography

In parallel with the research and development carried out at Metropolitan Vickers, Manchester, the research team set up in 1947 at the AEI Research Laboratory, Aldermaston, began a comprehensive long-term research program into all aspects, theoretical and practical, of the factors limiting the resolving power of the electron microscope. The main results of this program have been described by Haine (1954), who directed the work, which led to a considerable revision of current ideas and concepts in electron microscope design and operation, and especially in introducing quantitative assessment of all aspects of instrumental performance. At the time, considerable controversy surrounded the optimum mode of operation of the thermionic electron gun, often accompanied by misleading experimental observations. A classic investigation by Haine and Einstein (1952) settled many, but not all, of these controversial points. A related topic of great emotional significance to many electron microscopists, namely the apparently random failure of expensive tungsten hairpin filaments, was elegantly analyzed by Bloomer (1957), who showed how the two chief causes of failure, namely, failure by evaporation or by chemical attack by water vapor, could be distinguished merely by inspecting the worn filament and how the filament life could be predicted with high accuracy. Ennos (1953) carried out a definitive study of the causes of specimen contamination by hydrocarbon vapors present in the vacuum system and developed an efficient decontaminator. Liebmann attacked the problem of the direct calculation of the axial field distribution in magnetic electron lenses by a resistance analog of the finite-difference method for solving the Laplace and Poisson equations. The accuracy was sufficiently great to allow the third-order aberrations of real lenses to be calculated to an accuracy of a few percent (Liebmann & Grad, 1951), a significant advance over the previous approximate methods of assuming a model field distribution. Bradley (1954) invented the carbon replica technique and the thin-carbon support film, still in universal use. Haine and Jervis (1955) developed a pair of electron velocity spectrometers for stabilizing and monitoring the 100-kV accelerating voltage to an accuracy of the order of a few parts per million for high-resolution work calling for long exposures. Haine and Mulvey (1956) developed quantitative methods for measuring instrumental resolving power and the diagnosis of imaging faults by the use of Fresnel fringes. Sturrock (1951), as a visiting scientist to the laboratory, worked out the theory of the machining tolerances of polepiece bores and faces and the alignment tolerances required to reach atomic resolution. First estimates showed that they were unbelievably small! Archard (1953) translated the difficult mathematical terminology into realistic computer programs which completely verified the astonishing degree of accuracy which had to be achieved in a high-resolution objective lens. Archard (1956) also looked very closely into the mechanical and electrical feasibility of constructing and correcting the spherical aberration of the objective lens of the electron microscope by the use of nonrotationally symmetric elements such as quadrupoles, and octupoles as first proposed by Scherzer in the early days of electron microscopy. Although Archard came up with many ingenious ideas, the mechanical tolerances and effects of iron hysteresis as judged by preliminary experimental trials seemed to us to preclude this method of improving the resolving power of the electron microscope, even in an experimental instrument, let alone a commercial prototype.

On the light-optics side, J. Dyson produced a remarkable series of optical interference microscopes which were not strictly in line with the official product range of an electrical engineering company but which were so impressive that the broadminded director of the laboratory, Dr. T. E. Allibone, was able to subsume them into his general research program. The strong laboratory background in optical interference microscopy served us well when, one day, Gabor came over from the BTH Company at Rugby to introduce the idea of electron holography as the key to atomic resolution in the electron microscope, an eye opener indeed to any electron microscopist brought up in the tradition of the cathode-ray oscillograph.

Gabor had left Hungary in the summer of 1934, having sought a position with the Metropolitan Vickers Company at Manchester through the good offices of Dr. Allibone (1980). Gabor was immediately offered a 2-year contract with the British Thompson Houston Company at Rugby, sister company within the AEI group of companies. In the event he remained with the company until 1948, the year in which he invented holography, for which he later received the Nobel Prize.

Gabor had always openly admitted that he was disappointed with himself for failing to produce the theory of his ironclad lens before Busch beat him to it. However, when he later realized that such lenses have inevitable spherical aberration he constantly exercised his mind on inventing ways and means of carrying out electron microscopy without this drawback, in order to attain atomic resolution.

In 1946, after thinking about magnetrons, he proposed a "zonally corrected electron lens" (Gabor, 1946) in which a thin wire placed on the axis of a lens emits electrons and so creates an "electron swarm" which corrects the spherical aberration of an incident extra-axial bundle of incident electrons confined to a well-defined axial zone of the lens. He wanted Metropolitan Vickers to build this lens and try it out. Before agreeing to embark on such a project, Haine wisely insisted that estimates be made of the necessary mechanical tolerances. The results were discouraging, even for Gabor, who was not normally deterred by near-impossible tolerances, and the project was quietly dropped.

However, during the Easter holiday of 1947, while idly awaiting his turn for a game of tennis, the idea for an electron holographic microscope suddenly sprang into Gabor's mind (Allibone, 1980) "effortlessly" as if by supernatural intervention.

The basic idea seemed remarkably simple but highly improbable to the majority of electron microscopists: Place a small object, ideally a single atom, but preferably without using a support film, in a shadow electron microscope of the type pioneered by Boersch (but never produced commercially). On the photographic plate there would be projected the unscattered background (reference) wave together with a (weakly) scattered wave from the specimen. The resulting interference pattern (hologram) would thus contain all the information about the specimen in amplitude and phase since the background wave would act as a reference. The image from the hologram could be reconstructed by projecting a light beam through the hologram onto a photographic plate. By suitably adjusting the spherical aberration, astigmatism, and focus of the reconstructing system, these corresponding image defects of the electron microscope would be eliminated and hence atomic resolution could be realized!

Gabor tried out the idea optically at BTH Rugby with the help of Ivor Williams, a gifted research worker. It was not an easy experiment to carry out as, unfortunately, the laser had not yet been invented at that time. However, success was achieved and the results were published in 1948 in *Nature* (Gabor, 1948) and in more detail in 1949 (Gabor, 1949).

Work on electron holography was under way at the Aldermaston Laboratory in 1949 and the first electron holograms were quickly obtained by converting a transmission microscope into a projection (shadow) electron microscope. J. Dyson, A. E. Ennos, and J. Wakefield designed, constructed, and commissioned the optical reconstruction apparatus in which the spherical aberration and astigmatism could be varied, since it was not possible to predict the aberrations of the electron microscope with sufficient accuracy.

Unforeseen difficulties arose with the projection microscope method, especially with the severe limitation of the field of view caused by chromatic aberration. This was overcome when Haine and Dyson (1950) realized that holography could be carried out more conveniently in a standard transmission microscope provided with a coherent source (the transmission method). However, in both methods, a more fundamental difficulty arises. In this method, now called "in-line" holography, two images appear in superposition. One is the desired reconstruction, in focus and free from aberration; the other is out of focus and is afflicted with double the original aberrations. There are ways of reducing its disturbing effects but only at the cost of increasing the exposure.

The important thing now was that the taking of holograms and the taking of standard high-resolution micrographs could both be carried out in the same instrument (Haine & Mulvey, 1952). In fact, we succeeded in improving the resolution of both methods. However, we did not succeed in obtaining better resolution by holography than we did with conventional microscopy. The reason probably lay in the fact that in our electron microscope the limitation was probably not set by the spherical aberration of the objective lens. Nevertheless the encounter with Gabor had been wonderfully stimulating and had changed considerably our way of working at high-resolution microscopy.

These wide-ranging researches at Aldermaston and the manufacturing experience of the production department at Manchester placed the design of AEI's electron microscopes on a firm scientific basis, and culminated in 1953 in the design and construction of an experimental high-resolution microscope (Haine & Mulvey, 1956) in which a resolution of better than 10 Å could be regularly attained. This instrument paved the way for the EM6 series of microscopes, which were produced as soon as the commercial demand for such instruments arose. At that time, few electron microscopists needed such a resolution. The first fruit of this research program was, in fact, the EM3A shown in Fig. 8, a greatly improved version of the EM3 microscope with a guaranteed resolving power of 25 Å and a modest price. The improvement in resolution was made possible mainly by technological improvements and an elimination of all unnecessary components. Close attention was paid to the construction of the lenses themselves. The traditional polepieces were abandoned and replaced by an integral lens construction. Such lenses proved extremely easy to align. In addition, a new tilting section was inserted so that the illuminating beam could be tilted mechanically through an angle of 10° with respect to a solid specimen, as shown in Fig. 8, to enable a "reflection" image to be taken. This technique was widely used for studying wear and abrasion of metal surfaces (Bradley et al., 1956), but went out of fashion when scanning electron microscopes became generally available.

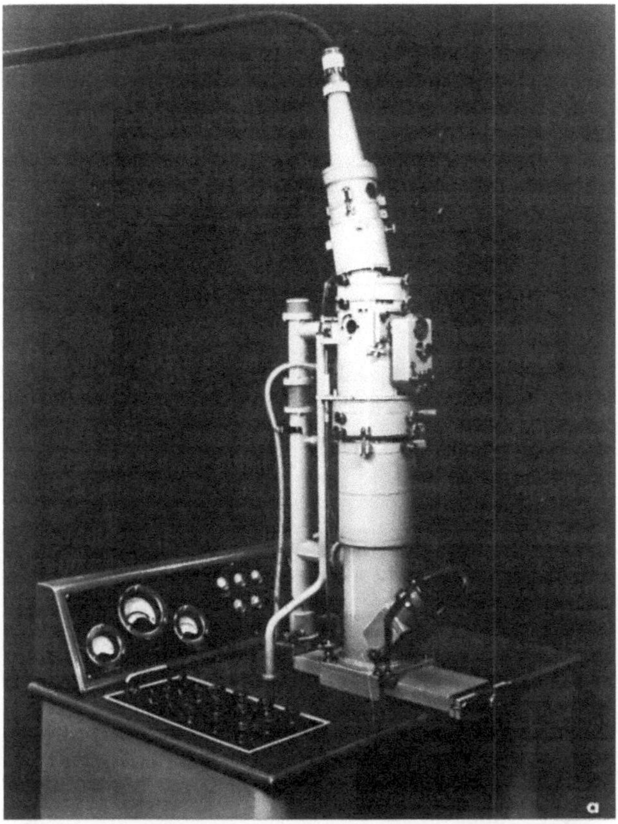

Figure 8 (a) The Met-Vick EM3A electron microscope with large-angle (10°) illumination tilt for reflection microscopy, (b) Diagram of inner and outer construction.

10. High-resolution microscopes—EM6

During the period 1947–1954, great progress had been made in Europe and the United States in improving instrumental resolving power. By 1954, in several laboratories in different parts of the world, there were specially "tuned-up" electron microscopes capable of a resolution of 10 Å or better in capable hands. A demand was building up for a commercial instrument of high resolution. It was perhaps fitting that Ruska and his colleagues were able in 1954 to introduce the Siemens Elmiskop I in which this performance could be obtained without difficulty in a production instrument (Ruska, 1956). The Elmiskop I thus became a landmark in the commercial world. The Elmiskop I was not without its shortcomings, but in the future all manufacturers were compelled to produce a matching resolution and re-

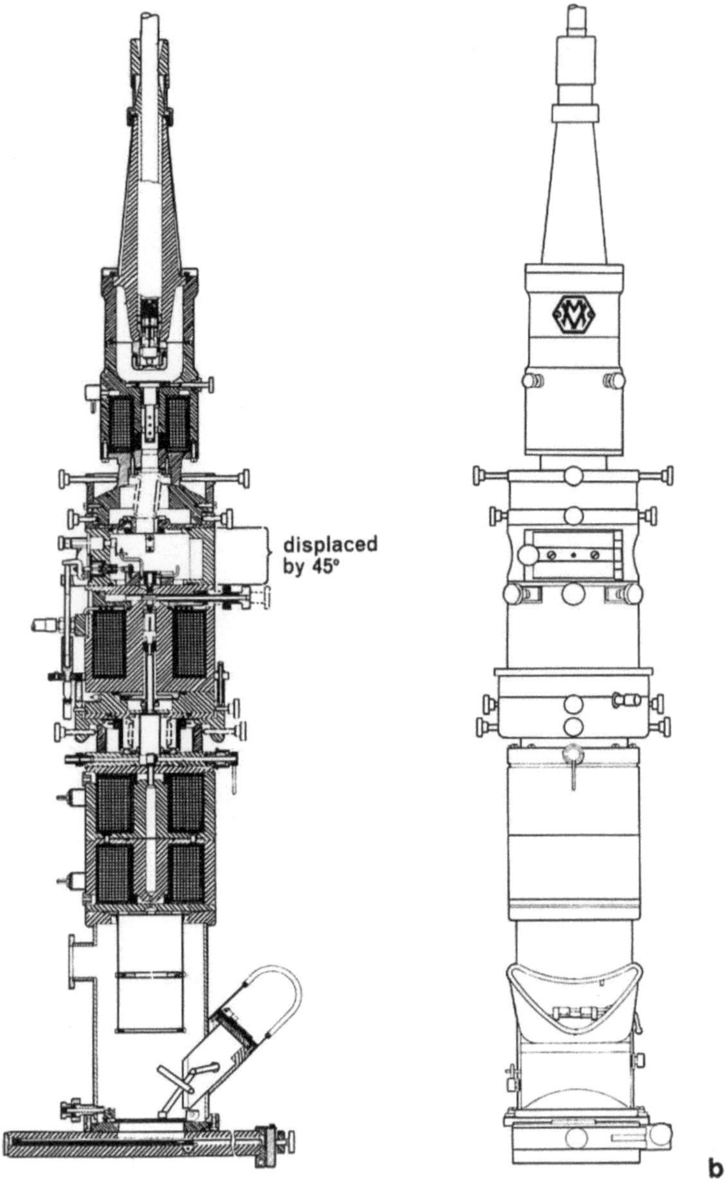

Figure 8 (*continued*)

liability. This transformation of the scene took place with remarkable speed. The EM6, with a similar high-resolution specification, appeared soon after (Haine & Page, 1957), thanks to the previous intensive period of research

Figure 9 The Met-Vick EM6 high-resolution electron microscope (Haine & Page, 1957).

and development at the AEI Research Laboratory. A general view of the EM6 is shown in Fig. 9. A double-condenser system was fitted; this has the big advantage that the operator can control the illuminated area of the specimen and its intensity independently. Consequently, a delicate specimen can be preserved for the maximum time. On the other hand, local thermal stressing is also possible when examining dislocation movements in thin-metal foils. It is not easy to tilt such a condenser mechanically so an electrical tilt system was devised that could tilt the illuminating beam by angles of up to 12°. This is a difficult electron-optical task, involving high-order aberrations, but it was made easy by previous investigations carried

Figure 10 An EM6 electron microscope fitted with an image intensifier transmitting high-resolution electron microscope images at TV rates directly into the BBC television network, without the aid of a studio camera and without the formation of an image on the fluorescent screen (December, 1959).

out by Archard on quadrupole and octupole systems. The resolution and image quality of the EM6 was checked by Fresnel fringe methods, which are completely objective and independent of the personal judgment of the operator. The first production model was installed at the Royal College of Surgeons in May, 1958.

11. The image intensifier

Even when a high instrumental performance can be guaranteed, the taking of a successful micrograph still depends largely on the skill and judgment of the operator. Even with dark adaptation and the use of a binocular telescope to improve the light-gathering power of his eyes, an operator can only barely compete with the photographic plate, which is sensitive to the impact of one electron. As early as 1955 Haine (1955) suggested that an electron image intensifier would be of great assistance and with the assistance of Ennos and Einstein, proceeded to develop one for the electron microscope (Haine et al., 1958). In this intensifier the normal fluorescent screen of the microscope was replaced by a thin layer of selenium (Haine & Einstein, 1962). The final image is projected onto this, as in the Vidicon television camera tube, so that electron intensity variations are converted into corresponding variations of charge in the layer. The charge is swept off by a low-voltage scanning beam, thereby creating a video signal which can modulate a cathode-ray display tube scanned in synchronism. The image intensifier eliminates the need for dark adaptation and is the only convenient way of demonstrating high-resolution microscopy to large audiences. Thus it proved possible in December, 1959, to display high-resolution images at one-million times magnification during BBC television broadcasts (Fig. 10) in which the video signal from the intensifier on the EM6 microscope modulated the television transmitter directly at TV rates, without the aid of a studio camera and without an image being formed on the fluorescent screen. This perhaps illustrates the point that the primary purpose of the image intensifier in high-resolution microscopy is to transmit an enormous quantity of data rapidly, thereby extending the visual powers and memory of the operator; these include not only the image-forming system of his eyes but also of the pattern-recognition system of the brain. It was realized early on in the development of electron beam holography that the storage of the original signal produced by the microscope was only the first stage in a complex process of extracting the information contained in the specimen from the inevitable instrumental perturbations superimposed on it. In the late 1950s, however, the dedicated computing power needed for this task was simply not available, and so a highly skilled operator could usually gain only minor benefits from the image intensifier.

12. The electron microscope in biology

In general, specimen preparation techniques rather than instrumental performance has been the dominating factor limiting the resolution of biological micrographs. However, in the 1960s great improvements were made in the cutting of thin sections and in negative-staining techniques. These made a high-resolution microscope essential in branches of the subject such as virology. The EM6B was designed to meet this need. A high-resolution "pinhole" objective lens initially suggested by Liebmann (1955) in the early 1950s but never before incorporated into a production instrument formed the basis of this new microscope. This lens has the low spherical aberration of the well-known condenser–objective lens first proposed by Glaser and realized by Riecke and Ruska, but without the operational inconvenience of the strong prefield. The controls of this microscope were therefore extremely simple. The top magnification was 250,000×. The resolution of 5 Å readily achieved in this instrument was a notable improvement over the 100 Å obtained on the EM2 only 17 years previously.

13. Very high-resolution microscopy

Although commercial production of transmission electron microscopes has now unfortunately ceased in the United Kingdom, brief mention should perhaps be made of the technology transfer that has been possible in the design and construction of the 600-kV high-resolution electron microscope at the University of Cambridge. This is a joint project by the Department of Engineering and the Cavendish Laboratory, with major grant support from the Science and Engineering Research Council, with some additional support from the Department of Industry and AEI Scientific Instruments Ltd. The basic arrangement and layout of the microscope is essentially that of the AEI EM7 1000-kV microscope including the Haefely generator supplying the accelerating voltage, but with many crucial refinements of design. The accelerating voltage of 600 kV was carefully chosen on technological grounds to maximize instrumental resolution (Cosslett, 1980). Important features are the high-brightness lanthanum hexaboride gun, unusual in a high-voltage microscope, a condenser–objective lens of the Riecke–Ruska type, a remarkably stable mechanical stage, an ingenious girder suspension system to reduce the effects of external vibration, and an image intensifier system interfaced to a dedicated on-line computer. This microscope represents the culmination of a long period of industrial development brought to perfection by the combined innovative forces of the Department of Engineering and the Cavendish Laboratory at Cambridge University. It was perhaps fitting that M. E. Haine was closely associated with the project from its early conception to the realization of atomic resolution. The mechanical and electrical stability of the instrument is about 0.1 nm (1 Å) and resolutions in the range 0.15–0.2 nm have already been achieved, as verified by optical diffractometry and computer image analysis. It is of great technical and scientific interest that this has been achieved without the necessity of correcting the spherical aberration or invoking the aid of holography, thereby perhaps vindicating Ruska's early faith that his polepiece lenses would one day succeed in achieving atomic resolution.

14. Afterword by Peter Hawkes

Oxford University Press has kindly allowed me to reproduce the article on Tom Mulvey that was published in the Oxford Dictionary of National Biography (2015).

Mulvey, Thomas [Tom] (1921–2009), physicist, was born on 26 July 1921 at 124 Greenhays Road, Chorlton upon Medlock, Manchester, the eldest son of Thomas Mulvey (1888–1951), cashier for an iron and steel merchant, later a schoolmaster, and his wife, Margharetha Magdalena, *née* Neethling (1888/9–1934), an Afrikaans-speaking South African. He had two brothers, Paul (*b.* 1923) and David (*b.* 1928), and a sister, Joan (*b.* 1926). After the early death of their mother, in 1936 Thomas Mulvey senior married Bridget Mary Teresa (Bridie) O'Regan (1898/9–1958), with whom he had a son, Joseph (*b.* 1940).

From 1934 to 1938, Tom Mulvey attended St Bede's, an independent Catholic grammar school in Manchester, where he became a wireless fan and came across *Television Reception* (1936), the English translation of *Fernsehempfang* by Baron Manfred von Ardenne, a pioneer of electron microscopy. The system of college apprenticeships took Mulvey to Metropolitan–Vickers in 1939, after which he gained a first degree in engineering at Manchester University (1940–43). In May 1943 he joined the anti-submarine branch of the navy as an acting sub-lieutenant in the Royal Naval Volunteer Reserve. After a spell of responsibility for anti-submarine equipment in Plymouth and Devonport, he was sent to Ceylon to oversee anti-submarine detection material.

After VJ-day, Mulvey was demobilized and rejoined Metropolitan–Vickers, where he encountered his first electron microscope. In these instruments, the image is formed by means of electrons, not light as in the more familiar kind of microscope. The electrons are focused by passing them through a circular opening in an electromagnet (magnetic lenses) or through round openings in metal electrodes held at different voltages (electrostatic lenses). At that time, a resolution of a few nanometres could be attained. While working at Metropolitan–Vickers, he studied for an MSc at Manchester, awarded in 1949 for a dissertation on 'The symmetrical electron lens as an element of the electrostatic microscope'; this led to his first publication, 'Distortion-free electrostatic lenses', which appeared in *Nature* in 1949. Meanwhile Metropolitan–Vickers, now part of Associated Electrical Industries (AEI), had set up a research laboratory at Aldermaston Court, in Berkshire, and there Mulvey remained until 1963 when the laboratory was closed down. On 16 April 1955, at the Roman Catholic Church of Our Lady and Saint Anne, Reading, he married Rita Bond (1927–2002), schoolteacher, and daughter of William Walter John Bond, gardener. They had two sons, David (*b.* 1956) and Nicholas (*b.* 1959).

Mulvey's years at Aldermaston Court were very happy and productive years, marked in particular by his meeting with Dennis Gabor, inventor of holography in 1948 as a way of correcting a severe defect of electron lenses. Mulvey was one of the small group of scientists who performed the first tests of holography. His 'Initial results in the practical realisation of Gabor's diffraction microscope' were presented at the international congress on electron microscopy held in Paris in 1950. But the time was not yet ripe for practical holography, which requires a laser or a powerful computer, neither of which existed, and a brighter electron source than those then available. Mulvey also worked on many other aspects of electron optics, notably on magnetic circuit design; this study was much admired by a celebrated German electron optician, Dieter Riecke (who had been trained as a U-boat officer but fortunately did not encounter Mulvey during the war). This work was important because the resolution of an electron microscope is governed by the quality of the objective (the lens situated immediately beyond the specimen), the properties of which are determined by the shape of the magnetic field and hence by the geometry and homogeneity of the magnet used for focusing.

After the closure of Aldermaston Court, Mulvey spent two uninteresting years at AEI, Harlow before accepting a readership in electron physics at Birmingham College of Advanced Technology in 1965; the college became Aston University the following year. In 1971 his alma mater, Manchester University, awarded him a DSc, which pleased him particularly, since it recognized the importance of his research. Three years later he was appointed professor of physics at Aston, a post which he occupied until retirement in 1986, when he was granted emeritus status. His Aston years were remembered for the miniature electron lenses that he would bring out from his pockets at congresses and flourish before his audience and also for the close personal links he established with his young research students from North Africa and the Middle East and from China. Already a considerable linguist, he added Arabic to the languages of which he had a working knowledge: German (in which he was fluent), French, Chinese, Japanese, and Russian. He was assured that he need never sleep in a hotel when travelling in the Arab world. He also established close relations with groups of scientists behind the iron curtain, notably in Brno in Czechoslovakia and Halle in the German Democratic Republic, which he visited regularly from 1981 onwards, bringing books and scientific papers, and above all a breath of air from the free world to colleagues oppressed by the communist tyranny. The

Czech Academy of Sciences awarded him the Křižík Gold Medal in 1995 in recognition of these services.

Mulvey also contributed extensively to the history of the electron microscope. As early as 1962, when little had been published about the beginnings of the subject, he wrote 'Origins and historical development of the electron microscope' (an article published in the *British Journal of Applied Physics*) and several more such studies appeared, as well as biographies of Ernst Ruska, builder of the first electron microscope, Jan Le Poole, who launched the Dutch school of electron microscopes, Ellis Cosslett, and Dennis Gabor. In 1966, Ernst Ruska was elected to the Deutsche Akademie der Naturforscher Leopoldina and later published a fully documented history of the electron microscope in their *Acta*. This essential document for the history of the electron microscope was translated by Mulvey as *The Early Development of Electron Lenses and the Electron Microscope* (1980). When Cosslett could no longer continue editing *Advances in Optical and Electron Microscopy*, Mulvey took over and remained as associate editor when this serial amalgamated with *Advances in Electronics and Electron Physics* to form *Advances in Imaging and Electron Physics*; for the latter, he edited a stout volume, 'The Growth of Electron Microscopy' (1996).

Tom Mulvey, a man of great kindness and generosity, devoted much time to the charitable activities of the Society of St Vincent de Paul. He also had a highly developed sense of humour and an inexhaustible fund of jokes. The communist regimes were the source of some of the best of these and a few were recalled by Eric Munro in his contribution to the eightieth birthday issue of the *Proceedings of the Royal Microscopical Society*. Whenever possible, he travelled light, preferably by the humbler forms of transport (notably bus and tram); on one visit to Brno, he arrived at the railway station with just one run-down piece of luggage and a conference bag. 'Our driver', recalled Bohumila Lencová, 'walked past him several times, because he did not believe that the visitor would come for four weeks with so little luggage and recognised him only when everyone but Tom had left' ('Tom Mulvey, a celebration', 217). In 1984 he was made an honorary fellow of the Royal Microscopical Society, of which he had been a fellow since 1966; he received the 2001 distinguished scientist award of the Microscopy Society of America, which elected him a fellow in 2009.

For the closing years of her life, Rita Mulvey needed constant attention and Tom Mulvey set aside electron microscopy to care for her. After her death in 2002, he picked up the threads and that same year participated in a seminar, 'Recent Trends in Charged Particle Optics', appropriately enough in the Czech Republic, not far from Brno. In his later years, he moved from Birmingham to Oxford, close to one of his sons. Afflicted with Alzheimer's disease, he died of an abdominal tumour at Heathfield House Nursing Home, Bletchingdon, Oxfordshire, on 16 July 2009. Both sons survived him.

References

Allibone, T. E. (1980). Dennis Gabor: a biographical memorial lecture. *Israel Journal of Technology, 18*, 201–208.

Archard, G. D. (1953). Magnetic lens aberrations due to mechanical defects. *Journal of the Institution of Electrical Engineers, 30*, 352–358.

Archard, G. D. (1956). *Electron-optical properties of electrode systems of four- and eight-fold symmetry. London, 1954*. (pp. 97–105).

Bloomer, R. N. (1957). The lives of electron microscope filaments. *British Journal of Applied Physics, 8*, 83–85.

Boersch, H. (1936). On the primary and secondary image in the electron microscope. Concepts of the diffraction image and its effect on imaging. *Annalen der Physik (Leipzig) [5], 26*, 631–644.

Bradley, D. E. (1954). *British Journal of Applied Physics, 5*, 65 and 96.

Bradley, D. E., Halliday, J. S., & Hirst, W. (1956). Stereoscopic reflection electron microscopy. *Proceedings of the Physical Society of London, Section B, 79*, 484–485.

Burch, F. P., & Whelpton, R. V. (1932). The technique of the high speed cathode ray oscillograph. *Journal of the Institution of Electrical Engineers, 71*, 380–388.

Busch, H. (1927). On the action of the concentration coil in a Braun tube. *Archiv für Elektrotechnik (Berlin), 18*, 583–594.

Cosslett, V. E. (1961). New electron microscopes of very high voltage. *New Scientist, 11*, 387–390.

Cosslett, V. E. (1980). Principles and performance of a 600 kV high resolution electron microscope. *Proceedings of the Royal Society of London. Series A, 370*, 1–16.

Driest, E., & Müller, H. O. (1935). Electron microscopic photographs (electron micrograms) of 'Chitin' specimens. *Zeitschrift für Wissenschaftliche Mikroskopie, 52*, 53–57.

Dupouy, G., Perrier, F., & Fabre, R. (1961). An electron microscope operating at very high voltage. *Comptes Rendus Hebdomadaires des Séances de L'Académie des Sciences, 252*, 627–632.

Ennos, A. E. (1953). The origin of specimen contamination in the electron microscope. *British Journal of Applied Physics, 4*, 101–106.

Franks, A., & Halliday, J. S. (1961). Plans for a large electron microscope. *Engineering (London), 21*.

Gabor, D. (1927). *The cathode ray oscillograph*. Forschungsh. Stud. Ges. Höchstspann Anlag. No. 1. Doctoral Dissertation. Berlin: Technische Hochschule.

Gabor, D. G. (1946). A zonally corrected electron lens. *Nature (London), 158*, 198.

Gabor, D. G. (1948). A new microscopical principle. *Nature (London), 161*, 111.

Gabor, D. G. (1949). Microscopy by reconstructed wavefronts. *Proceedings of the Royal Society of London. Series A, 197*, 457–487.

Haine, M. E. (1947). The design and construction of a new electron microscope. *Journal of the Institution of Electrical Engineers, 14*, 447–462.

Haine, M. E. (1954). The electron microscope—a review. *Advances in Electronics and Electron Physics, 6*, 295–370;
Also see Haine, M. E., & Cosslett, V. E.. *The electron microscope*. London: Spon, (1961).
Haine, M. E. (1955). Some observations on the present and future possibilities of the transmission electron microscope. In *C. R. C.N.R.S. colloq. recent techn. electron corpuscular microsc.*
Haine, M. E., & Dyson, J. (1950). A modification to Gabor's proposed diffraction microscope. *Nature (London), 166*, 315.
Haine, M. E., & Einstein, P. A. (1952). Characteristics of the hot cathode electron microscope gun. *British Journal of Applied Physics, 3*, 40–46.
Haine, M. E., & Einstein, P. A. (1962). Intensification of the electron microscope image using cathode-conductivity in selenium. *Proceedings of the Institution of Electrical Engineers. Part B, 109*, 185–195.
Haine, M. E., & Jervis, M. W. (1955). The use of an electron velocity analyser to stabilise a 50 kV D. C. voltage source to a few parts in a million. *Proceedings of the Institution of Electrical Engineers. Part B, 102*, 265–274.
Haine, M. E., & Mulvey, T. (1952). The formation of the diffraction image with electrons in the Gabor diffraction microscope. *Journal of the Optical Society of America, 42*, 763–773.
Haine, M. E., & Mulvey, T. (1956). *The regular attainment of very high resolving power in the electron microscope. London, 1954*. (pp. 698–705).
Haine, M. E., & Page, R. S. (1957). *A new universal electron microscope of high resolving power: Metro-Vick type EM6. Stockholm, 1956*. (pp. 32–37).
Haine, M. E., Page, R. S., & Garfitt, R. G. (1950). A three-stage electron microscope with stereographic, dark field and electron diffraction facilities. *Journal of Applied Physics, 21*, 173–182.
Haine, M. E., Ennos, A. E., & Einstein, P. A. (1958). An image intensifier for the electron microscope. *Journal of the Institution of Electrical Engineers, 35*, 466–467.
Knoll, M., & Ruska, E. (1932). A contribution to geometrical electron optics. *Annalen der Physik (Leipzig) [5], 12*, 607–640 and 641–661.
Kobayashi, K., Hashimoto, H., Suito, E., Shimadzu, S., & Iwanaga, M. (1963). Construction of 300 kV electron microscope and its electron microscopy. *Journal of Applied Physics, 2*, 47–57.
Le Poole, J. B. (1947). A new electron microscope with continuously variable magnification. *Philips Technical Review, 9*, 33–46.
Liebmann, G. (1948). A new experimental electron microscope. *Journal of the Institution of Electrical Engineers, 25*, 37–43.
Liebmann, G. (1955). The magnetic pinhole lens. *Proceedings of the Physical Society of London, Series B, 78*, 682–685.
Liebmann, G., & Grad, E. M. (1951). *Proceedings of the Physical Society of London, Section B, 64*, 956–971.
Martin, L. C., Whelpton, R. V., & Parnum, D. H. (1937). A new electron microscope. *Journal of the Institution of Electrical Engineers, 14*, 14–24.
Marton, L. (1934). The electron microscope. First attempts at application to biology. *Annales et Bulletin de la Société Royale des Sciences Médicales et Naturelles de Bruxelles, 5 and 6*, 92–106.
Marton, L. (1935). The electron microscope and its applications. *Revue d'Optique, Théorique et Instrumentale, 14*, 129–145.
Marton, L. (1945). A 100 kV electron microscope. *Journal of Applied Physics, 16*, 131–138.
Page, R. S. (1954). A compact console type electron microscope. *Journal of the Institution of Electrical Engineers, 31*, 200–205.
Popov, N. M. (1959). An electron microscope-electron diffraction camera with an accelerating voltage of 400 kV. *Izvestiâ Akademii Nauk SSSR. Seriâ Fizičeskaâ, 23*, 436–441.

Ruska, E. (1934). Progress in the construction and performance of the magnetic electron microscope. *Zeitschrift für Physik, 87*, 580–602.
Ruska, E. (1956). *A high resolution 100 kV electron microscope with small area illumination. London, 1954.* (pp. 673–693).
Ruska, E. (1980). *The early development of electron lenses and electron microscopy*. Stuttgart: Hirzel Verlag.
Sturrock, P. A. (1951). The aberrations of magnetic electron lenses due to asymmetries. *Philosophical Transactions of the Royal Society of London, Series A, 243*, 387–429.
Tadano, B., Sakaki, Y., Maruse, S., & Morito, S. (1956). New 300 kV electron microscope. *Journal of Electron Microscopy, 4*, 5–9.
von Borries, B., & Ruska, E. (1939a). An electron microscope for research institutes. *Naturwissenschaften, 27*, 577–582.
von Borries, B., & Ruska, E. (1939b). Construction and performance of the Siemens electron microscope. *Zeitschrift für Wissenschaftliche Mikroskopie, 56*, 317–333.
Wiechert, E. (1899). Experimental investigations into the velocity and the magnetic deflection of cathode ray rays. *Annalen der Physik und Chemie, 69*, 739–766.
Zworykin, V. K., Hillier, J., & Vance, A. W. (1940). An electron microscope for practical laboratory service. *Transactions of the American Institute of Electrical Engineers, 60*, 157–162.

CHAPTER NINE

The development of the scanning electron microscope

C.W. Oatley[a], D. McMullan[a], K.C.A. Smith[a], and Peter Hawkes (Afterword)[b,*]
[a]Formerly Engineering Department, Cambridge University, Cambridge, United Kingdom
[b]CEMES-CNRS, Toulouse, France
*Corresponding author. e-mail address: hawkes@cemes.fr

Contents

1. Introduction 255
2. The initial aims of the three groups 256
3. The Cambridge microscope 260
4. Uses of the microscope 275
5. Afterword by Peter Hawkes 283
Bibliography including cited references 283

1. Introduction

In the development of the scanning electron microscope for the examination of solid surfaces three groups have made significant contributions. Following earlier work by M. Knoll (1935), who was interested in studying secondary emission and did not use demagnifying lenses to produce a fine electron probe, M. von Ardenne (1938a) published an important theoretical paper in which he clearly enunciated the principles underlying the scanning microscope as we know it today. He introduced the use of demagnifying lenses, examined the aberrations that they would produce, and made calculations of the resolution that might be obtained under various conditions of operation. He constructed a scanning microscope and published details of his experimental work (1938b). The research was interrupted by the war and the microscope was destroyed in an air raid in 1944. von Ardenne has himself published his own account of this early work (von Ardenne, 1978), so it is unnecessary to say more about it here other than to pay tribute to the great importance of his pioneer research.

During the early part of the war a scanning microscope was constructed in the United States in the RCA Laboratories and a description of this

Advances in Imaging and Electron Physics, Volume 221
ISSN 1076-5670
https://doi.org/10.1016/bs.aiep.2022.03.009

instrument was published by V. K. Zworykin, J. Hillier, and R. L. Snyder (1942). The results obtained with this instrument, although by no means negligible, did not seem to justify commercial exploitation, and the project was discontinued.

After this there was a period of 6 years during which little or no work on scanning microscopy was done. Then, in 1948, C. W. Oatley decided that the construction of such a microscope would be a suitable subject for a research student working in the Engineering Department of the University of Cambridge. (Several experts expressed the view that this would be a complete waste of time!) Thus began a project which lasted for 13 years till 1961, when the commercial manufacture of scanning microscopes was assured. During this time the work progressed continuously and was carried out by a succession of research students, all supervised by Oatley.

A review of scanning electron microscopy published shortly after commercial instruments had become available (Oatley et al., 1965) contains a detailed account of the technical achievements of the Cambridge group. A more recent article (Oatley, 1982) gives a full account of the work of this group from a more personal point of view. It explains the financial constraints under which the group operated and assigns credit for the advances made as the project progressed. It also describes how the microscope finally went into commercial production.

From the above it will be clear that the historical facts relating to the development of the scanning microscope are already on permanent record. No good purpose would be served by repeating them here and the object of the present contribution is different. It is to consider the historical facts in the light of present-day knowledge, but also in the light of knowledge existing at the time in question and to attempt to explain why the development took the course that we know.

2. The initial aims of the three groups

When the leader of a group of research workers initiates a project, he has in mind, at least in broad outline, the ultimate object that he hopes to achieve. In the case of the scanning microscope, it is important to realize that the objects which led the three groups under consideration to begin work were quite different from each other.

To von Ardenne, cognizant of the great success being achieved by the transmission electron microscope, the scanning system offered an escape from the restriction to very thin specimens that chromatic aberration imposed on the conventional instrument. He was primarily interested in building a scanning transmission microscope. He fully appreciated that the instrument might be used for the examination of surfaces of solid objects and even foresaw the possibilities of electron beam microfabrication of small objects. When time permitted, he conducted a few experiments along these lines but, with the technique available at that time, these were not particularly successful. At all times, however, the transmission work had priority.

When, a few years later, work was started in the RCA Laboratory, the outlook was quite different. Improvements had been made to the transmission microscope and it seemed unlikely that a scanning transmission instrument would have any additional advantages to offer. On the other hand, a microscope for the examination of solid surfaces, giving better resolution than optical instruments, would be useful. A scanning electron microscope might meet this need if the output signal were obtained from secondary electrons, with contrast resulting from the variation of secondary-emission coefficient from one element to another in the specimen.

At that time little had been published about secondary-emission coefficients for electrons with high incident energy. It was known that, for most metals, the emission reached a maximum for incident energies of a few hundred electron volts and thereafter fell steadily with increasing energy. To obtain the maximum variation of secondary emission from one surface element to another, therefore, the incident probe should strike the surface with relatively low energy, and a figure of about 800 eV was chosen. This choice determined many of the remaining features of the microscope since the current that can be obtained in a focused probe of given diameter is roughly proportional to the energy with which the electrons strike the specimen. The low value chosen for this energy placed a severe restriction on the probe current and both theory and practical experience showed that it would be very difficult to obtain an adequate ratio of signal to noise in the output signal.

To overcome this handicap three different techniques were explored. Since the probe current, for a given spot diameter, is directly proportional to the current density which can be drawn from the cathode, it seemed worthwhile to consider the use of a field emission cathode. Experiments along these lines showed promise, but it was not found possible to keep the emission current sufficiently steady for use in the microscope. To obtain relatively noise-free amplification of the very small secondary-emission current, a new technique was introduced. The electrons were accelerated and caused to fall on a fluorescent screen: the light so generated was then directed onto the cathode of a photomultiplier, which had recently been developed in the laboratory (Janes & Glover, 1941). This scheme worked fairly well, though the collection of both electrons and light must have been far from perfect. Finally, to reduce the frequency bandwidth required, and hence the noise generated in the probe current, the time taken to build up a picture was increased to 10 min. This was done by modulating the probe current at 3 kHz and making use of a facsimile recorder which had earlier been developed in the laboratory.

These basic techniques have all been used successfully in later years, but, at the time in question, the state of the art had not advanced sufficiently to permit the construction of a satisfactory microscope which could compete with the replica techniques then coming into use.

When work on the scanning microscope was started in the Cambridge University Engineering Department in 1948, the initial outlook was quite different from that of von Ardenne or of the RCA team. One major difference was that no one in the Cambridge Engineering Department had worked with any electron-optical instrument other than a cathode-ray tube, so there were no preconceived ideas about the way in which development should proceed. There was a desire to gain experience in the general field of electron optics, since this seemed a promising subject for a department which was setting out to build up, from scratch, a research group in electron physics. The choice of the scanning microscope as a project was determined by the knowledge of two technical advances which had taken place since the cessation of the RCA work. The first of these was the development of electron multipliers with beryllium-copper dynodes, which could be exposed to the atmosphere (Allen, 1947; Baxter, 1949). A multiplier of this type could be used in a demountable system and seemed to offer a more efficient means of detecting electrons leaving the specimen than the arrangement used by Zworykin and his colleagues. Moreover, A. S. Baxter was working in the Cavendish Laboratory and he very kindly gave one of his multipliers to the Engineering Department. The second technical advance was concerned with improvements in cathode-ray tubes generally, and particularly in tubes with long-afterglow screens, which had been perfected for radar purposes. If the new electron multiplier really did provide a much more efficient detection of electrons, with a corresponding gain in signal-to-noise ratio, it might be possible to reduce the scan period to an extent permitting direct viewing of the specimen on a cathode-ray tube. This would have major advantages over the use of a facsimile recorder since,

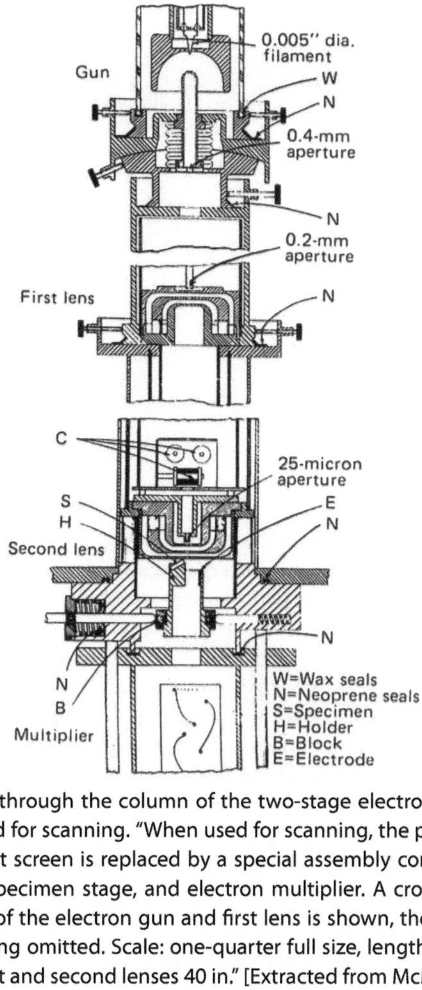

Figure 1 Cross section through the column of the two-stage electrostatically focused microscope that was modified for scanning. "When used for scanning, the projector lens is removed and the final fluorescent screen is replaced by a special assembly comprising deflection coils (C), electrostatic lens, specimen stage, and electron multiplier. A cross-sectional elevation of this assembly and also of the electron gun and first lens is shown, the interposing part of the microscope column being omitted. Scale: one-quarter full size, lengths between gun and first lens 8.5 in., between first and second lenses 40 in." [Extracted from McMullan (1953a).]

apart from being much less tedious, it would greatly assist focusing and the selection of interesting areas of the specimen.

These, then, were the thoughts leading to the idea that the construction of a scanning microscope might be a suitable Ph.D. project for a new research student. It might, or might not, succeed, but it would certainly provide good training in research and would allow ample scope for the student to demonstrate originality.

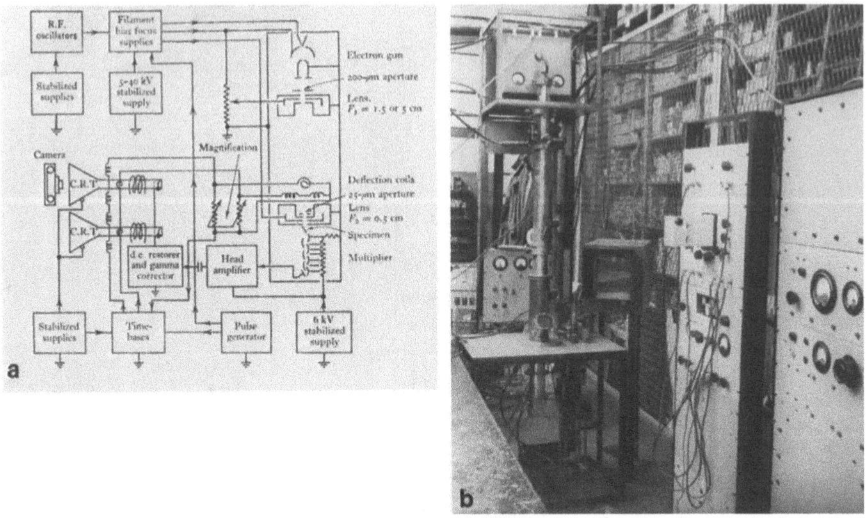

Figure 2 (a) Block schematic diagram of the scanning electron microscope. "The electron optical system is shown on the right. The diameter of the cross-over in the electron gun is about 50 microns and the total demagnification of the lenses can be set to between 500 and 2500 times by adjusting the voltage on the central electrode of the first lens. There are 405 lines (interlaced) per picture and about 0.9 pictures per sec for direct viewing and 550 lines and 1 picture in 300 sec for recording when the spot diameter (and hence the beam current) is reduced to obtain maximum resolving power. The electron current from the specimen is amplified in an electron multiplier having beryllium oxide-coated dynodes which can be exposed to the air without damaging them. The output of the electron multiplier is further amplified by a two-stage head amplifier and after passing through a gamma corrector modulates the cathode-ray tubes." [Extracted from McMullan (1953b).] (b) The original microscope constructed at the Cambridge University Engineering Department.

3. The Cambridge microscope

The project having been assigned in October, 1948, to D. McMullan, he was allowed considerable latitude as to the manner in which he tackled it, though certain constraints were imposed by the fact that almost all apparatus would necessarily be designed and constructed in the department. Thus it had already been decided that an electrostatic system would be used since this would greatly ease the stability requirements on the homemade power supplies. In fact, K. F. Sander, who had also been a research student in the department, had started the construction of an electrostatically focused transmission electron microscope, and some of the parts he had made (lens electrodes, specimen stage, and column) were used in the new microscope.

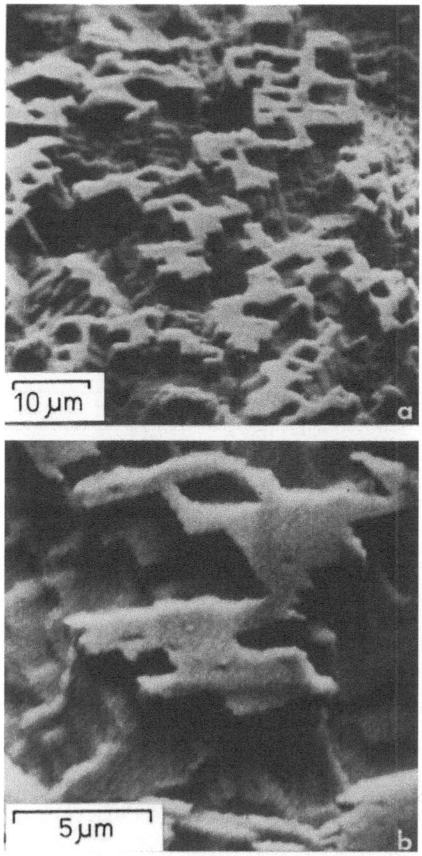

Figure 3 "Micrographs of etched aluminium specimen. Angle of incidence of 16-keV electrons = 25°. (a) Fair indication of the quality of the picture visible on the cathode-ray tube when scanning at the rate of 0.95 pictures per sec with a beam current of 1.5×10^{-10} amp; exposure 5 sec. (b) Recorded over a longer period (5 min) with a smaller spot diameter and beam current (10^{-12} amp)." [Extracted from McMullan (1953a).]

As a first step the transmission microscope was completed and the first images were produced in January, 1950; the resolution of about 200 Å was limited by astigmatism in the objective lens and insufficient magnetic shielding. This resolution was thought to be adequate for a scanning microscope and efforts were now devoted to providing the necessary auxiliary equipment: scan generators, video channel, and monitors. McMullan had had several years experience in industry on the design of radar and television circuitry and it was a comparatively simple task to build a slow-scan version of a television camera chain; it was, however, time consuming as

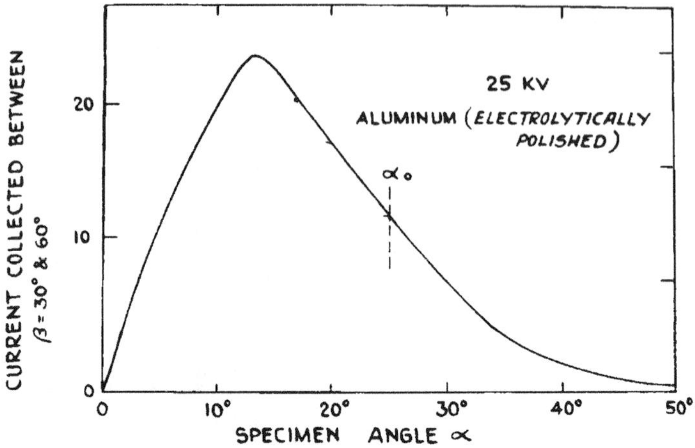

Figure 4 Variation with specimen angle of the high-energy electron current collected by an electrode subtending an angle of 30° at the specimen surface (McMullan, 1952).

everything including the power supplies had to be built from scratch. The microscope column was modified for scanning (Fig. 1) and in March, 1951, the first images were produced from a thin specimen in transmission.

In the meanwhile thought had been given as to how images from surfaces might be obtained. The results achieved by Zworykin were not very convincing, the main defects being a rather limited contrast range and excessive random noise. The reasons for these deficiencies were not identified by Zworykin in his paper but the most likely explanation appeared at the time to be that the incident electron energy was too low; after acceleration to 10 keV and focusing by two or three electrostatic lenses the beam was retarded by an inverse field so that the electrons struck the specimen with only 800 eV energy. According to Zworykin et al. (1945) "... this relatively low voltage (is) favorable for obtaining a large difference in the secondary emission currents from different parts of the object." However, the work of H. Bruining and J. H. de Boer (1938) had shown that the secondary-emission ratio for electrons of less than 1 keV depends on the state of the surface; in Zworykin's microscope, pumped by an untrapped oil diffusion pump using the fluids then available, the specimen surface must have been covered by a film of contamination which would have masked the variations in the emission ratio that would otherwise have been evident. What was missed at the time in reading Zworykin's paper was that all the micrographs published were of etched or abraded specimens and contrast was being produced by the topography of the specimen rather than by vari-

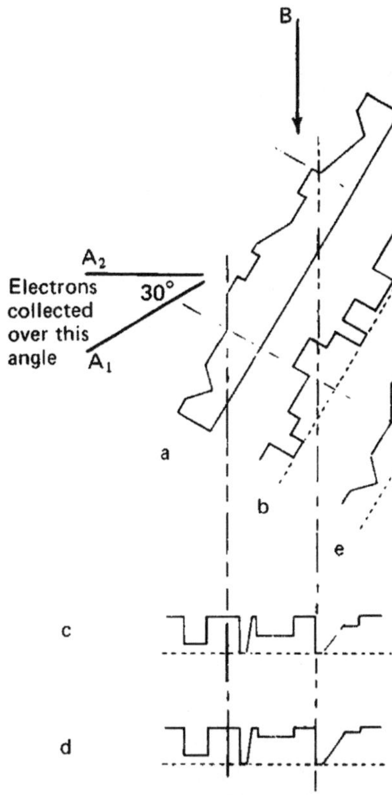

Figure 5 "Formation of contrast with oblique specimens in the scanning electron microscope compared with the optical microscope and replica techniques. (a) Cross section of a metal specimen with large irregularities but otherwise smooth and polished. (b) Brightness in different parts of the image formed by an optical microscope perpendicular to the surface which is lit at an angle of 30° from direction A. If the optical microscope was also at an angle of 30° (in direction B) and the specimen lit by a source subtending an angle A_1A_2 (30°), the brightness variations would be as in (c). The variation in image brightness with the scanning electron microscope is shown in (d). The scanning beam is at an angle of 30° to the specimen (from direction B) and all the electrons are collected from the specimen over the angle A_1A_2. As can be seen the variations in brightness are almost identical to those in (c) and it can be said that the image is of the same form as would be obtained if the specimen were lighted at an angle of 30° and viewed from the other direction at 30°. (e) Brightness variations using a negative replica." [Extracted from (McMullan, 1952).]

ations in the true secondary-emission ratio; this point was not commented on by Zworykin.

One way of reducing surface contamination which was considered was heating the specimen, but more attractive was the idea of using a

Figure 6 "A micrograph of a mild steel specimen mounted normal to the scanning beam. Specimen electrolytically polished (but unetched). Recording time 5 min; the beam voltage is 15 kV. Examination of the specimen with an optical microscope showed a very similar picture although much of the structure visible is due to scratches and staining during the electropolishing process." [Extracted from McMullan (1953b).]

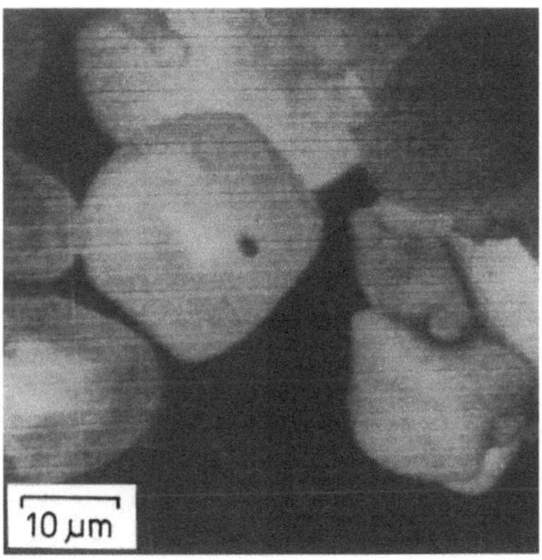

Figure 7 Grains of silver-activated zinc sulfide phosphor observed in the cathodoluminescent mode (D. McMullan, unpublished; Smith & Oatley, 1955).

much higher incident electron energy so that any contaminating layers on the specimen surface would be penetrated; however, it was unclear how contrast would then be produced since it was known that the secondary-emission ratio decreases as the primary energy is raised. Measurements reported by P. Palluel (1947) showing that the backscattering coefficient

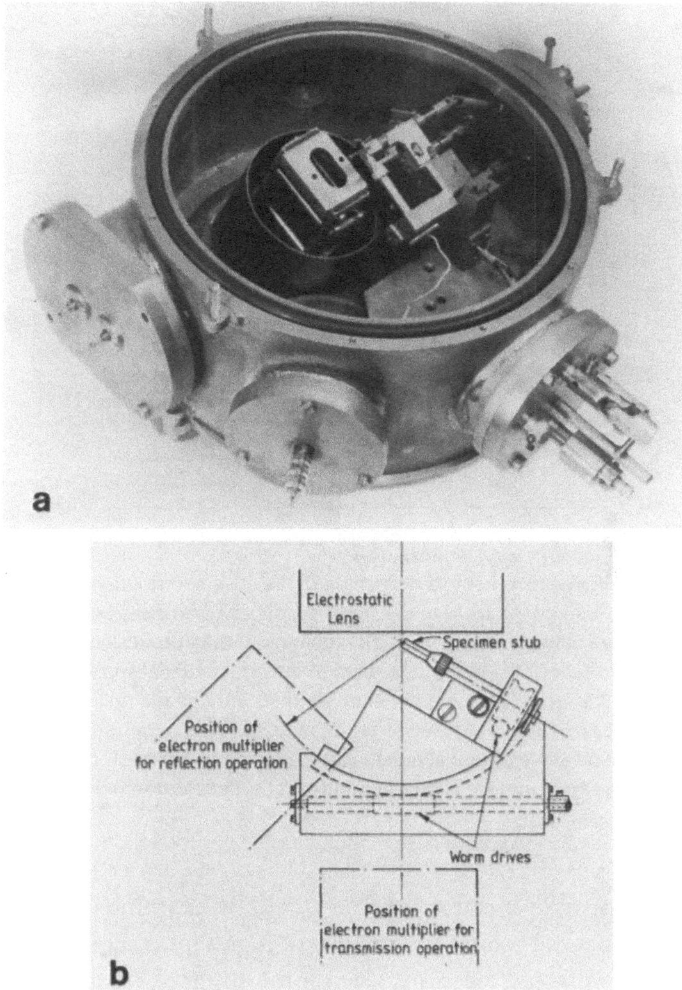

Figure 8 Improved specimen facilities and electron collection arrangement in the second phase of the development of the microscope shown in Figs. 1 and 2. (a) Photograph of new specimen chamber with electron multiplier in position for reflection operation. (b) Diagram of specimen stage and positions of electron multiplier for reflection operation and transmission operation. In reflection operation, secondary electrons can enter the electron multiplier directly (Smith, 1956).

for high-energy (20-keV) electrons depends on the atomic number of a metal had not then been seen.

In 1950, in spite of progress with replica techniques, there was still some interest in using conventional transmission microscopes in the reflection

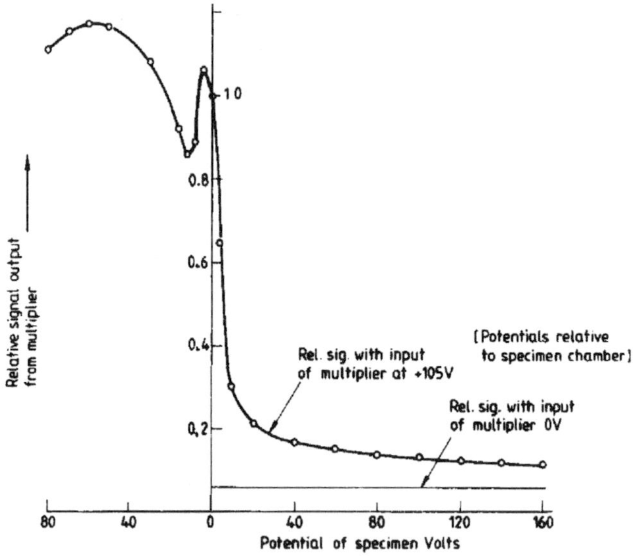

Figure 9 Results of an experiment to determine the relative contributions of true secondary electrons and high-energy electrons to the signal produced by an electron multiplier detector. For the arrangement shown in Fig. 8 (reflection operation) the output signal was found to depend strongly on the specimen potential and upon the bias potential applied to the input (first dynode) of the multiplier. This showed conclusively that, with the multiplier input at a positive potential, the major part of the signal was attributable to low-energy secondaries. The shape of the curve in the negative potential region was explained on the basis of the production of indirect secondaries at the surface on the lens and other surfaces in the vicinity of the multiplier (Smith, 1956).

mode so that opaque metal specimens could be examined. The specimen was placed so that the beam from the electron gun hit it at grazing incidence (∼4°) and the reflected electrons leaving the surface at a similar angle entered the objective lens and were focused in the usual way. The images were very elongated, with a ∼14:1 difference in the magnification along the two axes, but the method enabled etched metal surfaces to be examined without the use of replicas. The first micrographs had been obtained by B. von Borries (1940), and it is interesting to note that some of those reproduced in his paper show the striking three-dimensional effect now commonplace with scanning electron micrographs, although the foreshortening is excessive. In the early 1950s J. W. Menter was reevaluating the method (Menter, 1952), and seeing his results led to the thought that contrast could be produced in a similar way if the specimen were placed at an angle and scanned with a high-energy beam (∼20 keV), the high-energy backscattered electrons

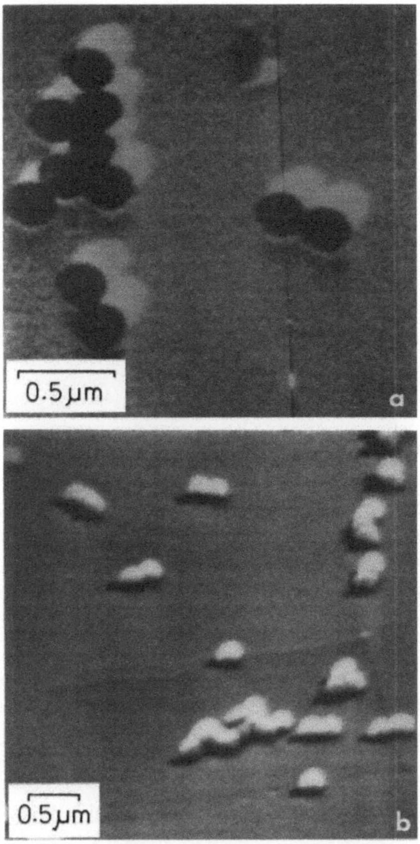

Figure 10 Latex particles, 0.25 μm in diameter on collodion film shadowed with gold-palladium. Micrographs obtained during tests of the specimen chamber arrangement shown in Fig. 8. (a) Transmission operation; (b) reflection operation. The resolution obtainable at this stage of the development, using the electron multiplier detector, was estimated to be in the region of 25 nm (Smith, 1956).

from the specimen being used to provide the output signal. Low-energy secondaries which were thought to be detrimental were to be excluded from the detector. Moreover, the incident and exit angles (measured to the surface) could be made much larger because the electrons leaving the surface would not have to be focused (chromatic aberration in the objective lens prevented the angles being set at greater than ∽4° with von Borries' method). It was expected that an angle of about 45° could be used and that the aspect ratio could be corrected by varying the amplitudes of the *X* and *Y* scans in the microscope; the possibility of a three-dimensional

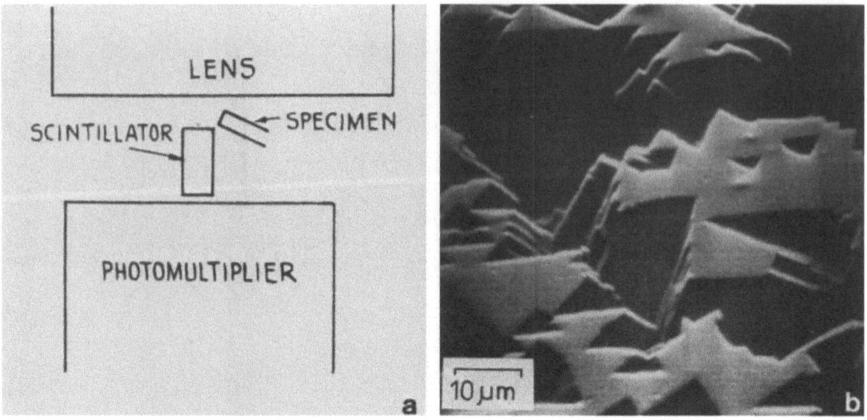

Figure 11 Electron collection using a scintillator-photomultiplier combination. (a) Experimental arrangement. The plastic scintillator (Pamelon: decay time 10^{-8} sec) took the form of a $\frac{1}{4}$-in. diameter × $\frac{1}{4}$-in. long cylinder. Light from the scintillator was detected by an EMI type 6094C photomultiplier. (b) Micrograph of a surface of etched aluminum obtained using the arrangement shown in (a). The absence of detail within the etch pits indicates that the signal was predominantly due to the high-energy reflected electrons taking direct paths to the scintillator (Smith, 1956).

impression in the displayed picture was not expected although a closer examination of von Borries' and Menter's micrographs would have pointed to it. The choice of a high-energy incident electron probe had also the very important added advantage of giving a much larger current density in the focused spot than had been possible with the Zworykin arrangement. The specimen chamber of the microscope was therefore designed so that, as shown in Fig. 1, the specimens could be placed at an angle close to the final lens. Fig. 2a is a schematic diagram of the complete microscope shown in Fig. 2b.

The first images of angled solid specimens were seen in June, 1951, and the three-dimensional effect was so clear that it was immediately realized that aspect ratio correction was not only unnecessary but would be undesirable. The micrographs in many cases looked very similar to those produced by replica methods (Fig. 3). Measurements were made on the reflection of high-energy electrons from metal surfaces as a function of angle and it was shown that when the incident probe strikes a metal surface at a glancing angle of about 25° the number of electrons reaching the detector is very sensitive to small variations in the angle of incidence (Fig. 4). Thus an etched metal surface examined in this way exhibits contrast in a manner similar to that in which the same surface, when viewed under an optical

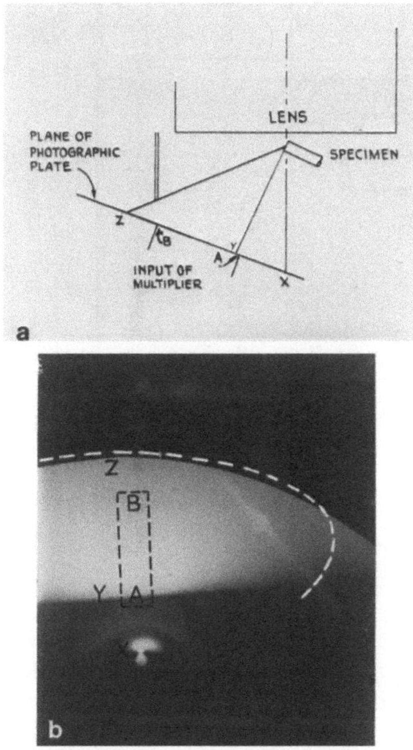

Figure 12 Determination of the spatial distribution of the high-energy reflected electrons. (a) Experimental arrangement: photographic plate placed over entrance aperture of the electron multiplier. (b) Contact print of exposed plate. Exposure time 15 min; beam energy 15 kV, beam current falling on specimen 10^{-10} A. The lower boundary of the exposed area, *Y*, was defined by the intersection with the plate of the plane containing the surface of the specimen; the upper boundary, *Z*, was defined by a Mumetal screening tube. The exposure at *X* was produced by removing the specimen from the path of the beam. The regions corresponding to the areas intercepted by the input aperture of the multiplier and by the scintillator arrangement of Fig. 11a are indicated by dashed lines. From this information, and from densitometric measurements of the plate, it was estimated that the collection efficiency of the scintillator arrangement shown in Fig. 11a was greater than that of the multiplier by a factor of about eight (Smith, 1956).

microscope, generates contrast by reflecting differing amounts of light into the objective from different crystal planes. It was realized that reciprocity applies in that the illumination appears to come from the direction of the detector and that the observer looks along the scanning beam (Fig. 5).

It was still thought that low-energy electrons should be excluded from the detector and in fact in order to prevent loss of resolution caused by electron penetration a low-loss image was proposed: "The beam from the

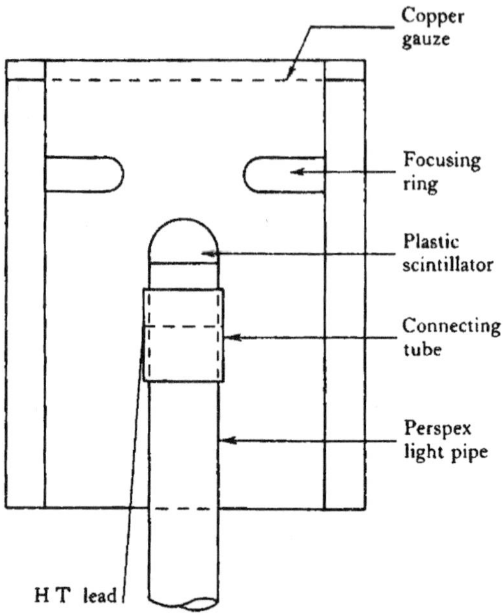

Figure 13 Secondary-electron collector devised by Everhart and Thornley. Secondaries emitted from the surface of the specimen are accelerated toward the collector, which is at a positive potential, with respect to the specimen, of a few hundred volts. The electrons pass through the gauze at the entrance of the collector and are then accelerated through a potential of several kilovolts onto a scintillator. Light from the scintillator passes along a light pipe to the photomultiplier (Everhart, 1958; Everhart & Thornley, 1960).

specimen could be restricted to the electrons which have lost small amounts of energy and which have therefore travelled only small distances through the specimen" (McMullan, 1952).

A reading of the paper by Palluel (1947) on the backscattering ratio for high-energy primary electrons had indicated that contrast would be produced by variation in atomic number over the surface. Some of his measurements were repeated and it was confirmed that ordinary surface contamination did not affect the results. Perpendicular scanning of a specimen was therefore tried but the short working distance of the objective lens made electron collection highly inefficient: even so, clear images were obtained with a high probe current and lower magnification (Fig. 6).

During the following 2 years detailed improvements were made to the microscope including shortening the column, double-deflection scan coils, and beam deflection blanking. Cathodoluminescence contrast with zinc sulfide specimens was also tried successfully (Fig. 7).

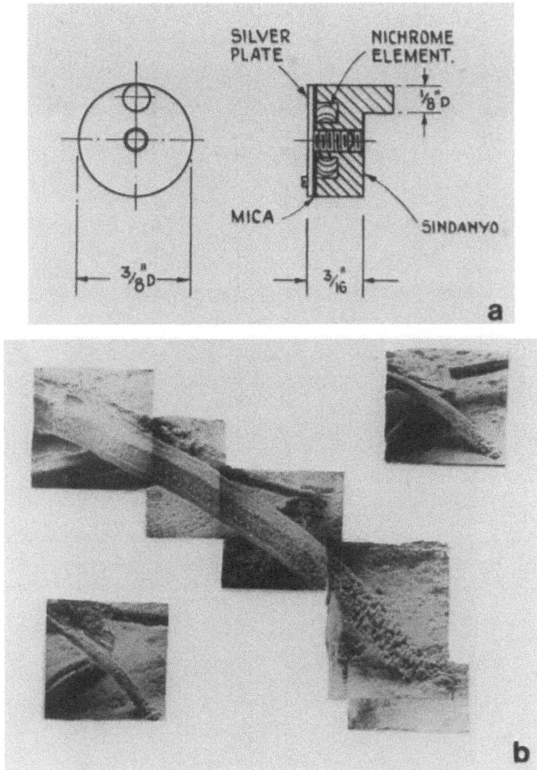

Figure 14 Thermal decomposition of silver azide. (a) Hot plate. (b) Needle of silver azide decomposing from the end in contact with the hot plate. Needle diameter approximately 10 μm (McAuslan & Smith, 1957; Smith, 1956).

Looking back on these 5 years it is plain that although substantial progress was made, there were a number of misjudgments. One of the most serious was probably the emphasis that was placed on attaining high resolution, while other advantages such as depth of field were not considered to be of such great importance. This was partly a consequence of using electrostatic lenses: the focal length of the einzel objective lens could not be increased as easily as a magnetic lens to give a longer working distance and lower magnification. The great advantages of using a scanning microscope at low magnification in place of a light microscope were not realized.

At this point, the outlook at Cambridge changed. The project, which had started as a Ph.D. exercise which might, or might not, lead to interesting results, now became a serious attempt to develop a satisfactory scanning microscope. In 1953 the original instrument was taken over by K. C. A.

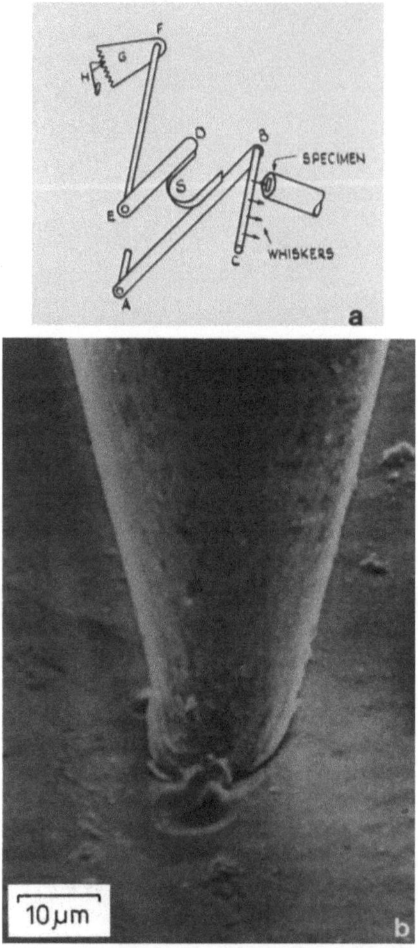

Figure 15 Experiments with a point-contact rectifier. (a) Whisker manipulation mechanism. (b) Tungsten whisker placed in contact with the surface of a germanium crystal. To "form" the rectifier a pulse of current was passed through the contact, giving rise to local melting in the region of the contact (Oatley et al., 1965; Smith, 1956).

Smith, who began by making a number of improvements: to increase the efficiency of electron collection, to provide for varying the tilt of the specimen, and so forth (Fig. 8). In the course of this work the importance of collecting secondary as well as high-energy electrons from the specimen became apparent (Fig. 9). Not only did the secondaries very greatly increase the signal current, but they also gave rise to contrast which did not depend primarily on the secondary-emission coefficient and so was not af-

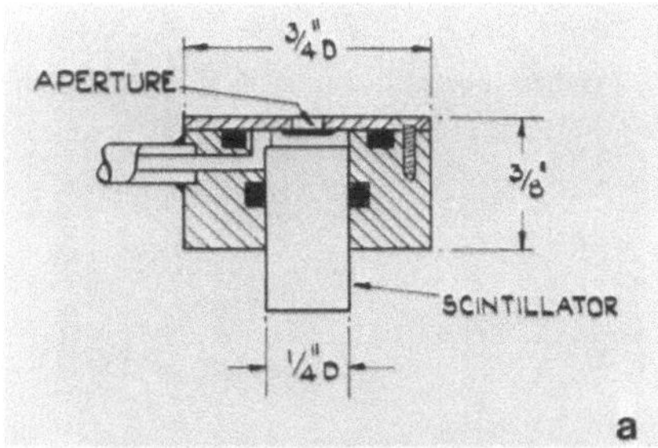

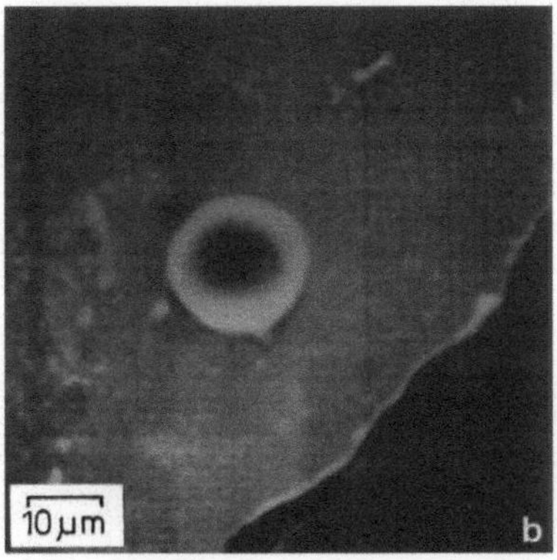

Figure 16 Observation of specimens in water vapor. (a) Water-vapor cell using a 50-μm aperture covered by a collodion film to separate the microscope column vacuum from the specimen, which is mounted on the film. The transmitted electrons are detected by the scintillator and photomultiplier in a way similar to that shown in Fig. 11a. (b) Fungus spore (*Aspergillus nidulans*) observed in the cell at water-vapor pressure (Smith, 1956).

fected by contamination. This contrast was caused by topographical surface irregularities (Fig. 10) which determined the number of secondaries reaching the collector, in much the same way that McMullan had envisaged for high-energy electrons. Its discovery opened the way to the examination of nonmetallic specimens, coated when necessary with a thin film of metal to overcome charging.

From the outset of the Cambridge work the possibility of replacing the electron multiplier by a scintillator/photomultiplier combination, as used in the RCA microscope, had been kept constantly in mind. A fundamental difficulty was that the decay of the inorganic phosphors then available is very much slower at the extremely low excitation achieved in a scanning microscope than it is in, for example, a cathode-ray tube. This slow decay could perhaps be tolerated in the RCA instrument, with a scan period of about 10 min, but not in McMullan's microscope, where direct visual

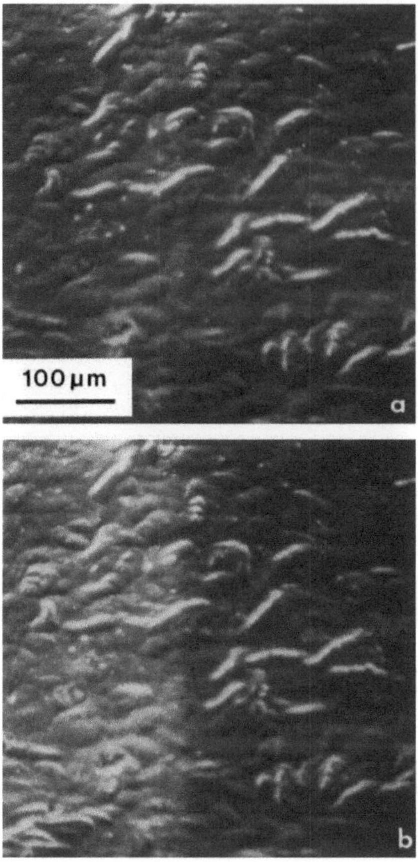

Figure 17 Voltage contrast observed on a grown p-n junction: (a) before the application of a bias voltage, (b) after the application of 4-V reverse bias (Oatley & Everhart, 1957).

presentation on a cathode-ray tube was required. When plastic phosphors with very short decay times became available, a scintillator/photomultiplier detector was constructed to provide an output signal from high-energy electrons leaving the specimen (Figs. 11 and 12). A little later, T. E. Everhart built a detector in which slow secondaries were accelerated to strike the scintillator with energies of about 10 keV, and the detailed design and assessment of this were completed by R. F. M. Thornley (Fig. 13). Thereafter the electron multiplier, which had served its purpose, was superseded by the much more convenient and efficient new detector.

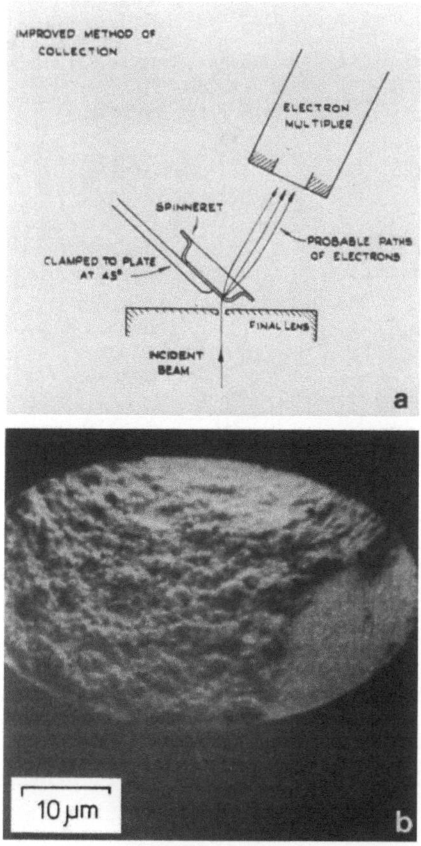

Figure 18 Examination of nylon spinneret holes. (a) Geometry of specimen and collector arrangement; the secondary electrons are collected on the side of the spinneret hole remote from the incident beam. (b) Bore of spinneret showing accumulation of debris (Wells, 1959).

4. Uses of the microscope

At about this time, the outlook at Cambridge changed once more. With the improvements recorded above, it was clear that the scanning microscope was a valuable instrument which was ripe for commercial production. The trouble was that instrument manufacturers did not see it in this light. In their view resolution was all important and, judged by this criterion, the scanning microscope could not compete with the transmission instrument using replicas. It has been pointed out by Everhart that, as late as 1963, a well-known research institute made a market survey to assess the number of scanning microscopes that might be sold in the United States.

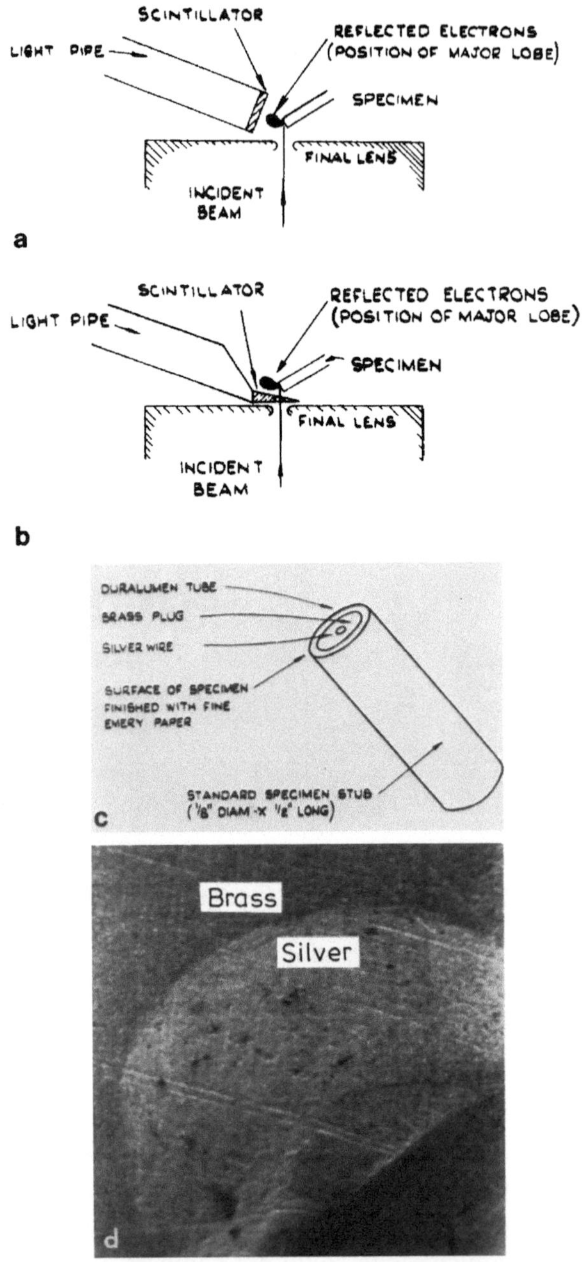

Figure 19 Atomic number contrast observed in the SEM. (a and b) Experimental specimen-collector geometries. (c) Composite test specimen. (d) Micrograph of composite test specimen, obtained with the geometry shown in (a), using 15-keV electrons (Wells, 1957).

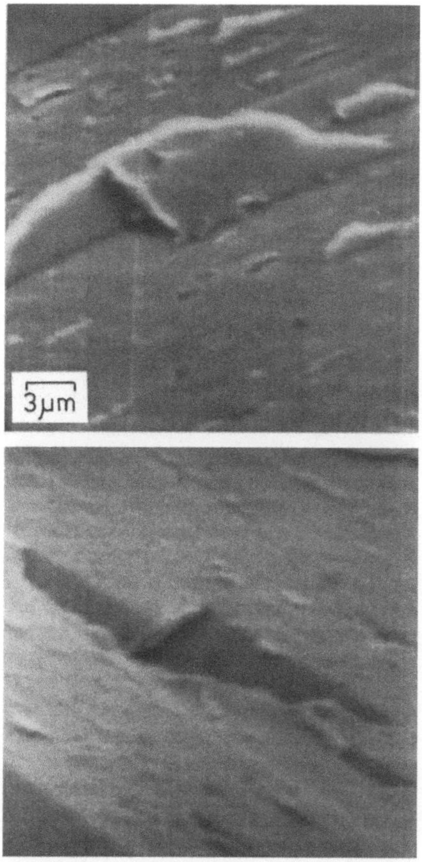

Figure 20 Early work on the topographical analysis of rough surfaces. The micrograph shows the two mating halves of the fracture surface in crystal of α-iron. Quantitative analysis of such surfaces were carried out by means of stereomicroscopy (Tipper et al., 1959; Wells, 1960).

Figure 21 Examination of particulate specimens in the SEM. Early example of the use of the SEM in soils structure research: specimen of kaolin (chalk). (Unpublished micrograph taken by L. R. Peters for M. Black in 1956.)

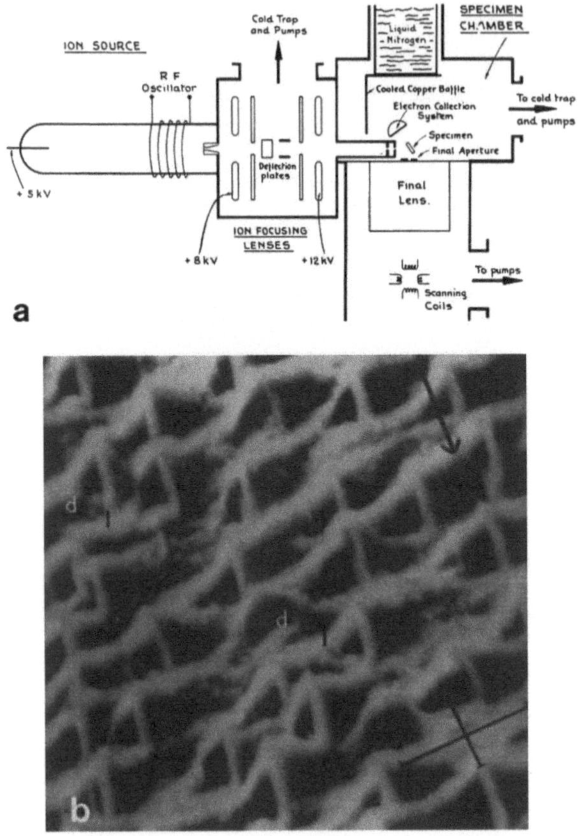

Figure 22 *In situ* etching of surfaces by ion bombardment. A scanning microscope was modified by A. D. G. Stewart for the study of the etching of specimens by bombardment with argon ions. A schematic diagram of the modified portion is shown in (a). The ion gun, the specimen chamber, and the remainder of the microscope column were separately pumped, so that micrographs of the specimen could be obtained while it was undergoing bombardment. Alternatively, bombardment and observation could proceed in sequence, without exposing the specimen to air. (b) Micrograph of a specimen of *Macropus* dentine, prepared by A. Boyde, after bombardment. The direction of bombardment is indicated by the arrow. Markers are 20 μm in both directions. (Boyde & Stewart, 1962a, 1962b; Stewart, 1962).

The prediction was that 3–5 might be sold in the first year that a commercial instrument became available and that an overall total of 10 would saturate the market.

The Cambridge work was therefore directed away from improvements in the microscope itself (though numerous improvements were in fact made) and toward the use of the microscope to do things which could

Figure 23 Examination of insulating specimens without metallization by means of low-voltage operation. (a) Fractured surface of sintered alumina: accelerating voltage 1.5 kV. (b) Early example of the examination of freeze-dried biological specimens in the SEM: silver-stained endothelial surface, accelerating voltage 1 kV (Thornley, 1961).

not be done with replicas. Thus the examination of biological and other specimens, with highly irregular surfaces, gradually brought realization that depth of field and a pronounced three-dimensional effect are often more important than high resolution. Another line of attack was the examination of specimens which underwent change while in the microscope.

A representative collection of micrographs resulting from work undertaken during this phase, which lasted until 1965, when commercial production of the microscope began, is shown in Figs. 14–25. Further scanning instruments, designated SEM 2 (Wells, 1957), SEM 3 (Smith, 1961a), SEM 4 (Stewart, 1962), and SEM 5 (Pease & Nixon, 1965a), were constructed

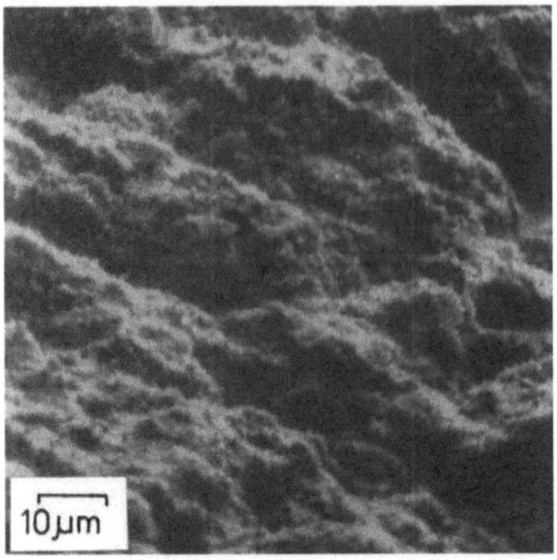

Figure 24 The examination of hot and emitting thermionic cathodes in the SEM. In this technique a scintillator light-pipe photomultiplier detector was used to detect the high-energy reflected electrons. The normal plastic scintillator was replaced by lithium glass to withstand the high temperature of the nearby specimen. All surfaces of the detector were coated with aluminum, and an optical filter was interposed between the light pipe and the photomultiplier to prevent light from the specimen entering the photomultiplier while transmitting the radiation from the scintillator. The micrograph shows the surface of a bariated-nickel dispenser cathode at a temperature of 1200 K emitting a total current of 25 mA (Ahmed, 1962).

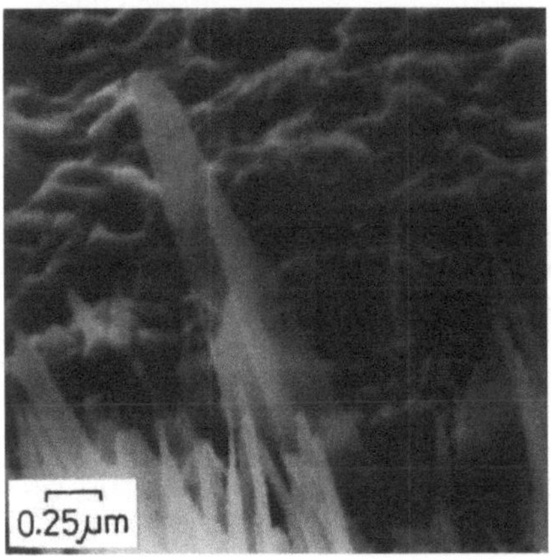

Figure 25 Early high-resolution scanning electron miroscopy. Micrograph shows oxidized surface of iron, demonstrating a resolution of about 100 Å. Accelerating voltage 30 kV (Pease & Nixon, 1965a).

and used in the course of this work. From 1961 some of the research was supervised by other members of the staff of the department.

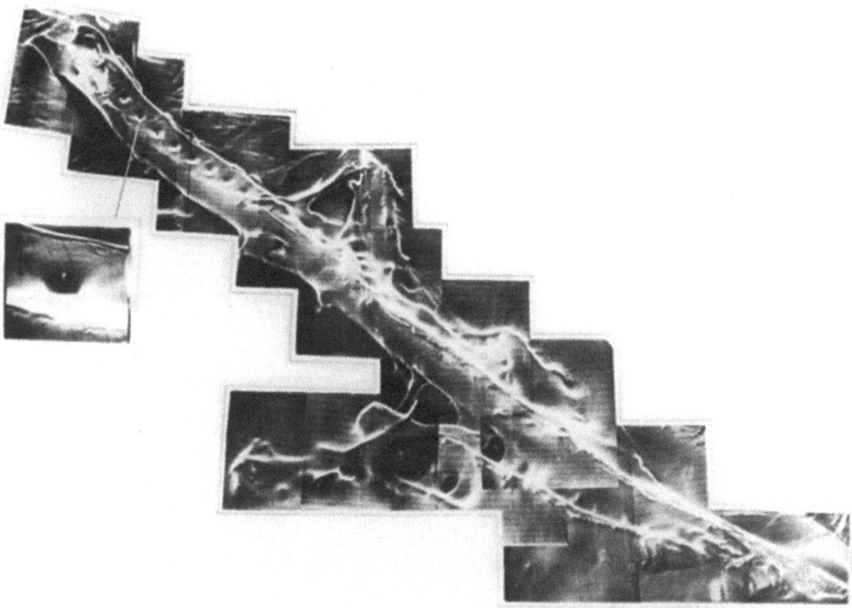

Figure 26 Application of the SEM to the study of wood fibers. A montage of a spruce fiber (diameter approximately 40 μm) from a groundwood pulp, prepared in collaboration with D. Atack, 1955 (Atack & Smith, 1956).

Gradually, reports of the above work at conferences and in scientific journals aroused interest in the advantages of the scanning microscope. One person who clearly saw its possibilities was J. Sikorski, who since 1958 (Sikorski, 1960) had been strongly advocating the use of the instrument for research on fibers. Another was D. Atack from the Pulp and Paper Research Institute of Canada, who was spending sabbatical leave in Cambridge during 1954–1955. With his collaboration the montage of a spruce fiber shown in Fig. 26 was prepared and this so impressed the director of the institute, L. B. Thiesmeyer, that on his next visit to Cambridge he arranged for Smith to spend 2 years in the Engineering Department to build a new instrument which he would then take to Canada. This provided time and money for the construction of the first properly engineered scanning microscope (Fig. 27). It was completed in 1958 and gave excellent service for some 10 years before being replaced by a more modem instrument.

The story of the somewhat involved set of circumstances which finally led to the first commercial production of a scanning microscope by the Cambridge Instrument Company has been related elsewhere (Oatley,

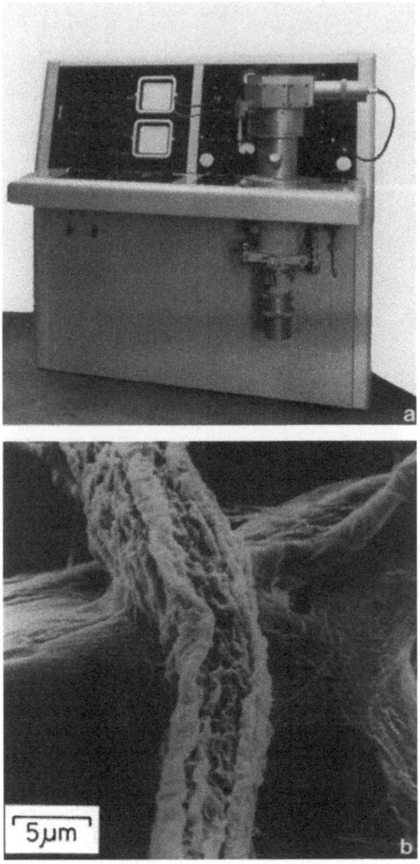

Figure 27 Scanning microscope, SEM 3, constructed for the Pulp and Paper Research Institute of Canada. (a) SEM 3 console: power supplies were housed in a separate remotely located cabinet. (b) Application of SEM 3: fibers used in the manufacture of paper. This instrument, the third to be constructed in the Cambridge University Engineering Department, incorporated magnetic lenses and double-deflection scanning coils. The scanning objective lens was of the pinhole type (Liebmann, 1955a, 1955b) with three adjustable apertures, and was designed in collaboration with R. S. Page. The condenser system consisted of a modified Metropolitan Vickers EM4 transmission-type instrument (Page, 1954). By replacing the portion of the column above the desk of the console (see figure), with the original EM4 viewing chamber, transmission microscope facilities could be retained. A vacuum-tight thin foil target could be mounted in the pinhole of the objective lens to obtain X-ray projection facilities (Smith, 1961a).

1982). The design of the instrument was largely the work of A. D. G. Stewart. The microscope was placed on the market in 1965, under the trade name Steroscan, and was an immediate success.

To end this account of the work carried out at Cambridge, it is appropriate to put on record the very important part played there by Leslie Peters. As a young assistant he became involved with the scanning microscope project at the outset, and he has had a hand in every subsequent development described in this article. The evolution of the scanning microscope owes much to his skill as instrument maker, photographer, and technician.

5. Afterword by Peter Hawkes

Much information about Sir Charles Oatley, Dennis McMullan and Ken Smith is available, especially in other volumes of these advances. Volume 133 contains contributions by many of Sir Charles's former research students and members of the Cambridge University Engineering Department. Auto-biographical articles by Ken Smith appear in Volumes 159 and 177. Dennis McMullan's career is described in 'Dennis McMullan (1923–2015)', which appeared in *In Focus* No. 38 (2015) 116–118 and in an article in the Pioneers series in *Microscopy Today* **28**(5) (2020) 60–61 on both Dennis McMullan and Ken Smith. We also draw attention to a key publication by Oatley in the *Journal of Applied Physics* **53** (1982) R1–R13 and to the seminal account of the scanning electron microscope by C.W. Oatley, W.C. Nixon and R.F.W. Pease in *Advances in Electronics and Electron Physics* **21** (1965) 181–247.

Bibliography including cited references

A very full bibliography of papers on the scanning microscope and its applications, compiled by Dr. Oliver C. Wells, was published in the *Proceedings of the Fifth Symposium on the Scanning Microscope* (O. Johari, ed.), IIT Research Institute, Chicago, Illinois, pp. 485–525 (1972).

The papers listed below are those which, prior to 1965, when the first commercial scanning microscope became available, deal with the theory, construction, and development of the instrument and with first applications to particular classes of specimen.

Ahmed, H. (1962). *Studies on high current density thermionic cathodes*. (Ph.D. Dissertation). Cambridge University.

Ahmed, H., & Beck, A. H. W. (1963). Thermionic emission from dispenser cathodes. *Journal of Applied Physics, 34*, 997–998.

Allen, J. S. (1947). An improved electron multiplier particle counter. *Review of Scientific Instruments, 19*, 739.

Allen, J. W., & Smith, K. C. A. (1956). Electron microscopy of etched germanium surfaces. *Journal of Electronics, 1*, 439–443.

Atack, D., & Smith, K. C. A. (1956). The scanning electron microscope—a new tool in fibre technology. *Pulp and Paper Magazine of Canada, 57*, 245–251.

Baxter, A. S. (1949). *Detection and analysis of low energy disintegration particles*. (Ph.D. Dissertation). Cambridge University.

Bernard, R., & Davoine, F. (1957). The scanning electron microscope. *Annales de l'Université de Lyon, Science Section B [3], 10*, 78–86 (in French).
Bernard, R., & Davoine, F. (1960). Scanning electron microscopy. *Berlin, 1958.* (pp. 273–276) (in French).
Bernard, R., Davoine, F., & Pinard, P. (1959). Extension of the technique of scanning electron microscopy to the observation of fluorescent substances. *Comptes Rendus Hebdomadaires des Séances de L'Académie des Sciences, 248*, 2564–2566 (in French).
Bowden, F. P., & McAuslan, J. H. L. (1956). Slow decomposition of explosive crystals. *Nature (London), 178*, 408–410.
Boyde, A., & Stewart, A. D. G. (1962a). A study of the etching of dental tissues with argon ion beams. *Journal of Ultrastructure Research*, 7, 159–172.
Boyde, A., & Stewart, A. D. G. (1962b). Investigation of the erosion of tooth sections with argon ion beam. *Philadelphia, 1962* Pap. QQ9.
Brachet, C. (1946). Note on the resolution of the scanning electron microscope. *Bulletin de l'Association Technique Maritime et Aéronautique, 45*, 369–378 (in French).
Broers, A. N. (1965a). Combined electron and ion beam processes for microelectronics. *Microelectronics and Reliability, 4*, 103–104.
Broers, A. N. (1965b). Micromachining by sputtering through a mask of contamination laid down by an electron beam. In *Proc. 1st Int. Conf. Electron Ion Beam Sci. Technol.,* 1964, Toronto (pp. 191–204).
Broers, A. N. (1965c). *Selective ion beam etching in the scanning electron microscope.* (Ph.D. Dissertation). Cambridge University.
Bruining, H., & de Boer, J. H. (1938). Secondary emission. *Physica (Amsterdam), 5*, 17.
Cosslett, V. E., & Duncomb, P. (1957). A scanning microscope with either electron or x-ray recording. *Stockholm, 1956.* (pp. 12–14).
Davey, J. P. (1965). *Electron microprobe scanning system for the study of luminescence.* (Ph.D. Dissertation). Cambridge University.
Davoine, F. (1957). *Secondary electron emission of metals under mechanical strain.* (Ph.D. Dissertation). L'Université de Lyon (in French).
Davoine, F., & Pinard, P. (1965). Study of cathodoluminescence by scanning electron microscopy. *Prague, 1964.* (pp. 113–114) (in French).
Davoine, F., Pinard, P., & Martineau, M. (1960). Observation of cadmium sulfide bombarded with electrons of 5 kev energy. *Journal de Physique et Le Radium [8], 21*, 121–124 (in French).
Davoine, F., Bernard, R., & Pinard, P. (1961). Fluorescence of alkali halides observed with a scanning electron microscope. *Delft, 1960.* (pp. 165–168) (in French).
Drummond, I. W., & Thornton, P. R. (1965). Studies of electrical high-field regions in high resistivity GaAs with the scanning electron microscope. *Prague, 1964.* (pp. 287–288).
Dwornik, E. J. (1965). Use of the scanning electron microscope in geologic studies. [US] *Geological Survey Professional Paper* 550-D (pp. D209–D213).
Everhart, T. E. (1958). *Contrast formation in the scanning electron microscope.* (Ph.D. Dissertation). Cambridge University.
Everhart, T. E., & Thornley, R. F. M. (1960). Wide-band detector for micro-microampere low-energy electron currents. *Journal of Scientific Instruments, 37*, 246–248.
Everhart, T. E., Wells, O. C., & Oatley, C. W. (1959). Factors affecting contrast and resolution in the scanning electron microscope. *Journal of Electronics and Control*, 7, 97–111.
Everhart, T. E., Smith, K. C. A., Wells, O. C., & Oatley, C. W. (1960). Recent developments in scanning electron microscopy. *Berlin, 1958.* (pp. 269–273).
Everhart, T. E., Wells, O. C., & Matta, R. K. (1963). Evaluation of passivated integrated circuits using the scanning electron microscope. *Extended Abstract - Electrochemical Society. Electronics Division. Meeting 1963, 12*(2), 2–4.
Green, D., & Nathanson, H. C. (1965). Observation of inversion layers under insulated-gate electrodes using a scanning electron microscope. *Proceedings of the IEEE, 53*, 183–184.

Higuchi, H., & Tamura, H. (1965). Measurement of the lifetime of minority carriers in semiconductors with a scanning electron microscope. *Japanese Journal of Applied Physics, 4*, 316–317.

Janes, R. B., & Glover, A. M. (1941). Recent developments in phototubes. *RCA Review, 6*, 43.

Jaques, W. E., Coalson, J., & Zervins, A. (1965). Application of the scanning electron microscope to human tissues—a preliminary survey. *Experimental and Molecular Pathology, 4*, 576–580.

Knoll, M. (1935). Static potential and secondary emission of bodies under electron irradiation. *Zeitschrift für technische Physik, 16*, 467–475 (in German).

Knoll, M., & Theile, R. (1939). Scanning electron microscope for determining the topography of surfaces and thin layers. *Naturwissenschaften, 29*, 335–336 (in German).

Kushnir, Yu. M., Fetisov, D. V., Raspletin, K. K., et al. (1961). Scanning electron microscope and x-ray microanalyzer. *Bulletin of the Academy of Sciences of SSSR, Physics Series (English Translations), 25*, 709–714.

Kushnir, Yu. M., Fetisov, D. V., Der-Shvarts, G. V., et al. (1963). Scanning electron microscope-x-ray microanalyser with magnetic electron optics. *Bulletin of the Academy of Sciences of SSSR, Physics Series (English Translations), 27*, 1146–1151.

Liebmann, G. (1955a). The field distribution in asymmetrical magnetic electron lenses. *Proceedings of the Physical Society. Series B, 68*, 679–681.

Liebmann, G. (1955b). The magnetic pinhole electron lens. *Proceedings of the Physical Society. Series B, 68*, 682–685.

Liebmann, G. (1955c). A unified representation of magnetic electron lens properties. *Proceedings of the Physical Society. Series B, 68*, 737–745.

Liebmann, G., & Grad, E. M. (1951). Imaging properties of a series of magnetic electron lenses. *Proceedings of the Physical Society. Series B, 64*, 956–971.

Matta, R. K., Green, D., & Larkin, M. W. (1965). The use of the scanning electron microscope in the fabrication of an integrated circuit. *Extended Abstract - Electrochemical Society. Electrothermal Metallurgy Division, 3*(1), 32–33.

McAuslan, J. H. L., & Smith, K. C. A. (1957). The direct observation in the scanning electron microscope of chemical reactions. *Stockholm, 1956*. (pp. 343–345).

McGrath, J. T., Buchanan, J. G., & Thurston, R. C. A. (1962). A study of fatigue and impact fractures with the scanning electron microscope. *Journal of the Institute of Metals, 91*, 34–39.

McMullan, D. (1952). *Investigations relating to the design of electron microscopes*. (Ph.D. Dissertation). Cambridge University.

McMullan, D. (1953a). An improved scanning electron microscope for opaque specimens. *Proceedings of the Institution of Electrical Engineers, 100*, 245–259.

McMullan, D. (1953b). The scanning electron microscope and the electron-optical examination of surfaces. *Electronics Engineering, 25*, 46–50.

Menter, J. W. (1952). Direct examination of solid surfaces using a commercial electron microscope in reflection. *Journal of the Institute of Metals, 81*, 163–167.

Nixon, W. C. (1964). Scanning electron microscopy. *Journal of the Royal Microscopical Society, 83*, 213–216.

Oatley, C. W. (1958). The scanning electron microscope. *New Scientist, June Issue*, 153–155.

Oatley, C. W. (1982). The early history of the scanning electron microscope. *Journal of Applied Physics, 53*, R1–R13.

Oatley, C. W., & Everhart, T. E. (1957). The examination of p-n junctions with the scanning electron microscope. *Journal of Electronics, 2*, 568–570.

Oatley, C. W., Nixon, W. C., & Pease, R. F. W. (1965). Scanning electron microscopy. *Advances in Electronics and Electron Physics, 21*, 181–247.

Page, R. S. (1954). A compact console-type electron microscope. *Journal of Scientific Instruments, 31*, 200–205.

Palluel, P. (1947). Rediffused component of secondary electron emission from metals. *Comptes Rendus Hebdomadaires des Séances de L'Académie des Sciences, 224*, 1492 (in French).
Pease, R. F. W. (1963). *High resolution scanning electron microscopy*. (Ph.D. Dissertation). Cambridge University.
Pease, R. F. W. (1965). The determination of the area of emission of reflected electrons in a scanning electron microscope. *Journal of Scientific Instruments, 42*, 158–159.
Pease, R. F. W., & Nixon, W. C. (1965a). High resolution scanning electron microscopy. *Journal of Scientific Instruments, 42*, 81–85.
Pease, R. F. W., & Nixon, W. C. (1965b). Microformation of filaments. In *Int. conf. electron ion beam sci. technol. [pap.], 1st, 1964* (pp. 220–230).
Sikorski, J. (1960). Studies of fibrous structure. *Berlin, 1958*. (pp. 686–707).
Smith, K. C. A. (1956). *The scanning electron microscope and its fields of application*. (Ph.D. Dissertation). Cambridge University.
Smith, K. C. A. (1959). Scanning electron microscopy in pulp and paper research. *Pulp and Paper Magazine of Canada, Technical Section, 60*, T366–T371.
Smith, K. C. A. (1961a). A versatile scanning electron microscope. *Delft, 1960*. (pp. 177–180).
Smith, K. C. A. (1961b). Scanning. In G. L. Clark (Ed.), *Encyclopedia of microscopy* (pp. 241–251). Princeton, New Jersey: Van Nostrand-Reinhold.
Smith, K. C. A., & Oatley, C. W. (1955). The scanning electron microscope and its fields of application. *British Journal of Applied Physics, 6*, 391–399.
Spivak, G. V., Saparin, G. V., & Pereverzev, N. A. (1962). Potential distribution in a p-n junction as observed by electron-optical scanning. *Bulletin of the Academy of Sciences of SSSR, Physics Series (English Translations), 26*, 1362–1365.
Stewart, A. D. G. (1962). Investigation of the topography of ion bombarded surfaces with a scanning electron microscope. *Philadelphia, 1962*. (pp. D12–D13).
Stewart, A. D. G., & Boyde, A. (1962). Ion etching of dental tissues in a scanning electron microscope. *Nature (London), 196*, 81–82.
Stewart, A. D. G., & Snelling, M. A. (1965). A new scanning electron microscope. *Prague, 1964*. (pp. 55–56).
Stintzing, H. (1929). *Method and device for automatically assessing, measuring and counting particles of any type, shape and size*. German Patent 485,155.
Te Gude, H. (1945). The scanning electron microscope—an imaging method using secondary electrons. *Funk Ton, 7*, 373–383 (in German).
Te Gude, H. (1965). Automatic inspection and fabrication of integrated circuits with the scanning electron microscope. *Internationale Elektronische Rundschau, 19*, 482, 567 and 570 (in German).
Thiele, R. (1938). Resistance-sensitive scanning electron tubes. *Telefunken-Röhre, 13*, 90–126 (in German).
Thornhill, J. W., & Mackintosh, I. M. (1965). Application of the scanning electron microscope to semiconductor device structure. *Microelectronics and Reliability, 4*, 97–100.
Thornley, F. R. M. (1960). *New applications of the scanning electron microscope*. (Ph.D. Dissertation). Cambridge University.
Thornley, F. R. M. (1961). Recent developments in scanning electron microscopy. *Delft, 1960*. (pp. 173–176).
Thornley, F. R. M., & Cartz, L. (1962). Direct examination of ceramic surfaces with the scanning electron microscope. *Journal of the American Ceramic Society, 45*, 425–428.
Thornton, P. R. (1965). The scanning electron microscope. *The Science Journal, 1*, 66–71.
Thornton, P. R., Culpin, M. J., & Drummond, I. W. (1963). Direct observation of the high-field regions in GaAs. *Solid-State Electronics, 6*, 532–533.
Tipper, C. F., Dagg, D. I., & Wells, O. C. (1959). Surface fracture markings on alpha iron crystals. *Journal of Iron and Steel Institute, London, 193*, 133–141.

von Ardenne, M. (1938a). The scanning electron microscope. Theoretical fundamentals. *Zeitschrift für Physik, 109*, 553–572 (in German).
von Ardenne, M. (1938b). The scanning electron microscope. Practical construction. *Zeitschrift für technische Physik, 19*, 407–416 (in German).
von Ardenne, M. (1978). The history of scanning electron microscopy and of the electron microprobe. *Optik, 50*, 177–188 (in German).
von Borries, B. (1940). High resolution images from the electron microscope used in reflection. *Zeitschrift für Physik, 116*, 370–378 (in German).
Wells, O. C. (1957). *The construction of a scanning electron microscope and its application to the study of fibres*. (Ph.D. Dissertation). Cambridge University.
Wells, O. C. (1959). Examination of nylon spinneret holes by scanning electron microscopy. *Journal of Electronics and Control, 1*, 373–376.
Wells, O. C. (1960). Correction of errors in electron stereomicroscopy. *British Journal of Applied Physics, 11*, 199–201.
Wells, O. C. (1962). Electron beams in microelectronics. In R. Bakish (Ed.), *Introduction to electron beam technology* (pp. 354–381). New York: Wiley.
Wells, O. C. (1964). Scanning electron microscope. In J. Kurshan (Ed.), *Analytical characterization of materials* (pp. 32–34). Princeton, New Jersey: R. C. A. Labs.
Wells, O. C., Everhart, T. E., & Matta, R. K. (1963). Automatic positioning of device electrodes using the scanning electron microscope. *Extended Abstract - Electrochemical Society. Electronics Division, 12*(2), 5–10.
Zworykin, V. K., Hillier, J., & Snyder, R. L. (1942). A scanning electron microscope. *ASTM Bulletin, 117*, 15–23.
Zworykin, V. K., Morton, G. A., Ramberg, E. G., Hillier, J., & Vance, A. W. (1945). *Electron optics and the electron microscope*. New York: Wiley.

CHAPTER TEN

Some recollections of electron microscopy in Britain from 1943 to 1948

R. Reed[a] and Peter Hawkes (Afterword)[b,*]
[a]Formerly Pudsey, West Yorkshire, United Kingdom
[b]CEMES-CNRS, Toulouse, France
*Corresponding author. e-mail address: hawkes@cemes.fr

Contents

1. Afterword by Peter Hawkes 307
References 309

These recollections should not be taken as being a full or indeed accurate account of the early days of electron microscopy in Britain. They are intended merely to convey something of the background against which the main lines of development were pursued. Although every attempt has been made to achieve factual accuracy, it must be remembered that the main impetus to the prosecution of electron microscopy on a fairly wide scale, namely, the availability of instruments produced on the commercial scale in America, occurred in the middle of World War II, when most scientists and engineers were engaged on more immediate tasks while the extent to which electron microscopists could collaborate and exchange views was severely limited.

The major incentive to practical electron microscopy in Britain can be said to have come with the acquisition of the so-called RCA Type B electron microscopes commercially produced by the Radio Corporation of America and sent over to Britain early in 1943 under lend–lease arrangements, which covered the wartime exchange of ideas and materials between Britain and the United States. However, it should be noted that the British interest in electron optics was of long standing. The work of J. J. Thomson (1897, 1899) in integrating the findings of earlier researches into the nature of cathode rays and putting them on a firm quantitative footing by considering them as rays of particles (later termed electrons) is well known. Soon after J. J. Thomson, Kaufmann (1897), Wiechert (1899),

Advances in Imaging and Electron Physics, Volume 221
ISSN 1076-5670
https://doi.org/10.1016/bs.aiep.2022.03.010

and Lenard (1904) obtained similar results. In particular, Wiechert in 1899 had actually concentrated a beam of cathode rays using a magnetic coil and he is usually credited with the first demonstration of magnetic focusing. Apparently, however, a few years previously the British electrical engineer Macgregor-Morris and Professor Clinton had, at the instigation of Sir Ambrose Fleming (1897), mounted a circular coil on a Crookes tube and found that by altering the coil current, the shadow of the Maltese cross at the end of the tube could be varied in size and degree of rotation. This clearly demonstrated the image-forming properties of a short magnetic coil.

Busch (1926) has rightly been called the founder of electron optics. He realized that the long solenoidal lens such as Wiechert and others had used for focusing cathode rays was useless for high-magnification optics, and the merit of his work lay in demonstrating and also proving mathematically that an inhomogeneous magnetic field of rotational symmetry produced by a shorter solenoid has the properties of a lens, and further that the relationship between focal length, magnetic field strength, and electron velocity fitted well-known formulas in light optics.

Busch's derivation of the focal length of a magnetic lens naturally provided a great stimulus for the development of various electron-optical devices. Thus many countries around the world contributed toward improving the cathode-ray oscillograph (or cathode-ray tube) as a highly versatile instrument for the study of periodic phenomena. Many natural effects having a periodic basis may be converted into an oscillatory voltage which may then be applied to the deflecting system of a calibrated oscillograph and displayed visually on a screen. Thus the cathode-ray oscillograph came to be widely used in the study of voltage waveforms developed at the various parts of electrical circuits, being still used today in frequency analysis and in fault finding. In this surge of applied electron optics Britain was in the forefront of many developments. Thus it was one of the first countries to produce double-beam cathode-ray tubes for the direct comparison of waveforms from two different sources. Britain also pioneered the development of the high-speed oscillograph, capable of handling transient waveforms of high frequency. As an example one may cite its use in studying the characteristics of high-voltage switch-gear assemblies whereby much information for design improvement was obtained (Macgregor-Morris & Henley, 1936).

During the 1930s Martin et al. (1937), working at Imperial College, London, under the auspices of Metropolitan Vickers Electrical Company Limited, had produced the first British electron microscope. It used a cold-cathode discharge tube as the source of 60-kV electrons and with magnetic lenses it occasionally achieved a resolution of about 100 nm with favorable test specimens (holes in a gold film) with a maximum magnification of 2000×. However, it was highly prone to image instability (Martin et al., 1939) and its erratic performance limited its use as a routine instrument. Thus it made little impact as a new tool for general research purposes.

At this time the possibility of using radio waves for the detection and ranging of objects at a distance (or radar, as it is now called) was foreseen and many scientists, including those familiar with electron-optical principles, were engaged in developing this novel and vital contribution to the military defense of Britain. Various ancilliary devices needed to display the detection pattern on the screen of a cathode-ray tube were quickly produced while electron-optical ideas were pressed into service to develop the means of generating high-frequency radio beams of high power output such as the magnetron and the klystron.

Over the same period the future importance of television, with all its essential auxiliary equipment, was also apparent. Here, along with RCA in America, Philips in Holland, and Telefunken and AEG in Germany, the research laboratories of Electrical and Musical Industries (EMI) in Britain played a prominent part in developing the science of electron optics. Indeed, some idea of the vast range of research topics covered by the EMI Laboratories may be gained from their publication "Electron Optics" edited by O. Klemperer (1939). The subject of electron optics was also well covered in many British universities and polytechnics at final-degree level and beyond. Here the well-known volume of L. M. Myers (1939) was outstanding. It provided a full and excellent treatment of electron optics and is valuable for its historical account of the various applications of electron-optical principles. Contemporary with Myers and more effective in fostering electron optics in Britain was D. Gabor, later to become Professor at University College, London. Soon after the work of Busch, Gabor (1927), while working for his Dr.-Ing. degree at the University of Berlin, made an important contribution to practical lens design by recognizing that the magnetic field could be compressed into a smaller distance along the axis of the solenoid if the latter were encased in soft-iron sheeting. This fundamental discovery later enabled the production of short-focus magnetic lenses which were essential for high-resolution transmission electron microscopes. In 1937 Gabor joined the research staff of the British Thomson–Houston Company Limited at Rugby, so starting his immense contribution to electron optics in Britain, both in the theoretical and practical fields. Apart from producing a stream of electron-optical devices (including those based on the holographic principle—a topic which he largely pioneered), Gabor was distinguished for so generously placing his vast knowledge at the disposal of younger scientists. Some measure of his erudition and enthusiasm may be gained from reading his monograph "The Electron Microscope" (Gabor, 1946).

Hence it could be said that by 1943, when the seven RCA Type B electron microscopes were sent over from America, there was already a deep and growing interest in electron optics and its applications.

The Scientific Advisory Committee of the British War Cabinet allocated these seven instruments to various institutions around the country, namely, The National Physical Laboratory (NPL), Metallurgy Division, Teddington, Middlesex; The Cavendish Laboratory, Cambridge; The National Institute for Medical Research, Mill Hill, London; The Rothamsted Experimental Station, Harpenden, Hertfordshire; The Shirley Institute, Didsbury, Manchester; The Medical Research Council (who then offered the instrument to Professor M. J. Stewart, The University of Leeds); and an undisclosed Scientific Research and Developmental Establishment run by the Ministry of Supply.

In later years there was much speculation among electron microscopists regarding these allocations. Cynics asserted that the NPL would naturally obtain one of the microscopes because Sir Charles Darwin, Director of the NPL, was a member of the allocating committee. It is not generally known, however, that under the leadership of Dr. G. D. Preston (an X-ray crystallographer, later to become Professor of Physics at the University of Dundee) the Metallurgy Division of NPL had taken over the first British instrument constructed by Martin et al. (vide supra) at Imperial College and had ironed out many of its innate troubles. In personal conversations with the author, Preston claimed he was the first ever to bring this microscope into routine working order.

The NPL Type B instrument was assembled and brought into working order by Mr. F. W. Cuckow assisted by Mr. J. Wilson and Mr. J. Trotter. It was used mainly for studying the structure of metals but it was also available to workers from other divisions within the National Physical Laboratory.

The allocation of an instrument (under the charge of Sir Lawrence Bragg) to the Cavendish Laboratory, the most prestigious home of physics research in Britain, was understandable. It is strange, however, that the Clarendon Laboratory at the University of Oxford was *not* offered one, in spite of the fact that Professor Lindemann (later Lord Cherwell), a personal advisor in scientific matters to the Prime Minister (Mr. Winston Churchill), was on the allocating committee. This microscope was housed in a small basement laboratory in the older part of the Cavendish, being assembled under the supervision of Mr. G. R. Crowe.

The instrument at the National Institute for Medical Research was one of two originally given to the Medical Research Council. This one was under the charge of Dr. A. S. McFarlane, being primarily intended for the study of pathogenic bacteria and viruses. The initial assembly was under the immediate charge of Mr. J. Smiles (also Hon. Sec. of the Royal Microscopical Society).

During World War II the Rothamsted Experimental Station, one of the Commonwealth Agricultural Bureaus, dealt with various problems in agricultural science but was allocated an instrument for the prime purpose of studying plant viruses. Dr. F. M. L. Sheffield and Dr. E. M. Crook were largely responsible for the initial assembly of this microscope.

The granting of an instrument to the Shirley Institute (The British Cotton Industry Research Association) came as a great surprise. Dr. D. G. Drummond, who was in charge of the microscope, always maintained in a marked Northumbrian accent that it was given by the Southerners as a sop to the rough tribesmen living north of Watford. More likely, the Cabinet Committee were well aware that many technical developments important to the war effort were being pursued in the Manchester area, including the impregnation of fiber assemblies with polymer compositions; hence possibly they assigned one for this very purpose. The assembly of the instrument was carried out by Dr. D. G. Drummond and Mr. J. B. Todd.

Of the sixth instrument allocated by the Cabinet Committee, nothing is known except that it was taken over by one of the research establishments controlled by the (then) Ministry of Supply.

The second of the two microscopes originally allocated to the Medical Research Council was placed under the joint charge of Professor M. J. Stewart, Department of Pathology, and Professor J. W. McLeod, FRS, Department of Bacteriology at the Medical School, The University of Leeds. Professor Stewart told me that the granting of an electron microscope to him came as a complete surprise. Probably the Medical Research Council wished to recognize his preeminence as a pathologist, for he was known nationally and was editor of the prestigious *Journal of Pathology and Bacteriology* published by the Pathological Society of Great Britain and Ireland. Professor McLeod was also a distinguished worker in his field, having gained distinction for his researches into diphtheroid bacilli.

In the event, this award of a microscope to the Leeds Medical School caused some slight consternation within the university. The commissioning of the instrument was held up since Professor Stewart was more familiar with autopsies than with electron optics and he had difficulty in finding someone to assemble it. Hence a university committee was established and at its first meeting Dr. W. T. Astbury, then Reader in Textile Physics, vigorously suggested that the microscope should be placed in the hands of a physicist like himself, "someone who knows how to operate such a new and complex instrument." Permission for this change in immediate ownership was granted with the approval (and relief) of Professor Stewart, Professor McLeod, and the Medical Research Council. Thus Astbury was given charge of the instrument, which was to be available to all departments within the university, including those in the Medical School.

Actually Astbury had taken on more than he bargained for. Even after advertising and asking around personally, he found, as had Professor Stewart, that no one was readily available in wartime to take on the duties of assembling the instrument and running his electron microscope section. As a Junior Scientific Officer working from 1942–1943 at the Chemical Research Laboratory, Teddington, within the same grounds as the National Physical Laboratory, I had made some use of the Type B microscope in the Metallurgy Division to study the surface structure of materials such as leather, glass, and paper, although my knowledge of electron optics was largely confined to watching Mr. Cuckow examine my preparations while lounging on the numerous large packing cases which literally surrounded the instrument. On finding that these cases housed an electron microscope destined for the University of Leeds, in October, 1943, I wrote to Astbury applying for the post of Research Assistant in Electron Microscopy—which he immediately offered me on the grounds that I was the only applicant who had both seen and worked with an electron microscope.

To change one's job in wartime was not easy, it being subject to the control of the Central Register of Scientists (administered by the Cambridge don, Dr. C. P. Snow). To jump this hurdle Astbury used the good offices of Sir Charles Darwin to arrange matters with Dr. Snow and to secure my transfer to Leeds. I joined his laboratory in December, 1943, and by April, 1944, I had assembled the microscope and brought it into reasonable working order.

Personally I found Astbury extremely pleasant and stimulating to work with; however, he had well-developed powers of self-aggrandizement and he never disclosed to me the true basis on which he had acquired the instrument—always maintaining that he had been given the microscope personally by the Medical Research Council. He expected total loyalty to his wishes and indicated that I must have nothing to do with anyone else in the Textiles Department or in the university as regards electron microscope projects. It was only later, in about 1950, that I learned the true facts regarding the Leeds Type B instrument and it embittered my view of Astbury. Only then could I understand the large number of medical people, some of high reputation and standing, who, too frequently for my liking, visited my laboratory armed with sizeable specimens of human biopsy and autopsy tissue or with tissue extracts hopefully containing viruses or with bacterial cultures, wanting "to see what they looked like under this new microscope" and who were at a loss to understand the chilly reception they mostly received.

On the lighter side, I recall the visit to my laboratory of Brigadier J. N. Tetley, DSO, TD, DL, Treasurer of the University and head of the famous brewery in Leeds. He confided to me that he was developing a new beer so that the forthcoming victory in the war could be worthily celebrated by the citizens of Leeds, but he was having trouble with the brew—it was cloudy in appearance. Could he send me a sample so that I could examine it in the new microscope and tell him the cause of the cloudiness? Of 24 pints of the new beer which he provided, I estimate that 0.5 ml formed the sample for the electron microscope. The remainder was "sampled" by members of the Textile Physics Laboratory, who declared it perfect in every way. The cloudiness was caused by small particles of oak wood from old vats which had recently been brought into commission; once the wood had been thoroughly rewetted, the problem disappeared.

On arrival in this country from America, the seven instruments were sent to the National Physical Laboratory at Teddington, which acted as the distribution center and which arranged for their onward journey to the various institutions to which they had been allocated. This work was made easy for the Metallurgy Division of NPL by the efficiency of Sir Charles Darwin. Being close to the War Cabinet and knowing many of the important people, he knew how to wade through wartime red tape by dealing directly with the highest officials. Thus by late January, 1943, six of the seven microscopes were in the hands of their intended recipients.

The recipients soon established that the eight huge wooden packing cases merely contained the parts for assembling the microscope. (Here I recall that in wartime, timber was extremely hard to come by; I can assure readers that the speed with which these packing cases disappeared for private use was close to that of light itself.) There were no instructions regarding the mode of assembly, merely a list of the contents of each packing case. Clearly much time would be needed to assemble the instruments and bring them into routine use. This meant that lots of ingenuity and judgment had to be exercised, and it says much for the people in charge and for the quality of the technical assistance they were able to call upon that by August, 1943, about three of the instruments had been built and a few electron-optical images obtained.

A handbook showing the appearance of the completed instrument and recommended methods for its assembly and operation was not forthcoming until late in 1944. Certain points in the assembly meanwhile remained obscure so NPL took on the task of referring these to RCA in Trenton, New Jersey, for elucidation. To these requests the Americans responded quickly and by July, 1943, a body of information regarding the Type B assembly and the working characteristics of the various electrical circuits plus a limited supply of spare components likely to be needed were acquired by NPL. Thus the National Physical Laboratory, as distributor, stockist, and a rich source of information on the lend–lease instruments, naturally became the oracle to which recipients of the microscopes turned for help.

At the time it became commercially available (April, 1941) the RCA Type B instrument could be said to be well engineered, being built on robust lines and able to withstand heavy and often inexpert usage. The various electrical circuits were easily accessible, well designed on conservative principles, and since they had robust components there were relatively few circuit failures. In the early stages of testing the instruments, however, the overall design of the electrical gear came in for much criticism, due to a crop of erratic circuit failures. These were due to the sudden and unforeseeable demise of wire-wound resistors coated with blue ceramic—indeed, among other names they were called blue devils. Once it was realized that *all* these resistors needed to be replaced (the British equivalents proved entirely satisfactory) the circuitry worked extremely well. The RCA parent company in New Jersey were well aware of the devilish behavior of this batch of blue resistors, and because their wholesale failure caused much friction with their main customer (the U.S. Armed Forces) they took the drastic step of banning the manufacturers of this particular component from doing future business with them.

The Type B microscopes also allowed research workers in wartime Britain to see something of contemporary American high-vacuum practice. The comparison was distinctly in America's favor. The vacuum system was efficient and easy to manage and monitor. Though noisy, the backing pumps were very large by British standards and of high pumping speed. As this was before the era of air-ballasted pumps, the backing system was very susceptible to moisture levels (mostly determined by the gelatin in the photographic plates), so that regular and frequent changes of drying agent (phosphorus pentoxide or calcium chloride) were necessary. The oil diffusion pumps were of a well-designed metal construction and had high pumping speeds, using newly developed pump fluids such as Octoil or silicone oil—which were superior to the British Apiezon oils pioneered by Professor C. R. Burch for Metropolitan Vickers Electrical Co. Ltd. some 20 years earlier.

Although the microscopes were soundly built and well designed from the visual standpoint (they retained a surprisingly modern look for a decade or more), there were certain irksome aspects in running and maintaining them, as their early operators soon discovered.

For example, it was difficult to service the high-tension tank, weighing nearly 200 kg and located in a tight position at the top of the assembly, unless overhead lifting gear was available. Also, when lower parts of the instrument such as the photographic chamber required attention, the upper and quite heavy part of the column needed to be lifted both quickly and in one piece. Otherwise, to dismantle the column in sections by loosening and storing over 100 screws and to provide dust protection for the several parts was time consuming and involved about 4 hours' labor. Some of the microscopes installed later than the NPL instrument benefited greatly from NPL's advice to include overhead mobile lifting gear right from the start.

Specimen movement was by means of two hand-operated wheels set at right angles and located at specimen chamber level some 1.8 m above the floor, but the viewing screen was about 0.8 m from floor level. Thus the simultaneous use of hand and eyes at a separation of 1.0 m placed great physical strain on the operators, and people with short arms and bodies were at a disadvantage. To obviate this awkward physical requirement, specimen movement was later actuated by worm/pinion devices connected by flexible Bowden cables to hand wheels at viewing-screen level.

The Type B instruments used photographic plates 10 × 2 in. in size. At first these were regarded as being too small for regular research purposes but in fact proved quite versatile in use because five separate recordings each 2 in. square could be taken on one plate. Early users of the microscope soon realized that photographic plate gelatin can absorb appreciable amounts of water, which meant long pumping-out time before a working vacuum could be established in the microscope column. Even by the end of 1943 operators had adopted prepumping of plates (in a vacuum system separate from the microscope), which led to considerable saving of time.

During these early days some amusing situations arose from the fact that in some parts of the microscope column the Mumetal screening was poor; hence stray magnetic fields could easily deflect the electron beam. For a long time it was thought that many of the visitors who wished to see the microscope in operation had distinctly magnetic personalities until it was realized they might be carrying iron objects in their pockets, handbags, or garments.

By April, 1944, four of the original seven microscopes had been assembled and brought into some degree of working order. However, Cuckow's enquiries from NPL had established that some owners of microscopes were clearly having difficulty in bringing them into operation and he realized that much might be gained by having a get-together of all those involved with electron microscopy. Sir Charles Darwin, both as Director of NPL and as a member of the allocating committee of the War Cabinet, immediately saw the value of Cuckow's suggestion and bent all his energy and resource in arranging the first meeting of users of electron microscopes in Britain. This took place in April, 1944, in the rooms of the Royal Society (then at Burlington House, off Piccadilly, London) in a pleasant and exciting atmosphere.

In addition to scientists from civilian institutions (including the owners and operators of the RCA Type B instruments) there were many representatives from the War Ministry and various defense establishments—indicating perhaps a greater wartime use of electron microscopes than previously thought. I recollect that few if any representatives of British manufacturers of electron-optical devices seemed to be present, and it is unclear why Sir Charles chose not to invite them on this occasion. One or two American service people were present but I do not recall any visitors from France, Holland, or other European countries attending the meeting.

Even during the dark days of the war, one never met a scientist who doubted that victory would eventually come to the Allies, but at this meeting an atmosphere of optimism and youthful exuberance was unmistakably present. No doubt the scientists from the various service establishments were well aware that the D-Day landings, so long in preparation, were soon to be mounted. With absolute certainty as to the outcome, they displayed a highly infectious gaiety. Thus on what really was nothing more than a delightful social occasion within a splendid and historic setting, perhaps it was this exhibition of national fervor and confidence which hid the meager scientific content of the proceedings.

Sir Charles was an outstanding chairman of great urbanity and charm. Dr. W. T. Astbury, always enthusiastic, expounded the view that the electron microscope would prove to be the greatest single advance in helping to elucidate the structure of biological materials. Mr. Cuckow, in mentioning some useful working hints, stated that much instability in operation was due to the bad state of the dry reference batteries—they had not traveled well over the Atlantic. Nothing was said of electron-optical principles or of methods of specimen preparation for the new kind of microscope.

The next meeting of users of the electron microscope was again initiated by Sir Charles Darwin, being eventually held toward the end of 1944 (December, I think). As before, it was intended to have the meeting in the rooms of the Royal Society but at very short notice the venue was switched to the lecture theater of the Institution of Electrical Engineers in Savoy Place alongside the Thames.

The arrangements for this meeting were somewhat ad hoc in character because Sir Charles' war commitments were onerous and continually changing (by this time the Allies were fighting on the very soil of Nazi Germany and victory was not far off) and so he devolved much of their detail upon Mr. Cuckow in the NPL Metallurgy Division. About 30 people attended the meeting; these were mostly from civilian laboratories and on this occasion the representatives from the scientific establishments of the Armed Services were not so numerous.

Sir Charles Darwin opened the proceedings with a short speech of welcome but he immediately left the building, with Professor W. T. Astbury taking over the duties of chairman. There was no fixed agenda; anyone with a contribution to make could come forward and speak. It was still evident, as at the earlier meeting, that the routine use of the instrument was a long way off and that people were finding it difficult to prepare specimens for the electron microscope.

Dr. Drummond showed a few photographs of cotton dust particles taken on the RCA instrument at Manchester, while Mr. Cuckow again gave generously of his experience and supplied hints on operational matters and on specimen preparation. Professor Astbury, using his prerogative as chairman, provided the longest contribution, and though he exhibited his customary erudition and enthusiasm, much of the audience found his thoughts on the clotting of blood, the fibrinogen–fibrin conversion, a little above their heads. A few months later I recalled this lecture with some shame, for the slides which I had prepared for Astbury, and purporting to be fibers of the blood protein known as fibrin, were artifacts—simply being folds in the nitrocellulose films used as specimen supports and about which little was known in Britain at this time.

This occasion was important in disclosing that an electron microscope of native British construction, i.e., *not* of the RCA Type B variety, was being operated successfully. Toward the end of the meeting, Mr. N. D. P. Smith of the Paint Research Station, Teddington, Middlesex, astonished the audience by saying that he had constructed his own electron microscope at Teddington and that it seemed to work reasonably well. With his colleague Dr. Dorothy L. Tilleard he had already photographed particles of paint pigments with some success. This came as a surprise, especially as the instrument had been built only about half a mile from the laboratory housing the RCA Type B microscope at NPL. Mr. Smith had based his design on published accounts of microscopes constructed in Germany, Canada, and America and modestly claimed that the bulk of the work was within the competence of a physicist who had some skill in metalworking and practical electronics. He endeared himself to his audience by saying that the glass insulation sleeve for the electron gun assembly had been constructed from a humble pint bottle of English beer, so setting a precedent for later builders of electron microscopes, e.g., Challice (1948) of Imperial College, London. Tests later carried out at NPL, The Shirley Institute, and the University of Leeds showed that in performance this homemade instrument compared favorably with the Type B microscopes except at higher magnifications (of the order 10,000×), when marked aberrations in the image became apparent.

The third meeting on electron microscopy was held in June, 1945 (i.e., immediately after the end of the war in Europe), in a schoolroom close to the National Institute for Medical Research, Mill Hill, London. Again, Sir Charles Darwin was the prime mover, with Dr. A. S. McFarlane being responsible for the local arrangements. About 30 people attended, including, for the first time at these meetings, Sir Lawrence Bragg (The Cavendish Laboratory, Cambridge), Dr. V. E. Cosslett (The Electrical Laboratory, Oxford), Dr. D. Gabor, and a most welcome visitor from Delft in Holland, Dr. J. B. Le Poole.

Apart from a few micrographs of bacterial and smallpox virus preparations shown by Mr. J. Smiles of the host institute, few other slides were presented. On the contrary, for the first time some theoretical aspects of electron optics were discussed, including the theory of image formation and contrast. I recall showing densitometer recordings across a collagen fibril and their variation with changes in beam aperture, which I considered to represent Fresnel-type diffraction fringes (Reed, 1946). However, Dr. Gabor and others thought this was highly fanciful and regarded the fringes as chromatic in origin, arising from the inelastic scattering of electrons as they traversed the specimen.

At this particular meeting it became apparent that lack of knowledge of the practical procedures in electron microscopy would delay future progress and thus it was decided that a compilation of practical methods, including those of specimen preparation, should be made by the members present, sending any information to Dr. D. G. Drummond, The Shirley Institute, Manchester, who would also act as editor. (This venture was completed in 1947 and was first published in typewritten form by the Institute of Physics in 1948 and as a full, printed book in March, 1950, by the Royal Microscopical Society, Tavistock Square, London.)

For the first time, the desirability of electron microscopists forming a special group under the aegis of a learned society or institute was mooted. Dr. A. S. McFarlane and others suggested that perhaps the Royal Microscopical Society of London would be ideal for this purpose but Professor W. T. Astbury and Sir Charles Darwin thought that because applied electron optics relied so heavily on complex physical principles, only a body such as the Institute of Physics should be considered. However, the meeting decided to defer taking any action on this matter.

The fourth conference on electron microscopy was held in March, 1946, at the Midland Hotel, St. Peter's Square, Manchester. Sir Charles Darwin again convened the meeting with Dr. D. G. Drummond, The Shirley Institute, Manchester, in charge of the local arrangements. About 120 people attended, including visiting scientists from the Netherlands, France, America, and the USSR.

Now that victory in World War II had been won, the occasion was more of a friendly get-together than a serious scientific conference. Only a small number of slides were presented by various speakers, but each electron micrograph was discussed seriously and the interesting contributions made by the foreign visitors were most welcome. The first micrographs of metal-shadowed specimens were presented, prepared by the method of Williams and Wyckoff (1944), details of which had recently become available in Britain. Dr. D. G. Drummond had used gold to shadow samples of airborne dust from the carding rooms of cotton mills, while Dr. R. Reed showed rodlets of tobacco mosaic virus shadowed with silver atoms. Even from these early micrographs it was apparent that the high contrast and the illusion of three-dimensional realism afforded by this type of preparation had great future potential, especially if the technique could be adapted to replica studies of surfaces.

With the war now won, British scientists were keen to return to normality and to open the floodgates of scientific research and development. By 1946 it was also apparent that electron microscopy had a promising future and would involve scientists from many diverse disciplines. Already Metropolitan Vickers in Britain, Philips in The Netherlands, RCA in America, and even Siemens in postwar Germany were planning the commercial production of new microscopes and it was now timely to set up an Electron Microscopy Group to coordinate the activities of those working in this growing field. The question of how this should be realized formed the main business of the conference.

From the Hampstead meeting in 1945 it was clear that two professional bodies would be interested in administering a group of electron microscopists, namely the Royal Microscopical Society, then at BMA House, Tavistock Square, London, and the Institute of Physics, 47, Belgrave Square, London. Both these parties were well represented at the meeting, having given much thought to this matter.

Professor Astbury and Mr. Cuckow favored going under the Institute of Physics banner, having the support of Dr. F. C. Toy (Director of the Shirley Institute, the host for the meeting and President of the Institute of Physics) and Dr. H. R. Lang, full-time Secretary of the Institute of Physics. Astbury was aware that the chairman, Sir Charles Darwin, no mean physicist himself, also favored the aegis of the Institute of Physics, but Sir Charles, with his customary urbanity and courtesy, conducted the meeting with scrupulous fairness to all. Dr. A. S. McFarlane, Dr. R. J. Ludford (President, Royal Microscopical Society), Mr. J. Smiles (Hon. Sec., Royal Microscopical Society), and Professor I. Manton (newly appointed to the Chair of Botany, University of Leeds), on the other hand, ably put the case for going with the Royal Microscopical Society.

Although the two parties debated fiercely, the argument was conducted in a friendly spirit and no rancor or resentment was discerned. When the vote was taken 60% of those attending voted in favor of going with the Institute of Physics, and instructed the chairman and Professor Astbury to approach the institute. With hindsight it can now be said that both professional bodies were much keener to take on the task of looking after the interests of electron microscopists than appeared at the Manchester meeting. The Royal Microscopical Society, for so long the famous upholders of the light microscope—the "brass and glass brigade" as they were affectionately known, were worried for their future and feared a dramatic fall in membership with the advent of the new type of microscope. The Institute of Physics had similar problems. Having come through the war, when scientific conferences were difficult if not impossible to organize, they now wished to expand their activities and recruit the growing number of young physicists in training. They feared rivalry from the Physical Society, the other large body of professional physicists and which boasted a highly prestigious journal. In 1946 the task of providing a special grade of membership to accommodate people not well qualified in physics and who were ineligible for the Associate or Fellowship grades did not seem to pose too many problems, but in later years Dr. Lang admitted (privately to the author) that looking after the needs of nonphysicists did at times place a severe strain on the administrative staff of the institute. Although the Royal Microscopical Society failed in its aspirations to organize scientists in the new field of microscopy, it should be stated that no member of its council was ever heard to express ill-feeling over this matter. In fact, over the years its council collaborated with the Electron Microscopy Group of the Institute of Physics in a most friendly manner, particularly in arranging large and important conferences (such as the Third International Conference held in London, 1954), and in publishing Group proceedings and abstracts of papers in electron microscopy.

In the immediate postwar period, with the return of international trading practices, The Radio Corporation of America had paid some attention to the possibility of selling their electron microscopes to European countries. Thus they had introduced some EMU models (developments of the Type B instruments) and also some small table models into France, Spain, Britain, and possibly even other countries in Europe and had sent over one of their best service engineers (and linguists), Mr. Holmes Halma, to oversee their installation and maintenance. As the sole RCA microscope representative in Europe, he also had the task of servicing the older Type B instruments which had been sent over in 1943 under the lend–lease arrangements.

Mr. Halma found this aspect of his work very trying because Type B instruments were by now somewhat out of date and in any case it was difficult to maintain a stock of the necessary spare parts across the Atlantic, so far removed from RCA headquarters in New Jersey. Even so, over the years 1946–1948, Mr. Halma did remarkable work in servicing many of the seven original Type B instruments and in modifying them to enhance their performance. He was a most capable service engineer and indefatigable in the cause of his company and those who used its products. He did much to foster Anglo-American relationships but sadly he did not always receive the encouragement his abilities and friendliness deserved. Some owners of Type B instruments would have little to do with him, perhaps regarding his intrusion as a slight on their ability to maintain and service an electron microscope, and at times he had much to contend with. In any case he was at a disadvantage when surrounded by so many of his business rivals such as Metropolitan Vickers, Philips, TSF (in France), and Siemens (in Germany).

I recall his telling me of a routine visit to the National Institute for Medical Research, Mill Hill, and finding not the RCA Type B instrument he had come to service but a fairly new-looking Siemens microscope, which by some strange means had come from a recently defeated Germany. To his horror, closer inspection revealed that the high-tension supply to this alien newcomer was none other than the HT tank assembly from the RCA Type B instrument "since it works much better than the German one."

This particular Siemens instrument had an interesting history and always seemed to be at the center of some high drama. Actually it had belonged to Dr. Carlheinrich Wolpers, a highly competent physician with a lucrative medical practice in far-away Hamburg, who had bought it for his own personal use from Siemens during the war. Wolpers was an accomplished electron microscopist and was extremely well skilled in the techniques of specimen preparation. Indeed, after the war, when German scientific journals again became available, it appeared that he had pioneered much important work in the biological field; e.g., he was the first to demonstrate cross-striations in the fibers of animal connective tissue. In 1945, not wishing to leave himself or his microscope in the Russian Zone, he was granted the use of a German army lorry to transport the instrument westward toward the advancing troops of Lt.-General Sir Brian Horrocks. His captors, disbelieving the suggestion that the contents of the lorry were his personal property, promptly took them over and "liberated" them to the United Kingdom, where, for the second time, a benevolent authority passed it freely to the charge of the National Institute for Medical Research at Hampstead. (Other laboratories, e.g., the Chemical Defence Research Establishment at Porton Down, Wiltshire, similarly were the grateful recipients of Siemens-type microscopes.)

In my correspondence with Dr. Wolpers on scientific matters, he invariably mentioned this microscope, which he asserted had been stolen from him, and he implored me to keep an eye open for it in the United Kingdom. It was in 1954 on the occasion of the Third International Conference on Electron Microscopy that Dr. Wolpers dramatically identified the Hampstead instrument as his own when visiting the National Institute. After representation had been made to the German Embassy in Britain it was soon agreed that this instrument was indeed his personal property and belatedly he regained possession and had it sent back to Germany.

The fifth conference on electron microscopy, the first arranged by the Electron Microscopy Group of the Institute of Physics, was organized by Dr. V. E. Cosslett and took place in September, 1946, at the Clarendon Laboratory, The University of Oxford. Being held in that pleasant city, with an interesting program of papers for discussion, including a general lecture on the work of the Clarendon Laboratory and a visit to the Atomic Energy Research Establishment at nearby Harwell, this conference was well attended by people from Britain, France, and The Netherlands. Nine papers were devoted to the theoretical principles underlying the construction of electron microscopes (the session being chaired by Lord Cherwell), followed by a further nine covering specimen preparation techniques and their application. [For the summarized proceedings see Cosslett (1947).] For the first time in Britain the so-called Small Electron Microscope of RCA Laboratories, a compact table model, operating at fixed magnifications of 500 or 5000× using 30-kV electrons and developed by Zworykin and Hillier (1943), was available for inspection (being in use at Harwell), while the prototype of a new Philips microscope was also on display.

At this meeting Sir Charles Darwin reported that the Board of the Institute of Physics had formally agreed to allow the formation of an Electron Microscopy Group of the institute. After members had unanimously approved this development, Sir Charles Darwin was elected as chairman and Dr. V. E. Cosslett as honorary secretary of the newly established group. Hence this fifth conference on electron microscopy can also be regarded as the first arranged by the Electron Microscopy Group of the Institute of Physics. Shortly after this conference Dr. Cosslett left Oxford to take up a university post at the Cavendish Laboratory, Cambridge, so beginning his long and fruitful association with this famous laboratory.

The sixth conference on electron microscopy, held at BMA House, Tavistock Square, London, in March, 1947, was arranged jointly by the Electron Microscopy Group of the Institute of Physics and the Royal Microscopical Society. The chairman of the group, Sir Charles Darwin, presided over an attendance of about 120 people, including visitors from the Netherlands, France, and the United States.

There is no need to detail the papers that were presented, for they were fully reported by Sayer et al. (1948), but it may be said that at this second conference of the Electron Microscopy Group, the format of a typically British electron microscope meeting finally and fully emerged.

Authoritative speakers both from Britain and overseas presented papers on topics of immediate interest and ample time was allowed to discuss them with frank and full exchanges of information. Thus Dr. J. B. Le Poole (Institute for Electron Microscopy, Delft) introduced a fundamental treatment of the resolving power of the electron microscope which was discussed with enthusiasm. Mr. M. E. Haine and Dr. G. Liebmann (Associated Electrical Industries Laboratories, Aldermaston) spoke of developments in the production of new models with magnetic lenses, while Professor P. Grivet and Professor H. Brück (TSF Laboratories, Paris) discussed design features of their newly constructed electrostatic electron microscope. The techniques of specimen preparation were now being widely investigated. Various workers presented papers on metal-shadowed preparations of biological structures such as collagen and keratin fibers, and the marked granularity often seen in such preparations was fully discussed. Likewise the flow properties of evaporated silica during the preparation of surface replica films was also considered.

The clear success of the London conference in March, 1947, was followed by the Electron Microscopy Group of the Institute of Physics holding a further meeting on September 16–17 of the same year at the University of Leeds. For details of this meeting see Reed (1947) and Cosslett (1948). Over 150 people attended, including welcome visitors from the Netherlands, France, and the United States, and the proceedings followed the now well-established format.

In the session devoted to instrument design and construction the main feature was the introduction of the Met-Vick experimental EM3A microscope by Mr. M. E. Haine and colleagues of the AEI Research Laboratories, Aldermaston. A new double-projector lens system permitted a wide range of magnification (1000–50,000× at 60 kV), while an improved electron gun gave usable image brightness at the highest magnification. The instrument was enthusiastically examined by a large number of visitors.

From papers presented in the Applications session it was clear that the methods of specimen preparation for the electron microscope were now better understood and were being freely applied. For the first time also, papers on the choice of photographic recording material for use in the electron microscope were presented. Dr. R. Reed (University of Leeds) gave the results of testing a large number of plates and films, preferring for general use a plate emulsion which was very fast to electrons, while Mr. I. M. Dawson (National Institute for Medical Research, Hampstead) described a successful roll-film adaptor for use in the Siemens microscope designed by von Borries and Ruska (1940).

This meeting, the third organized by the Electron Microscopy Group, was highly successful, and in reporting it Reed (1947) could say

> *In conclusion, it is gratifying to report that teething troubles in British electron microscopy are now more or less over, and that, so far as present conditions permit, the outlook is full of promise.*
> *Not only have the new techniques been learned; they are being actively applied and developed in the elucidation of many fundamental problems, particularly in the biological field.*
> *It may be claimed that the new Electron Microscopy Group of the Institute of Physics is now safely launched, and the growing membership and increasing vigour of its discussions augur well for the future.*

This optimism for the future of electron microscopy in Britain was fully justified, for following highly successful conferences at Delft in 1948 and at Paris in 1950, the aspirations of the Electron Microscopy Group, now a member of the Joint Commission on Electron Microscopy of the International Council of Scientific Unions, reached a zenith in the large-scale International Conference held in London, July, 1954. Here visitors from many parts of the world gathered to discuss virtually every aspect of electron microscopy. Perhaps those members of the lend–lease committee who first decided to ship seven RCA Type B instruments across the North Atlantic to a war-stricken Britain were wiser than we thought, and in language they would well understand they "sure started something."

1. Afterword by Peter Hawkes[1]

Ronald Reed was born on 15 November 1919 and died on 23 March 1990. According to its title, his article stops at 1948 though the last few sentences take it up to the International Congress on Electron Microscopy in London in 1954. During those years, he was author or coauthor of several papers on the instrument itself (Reed, 1947) and on early applications (Astbury et al., 1947; Passey et al., 1947, 1948; Preston et al., 1948; Reed &

[1] I am most grateful to Ira Rabin and John Hodgson for invaluable help.

Rudall, 1948a, 1948b). There is no hint that he was to become an important figure in a very different field of research: the study of the Dead Sea Scrolls—his advice on the preservation of the material was sought by the Vatican. The majority of these scrolls, many of which are very early versions of chapters of the Old Testament and the Apocrypha, had been found and deposited in the Rockefeller Museum in East Jerusalem by 1956. The scrolls consist of leather, papyrus and a leather-like parchment apart from an exceptional scroll on copper.

Ronald Reed was known to be an expert on leather and parchment. It was thus natural for John Allegro, a Dead Sea Scrolls scholar, to invoke his help in analyzing the fabric of the scrolls. For this, he was given an assortment of fragments of leather and parchment containing no text. Over the next few years, Reed and his doctoral student John Poole (Poole, 1959) examined these in great detail, showing that light and electron microscopy were capable of distinguishing between leather and parchment and elucidating the methods used to prepare these materials (Burton et al., 1959; Poole & Reed, 1961, 1962; Reed, 1965, 1972; Reed & Poole, 1962, 1964).

The John Rylands Library in Manchester houses the "Ronald Reed Archive" of Scrolls fragments, which it acquired in a dramatic fashion. In 1996, G.J. Brooke was delivering his invited speech on 'Manchester and the Dead Sea Scrolls" when a voice came from the audience: "My mother has some Dead Sea Scrolls in her attic". The speaker was the daughter of Ronald Reed and his wife Marion! These proved to be the samples that had been sent to Reed for study and, if possible, dating. "For much of his career at the University of Leeds and afterwards,", writes G.J. Brooke (2006), "Ronald Reed... was one of Britain's foremost leather analysts; many regarded him as the foremost. His time at Leeds seems not always to have been entirely happy; he suffered from the dwindling industrial interest in and support for leather studies at Leeds and his subject specialism was reorganized more than once so that, having started his career there in the Department of Biomolecular Structure in charge of the electron microscope laboratory, he eventually ended up in the University's Department of Food Science. Furthermore, perhaps because of the unfortunate circumstances of reorganization, he never gained the academic seniority and recognition at his University that was his without question on the international scene. But this has resulted in a happy circumstance for us, since he seldom seems to have entrusted the objects of his research and teaching to the University departments with which he was associated, but took home what was important to him. Rather than being lost or destroyed in the many departmental reorganizations, some of it ended up in his study, at the side of the house, and some of it in the attic. His experience was valuable when flooding in Florence damaged many works of Renaissance scholarship. In 1986, the ninth centenary of the Domesday Books, the Public Record Office requested him to study the parchment on which they are written during the rebinding of the Books".

Much more information about the scrolls and in particular, about the John Rylands Library holdings can be found in Rabin et al. (2007) and Brooke (2006).

References

Busch, H. (1926). *Annalen der Physik (Leipzig) [4]*, *81*, 974.

Challice, C. E. (1948). *Journal of Scientific Instruments*, *25*, 27.

Cosslett, V. E. (1947). *Journal of Scientific Instruments*, *24*, 113.

Cosslett, V. E. (1948). *Journal of Scientific Instruments*, *25*, 167.

Fleming, A. (1897). *Electrician*, *38*(972), 302.

Gabor, D. (1927) (Dr.-Ing. Thesis), University of Berlin.

Gabor, D. (1946). *The electron microscope*. London: Hulton Press.

Kaufmann, W. (1897). *Annalen der Physik (Leipzig) [3]*, *61*, 544.

Klemperer, O. (1939). *Electron optics*. London and New York: Cambridge Univ. Press.

Lenard, P. (1904). *Annalen der Physik (Leipzig) [4]*, *15*, 485.

Macgregor-Morris, J. G., & Henley, A. W. (1936). *Cathode ray oscillography*. London: Chapman & Hall.

Martin, L. C., Whelpton, R. V., & Parnum, D. H. (1937). *Journal of Scientific Instruments*, *14*, 14.

Martin, L. C., Parnum, D. H., & Speak, G. S. (1939). *Journal of the Royal Microscopical Society*, *59*, 263.

Myers, L. M. (1939). *Electron optics*. London: Chapman & Hall.

Reed, R. (1946) (Ph.D. Thesis), University of Leeds.

Reed, R. (1947). *Nature (London)*, *160*, 762.

Sayer, J., Brown, A. F., & Todd, J. B. (1948). *Journal of Scientific Instruments*, *25*, 23.

Thomson, J. J. (1897). *Philosophical Magazine [5]*, *44*, 293.

Thomson, J. J. (1899). *Philosophical Magazine [5]*, *48*, 547.

von Borries, B., & Ruska, E. (1940). *Siemens-Zeitschrift*, *20*, 217.

Wiechert, E. (1899). *Annalen der Physik (Leipzig) [3]*, *69*, 739.

Williams, R. C., & Wyckoff, R. W. G. (1944). *Journal of Applied Physics*, *15*, 712.

Zworykin, V. K., & Hillier, J. (1943). *Journal of Applied Physics*, *14*, 658.

Afterword references

Astbury, W. T., Perry, S. V., Reed, R., & Spark, L. C. (1947). An electron microscope and x-ray study of actin. *Biochimica et Biophysica Acta*, *1*, 379–392.

Brooke, G. J. (2006). The historical documents at the John Rylands University Library: The Reed Dead Sea Scrolls collection. *e-Preservation Science*, *3*, 3–40.

Burton, D., Poole, J. B., & Reed, R. (1959). A new approach to the dating of the Dead Sea Scrolls. *Nature*, *184*, 533–534.

Passey, R. D., Dmochowski, L., Astbury, W. T., & Reed, R. (1947). Electron microscope studies of normal and malignant tissues of high- and low-breast-cancer strains of mice. *Nature*, *160*, 565.

Passey, R. D., Dmochowski, L., Astbury, W. T., Reed, R., & Johnson, P. (1948). Ultracentrifugation and electron microscope studies of tissues of inbred strains of mice. *Nature*, *161*, 759.

Poole, J. B. (1959). *The nature, origins and techniques of manufacture of those of the Dead Sea Scrolls which are made from animal skins* (Dissertation). Leeds.

Poole, J. B., & Reed, R. (1961). The 'tannery' of Ain Feshkha. *Palestine Exploration Quarterly*, *93*, 114–123.

Poole, J. B., & Reed, R. (1962). The preparation of leather and parchment by the Dead Sea Scrolls community. *Technology and Culture*, *3*, 1–26.

Preston, R. D., Nicolai, E., Reed, R., & Millard, A. (1948). An electron microscope study of cellulose in the wall of Valonia ventricosa. *Nature*, *162*, 665–667.

Rabin, I., Brooke, G., Hodgson, J., Pantos, M., & Prag, J. (2007). *e-Preservation Science*, *4*, 9–12.

Reed, R. (1947). Construction and operation of an electron microscope. *Journal of Scientific Instruments*, *24*, 117.

Reed, R. (1965). Examination of ancient skin writing materials in UV light. *Proceedings of the Leeds Philosophical and Literary Society, Scientific Section*, *9*, 257–.276.

Reed, R. (1972). *Ancient skins, parchments and leathers*. New York: Seminar Press.

Reed, R., & Poole, J. B. (1962). A study of some Dead Sea Scroll and leather fragments from cave 4 at Qumran. Part I. Physical examination. *Proceedings of the Leeds Philosophical and Literary Society, Scientific Section*, *9*, 1–13.

Reed, R., & Poole, J. B. (1964). A study of some Dead Sea Scroll and leather fragments from cave 4 at Qumran. Part II. Chemical examination. *Proceedings of the Leeds Philosophical and Literary Society, Scientific Section*, *9*, 171–182.

Reed, R., & Rudall, K. M. (1948a). Electron microscope studies on the structure of earthworm cuticles. *Biochimica et Biophysica Acta*, *2*, 7–18.

Reed, R., & Rudall, K. M. (1948b). Electron microscope studies of muscle structure. *Biochimica et Biophysica Acta*, *2*, 19–26.

CHAPTER ELEVEN

Rudolf Rühle and the BOSCH electron microscope: another early commercial instrument

Hans R. Gelderblom[a,*] and Heinz Schwarz[b,†]
[a]Retired from Robert Koch Institute, Berlin, Seddiner See, Germany
[b]Retired from Max Planck Institute for Biology, Tübingen, Germany
*Corresponding author. e-mail address: gelderblom.seddin@t-online.de
†Alternative e-mail address: heinz.schwarz@tuebingen.mpg.de

Contents

1. Origins of electron microscopy in Berlin 311
2. First steps for the S&H instrument 312
3. The commercial Siemens & Halske magnetic electron microscope 314
4. The electrostatic electron microscope of the AEG 317
5. Competition between concepts, characters, and companies 317
6. The gifted entrepreneur: Baron Manfred von Ardenne 321
7. Development of electron microscopy inside the Bosch company 324
8. Rudolf Rühle (1907-2003): personal background and development 325
9. Rudolf Rühle: the technical engineer at Bosch 327
10. Walter Rentschler and the Bosch electron microscope 331
11. Rudolf Rühle's groundwork for the Bosch electron microscope 332
12. Rudolf Rühle's thesis: on the future Bosch electron microscope 333
13. Rebuilding of Germany after WWII 337
14. Foundation of the German society for electron microscopy 340
15. The commercial Bosch electron microscope 340
16. Promotion for the Bosch electron microscope 345
17. The Bosch electron microscope in Hohenheim 347
18. The Bosch-EM in Tübingen 354
19. Giving up the commercial electron microscope 364
20. A list of Rudolf Rühle's patents 366
Acknowledgments 368
References 369

1. Origins of electron microscopy in Berlin

Light microscopy, according to the Abbe law, had basically reached its limits at the end of the 1880s. There was a growing need for higher

Advances in Imaging and Electron Physics, Volume 221
ISSN 1076-5670
https://doi.org/10.1016/bs.aiep.2022.03.011

resolution, e.g., in the fight against infectious diseases caused by dreaded, invisible infectious agents: viruses, not blocked by regular bacterial filters. Fortunately new imaging solutions emerged when de Broglie had described the wave nature of electrons (De Broglie, 1925) and when Hans Busch showed that an electromagnet focuses electrons just as a glass lens focuses light (Busch, 1927). As early as 1928 Leo Szilard proposed to Dennis Gabor to build an electron microscope (Gabor, 1974). The implementation of this bright idea required much progress in technology, notably in high tension and vacuum techniques, also high curiosity and motivation. Activities began world-wide and the progress in electron optics fueled the development of the cathode ray tube, television, and the electron microscope [for reviews see Gabor, 1957; Haguenau et al., 2003; Ruska, 1979; van Gorkom, 2018].

The construction of such a complex instrument required ideas, perseverance and a strong background in academia, science, and industry. Despite the great depression and much political unrest in the 1930s, Berlin was a favorable place: engineering of cathode ray oscillographs was a major topic at the Technical University Charlottenburg (Gabor, 1957; Ruska, 1979, 1980) and was advanced mainly by two companies, at the Allgemeine Elektricitäts Gesellschaft (AEG) in Berlin-Reinickendorf and at Siemens, at the Siemens & Halske company (S&H) in Berlin-Siemensstadt. Berlin was a melting pot of ideas with a lively exchange between academia and industry and often fierce competition. The exciting working environment fostered progress, yet also produced long-lasting tough conflicts between the young scientists (Freundlich, 1963; Lin, 1995; Müller, 2021; von Buol, 1943).

2. First steps for the S&H instrument

During the 1930s in Berlin Max Knoll's team of young students at the Electro-Technical Institute, a part of the Technical College headed by Prof. A. Matthias, attempted to improve the performance of the cathode-ray-tube developed there earlier by Dennis Gabor (1900-1979) between 1924-1926 (Gabor, 1957). In the Knoll group the prototype of an electron microscope was created, when Ernst Ruska in 1931 achieved a 17-fold, two-step magnification of a wire-mesh by combining two magnetic lenses - an experiment implicitly honored in 1986 by one half of the Nobel Prize in Physics (Nobel Committee, 1986). This was a first step towards the building of an instrument surpassing the resolution of the light microscope - finally achieved late in 1933 (Ruska, 1934) using polepiece magnetic lenses with short focus length constructed by the two electrical engineers Ernst Ruska

Figure 1 The friends Bodo von Borries (left) and Ernst Ruska on May 26, 1931 during a cycling trip to the Baltic Sea on the island of Hiddensee, on the Dornbusch hill. The close relation was strained occasionally, their cooperation ended after World War II (Gelderblom, 2019; von Borries, 1991). *(Photo taken by Bodo von Borries with self-timer; by courtesy Archives of the Max Planck Society, Berlin.)*

(1906-1988) and Bodo von Borries (1905-1956) (Fig. 1) (von Borries & Ruska, 1932; for reviews see: Freundlich, 1963; Marton, 1968; Müller, 2008; Ruska, 1980; van Gorkom, 2018; von Borries, 1991). The further development of electron microscopy in Berlin was also stimulated by the early life science microscopy studies of Ladislaus Marton (1901-1980) in Brussels (Marton, 1934), but also in Berlin (Driest & Müller, 1935; Krause, 1937).

To interest industry with these developments and in the construction of a commercial research instrument was arduous, and achieved finally thanks to the expert opinion of Professor Richard Siebeck (1883-1965), the Head of the I. Medical Department of the Berlin Charité (Gelderblom & Krüger, 2014). One of his clinical assistants, Helmut Ruska (1908-1973), the younger brother of Ernst Ruska, had arranged a meeting of Siebeck with Ernst Ruska and von Borries. On September 29, 1936 they presented their achievements and expectations. On October 2, 1936, Siebeck, much impressed, wrote a positive expert opinion anticipating much scientific progress as the new instrument surpassed the resolution of the light microscope. He realized that medicine would benefit by deepened knowledge of the fine structure of human tissues and he foresaw particular progress in the field of the noncultivable, nonvisible infectious agents that cause smallpox, chickenpox, measles, influenza, rabies, and foot-and-mouth disease for example, but also in hematology and cancer research. Siebeck's opinion finally convinced industry (Siebeck, 1936).

3. The commercial Siemens & Halske magnetic electron microscope

On February 2, 1937 Ruska and von Borries joined the S&H in Berlin, in the Zentrallabor of the Wernerwerk, as senior engineers in the "Entwicklungsstelle für Elektronenmikroskopie" (S&H, intern: Z.L. 44) with equal rights. Together with the medical doctor Helmut Ruska, they formed a highly focused team executing a fair distribution of tasks, a successful family business (Gelderblom & Krüger, 2014; Karrenberg, 2019; Lin, 1995; Müller, 2008). The serious, matter-of-fact Ernst Ruska focusing mainly on development, Bodo von Borries with his additional communicative competence and extrovert nature took care also of administration and public and political relations. In December 1937 the first pre-serial instrument went into operation (Fig. 2). Their rapid progress was documented in two publications (von Borries & Ruska, 1938; von Borries et al., 1938a). Using a second pre-serial instrument Helmut Ruska, with his coworker Gustav-A. Kausche (1901-1960) and later with Carlheinrich Wolpers (1906-2003), investigated preparation and application problems in the life sciences, in particular in virology (Gelderblom & Krüger, 2014; Krüger et al., 2000).

In June 1938, the publication "Bakterien und Virus in übermikroskopischer Abbildung" (von Borries et al., 1938b) appeared and the novel instrument was presented to the press. Success and the enthusiastic reports led S&H to advance Z.L. 44 to the "Laboratorium für Elektronenoptik" (Laboratory of Electron Optics, LfE) on July 1, 1938 (Fig. 3).

Soon thereafter the need for more applications in the life sciences came up and when external scientists asked for access to the new instrument, Bodo von Borries and Ernst Ruska - with much input from Helmut Ruska - founded a guest laboratory, the "Laboratorium für Übermikroskopie" (LfÜ) (Siebeck, 1941). Initially equipped with three instruments and ample functional workspace this laboratory was headed by Helmut Ruska and also visited by foreign scientists (Gelderblom & Krüger, 2014). The LfÜ was inaugurated only on April 18, 1940 by an address of Dr. Hermann von

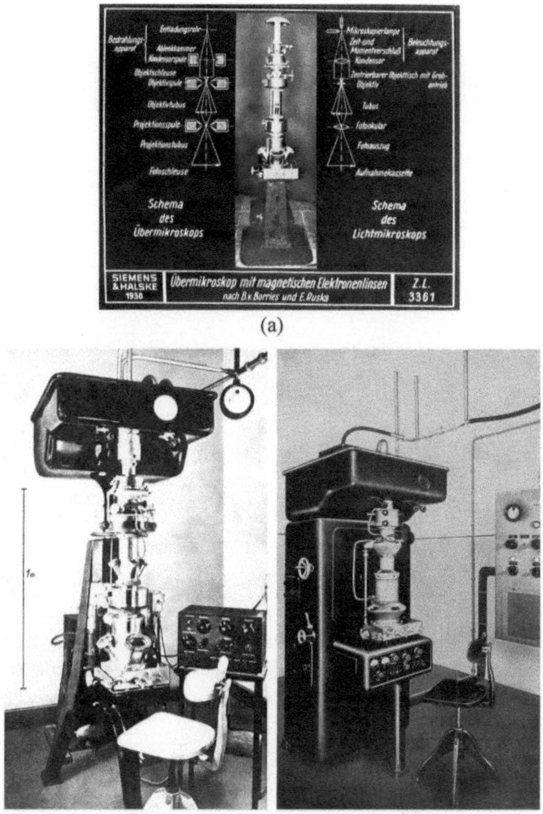

(a)

(b) (c)

Figure 2 The magnetic electron microscope of Siemens. Early instruments constructed in 1937-1938 at the S&H Laboratorium für Elektronenoptik (LfE). (a) Scheme of the pre-serial instrument accompanied by schematic diagrams of the electron microscope and the light microscope. (b) Pre-serial S&H instrument 1938. (c) The first serial Übermikroskop of S&H, one of a series of twelve instruments, from 1939 (*by courtesy of Siemens Corporate Archives*).

Siemens (1885-1986), Board Member then of S&H. He described the history of the instrument, followed by seven talks by well-known experts presenting the scientific progress made possible by the S&H Übermikroskop (ÜM) in the life and materials sciences. All the contributions including significant images and plenty of current references are included in the book of 104 pages *Das Übermikroskop als Forschungsmittel* (Siebeck, 1941). Still of interest today - it reveals progress and confidence - despite a number of contemporary misinterpretations and errors. The news about the ÜM as a research instrument spread widely. E.g., already in 1942 in France a com-

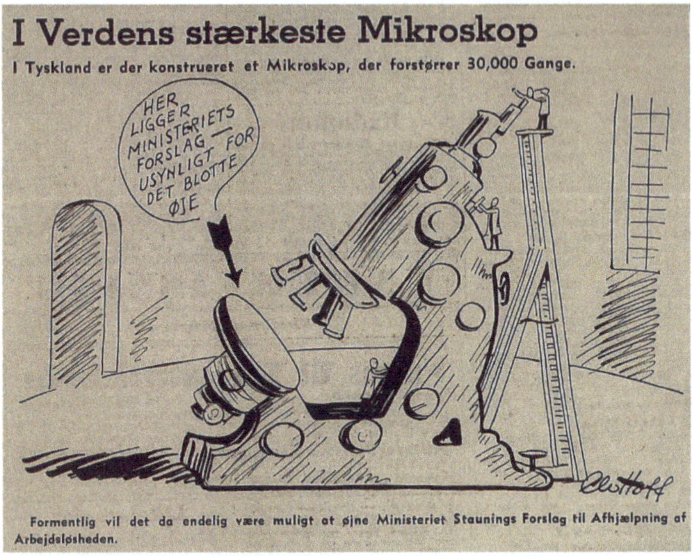

Figure 3 Much interest was raised in 1938 by the SUPERMICROSCOPE of S&H. The press clip from the Danish Berlingske Tidende makes use of the hype around the new instrument and its 30,000-fold magnification. The cartoon is aiming at the hard unemployment in Denmark and criticizes minister Staunings plans as insufficient. It would need the world's best microscope to see perhaps any benefits in the government's plans to cure the crisis (by *courtesy of Royal Danish Library, Copenhagen*).

prehensive review on the history and application of the instrument and on methods and results in virology appeared with 50 references (Levaditi & Bonét-Maury, 1942).

The first commercial ÜM still functioned at 70 kV acceleration voltage. It achieved 30,000-fold magnification, a resolution of 3 nm and was installed in 1938 at the Farbwerke Hoechst in Frankfurt. Up to the end of World War II, 37 more instruments were completed, among them, in 1940, a 220 kV experimental instrument. From 1943 onwards, all research ÜM could operate up to 100 kV (Wolpers, 1991). The Second World War limited production dramatically: only 13 ÜM had been delivered by January 1943, with 45 instruments still on the order book. In a letter to the influential Siemens Direktor Leifer, high up in the German War Production Organization, Bodo von Borries complained about shortages in materials and manpower due to the war conditions. He closed by mentioning that since 1942 the Radio Corporation of America (RCA) had produced more instruments than S&H (von Borries, 1943).

4. The electrostatic electron microscope of the AEG

The Forschungs-Institut der Allgemeinen Elektricitäts Gesellschaft (AEG) in Berlin Reinickendorf, the other early protagonist in electron optics, was led by Professor Carl Ramsauer (1878–1955) (Ramsauer, 1941, 1943). In Ramsauer's Physics Department Dr.-Ing. Ernst Brüche (1900-1985) together with a dedicated team conceived the "Geometrische Elektronenoptik" (Brüche & Scherzer, 1934). The term Elektronenmikroskop had already been coined in parallel in 1932 by the AEG and the later S&H teams (Brüche & Johannson, 1932; Knoll & Ruska, 1932a, 1932b). At AEG manifold theoretical (Brüche & Scherzer, 1934) and practical activities were pursued, e.g., emission microscopy (Brüche & Johannson, 1932) and the construction of the electrostatic einzel lens for high tension by Boersch and Mahl in 1938/39. This lens facilitated AEG's breakthrough to high resolution electron microscopy, to the shadow microscope (Schattenmikroskop), a very simple transmission instrument (Boersch, 1939) and to the AEG ÜM, built by Hans Mahl (Mahl, 1939a), to be developed further into a commercial instrument (Mahl, 1939b, 1940a) (Fig. 4).

Compared to the S&H ÜM in 1940 with magnetic lenses, an acceleration voltage of 75 kV, a 30,000-fold final magnification and a resolution of 2.5 nm, the AEG microscope with its electrostatic lenses operated only at 50 kV, though resulting in excellent contrast, and achieved a final magnification of only 5.000 times and a resolution of 10 nm at best (Jakob, 1941). Basically, it represented a technically less demanding and less expensive approach. It was applied then with success in bacteriology and cytology (Brüche & Haagen, 1939; Jakob & Mahl, 1940).

5. Competition between concepts, characters, and companies

Fierce competition existed between the AEG and the S&H groups regarding the priorities in theory and the advantages and disadvantage of electrostatic versus magnetic lenses - this rivalry was especially prominent between Ernst Brüche and Ernst Ruska (Lin, 1995; Müller, 2021; van Gorkom, 2018). When Ernst Ruska submitted his typescript "Abbildungsgeräte" on April 22, 1939 for the book *Das freie Elektron in Physik und Technik*, a compilation of lectures held during the winter 1938/39, to the editor Professor Ramsauer, an exchange of ten letters ensued. Ruska complained sequentially about falsifying footnotes and supposedly incor-

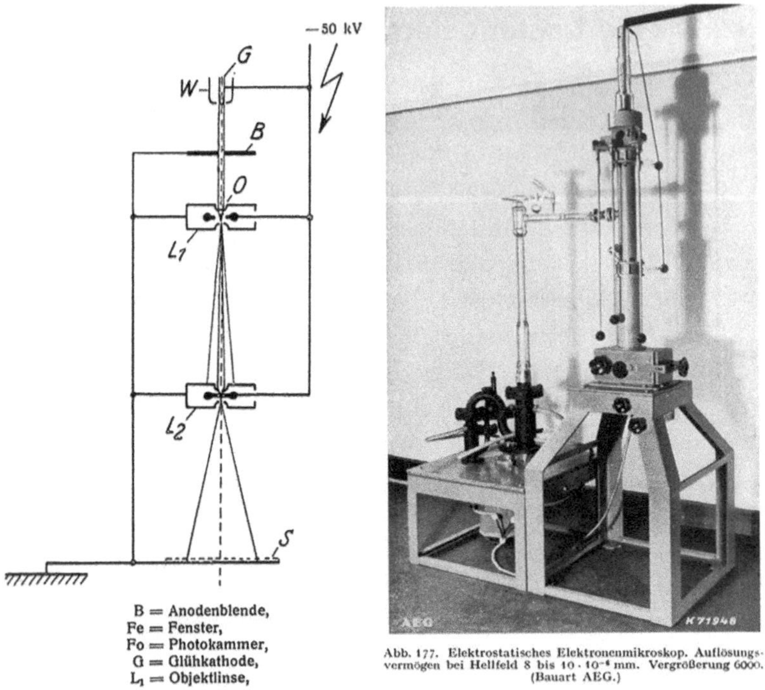

(a) (b)

Figure 4 The electrostatic Übermikroskop of the AEG. (a) Sketch of the earliest version of the AEG microscope (Mahl, 1940a). (b) Early version of the AEG instrument (from von Ardenne, 'Elektronen-Übermikroskopie' 1940, Fig. 177; *with permissions from SPRINGER NATURE*).

rect definitions, e.g., of the term "Übermikroskopie", as proposed by the AEG group. Ruska "corrected" his proofs, often provoking objections by Brüche. His chapter grew in volume beyond the set agreement by the inclusion of new results, but he was still afraid that Brüche's contribution was more attractively presented than his own. When Ruska threatened to withdraw his contribution, Ramsauer finally refused further negotiations ("Weitere Verhandlungen lehne ich ab") and wrote to Dr. Trettin, a neutral person in the Siemens management on July 29, 1940: "I would like to ask you, to instruct Dr. Ruska - also in the interest of your company - that all action has its limits." ("möchte Sie daher bitten, Herrn Dr. Ruska, ... auch im Interesse Ihrer Firma, dahin zu belehren, daß alles seine Grenzen hat) (Ramsauer, 1940). S&H even involved high-ranking copyright lawyers (Flir, 1940a). Nevertheless, the book still appeared in 1940. Yet priority quarrels, accusations of unfair citations and omissions continued and pro-

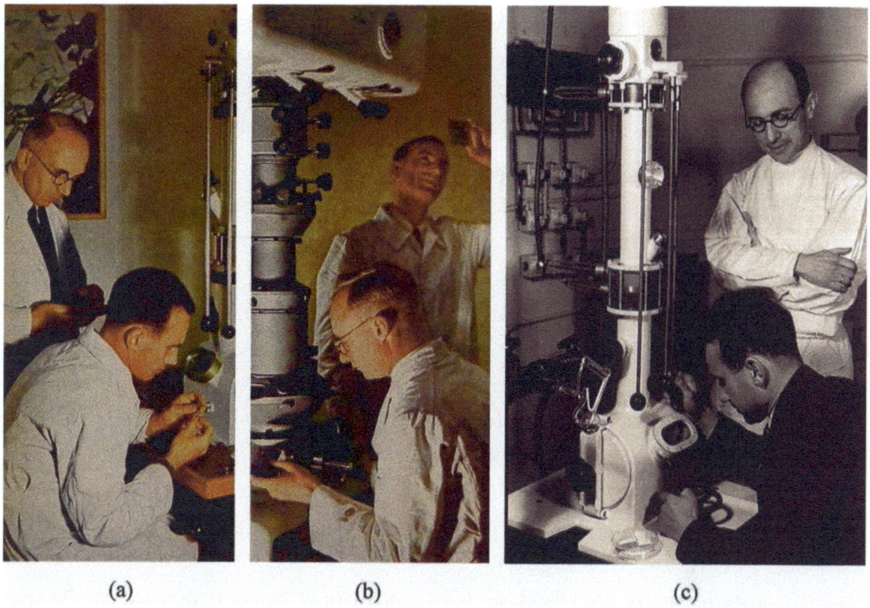

Figure 5 The protagonists of the two groups developing commercial electron microscopes in 1943. (a) Hans Mahl (sitting) and Ernst Brüche with the AEG instrument. (b) Bodo von Borries (sitting) and Ernst Ruska with the S&H ÜM [from SIGNAL, journal of the German Armed Forces, 1943, issue 1, p. 31; Signatur Bib. ZMSBw: Z 668]. (c) The electrostatic ÜM of the AEG with Hans Mahl (sitting) and Ernst Brüche in 1940/1942 (*photo courtesy Annelise Schulze, Mauritius*). AEG donated a microscope of this version to the Robert Koch Institute in Berlin resulting in a number of publications (Gelderblom, unpublished).

voked mutual recriminations and corrections (Brüche, 1940; von Borries & Ruska, 1944a, 1944b). (Fig. 5). The scientific public was not amused and in 1941 the editor of "Glasers Annalen" terminated publication of the heated discussions (Glasers Annalen, 1941).

The rivals were closely watching each other. When the book *Das Übermikroskop als Forschungsmittel* (Siebeck, 1941) appeared in August 1941, the General Manager of S&H in the Wernerwerk, Dipl.-Ing. Desiderius Flir and directly responsible for the LfE, criticized its low typographic quality and the weak image contrast compared to the appealing appearance of the AEG publication *Zehn Jahre Übermikroskopie* (Flir, 1941; Ramsauer, 1941). To calm down the public hostilities, but clearly also to demonstrate the global predominance of German research and engineering, the Prussian Academy of Science awarded the 1941 Silver Leibniz Medal to Max Knoll, Ernst Ruska, Bodo von Borries, Ernst Brüche, Hans Boersch, Hans Mahl and Manfred von Ardenne in recognition of "Leistungen bei der wissenschaftlichen Entwicklung des Elektronenmikroskops" (for their accomplishments in the construction of the electron microscope). Yet, Brüche was not satisfied as we can learn from his meticulously detailed essay on the origin of the instrument (Zum Entstehen des Elektronenmikroskops, Brüche, 1943). The two S&H engineers replied twice in the same detailed way (von Borries & Ruska, 1944a, 1944b), however, without causing much impression on their opponent (Brüche, 1944). Part of the dispute was published also in the VDI-Zeitschrift, the engineer's bread-and-butter newspaper in Germany, the discussions on priorities went on up to the end of 1944 (VDI, 1944). Interestingly and consequently S&H based its strong legal position on the early patents by Reinhold Rüdenberg (1883–1961), the Chief Electrical Engineer at Siemens and Director of the Scientific Department of Siemens-Schuckert Werke (Rüdenberg, 1931a, 1931b, 1931c, 1932). Manfred von Ardenne, the third party in the lively Berlin ÜM scene, also experienced critical relations with the dominant S&H company (see below).

The antagonisms also had serious economic consequences: In March 1943, S&H already owned 261 patent rights in the field of electron microscopy (Ruska, 1943), so that AEG was limited to only a small series of five commercial instruments (Müller, 2021; von Borries, 1944; Wolpers, 1991). One of them was donated to the Robert Koch Institute in Berlin (Ramsauer, 1941).

When, in October 1941, the AEG patent-office applied for permission to use S&H patents to fulfill research requests of the R&D Agency of the German Army, S&H delayed the negotiations. And in November 1942, when S&H learned about proposals for compulsory licences in favor of AEG and suggestions for future joint developments, Dr. Heinrich von Buol, Chairman of S&H, and his General Manager at the Wernerwerk of S&H Desiderius Flir excluded any collaboration as unacceptable ("das kann man uns nicht zumuten. Das Verhältnis der wissenschaftlichen Bearbeiter zu einander schließt jede nützliche Zusammenarbeit aus") (von Buol, 1942) and offered an apparently "poisoned" contract. On January 12, 1943, AEG complained about the levels of the license fees as "exceeding the sales prices of the instrument several-fold and asked them to set a decent limit for the fees" (AEG, 1943). S&H protracted the negotiations still further and on mid-1943, von Buol summarized the history of the dissonances in a nine-page letter to Ramsauer (von Buol, 1943). After regretting the splitting hairs feud ("die literarische Fehde") between Brüche and his AEG group and Bodo von Borries and Ernst Ruska on the S&H side, von Buol pointed to the Rüdenberg patents of May 1931 (Rüdenberg, 1931a, 1931b, 1931c) and presented a detailed list of historic facts emphasizing the fundamental priority of S&H in the field of transmission electron microscopy. Next, on four pages he accused AEG of a "literary campaign" (literarischer Feldzug), omissions, text manipulations and further misconduct, including fake information printed repeatedly in the press. Regarding the patent requests, von Buol finally offered concessions provided that AEG published proper corrections. When in autumn 1943 Allied bombing shattered Berlin, AEG evacuated the Brüche institute, with more than 100 members of staff, to the East, to Schönberg near Görlitz. This dislocation opened up further developments not only in electron microscopy but also in night vision devices, i.e. in equipment for the army. Early in 1945, when the east front approached from Russia, Brüche and his team went to the West for another restart (Rang, 1992; Schulze, 2001).

6. The gifted entrepreneur: Baron Manfred von Ardenne

The third developmental line in electron microscopy in Berlin was run by Baron Manfred von Ardenne (1907-1997) in Berlin-Lichterfelde. Inspired from his boyhood in tinkering with radio ("wireless") receivers, in 1922 he began a ten years collaboration with Dr. Siegmund Loewe (1885-1962), founder of the Loewe AG in 1923. Manfred von Ardenne left school without passing the Abitur (the qualification for university entrance) and became self-reliant at the age of 17 years: He developed the triode valve, broad band amplifiers and other pioneering radio technology. In 1923, he applied for the first of his more than 600 patents. Without the Abitur certificate, but specially recommended by the Nobel prize winner Professor Walter Nernst, he began studies at Berlin University in 1926 and 1927 in Physics, Chemistry and Mathematics. He was deeply impressed by his professors, among whom were Max Planck and Walter Nernst, but he left university after only four semesters and followed his research endeavors for the rest of his life. For his growing laboratory and for himself, he bought a villa in Berlin-Lichterfelde, Jungfernstieg 18, in 1928 resulting in a debt of 200.000 Reichs Mark, corresponding in today's value to 750.000 €. The house served him for the next 17 years to develop his contributions in electron microscopy, the electron microprobe, isotope separation and mass spectrometry - until May 21, 1945, when he left after World War II for the Soviet Union (von Ardenne, 1997).

His work depended very much on third-party funding, by the Reich Post office, a government institution, by the Krupp AG and other companies: To cope with his debts he produced and sold cathode-ray oscillographs. Next he turned to television engineering ("the flying spot scanner"). Together with the Loewe AG he presented the first fully electronic television during at the annual Radio Exhibition in Berlin in 1929 and 1931; in this he cooperated with the Fernseh AG. At the Olympic Games of 1936 in Berlin, TV transmission with 180 lines definition was offered to the public as a German achievement. Nazi propaganda concealed that part of the TV equipment was produced with the help of the Russian-American Jewish engineer Vladimir K. Zworykin (1888-1982). Early in 1937, von Ardenne thought about combining an electron beam of 50 nm in diameter as an electron probe with the line-by-line "flying spot technique". He created a scanning electron microscope (SEM), working in the scanning transmission mode, followed by another version, a real TEM in 1938 (Fig. 6a) (von Ardenne, 1938, 1985). The war stopped the further development of the SEM. A commercial SEM named "Stereoscan" was not produced until 1965 by the Cambridge Instrument Company in the UK (Oatley, 1982; von Ardenne, 1997).

In 1937, von Ardenne co-founded the Leybold und von Ardenne Oszillographengesellschaft - a company soon transferred to S&H. As early as 1937, S&H became interested in von Ardenne's ideas and developments and signed contracts with him. While S&H plunged into commercial, serial production of the ÜM, von Ardenne built his own transmission electron microscope (TEM) excellent in stability and versatility enabling also dark-field and stereo-imaging (von Ardenne, 1940a, 1985). Using colloidal gold and carbon black he achieved a 3 nm resolution. A special specimen stage allowed heating processes up to 1.500 °C to be studied and, with the help of a cine camera, he recorded melting processes. Also in 1940, he published his influential book on *Elektronen-Übermikroskopie* (von Ardenne, 1940b), translated during WWII in Japan, USA, and the Soviet Union. He imaged hemocyanin, tobacco mosaic virus and details of bacteria and started ambitious co-operations in the materials and life sciences: Together with the virologists Gerhard Schramm (1910-1969) and Hans Friedrich-Freksa (1906-1973) from the Kaiser Wilhelm Institutes for Biochemistry in Berlin he described cross-linking of tobacco mosaic viruses by virus-specific antibodies, the first immune-electron microscopy study on this side of the Atlantic (von Ardenne et al., 1941) - in parallel to similar achievements in the US (Anderson & Stanley, 1941). In 1944, von Ardenne published details

(a) (b)

Figure 6 Manfred von Ardenne and his universal electron microscopes. (a) The 1940 microscope (from his 1940 book 'Elektronen-Übermikroskopie', p 178; *with permission SPRINGER NATURE*), (b) von Ardenne and his 1944 universal electron microscope (taken from W. Gloede, "Vom Lesestein zum Elektronenmikroskop", p. 190; *by courtesy of Wolfgang Gloede, Stockelsdorf*).

of a new, advanced "Universal-Elektronenmikroskop" (Fig. 6b) equipped with an improved objective lens approaching 1.2 nm resolution and with eased operation modes (von Ardenne, 1944).

Von Ardenne possibly intended to switch to commercial production when the Kaiser Wilhelm Institutes (KWI) in 1940/41 thought about establishing a joint Virus Institute in Berlin, and later in Heidelberg, with an integrated EM department run by him and equipped with his instruments. Indeed, his contract with S&H allowed him to build three instruments (von Borries, 1940). Besides his recent 100 kV Universal-Elektronenmikroskop he planned for a 350 kV instrument and a third one with even higher acceleration voltage. On November 29, 1940, a high ranking S&H meeting with Hermann von Siemens, von Buol, Dr. Lüschen and the engineers von Borries and Ruska discussed relations with von Ardenne. S&H director Flir stated in his minutes that S&H would not grant any licence to von Ardenne as long as he was considered a candidate member of the putative

Virus Institute of the KWI. Initially, S&H had granted favorable licences for von Ardenne's in-house microscopy program but was now afraid of losing much higher patent fees if the private von Ardenne Institute was integrated into the KWI Virus Institute. S&H prevented him from following that idea and likewise did not allow him any rights of sale for a preparation device (Flir, 1940b, 1940c). Besides electron microscopy von Ardenne increasingly took up nuclear physics projects, notably isotope separation and a technique required to enrich uranium, and constructed a particle accelerator, a 60-ton cyclotron (Barkleit, 2006; von Ardenne, 1997).

In the 1945 Yalta Conference Agreement the USA, the UK, and the Soviet Union had mentioned the use of German workforce as one possible kind of war reparation. In July 1945, von Ardenne, like some other German scientists, left for the Soviet Union, together with his complete institute. Near Suchumi at the Black Sea he investigated the basic requirements for the hydrogen bomb and created efficient focused ion sources. When in 1954 he was allowed to return to the German Democratic Republic, he founded the private "Forschungsinstitut Manfred von Ardenne" in Dresden (East Germany). There, he developed a cancer multi-step therapy ("Krebs-Mehrschritt-Therapie", a combination of whole-body hyperthermia, hyperacidification and hyperoxygenisation) and in parallel electron probe, ion jet, welding and vacuum coating technologies. It is remarkable that he achieved his success under three different dictatorships. He died at an age of 90 on May 26, 1997, and today, his sons are pursuing his legacy separately in two independent business branches (Barkleit, 2006).

7. Development of electron microscopy inside the Bosch company

Progress in particular in the metal and the lacquer industries required quality control in the production lines. For this, industry also constructed and applied electron microscopes, for example the companies of Krupp in Essen, who ordered a S&H instrument, and Robert Bosch in Stuttgart. The latter company originated in the brain of the ambitious mechanic Robert Bosch (1861-1942). In 1886, he started a small workshop for Precision Mechanics and Electrical Engineering in Stuttgart, the origin of today's Robert Bosch Company with 395.000 employees worldwide, 92% owned by the Robert Bosch Stiftung, a charitable foundation. Bosch developed ignition devices for automobiles and became a leading automotive supplier of lighting systems, horns, wipers and battery powered ignition. Power tools followed in 1932, the Bosch refrigerator in 1933, injection pumps for diesel engines in 1927 and for gasoline engines in 1937.

In 1929, Bosch, together with the Scottish TV pioneer John Logie Baird and the companies Zeiss Ikon and Loewe, established the Fernseh AG and enabled public TV viewing during the Olympic Games in Berlin in 1936 (von Ardenne, 1985). Social responsibility is still a major mark of the foundation. Already in 1906, working hours were limited to 8 h per day and in 1936, on the occasion of his 75th birthday, Robert Bosch founded a significant hospital, the Robert-Bosch-Krankenhaus inaugurated in 1940 (Bähr & Erker, 2015). The Bosch choir and symphonic orchestra, both founded in 1934, still support the musical ambitions of the regional Bosch employees. They played a major cultural role in the Stuttgart area.

During the Third Reich (1933-1945), the Bosch company became part of the Nazi economic system. But compared to other big industry in Germany, Bosch had clearly kept a distance and had even facilitated civil resistance activities. Like other companies involved in the manufacture of armaments, Bosch was assigned prisoners of war and later also forced labor in order to boost production. Between 1939 and 1945, there were a total of around 20,000 forced laborers across the company, including prisoners of war and concentration camp inmates. On the other hand, Robert Bosch and senior Bosch executives supported resistance against Hitler and provided those facing persecution with money or help with emigration (Bähr & Erker, 2015; Kuhlgatz, personal communication, 2021). And as we will learn in Section 10, despite the Nazi organized workforce ("Gefolgschaft"), Bosch offered a "protected workplace" for opponents of the Reich, from which, e.g., the physicist Walter Rentschler benefited between 1936 and 1945. War-damage in the Stuttgart region was heavy, but after 1945, Bosch renewed its earlier automotive program, broadened its range of products by taking up more household technology, tools for professionals and do-it yourselfers, and entered the field of commercial electron microscopy. In the mid-1950s, Bosch started production of mechanical and electronic components for the automotive program - still its core business (Bähr & Erker, 2015).

8. Rudolf Rühle (1907-2003): personal background and development

Rudolf Rühle was born in Zuffenhausen, then a poor village near Stuttgart, on September 29, 1907 as the second of six children. His parents

Wilhelm (1877-1960) und Friederike Rühle (1878-1939) were peasants and wine-growers with a property so humble that father Wilhelm had to earn extra money as a Rottenmeister (foreman) at the Reichsbahn, the national railway. Family life was patriarchal, strongly influenced by the Protestant pietism typical of this Württemberg region. Religion, obedience, diligence and thrift dominated. Father Wilhelm was always busy and hard-working and imparted these characteristics to his children. As was common at that time, flogging was part of the education, but generally Wilhelm Rühle was tolerant and clear-cut against war and against any Nazism (Manfred Rühle, personal communication, 2021).

For his three sons father Wilhelm expected solid trades, and Rudolf was to become a carpenter. But after primary school, mother Friederike registered him clandestinely at the Realschule, the natural science grammar school (Gymnasium). She secretly paid his school fees, but finally, when Rudolf excelled, father Wilhelm also approved of these higher ambitions. In 1922, Rudolf entered the Oberrealschule in Feuerbach and excelled in Mathematics, Physics, and Sports, but not so much in English and German. As a young pietist in the CVJM (Christlicher Verein Junger Männer, member of the World Alliance of YMCA), he was active in the local brass ensemble (Posaunenchor, an institution typical for puritanistic German countries) playing the trumpet with great skill. He was talented also in writing and presenting his sketches during family parties.

On March 13, 1926, Rudolf Rühle passed the Abitur and, after half a year of compulsory practical work in an elevator and a special metal plant, he began studies in Technical Physics at the Technical College of Stuttgart in October 1926. He passed the intermediate examination (Diplomvorprüfung) in March 1929 with distinction and went for a semester to Göttingen to study Theoretical Physics, focusing mainly on solid-state physics. As a consequence of the World Financial Crisis, following Black Tuesday in the US (October 29, 1929), widespread unemployment and political unrest prevailed during the end of the Weimar Republic in Germany. The era was socially unsafe. Hence Rühle followed a recommendation of Professor Paul Peter Ewald (1888-1985) in Stuttgart (famous for his Ewald sphere) and in January 1930 took up the position of a Personal Research Assistant with Professor Fritz Emde (1873-1951) at Stuttgart Technical College. His task was to help with the preparation of a new edition of the book "Jahnke-Emde: Funktionstafeln mit Formeln und Kurven". The revision of these Tables turned out to be a very laborious work lasting three and a half years (Emde & Rühle, 1934). Rühle resumed his Physics studies in

October 1933 and passed his Diploma examination on May 17, 1934 again with an honors degree. On July 15, 1934, he entered the department "Versuche Angewandte Physik" (VAP, Experiments in Applied Physics), the R&D group of the Robert Bosch company as a Technical Physicist. Before joining Bosch, Rühle had established contacts there as a working student improving thus his meager financial situation (Manfred Rühle, personal communication, 2021).

Having reached a socially safe position, Rühle consolidated his family status: He married his fiancé Marta Margarete Motz on June 20, 1936. They have four gifted children. Among them was son Manfred, born on March 15, 1938 who became Director at the Max Planck Institute for Metal Research from 1989-2005 in Stuttgart. During this period, from 1994 to 1997, Manfred Rühle acted as President of the German Society for Electron Microscopy (DGE).

Rudolf Rühle loved being challenged by new tasks, he was hard-working, but not preoccupied with making a career. He was impulsive and pleasant, down-to-earth, often using his Swabian German dialect. He was loyal to Bosch all his life, although outspoken, and he was appreciated for that. Rühle spend his free Saturdays in the vineyard or on his land, loved solid Swabian food and his own wines. He used to take two or three weeks of annual vacation, often on the island of Sylt or in Switzerland, but otherwise he was not really interested in travel. He attended merely three or four early meetings of the German Society for Electron Microscopy (Deutsche Gesellschaft für Elektronenmikroskopie, DGE) and a few conferences on vacuum techniques in the USA and in Europe. Basically he preferred to spend his time working at Bosch. Aside from his serious work ethic, he enjoyed culture. He wrote sketches for parties at Bosch and himself went on stage. Rühle enjoyed the subscription concerts of the Stuttgart State Opera. After 1945, he played trumpet in the Bosch orchestra and became Leader of the orchestra (Fig. 7). The Bosch orchestra was led by the excellent conductor Hans Grischkat (1903-1977) who also became a friend of the Rühle family (Manfred Rühle, personal communication, 2021).

9. Rudolf Rühle: the technical engineer at Bosch

During the 1920s and 30s, Bosch with its narrow automotive program strived to return to its pre-war global position. To develop future products and new technical solutions, the R&D-Division VAP ("Versuche Angewandte Physik") had been founded in 1928, headed by the physicist

Figure 7 Rudolf Rühle and his endeavors in the world of arts. The conductor Hans Grischkat (left) and Rudolf Rühle, the spokesman of the Bosch symphonic orchestra on its 25th anniversary in 1959 (*from Rudolf Rühles's "Festschrift" 1971, courtesy Manfred Rühle*).

Dr. Walter Dorn. VAP focused on technical details of headlamps, ignition systems and capacitor membranes. The engineering of a commercial TEM at Bosch took a long time and the decision to construct it was taken only after the end of World War II. But the essential groundwork started around 1934, as documented in a selection of 31 internal VAP progress reports, bound in a "memoirs book" ("Bosch-Versuchsberichte 1934-44"), handed over to Rudolf Rühle on the occasion of his 75th birthday in 1982, kindly made available to the authors by Professor Manfred Rühle in 2021. Retrospectively, many of the reports point to cornerstones of the future electron microscope, though at that time they merely documented clear-sighted tasks set for the future automotive program.

Rudolf Rühle had entered the VAP division on July 15, 1934. The engineers at VAP fulfilled long-term projects independently but - wherever required - also in close co-operations. Rühle dealt successfully also with somewhat "abnormal" tasks. In December 1937 for example, he was asked to describe sunlight in a building ("Sonnenlicht in einem Gebäude"). He had to determine the angles of the incident light in a given building at a given time depending on the place and orientation of the house. Using astronomy and spherical geometry, the "path" of the sun "from dawn to dusk" and correlating local time and the position of the windows, Rühle calculated the incidence of the sun. Nevertheless, he also described an alternative procedure: placing a model of the house in proper orientation in an observatory and observing the incidence of light. This quest for reasonable alternatives was part of his alert in-depth intellectual character [VAP report no. 123 of 13.12.1937].

As his first project in 1934 Rühle studied the currents and fields of a copper coil. Further tasks focused on the control of cathode-ray oscilloscopes and the physics and techniques of electron scattering. In three VAP reports Rühle evaluated and improved electron scattering as a tool in analyzing evaporated metal layers [VAP no. 102 of 15.10.1936/VAP no. 122 of 24.03.1937/VAP no. 157 of 26.10.1938]. His apparatus worked at an accelerating voltage of 60 to 80 kV (Fig. 8). An electron beam limited by two apertures generated accurate Debye-Scherrer-diagrams comparable in resolution to or better that X-ray generated images. Next he planned to use apertures smaller than 10 μm in diameter, shielding from magnetic stray fields and a highly constant high tension supply.

Rühle's basic, long-term task was to improve the metal mirrors of searchlights. He developed a focused electron gun, protected by four sequential patents (see List of Patents) as a central prerequisite for efficient metal evaporation. To improve his skills in the preparation of thin metal films, he visited the physicist Dr. Max Auwärter (1908-1995). Originally, Auwärter had also been working at VAP, but since 1936 he led the physics laboratories at the W.C. Heraeus company in Hanau. In 1949, Auwärter established the Balzers company (Gerätebau-Anstalt Balzers, Balzers AG für Hochvakuumtechnik und Dünne Schichten) in Lichtenstein - highly esteemed also in electron microscopy for coating procedures. In 1976, Auwärter retired and sold his company to the Oerlikon-Bührle AB, now OC Oerlikon Balzers AG.

After 1933, the production of head-lamps had to be scaled up, not least for the re-armament of the Third Reich. Bosch was asked to reduce production costs by cutting down the cost per piece and to save foreign currency. So in August 1938, Rühle constructed a highly effective, half-automated unit for the metal-coating [VAP report no. 152 of 01.08.1938]. The device was patent protected as "Verfahren zur Herstellung von Spiegeln" by Dorn, Reger and Rühle on February, 25, 1937.

The production of improved mirrors had required major steps in technology: Rühle compared different materials, iron, brass, as materials for the bodies of the lamp and finally arrived at the conveniently to press Bakelit plastic. He tested different lacquer and metal coatings and ultimately combined the electron beam-generated evaporation of silver or aluminum with high vacuum technologies in order to guarantee a continuous production

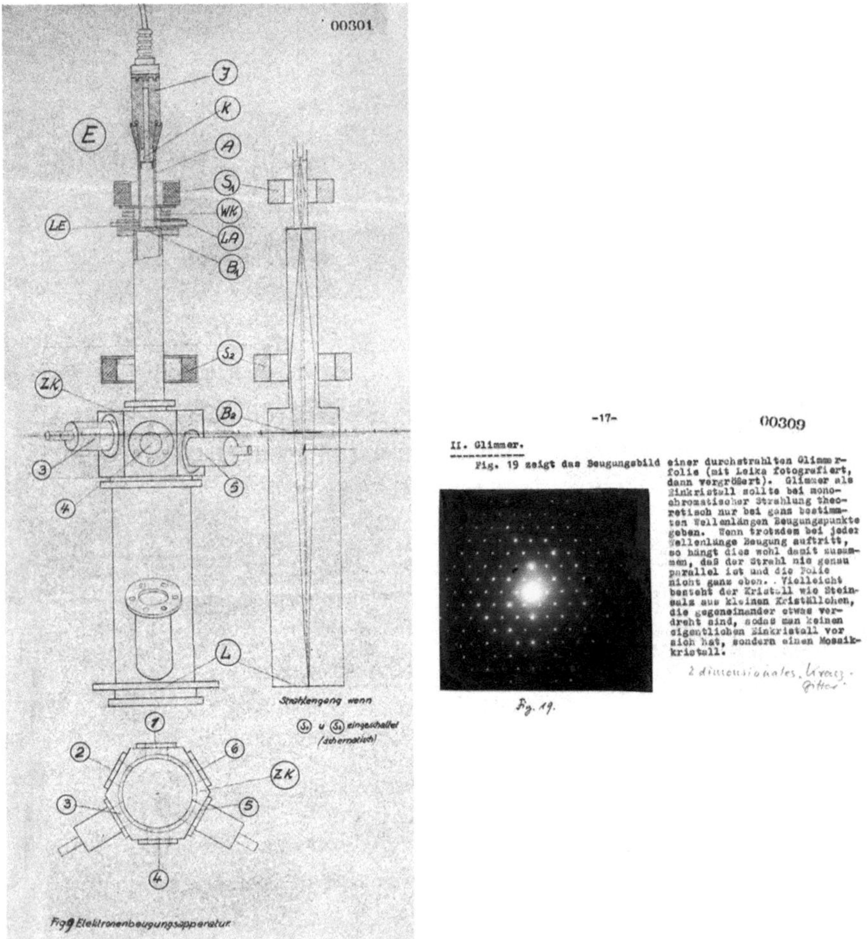

II. Glimmer.

Fig. 19 zeigt das Beugungsbild einer durchstrahlten Glimmerfolie (mit Leika fotografiert, dann vergrößert). Glimmer als Einkristall sollte bei monochromatischer Strahlung theoretisch nur bei ganz bestimmten Wellenlängen Beugungspunkte geben. Wenn trotzdem bei jeder Wellenlänge Beugung auftritt, so hängt dies wohl damit zusammen, daß der Strahl nie genau parallel ist und die Folie nicht ganz eben. Vielleicht besteht der Kristall wie Steinsalz aus kleinen Kriställchen, die gegeneinander etwas verdreht sind, sodaß man keinen eigentlichen Einkristall vor sich hat, sondern einen Mosaikkristall.

(a) (b)

Figure 8 Rudolf Rühle's electron diffraction apparatus presented in the Bosch VAP report No. 102 of 15.10.1936, "Elektronenbeugung": (a) Construction sketch, (b) a diffraction image of mica (*courtesy Manfred Rühle*).

process. There was also the need to control pre-products and product quality, and to do so, new diagnostic techniques based on electron diffraction and electron microscopy were developed step-by-step [VAP reports nos. 179 and 228 of 17.05.1939 and 21.03 1940]. In the latter report Rühle complained of a delay of 14 months due to the lack of time and manpower, yet he was apparently eager to report on advances in the construction of three analytical tools: electron scattering apparatus, electron microscope and cathode ray oscilloscope. In the meantime a dedicated specimen airlock, various polepiece lenses, means for the precise alignment of the components and magnetic shielding against stray fields also became available. They delivered reliable results up to 50 kV. For more ambitious work Rühle planned for a more stable high tension supply, improved magnetic lenses and a high resolution final screen.

To study zinc coatings, required in the production of capacitors, with regard to their crystallinity, state of oxidation and stability, Rühle applied electron diffraction [VAP report no.161 of 05.07.1939]. In February 1942, together with his coworker Suse Löffler, he also used electron microscopy in two ways: By examination from a shallow angle or by studying aluminum replicas of the Zn layers prepared according to Hans Mahl (Mahl, 1940b). Only the replicas yielded reliable data on the surface profile [VAP report no. 262 of 27.02.1942] (Löffler & Rühle, 1942).

10. Walter Rentschler and the Bosch electron microscope

The VAP report dealing explicitly with a TEM was signed not by Rühle, but rather by Rentschler [VAP report no.158 of 29.09.1938]. The author Dr. Walter Rentschler (1911-1984) was a physicist; he passed his doctoral examination with a thesis on X-ray absorption and chemical processes in the soft X-ray spectrum ("Röntgenabsorption und chemische Bindung im ultraweichen Röntgengebiet") with distinction at Tübingen University on November 8, 1934. He became an assistant professor at the Institute of Physics in Tübingen, but in April 1935 political pressure by the Nazis obliged him to move to the Physics Department of the Agricultural College ("Landwirtschaftliche Hochschule") in Hohenheim near Stuttgart. When the Physics Department of this University was suspended by a general decree of the Third Reich, he sought a safe haven on October 1, 1936 at the Bosch R&D Division, in VAP under Dr. Walter Dorn.

Rentschler took over a wide spectrum of tasks: he investigated polarized light for searchlights, secondary electron emission microscopy and improved an already existing TEM. Regrettably, the previous state of this TEM is not documented. Rentschler assembled a new instrument which was described briefly in the report "Elektronenmikroskop" on merely three pages [VAP report no. 158 of 29.09.1938, Rentschler, 1938; Seiler et al., 1981]. He compared cold emission versus emission from a heated tungsten wire and proposed improvements, e.g., protection against magnetic stray fields and stability of the high tension supply. Experimentally he compared the results of cold emission versus thermal emission microscopy of metal surfaces - iron, nickel, tungsten, zinc - at 30- to 50-fold magnifications. He also recorded technical progress: lenses allowing now for a more precise alignment, the Permalloy shielding and the relatively constant lens currents, but there was still the need for further improvements.

On February 9, 1942, Rentschler was drafted into military service but was already released on April 22, 1942 and ordered back to work again at Bosch, but this time in the central administration. After the war, and employed again in VAP, he turned once more to dazzle-free searchlights, but in parallel he renewed his contacts with the Hohenheim College to teach Physics and Mathematics. Rentschler left Bosch after a final year of half-day employment on August 31, 1947. During his stay at Bosch, between 1936 and 1947, he had filed two patents: in 1940 and 1941 for headlights for vehicles and for specially coated lenses, respectively (Kuhlgatz, 2021) and he had contributed nearly 30 VAP-reports (Denisa Kohan, UAH, personal communication, 2021).

11. Rudolf Rühle's groundwork for the Bosch electron microscope

In parallel to Rentschler's report "Elektronenmikroskop", Rühle reported in "Magnetische Linsen für das Elektronenmikroskop" on a number of basic constituents for a future Bosch microscope. He was aware of the innovative polepiece work by Ernst Ruska and Bodo von Borries (von Borries & Ruska, 1932) and described his version of iron clad magnetic lenses of very short focal length, the polepieces specifically shaped to facilitate the insertion of the specimen (Fig. 9) [VAP report no. 159 of 13.10.1938].

Rühle also calculated the impact of magnetic stray fields on the resolution of both the electron microscope and diffraction apparatus. To take care of proper magnetic shielding he scrutinized six different ferromagnetic materials [VAP report no. 160 of 04.02.1939].

To image the evaporated metal coats in the headlamp production in "more depth", Rühle turned to stereo-images. He engineered a sophisticated, though robust specimen stage, protected by patent on April 13, 1943 (Fig. 10). Its details were published only after World War II (Rühle, 1949c). In 1944, he examined stereo-images already acquired a year before - a remarkable delay in evaluation - and summarized his experiences in the report "Stereobilder am EM". He showed images of a finely machined iron

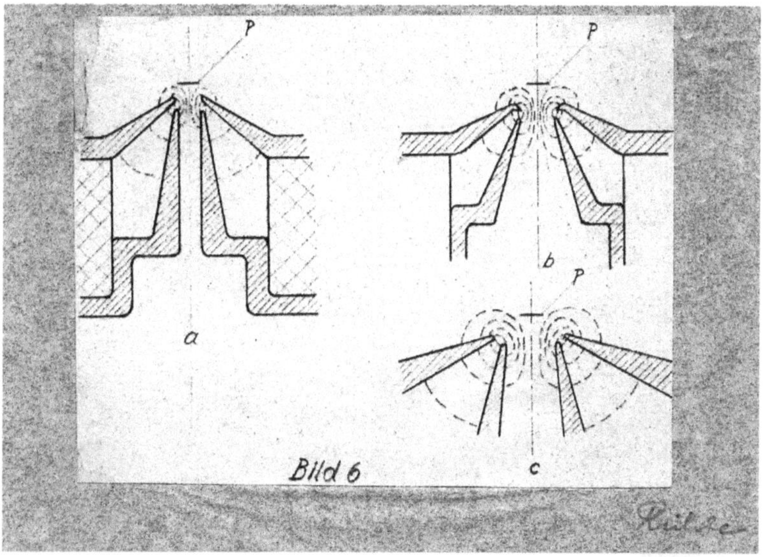

Figure 9 Specifically shaped pole pieces facilitating an easier insertion of the specimen. "P" is the position of the specimen (Bosch VAP report No. 159, 13.10.1938, signed "Rühle") (*courtesy Manfred Rühle*).

spindle at 2,500-fold magnification and of aluminum-oxide deposits on a paper substrate at 1,300-fold magnification.

While satisfied with the stereo-images, he was still displeased with the resolution of the microscope: "eine 3.000-fache Vergrösserung entsprechend einer Brennweite von 0.1 mm noch nicht erreicht" (a 3,000-fold magnification corresponding to the focal length of 0.1 mm is still not achieved) [VAP no. 327 of 28.06.1944].

12. Rudolf Rühle's thesis: on the future Bosch electron microscope

Having started in 1934 at Bosch, Rühle had accumulated extensive experience and many achievements in electrical engineering and in vacuum technology by 1943. The essential components for the complex instrument were proven: from high tension technology, electron gun, magnetic lenses down to vacuum measuring device, the tilting stage and photographic recording. The final instrument, still an in-house work-horse, reflected Rühle's concept (Fig. 11). Internally it was named "Drei-Wege-Apparatur" (three-option-device) and helped to generate knowledge in material sci-

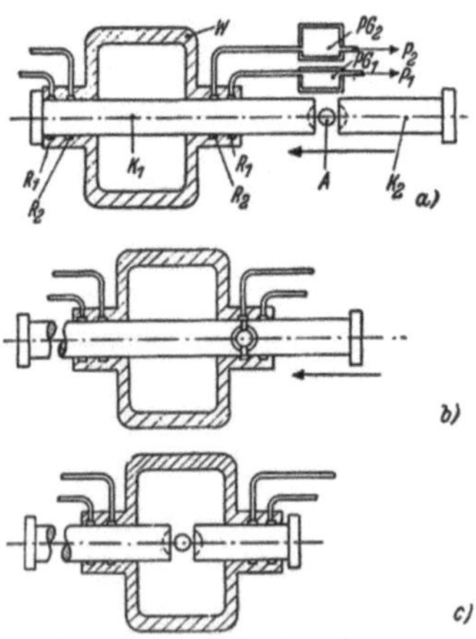

Abb. 3. Schematische Darstellung der Vakuumschleuse.

(a)

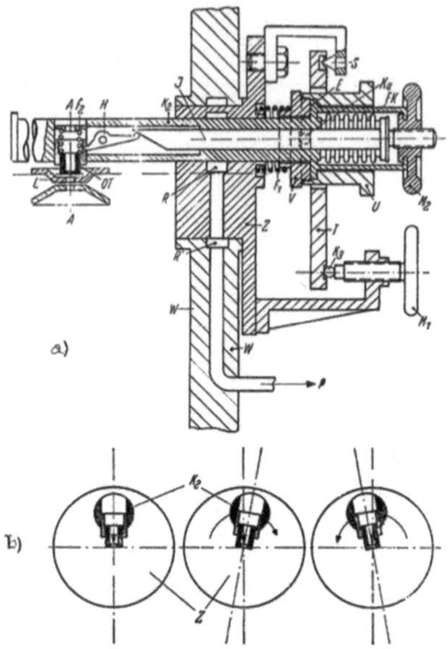

Abb. 4. Vakuumschleuse mit Objektbewegungsvorrichtungen, halbschematisch. a) Längsschnitt, b) Querschnitt längs AA.

(b)

Figure 10 Airlock and goniometer of the Bosch electron microscope (Rühle, 1949c). (a) Schematic drawing of the airlock. (b) Details of the specimen loading, reliable sealings and holder release (*with permission by ELSEVIER*).

ences by high resolution imaging. Stereo imaging promised additional 3D surface data and the integrated electron scattering device provided insight into the chemical composition of the specimen.

When Rudolf Rühle was allowed to "summarize" his experiences in a doctoral thesis, he put emphasis on the latter two capabilities, stereo imaging and chemical analysis. On January 6, 1944, Rühle submitted the thesis "Fortschritte in den Untersuchungsmethoden am Elektronenmikroskop" ("Progress in Electron Microscopic Imaging Techniques") to the Roentgen-Institute, one of the three Physics Institutes at the Stuttgart Technical College. The Reviewers for his thesis were Professor Ulrich Dehlinger and Professor Richard Glocker, the latter being the Head of the Roentgen-Institute. Initially, Glocker intended to act as the First Reviewer. But he withdrew when Rühle's thesis appeared excessively theoretical to him - not quite understandable - and his deputy Dehlinger took over. Both were also members of the Kaiser Wilhelm Institute (KWI) of the Physics of Metals, their institutes immediate neighbors with a strong focus on metallurgical research, a field of central interest to Rühle and his thesis. On

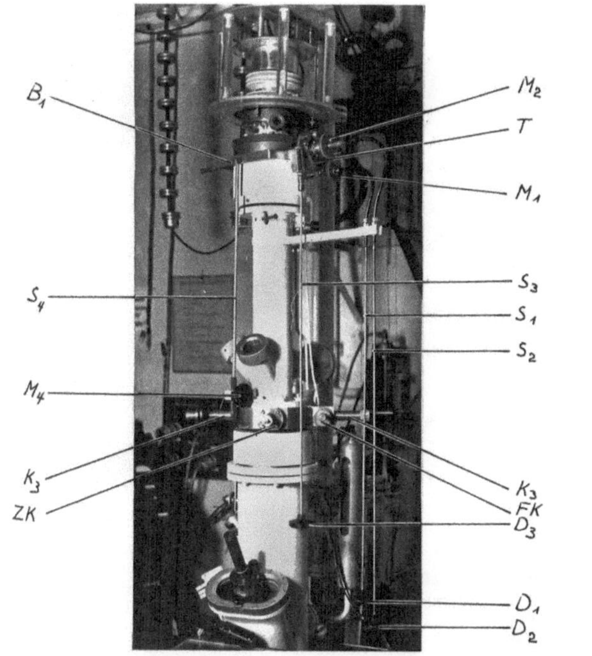

Abb.8 Elektronenmikroskop (zu S.16 und 21).

Längsverschiebung des Objektes	D_1 S_1 M_1	Stereoverstellung	M_4 S_4 B_1
Senkrechtverschiebg." "	D_2 S_2 M_2	Zwischenbildkamera	ZK
Querverschiebung " "	D_3 S_3	Fotokassette für ZK	FK
Träger	T	Kolben der ZK	K_3

Figure 11 The electron microscope presented in the dissertation of Rudolf Rühle, 1944 (*from Rudolf Rühles's "Festschrift" 1971, p. 22; courtesy Manfred Rühle*).

January 27, 1944, Rühle passed the examination with distinction (Fig. 12). He was awarded the title of Dr.-Ing. and, though basically humble and not overambitious, he took great pride in this achievement (Manfred Rühle, personal communication, 2021).

The thesis of 60 pages is written in a clear, straightforward manner and presents unique technical solutions different from those found in extant electron microscopes. The microscope's main purpose was the fine structure analysis of metal surfaces. He underlined the benefits of combining low and high magnification imaging in order to better understand the sample. To generate an undisturbed wide range of magnifications, he had arranged for three projector lenses to be activated and used in different combinations (Fig. 13).

Technische Hochschule Stuttgart, Stadt der Auslandsdeutschen.

Die Fakultät für Naturwissenschaften und Ergänzungsfächer der Technischen Hochschule Stuttgart verleiht unter dem Rektorat des ordentlichen Professors des Elektromaschinenbaus Dr.-Ing. habil. H. Hess und unter dem Dekanat des ordentlichen Professors der Technischen Physik Dr.-Ing. H. Reiher

Herrn

Rudolf Rühle

aus Stuttgart-Zuffenhausen

den Grad eines Doktor-Ingenieurs

nachdem er in ordnungsmäßigem Promotionsverfahren durch die Dissertation

„Fortschritte in den Untersuchungsmethoden am Elektronenmikroskop"

sowie durch die mündliche Prüfung seine wissenschaftliche Befähigung erwiesen und dabei das Gesamturteil „mit Auszeichnung" erhalten hat.

Stuttgart, den 27. Januar 1944.

Der Rektor: Der Dekan:

Hess

Figure 12 Rudolf Rühle's Doctoral Certificate of January 27, 1944 documenting "Progress in Electron Microscopic Imaging Techniques" and the assessment: "with honors degree" (mit Auszeichnung) (*courtesy Norbert Becker, Archive University Stuttgart*).

Rühle was aware that the surface texture of a workpiece determines both its function and durability and he also knew that simple transmission microscopy of a uni-directionally shadowed specimen does not lead to precise dimensional information. He considered the existing stereo image technologies - AEG, Ladislaus Marton, S&H or Manfred von Ardenne - as either imprecise or too complex (Rühle, 1949c) and developed his own goniometer design. It combined a precisely working specimen airlock and a specimen holder (Fig. 9), patented already in April 1943 (see separate List of Patents below). Importantly, during tilting the image did not change position or focus on the final screen. The holder allowed for high resolution stereo-images to be taken at +/− 10° angles. Rühle showed a number of

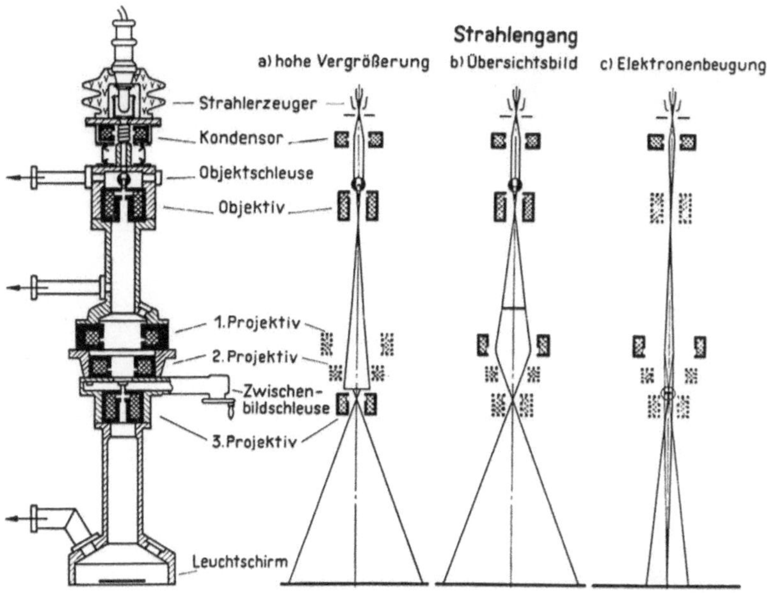

Figure 13 Three projector lenses are used to regulate the wide range of magnifications from 300 to 30,000-fold and the corresponding beam paths (from the brochure 'BOSCH Elektronen-mikroskop' 1950; *courtesy archives of the Robert Bosch GmbH, Stuttgart*).

examples, e.g., an image of a polished steel spindle (Fig. 14) and a thin aluminum foil prepared according to up-to-date replica methods (Mahl, 1940b). He put emphasis on pitfalls in the interpretation of stereo-images and indicated how they could be avoided. Finally, Rühle discussed the potential of electron scattering and presented an example of the electron probe analysis of zinc layers, the third high resolution diagnostic tool integrated in his microscope.

13. Rebuilding of Germany after WWII

Following the unconditional surrender of the Third Reich in Berlin-Karlshorst on May 08/09, 1945, social life and economy in Germany paused. After the bombing raids, factory buildings laid flattened and any surviving infrastructure was often removed as war reparations (Fig. 15).

The Western Allies helped with "denazification" as well as with food. With the Soviet expansion to the West, they saw also the need to stabilize life and economy in their three occupied zones. With the currency reform on June 20, 1948 and much support by the Marshall plan (from

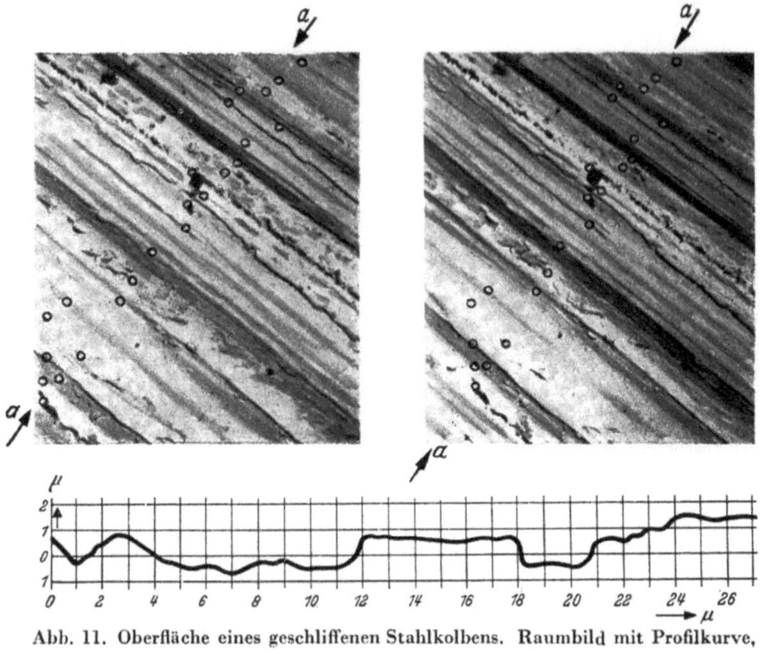

Figure 14 Stereo micrographs of a polished spindle (VAP report No. 327, 28.06.1944) (*from Rühle, 1949c; with permission by ELSEVIER*).

1948-1952, the European Recovery Program), reconstruction and development in Western Germany began to sprout and when on May 23, 1949 the Bundesrepublik Deutschland was founded, the way for an economic miracle was paved.

Electron microscopy also experienced a revival: Early in 1946, Ernst Brüche established the Süddeutschen Laboratorien (SDL) in Mosbach, Baden-Württemberg, in collaboration with AEG after April 1946. In 1947, the first post-war AEG-SDL-microscope was built, a precursor of the EM 8 (Fig. 16), but Brüche became disappointed with AEG's attitudes. In 1949 Brüche sold the SDL to AEG and renewed his contacts with the Carl-Zeiss-Foundation. AEG abandoned the construction of microscopes and when Brüche got together with Carl Zeiss, the advanced AEG-ZEISS EM 8 was built in 1949 (Fig. 16b) (Brüche, 1949; Rang, 1992; Schulze, 2001). In 1953, Zeiss withdrew from the joint venture and moved the further development of the ZEISS EM 8 to its center at Oberkochen, Württemberg, while Brüche founded a new research institute, the "Physikalische Laboratorium Mosbach" (Schimmel, 1998; Schulze, 2001). S&H had constructed

Figure 15 The Bosch-VAP team in summer 1944 recording the damage after the bombing of Stuttgart. From left: Walter Dorn, Alfred Ortlieb, Eberhard Traub, Rudolf Rühle, Karl Reger (*from Rudolf Rühle's "Festschrift" 1971, p. 24; courtesy Manfred Rühle*).

(a)

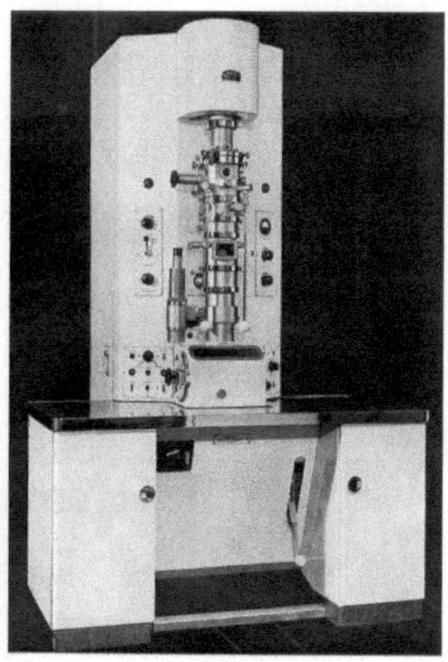

(b)

Figure 16 The first electron microscopes after World War II in Germany were built by the Brüche/AEG group in Mosbach. (a) SDL-EM on a local trade show in 1947 (Rang, 1992) (*with permission by ELSEVIER*), (b) AEG-Zeiss EM 8 (courtesy archives of ZEISS Jena).

their first post war instrument, the ÜM 100D with a resolution of 2-3 nm, in 1950 (Fig. 28) (Ruska, 1950).

14. Foundation of the German society for electron microscopy

Following the plans of S&H for the time after the war, Bodo von Borries had left Berlin on March 22, 1945 with parts of the LfE for the West (Gelderblom, 2019; von Borries, 1991). In Leopoldshöhe, near Herford, von Borries wrote his comprehensive book *Übermikroskopie* of 416 pages (von Borries, 1949). After three years of frustrating search he organized in 1948 in Düsseldorf a consortium consisting of industry and science that supported the Nord-Rhein-Westfälisches Institut für Übermikroskopie" (North Rhine-Westphalian Institute for Super Microscopy). It provided microscopy services in the life and the materials sciences using the surviving S&H ÜM 100, served also as a guest laboratory and allowed von Borries to engineer a low-cost electro-static 60 kV-instrument and an ultramicrotome (Karrenberg, 2019; von Borries, 1991). The institute was inaugurated on February 15, 1949, in the presence of Hermann von Siemens and scientists involved before in electron microscopy including Rudolf Rühle. During this meeting von Borries arranged the foundation of the German Society for Electron Microscopy (Deutsche Gesellschaft für Elektronenmikroskopie, DGE) with Ernst Ruska and Hans Mahl as the First and Second Chairmen respectively, and with himself as the Secretary: a remarkable good-will reconciliation of previously "difficult" colleagues. Initially, the DGE comprised 27 members - it met again only two months later, on April 23-24, 1949 in Mosbach at the invitation of Ernst Brüche for its first annual conference doubling the number of attendants (Schimmel, 1996).

15. The commercial Bosch electron microscope

In December 1945, VAP was converted into the bigger R&D-Division Development and Material Testing (Entwicklung, Stoffprüfung, Angewandte Physik, ESP) still with Walter Dorn as Head. It comprised 10 departments focusing on, e.g., magnets, metals, chemistry, electrochemistry, ceramics, with Rudolf Rühle's team serving as ESP 9. In this place he developed the 100 kV instrument, well aware of "other places". Though not completely independent of S&H patents (Siemens & Halske,

1955), the Bosch microscope and its precursors can be judged as a genuine Bosch products. An interesting question is whether or not Rühle socialized with his competitors in electron microscopy at S&H or AEG. At that time he was well known among the German electron microscopists. Professor Otto Scherzer described him as having a reliable, highly open, friendly, and helpful character (Scherzer, 1954). Rühle's position as a newcomer in the field of electron microscopy was probably not easy (Fig. 17). After World War II, Rühle was, however, still on cordial terms with Walter Rentschler.

Interestingly, Rühle and his longtime boss at Bosch, Dr. rer. nat. Walter Dorn, never became friends. This might be due to their different academic backgrounds: Technical College versus University (Müller, 2019). But presumably Dorn thought of the younger hard-working colleague as a growing competitor (Manfred Rühle, personal communication, 2021).

The Bosch product was a timely, robust, and dedicated TEM (Fig. 18). It allowed for high resolution imaging, for efficient tilting of the specimen, i.e. the image did not "disappear" during tilting, and included a powerful electron diffraction facility. Encased in a trendy whitish iron sheet it had on top the impressively protruding high tension generator, also in overall shape comparable to the ÜM 100 series of S&H.

The high tension could be selected between 40 and 90 kV, in the final version up to 100 kV. Electrons were emitted from a heated tungsten cathode and controlled by magnetic lenses. The illumination system, the "Strahlkopf" with a single condenser lens was centered by means of a complex manual shift and tilt system (Rühle, 1949b). See Fig. 19.

Rühle himself, the physicist and father of the construction, considered the alignment procedure logical, easy to perform (Rühle, 1949b). Some users, however, especially from the life science fields, found it difficult to handle (Hans-Joachim Gerth, Tübingen, personal communication, 2021; Schreil, 1955). Magnifications, from 300 to 30,000 times, were adjusted by merely switching between combinations of three different projector lenses, which had different polepiece bore-diameters: one high-magnification main lens for 30,000-fold magnification and two further auxiliary projector lenses (for 300- and 3,000-fold magnifications) mounted above it. To generate different ranges of magnifications, the lenses were electrically switched in different combinations. Such sequential projection lenses were not unique, SDL/AEG had adopted a similar solution with two different lenses (Rang, 1992).

The lens currents were supplied by a stack of car batteries. When changing magnification the image filled the final screen over the whole range of

(a)

(b)

Figure 17 Participants in the meeting of the German Society for Electron Microscopy in Tübingen 1952 with a number of being identifiable: (1) Karl-Heinz Steigerwald, Zeiss; (2) Richard Haas, Freiburg; (3) Rudolf Rühle, Bosch; (4) Christian Gerthsen, Karlsruhe; (5) Hans Mahl, Zeiss; (6) Walther Kossel, Physik Tübingen; (7) Ernst Zehender, Bosch; (8) Harald Volkmann, Zeiss; (9) Bodo von Borries, Düsseldorf; (10) Erich Menzel, Braunschweig; (11) Dierick Kossel, Leitz, son of Walther Kossel; (12) Walter Rollwagen, München; (13) Heinz Niehrs, Berlin; (14) Gottfried Möllenstedt, Tübingen; (15) Peter Schiske, Berlin. As special guests Ladislaus Marton (16) and his wife Claire (17) from the National Bureau of Standards, USA, were welcomed by the DGE members *(photos by courtesy of Albrecht H. Kossel, Tübingen, who identified a number of participants, while the identity of others was determined by help of Rudolf Rühle's "Festschrift", Christof Niehrs, Mainz, Peter Schattschneider, Vienna, and of the archives of ZEISS in Jena and the Technical University of Braunschweig).*

Figure 18 The Bosch TEM. (a) Overall aspect of the instrument encased by a modern covering (from the brochure 'BOSCH Elektronenmikroskop' 1950; *courtesy archives of Robert Bosch GmbH, Stuttgart*). (b) Details showing the robust construction of the column. (*Photo taken by Heinz Schwarz, August 12, 2021.*)

magnifications and remained stationary with the help of internal electric switching (Patent Nr. 972910, April 16, 1950), an advantage compared with hitherto "known instruments" ("bekannte Elektronenmikroskope"), for which a major change of magnification required a mechanical change of the polepieces, inconvenient and disturbing the image. The resolution of the Bosch microscope was 3 nm, while the resolution of the diffraction mode exceeded that of any other available apparatus. The peculiar combination of specimen lock and specimen holder was safe and enabled a continuous, quick change of tilt axes and taking stereo-images with convergence angles up to 20°.

The camera system housed 18 photographic plates of 6.5 × 9 cm and the automated camera air-lock allowed the instrument to function during plate change. The high vacuum was produced by a diffusion pump and controlled by a vacuum-gauge meter, developed in-house, with a measuring range of 10^{-2} down to 10^{-5} mm Hg. Interestingly, the Bosch brochure (Technische Daten 1950) presented twelve images, but only one biological

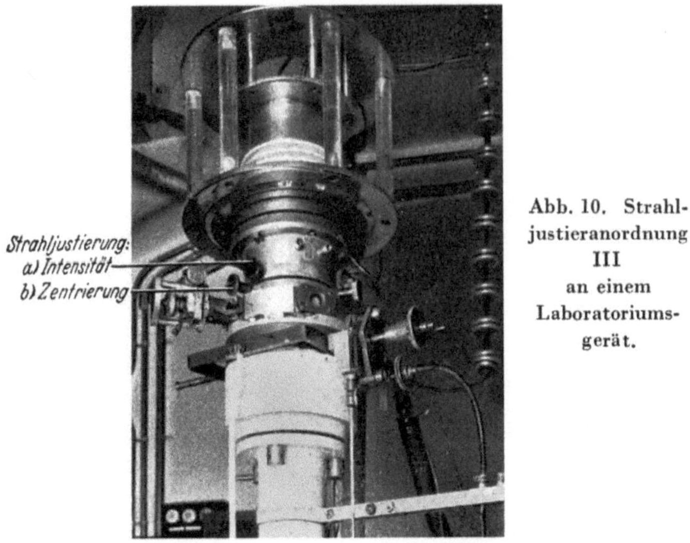

Figure 19 Details of the condenser of the Bosch microscope with the controls for beam intensity and alignment (from Rühle, 1949b; *with permission of ELSEVIER*).

(a) (b)

Figure 20 On the performance of the Bosch microscope in life and material sciences. (a) Q-fever bacteria after shadow cast preparation. (b) Evaporated zinc layers in diffraction (from the brochure 'BOSCH Elektronenmikroskop' 1950; *courtesy archives of the Robert Bosch GmbH, Stuttgart*).

sample, the agents of Q-fever (Fig. 20a), an occasionally deadly lung infection. At that time, the causative small *Rickettsia* bacteria were spreading in Germany.

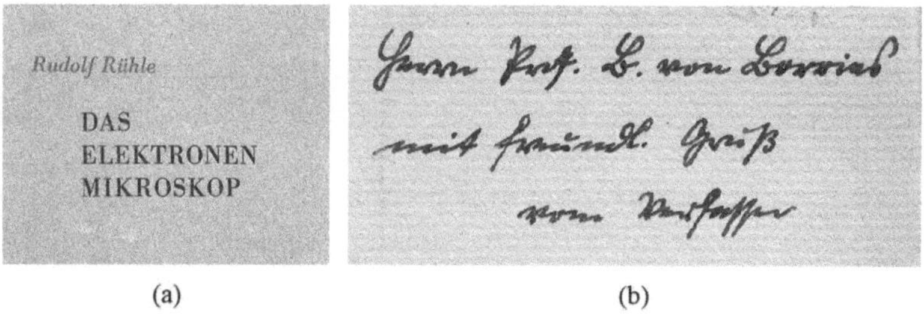

Figure 21 Rudolf Rühles book DAS ELEKTRONENMIKROSKOP: (a) A pedagogical treatise of 168 pages plus 26 figures (Rühle, 1949a). (b) Hand-written dedication in a copy of this book by Rudolf Rühle "Herrn Prof. B. von Borries mit freundl. Gruss vom Verfasser". (To Herr Prof. B. von Borries with kind regards from the autor.) (*Photos taken by Hans Gelderblom, February 26, 2021.*)

Instead of rubber sealing gaskets, Bosch introduced precision hard chromium plated sealing surfaces for a number of air locks (Rentschler, 1957).

16. Promotion for the Bosch electron microscope

Early on Rühle planned and carried out an intense promotion - proving his dedication to the instrument and his loyal commitment to Bosch. In 1949 he published a book of 180 pages *Das Elektronenmikroskop*. Written in a neat and systematic way, the fundamentals of electron optics are described, the diverse approaches of AEG and S&H, different geometries of the components and the basic features of the vacuum system. Sample preparation and applications in materials and life sciences are dealt with. Only 5 of the 25 micrographs included show results from the life science, bacteria and phages, the latter a gift from Helmut Ruska. This book is interesting for an educated layperson, but also for the expert of today (Rühle, 1949a). He dedicated one copy of the book to Bodo von Borries ("Herrn Prof. B. von Borries mit freundl. Gruß vom Verfasser") (Fig. 21). The wording might sound a little reserved, possibly because Rudolf Rühle was a relative newcomer.

Also in 1949, Rühle presented the new instrument at the First Meeting of the DGE in Mosbach in a lecture entitled "Einige spezielle Anordnungen am Beleuchtungs- und Objektivsystem eines Elektronenmikroskops / A special arrangement of illumination and specimen system of an electron microscope" (Schimmel, 1996). He published the key properties of the instrument in five parts in the new German journal *Optik*. In the first, on

6 pages with 10 figures, he described the centering process of the electron beam relative to the objective lens, involving tilting and shifting. The instrument offered two ways of aligning the beam, differing in complexity (Rühle, 1949b). Next he covered his favorite topic, the generation of stereo images ("Raumbilder am Elektronenmikroskop") (Rühle, 1949c). He stressed the need for an accurate tilting stage and presented evidence for the robust performance of the Bosch goniometer. Rühle advised on proper image interpretation and measurements of surface profiles and presented as a proof of performance five examples from his field in the materials sciences and just a single life science specimen, the rods of a typhoid bacterium after shadow cast preparation. This imbalance in favor of materials science reflects his training as a physicist and engineer and also reflects the course of his professional life and his interests. The paper contains nine references to the hitherto existing tilt mechanisms and was submitted in September 1949. The instrument was still not in a final stage, as Rühle announced a further improvement of the goniometer (Rühle, 1949c). At the end of 1949, the Bosch electron microscope was ready for the market. Bosch, the newcomer on the scene, built his product only "on demand" - in contrast to the serial productions of the established companies.

During the second meeting of the DGE in Bad Soden in May 1950, Rühle lectured on the simple and wide change of magnifications between 300-fold up to 30,000-fold by switching between three different projector lenses. He emphasized the advantages of this arrangement as allowing for low magnification examination and high resolution imaging and also electron diffraction (Rühle, 1950a). Next he analyzed systematically the principles of electron diffraction (Rühle, 1950b). A last contribution to electron microscopy was published by Rühle in 1953, together with his colleague Professor E. Zehender: "A detailed study of tilted objects in the electron microscope" (Die Beurteilung von Objekten im Elektronenmikroskop, die während der Beobachtung geschwenkt werden können (Rühle & Zehender, 1953). Rühle had covered the same topic a year before in his lecture at the 1952 annual meeting of the DGE in Tübingen.

Rühle had announced the instrument also in the Bosch house-magazine, the "Bosch-Zünder", well before it went onto the market. In "Ein Vorstoß in die Welt des Kleinsten mit dem Bosch-Elektronenmikroskop" (An approach to the world of the very small) he introduced the reader to the basics and limits of visualization and covered Abbe's law and Louis de Broglie's and Hans Busch's contributions. He stated that Bosch had started to develop its instrument early in the 1940s for internal use and described it as one of the most versatile instruments capable of solving a wide range of problems in R&D and in quality control. Besides engineers and physicists, physicians would also benefit from it, e.g., in the fight against infectious diseases. As an example he mentioned the case of Q-fever (Fig. 20a, see above on p. 344). Rühle was confident: To explore the range of the smallest structures still further will be an ongoing task for Bosch in order to improve products and production.

When Walter Dorn retired in 1960, Rühle took over as head of the R&D Division of Bosch. He felt relieved at being able to focus again on the key products of Bosch (Manfred Rühle, personal communication, 2021). He was promoted: in 1962 he became Head of the influential Technical Research Center (Entwicklung für Grundlagenforschung und angewandte Physik, EAP) and in 1969 Head of the 13 Departments of Bosch's Technical Research Center. As his last personal invention Bosch noted in 1964 "Laser als Informationsüberträger" (laser as a transmitter of information). As a manager he improved the Bosch repertoire establishing semiconductors and Peltier elements as steering components. In order to enhance basic research and the promotion of specific fields, like integrated circuits, he established two external Research Centers in Berlin, in 1965, and Lonay, Switzerland. To meet these challenges, the staff in the R&D group grew up from 36 persons in 1955 to 478 persons in 1961 (Bähr & Erker, 2015; Bosch, 2019; Rühle, 1971).

After his retirement in 1971, Rühle served as an external consultant to Bosch. Yet increasingly he spent his time farming, working in the vineyard (Fig. 22) and in the local brass ensemble. Attending concerts and the opera became his highest pleasure. Rudolf Rühle died on April 21, 2003 (Manfred Rühle, personal communication, 2021).

17. The Bosch electron microscope in Hohenheim

In September 1947, Walter Rentschler was back again full-time at the Agricultural College in Hohenheim, since 1967 Hohenheim University (Bosch, 1947). In 1949, he qualified as Privatdozent (Assistant Professor) with a habilitation on "Optics of multiple transparent layers" based on his achievements attained at Bosch ("Die Optik dünner, durchsichtiger Mehrfachschichten" (Seiler et al., 1981; UAH archive, 2021)). Ten Departments of the small, busy College wanted fine-structural capability for studies in the fields of infectious diseases, casein and milk studies, soil properties, paint and lacquers. Rentschler took care of the application for funding

(a) (b)

Figure 22 Rudolf Rühle: optimistic and down-to-earth. (a) Rühle around 1949. (b) Rühle in his vineyard 1971 (*from "Festsschrift"by courtesy of Manfred Rühle*).

an electron microscope and addressed the Deutsche Forschungsgemeinschaft (DFG, German Research Council). The acquisition was approved as a loan-gift and when the DFG and the University together compared the three instruments on the German market (ÜM 100 of S&H, EM 8 of Zeiss, the new Bosch instrument), the Bosch microscope was chosen. On November 15, 1952, the DFG ordered the instrument including the facilities for electron scattering, stereo-images and the local set-up at the cost of 96.000 DM. In the order form the DFG mentioned previous talks between Dozent Rentschler and Dr. R. Rühle (DFG, 1952) (Fig. 23). From February 12 to March 05, 1953, the instrument was mounted in the Physics and Meteorology institute headed by the later Professor Rentschler (Deutsche Forschungsgemeinschaft, 1952; Seiler et al., 1981).

The inauguration party for the instrument on March 6, 1953 brought together the Heads of the ten Departments of the Agricultural College interested in ultrastructural studies. Drs. Dorn, Rühle and Zehender were also among the guests (Fig. 24).

The resolution was shown to be 3 nm using gold crystals (Fig. 25a). As additional evidence of the instrument's performance fowl pest influenza virus was shown after shadow-cast preparation at 40,000-fold mag

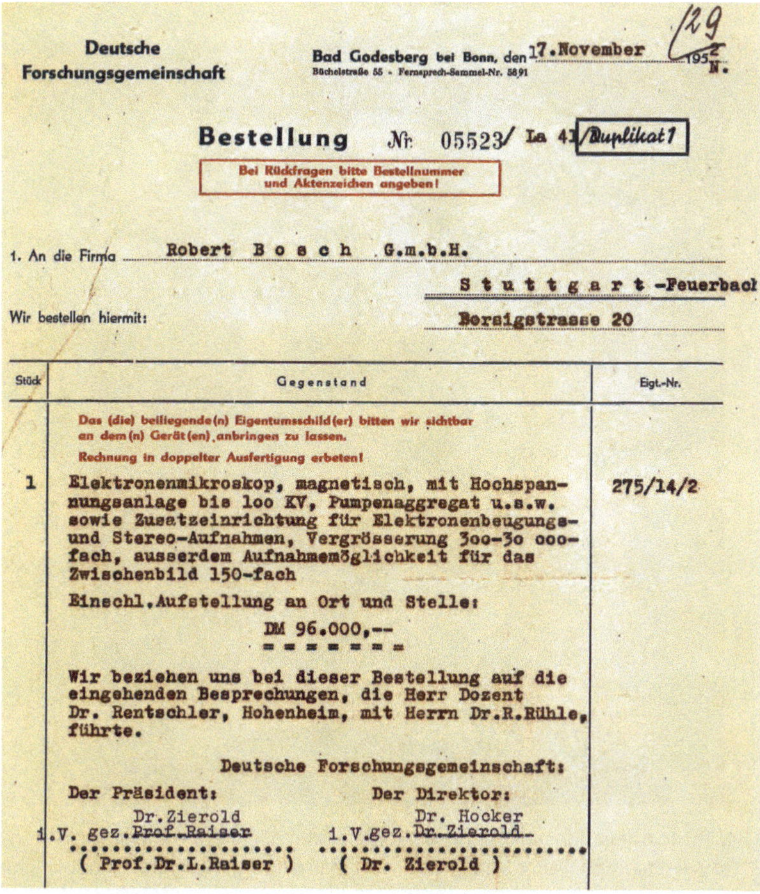

Deutsche Forschungsgemeinschaft

Bad Godesberg bei Bonn, den 17. November 1952

Büchelstraße 55 · Fernsprech-Sammel-Nr. 5891

129 N.

Bestellung Nr. 05523/ La 41/ Duplikat 1

Bei Rückfragen bitte Bestellnummer und Aktenzeichen angeben!

1. An die Firma Robert B o s c h G.m.b.H.

S t u t t g a r t -Feuerbach

Bersigstrasse 20

Wir bestellen hiermit:

Stück	Gegenstand	Eigt.-Nr.
	Das (die) beiliegende(n) Eigentumsschild(er) bitten wir sichtbar an dem(n) Gerät(en) anbringen zu lassen. Rechnung in doppelter Ausfertigung erbeten!	
1	Elektronenmikroskop, magnetisch, mit Hochspannungsanlage bis 1oo KV, Pumpenaggregat u.s.w. sowie Zusatzeinrichtung für Elektronenbeugungs- und Stereo-Aufnahmen, Vergrösserung 3oo-3o ooo-fach, ausserdem Aufnahmemöglichkeit für das Zwischenbild 150-fach	275/14/2
	Einschl. Aufstellung an Ort und Stelle: DM 96.000,--	
	Wir beziehen uns bei dieser Bestellung auf die eingehenden Besprechungen, die Herr Dozent Dr. Rentschler, Hohenheim, mit Herrn Dr. R. Rühle, führte.	

Deutsche Forschungsgemeinschaft:

Der Präsident: Dr. Zierold i.V. gez. ~~Prof. Raiser~~ (Prof. Dr. L. Raiser)

Der Direktor: Dr. Hocker i.V. gez. ~~Dr. Zierold~~ (Dr. Zierold)

Figure 23 Order document for the Bosch microscope at the University Hohenheim from November 17, 1952 with an additional note on exhaustive talks/discussions between Walter Rentschler und Rudolf Rühle [UAH3_08b.6(7)].

(Fig. 25b) as well as very fine electron diffraction spectra of clay minerals (Rentschler, 1953).

Bosch covered the cost of maintenance, repairs and in part also of spare parts like batteries (Rentschler, 1963). Bosch also provided a four-month training course (November 1952 to February 1953) for Ingeborg Günther, who became the research assistant in the electron microscope facility in 1956, when the preceding scientist E. Fischer had left (Rentschler, 1957).

The EM laboratory was strengthened by additional staff and equipment: in 1955, an ultramicrotome allowed ultrathin sectioning for TEM. In 1956 Dr. Günter F. Meyer came from Fritiof Sjöstrand's (1912-2011) laboratory

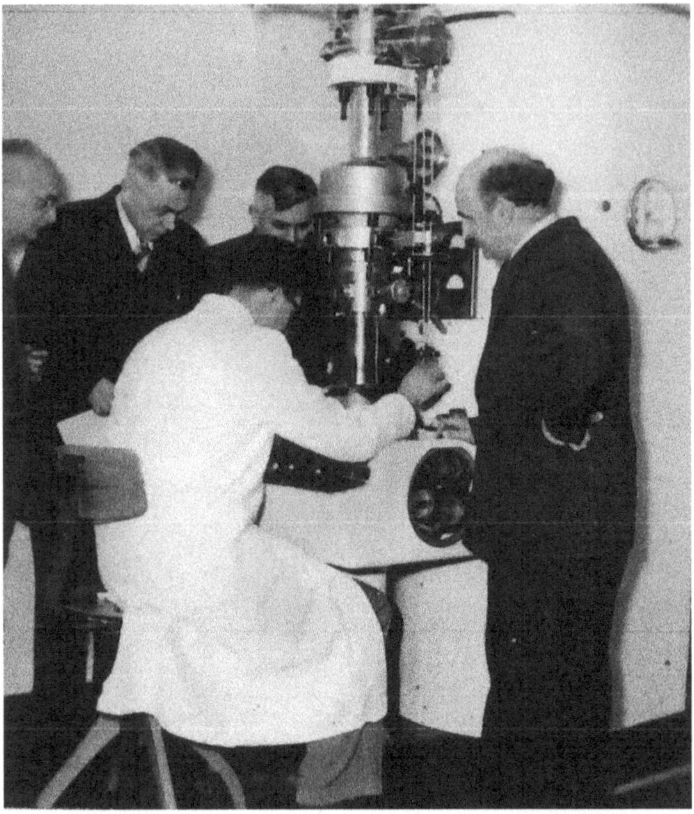

Figure 24 The inauguration party of the BOSCH electron microscope in Hohenheim on March 6, 1953. From left to right: Walter Dorn, Rektor Kurt Maiwald, Ernst Zehender, Rudolf Rühle, Walter Rentschler (*from Rudolf Rühle's 'Festschrift' 1971, p. 52; by courtesy of Manfred Rühle*).

at the Karolinska Institute to join the Hohenheim EM laboratory. Meyer later, during 1959-1986, headed the EM laboratory at the Max Planck Institute for Biology (Dept. Beermann) in Tübingen. The Bosch microscope was used to study the fine structure of plant chromoplasts, freezing processes of plants, texture of chicken organs, but also the action of plant-protection aerosols. It was run 70 hours per month on average and progress was made with eleven projects, e.g., in attempts to improve soil quality by the use of highly-polymeric polystyrene (Fischer & Rentschler, 1957), stability of paint and lacquer against environmental factors (Fischer & Hamann, 1957), fine structure of neurofibrills of *Hirudo medicinalis* (mainly achieved using the RCA microscope in Sjöstrand's laboratory) (Meyer, 1957) and of pigment granula of myxomycetae (Lieth & Meyer, 1957). A permanent

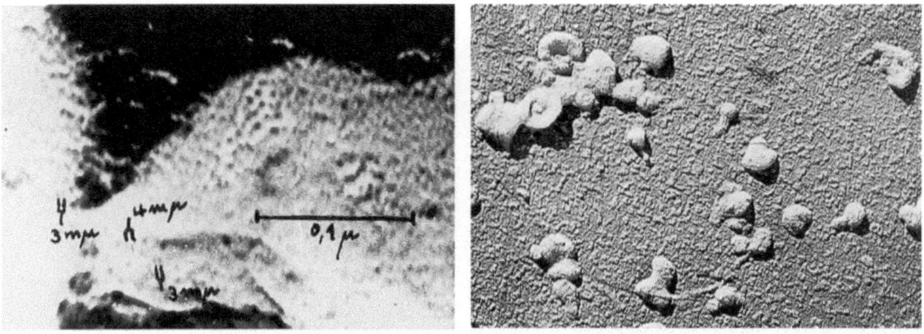

(a) (b)

Figure 25 Performance of the Bosch electron microscope in Hohenheim. (a) 3 nm resolution was proven using evaporated gold in February 1953. (b) Fowl plague virus: avian influenza virus particles, after shadow cast preparation at 40,000-fold magnification [Rentschler, 1. Jahresbericht, 18.12.1953. UAH3 08b.6(7)].

Figure 26 *Agrobacterium* after trypsin digestion and shadowing with Pt-Ir at 20,000 fold magnification [Rentschler, 4. Jahresbericht, 16.10.1957. UAH3 08b.6(7)].

challenge was the detection and morphological characterization of viruses and bacteria (Rentschler, 1957) (Fig. 26).

Despite its relative complex buildup (Fig. 27) and handling, the Bosch microscope helped to generate 20 fine-structural publications in Hohenheim (UAH, personnel files W. Rentschler, 2021).

In June 1963, the DFG invited Hohenheim University to take ownership of the loaned instrument, after ten years of application, for the price of 10.000 DM (Rentschler, 1963). Earlier, in 1962, the DFG had granted Rentschler funds for the acquisition of a "modern" Siemens Elmiskop I. The University gladly took over the Bosch microscope and promised to apply it both in future research and in electron microscope training courses

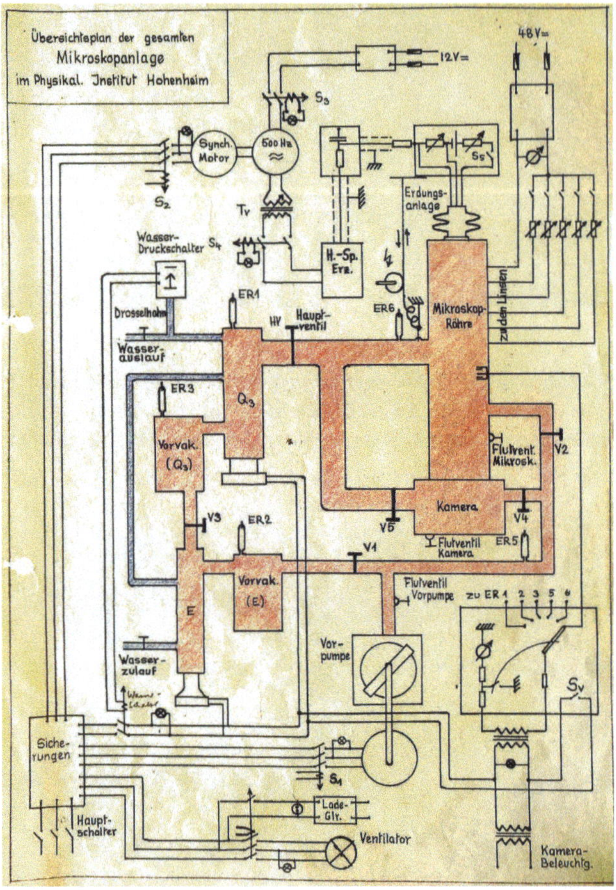

Figure 27 Circuit and vacuum diagram of the Bosch electron microscope at the Physics Institute in Hohenheim (by *courtesy of Anna Krupp, University Hohenheim*).

(Rentschler, 1963). It was shut down early in the 1990s and is exhibited now in a showcase as a historical instrument in front of the lecture hall B 13 of the Biology I Building of Hohenheim University (Fig. 28).

In 1963, Rentschler was appointed Full Professor and Head of the Institute of Physics and Meteorology. Due to his failing health, the tireless Rentschler retired at the end of September 1977. Until his death on October 21, 1984, the Emeritus still followed his strong scientific interests in optics and astronomy (Denisa Kohan, UAH, personal communication, 2021).

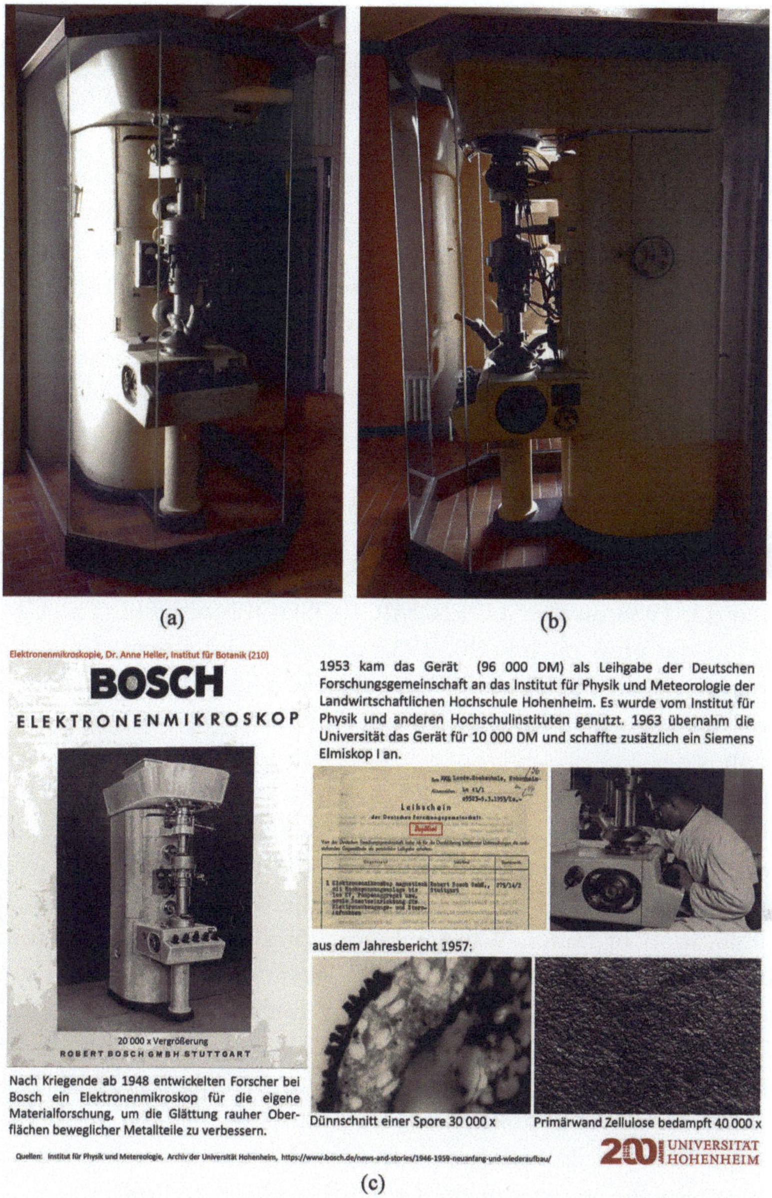

Figure 28 The appearance of the 1952 Bosch microscope in Hohenheim today. Instrument in its actual glass cabinet in the Biology I building. (a) Frontal view. (b) Side view (*Photos taken by Heinz Schwarz, August 12, 2021*). (c) Information Board in front of the displayed microscope with details to the history of the instrument (*by courtesy of Annerose Heller/Anne Krupp, University Hohenheim*).

18. The Bosch-EM in Tübingen

Already in 1938, significant insights into the microworld were presented. On June 22, Helmut Ruska gave a speech before the Berlin Medical Society on "Bakterien und Virus in übermikroskopischer Aufnahme" (bacteria and viruses under the electron microscope). In August he lectured at the Fifth International Congress on Cell Research in Zürich (V. Internationaler Zellforscherkongreß, 7.-13. August 1938) on Fine-structural observations: From cells down to viruses ("Übermikroskopische Darstellung organischer Struktur (vom Größenbereich der Zelle bis zum Ultravirus)") (Ruska, 1939). A powerplay of publicity started [for review see Gelderblom, 2019; Gelderblom & Krüger, 2014; Müller, 2008, 2009, 2021]. Bodo von Borries and this brothers-in-law Ernst and Helmut Ruska, the familial trio at S&H, published on progress in the construction of the Übermikroskop and observations on bacteria and viruses Ruska et al., 1939; von Borries & Ruska, 1938; von Borries et al., 1938a, 1938b. The instrument raised world-wide interest, e.g., the *New York Times* wrote on May 23, 1938 on the SUPER MICROSCOPE of S&H. The ÜM was shown to the press on July 20, 1938 in Berlin and attracted much public attention, e.g., the Danisch *Berlingske Tidende* used "Verdens staerkeste Mikroskop" to criticize the barely visible measures against unemployment (Fig. 3).

The news raised interest also at Universities. Early in 1939, the Natural Science and the Medical Faculties at the University of Tübingen started internal discussions on the acquisition of the novel instrument with the Rector of the University. He approached the Ministry of Cultural Affairs and Education of Baden-Württemberg asking for approval and funding. Professor Otto Stickl (1897-1951), then Dean of the Medical Faculty and Head of the Hygiene-Institute, a driving force at the Faculty and since 1933 also an influential member of the NSDAP, the Nazi party, took over the negotiations with S&H. From November 1, 1939 until May 7, 1945, Stickl had the powerful position of Rector. In July 1939, he went to Berlin and negotiated details of the acquisition with von Borries and Ruska, the two *pari passu* leading engineers at the LfE at S&H. Internally, at the University, Stickl emphasized the scientific advantages of the novel instrument in medical research. Indeed, it would become the first ÜM to be supplied to a University, one of the first series of ten instruments (Stickl, 1939a; Wolpers, 1991). On July 5, 1939, S&H submitted their offer. The costs of the ÜM would amount to 70.000 Reich Marks (RM) representing S&H's

manufacturing costs, i.e. the common costs for customers in Germany. Local installation was planned for an additional 5.000 RM, while the initial training of a technical assistant was already included. The price for the foreign acquisition was 95.000 RM plus installation and taxes (von Borries, 1939b). The delivery time for the instrument would amount up to 18 months. Dean Stickl forwarded the offer to the Rector on July 8, 1939 (Stickl, 1939b). On July 13, 1939, the Ministry of Cultural Affairs and Education ("Der Kultminister") with the backing of the Ministry of Finances approved the acquisition of the ÜM for both the Medical and the Natural Science Faculties (Kultminister, 1939). The Rector ordered the microscope the next day (Rektor, 1939), an unusually rapid acquisition, facilitated by Stickl's acquaintance with the Prime Minister Mergenthaler in Stuttgart, being himself a Parteigenosse (Wikipedia, 2021a, Otto Stickl and at http://www.leo-bw). S&H accepted the order on July 19, 1939. The confirmation was signed by "Lüschen" and per proxy "i.V. von Borries"(Siemens & Halske, 1939). Dr. Friedrich Heinrich Lüschen (1877-1945) was on the executive board of the S&H Wernerwerk, the home of the LfE where von Borries and Ruska developed the commercial ÜM.

In 1942, Lüschen became a member of NSDAP and the SS and a very high-ranking manager in the electrical industry. He was involved in the recruitment of 1500 (later 3000) forced foreign workers, later joined by inmates of concentration camps. Lüschen was accused retrospectively and committed suicide on June 18, 1945 (Wikipedia, 2021b, Friedrich Heinrich Lüschen). In this context it can be stated that the leading engineers at AEG, Bosch, and S&H, as well as Manfred von Ardenne, did not join the NSDAP, with the exception of Bodo von Borries (van Gorkom, 2018). Fostered by his conservative background, von Borries felt deutschnational (sic) and became a member of the NSDAP as early as May 1, 1933 (Gelderblom, 2019). He carried the party button, but in 1939, he did not participate in the mandatory NSDAP statistics ("Parteistatistische Erhebung 1939") (von Borries, 1939a). He had to negotiate with highest Nazi officials, e.g. with Professor Karl Brandt, the accompanying physician of the "Führer". Brandt was the supervising "Generalkommissar" for sanitation and health and responsible for euthanasia in the Reich (von Borries, 1939b). Yet, von Borries's many business trips, especially to Sweden and Switzerland, opened his eyes (von Borries, 1991). He focused on his tasks at S&H and protected his staff, including foreign workers and non-Aryans, against malicious aims of the Party, but he kept his membership and was therefore politically charged after the war. Accordingly, in his denazification process in summer 1946, Bodo von Borries came out as a D2, as a "follower", though with a clean character ("edle Gesinnung") (von Borries, 1946).

World War II was declared on September 1, 1939 and immediately caused shortages and much delay at S&H. On January 31, 1940, Ernst Ruska warned the University of delay owing to shortages in material and staff (Ruska, 1940). The delivery of the ÜM was postponed several times until the Reichsforschungsrat (German Research Council, executing also a central war-oriented planning) gave priority to the instrument for the war effort (Reichsforschungsrat, 1942). Nevertheless, the supply of the instrument still seemed uncertain as Reichsgesundheitsführer Dr. Leonardo Conti (1900-1945), high ranking SS-member, Reichsärztefführer (Leader of the Nazi Medical Association) and State Secretary in the Ministry of the Interior got suspicious. He had not been involved in the approval of the instrument. On April 20, 1942, his office questioned the purpose of the microscope (Schütz, 1942). The answer took a while. Only on June 15, 1942 did Stickl underline the need for fine structural studies in bacteriology, virus research and immunology at the Hygiene Institute and promised access to the S&H microscope for other potential user groups (Stickl, 1942). But in the meantime, on May 27, S&H had prepared the shipment of the instrument to Tübingen (Siemens & Halske, 1942).

The Tübingen ÜM is listed as no.15 in the Siemens list. While in 1941 only four microscopes had been built, this is one of seven microscopes delivered in 1942. It already reached an accelerating voltage of 85 kV (von Borries, 1943; Wolpers, 1991) (Fig. 29a). The Laboratory for Supermicroscopy of the Hygiene-Institute in Tübingen was inaugurated on July 1, 1942. Yet the instrument was not widely used, also due to the restrictive war times.

After World War II, the S&H ÜM went to France as a war reparation, a procedure covered by the agreements on German war reparations signed at the 1945 Yalta Conference. Here, at the Institut Pasteur, it was applied successfully for several years (Müller, 2021, p. 323, Gaudilliere, 2013).

Also in 1945, Stickl lost his positions as Rektor and Dean due to his NS history but after a short period of imprisonment, he was allowed back to take care of "his" Hygiene Institute activities (Wikipedia, 2021a, Otto Stickl). He eagerly attempted to replace the lost microscope (Stickl, 1946a). On March 6, 1946 S&H offered an ÜM 100B for the price of 96.000 Reich Marks; a personal letter of support from Ernst Ruska to Stickl accompanied the offer (Ruska, 1946). A delighted Stickl officially ordered the

(a) (b)

Figure 29 Early commercial Siemens microscopes: (a) A S&H ÜM from 1942 at the Hygiene-Institut in Tübingen (*source Universitätsarchiv Tübingen UAT-376-86-1*). In 1945, it was transferred by French Marine forces to Paris as a war reparation (Stickl, 1946a). During transport through the Black Forest it fall down into a ditch, later, however, it succeeded for several years at the Institute Pasteur (Giuntini et al., 1947; Müller, 2021, p. 247, p. 323). (b) In 1953, lecturer Hansjürgen Raettig sitting at "his" S&H ÜM 100D, Robert Koch-Institut, Berlin. Raettig worked then on the classification of bacteriophages and the physiology of bacteria (Raettig, 1954). The ÜM 100D was later transfered to Deutsches Museum in Munich (*by courtesy of Bärbel Raettig, Berlin*).

ÜM on March 21, 1946, and only then over-optimistically asked the Rektor for funding (Stickl, 1946b). The meager post-war circumstances also retarded the production at S&H in Berlin. There was no regular money transfer between the different occupational zones and apparently essential raw metals were lacking. In 1947, S&H had even asked Stickl to provide 30 grams of platinum, 79 grams of gold, 400 grams of silver plus 1 gram of a high-quality tin-solder as a pre-requisite for the making of the microscope (Siemens & Halske, 1947).

In October 1949, Stickl was mildly judged as a Mitläufer (follower) in his denazification process (Wikipedia, 2021a, Otto Stickl, see also Stickl at http://www.leo-bw). He was able to renew his professorship at the Hygiene-Institut and pursued the worrying question of a new electron microscope with renewed vigor. In April 1949, Stickl became aware of the lack of any financial resources (Erbe, 1949). Time passed and Stickl received new offers. In November 1949, Ernst Ruska informed Stickl that the first S&H post-war ÜM had been delivered to the Deutsche Forschungshochschule Berlin Dahlem, part of the former Kaiser-Wilhelm-Institut für Physikalische Chemie und Elektrochemie in Berlin. Ruska wrote: "Mein Bruder, der ebenfalls eine Forschungsgruppe in Dahlem hat, und ich sind sehr froh, nach so langer Zeit wieder mit einem Mikroskop arbeiten zu können" (my brother, who also has a research group in Dahlem, and I are so happy to be able to work again with an ÜM after so long a time) (Ruska, 1949).

In January and again in October 1950, Stickl applied for funds for a S&H ÜM (Stickl, 1950a, 1950b). When the Pro-Rektor Erbe consoled Stickl by informing him that Professor Butenandt from the Virus Research Group of the Kaiser Wilhelm/Max Planck Institute of Biochemistry, located in Tübingen from 1943-1956, had ordered in the meantime an SDL/Zeiss EM 8 microscope and that Butenandt would allow access for the colleagues from the Hygiene-Institut, Stickl was not impressed. He pointed to the need to have his own instrument inside the institute for biosafety reasons when working with highly infectious pathogens. He underlined also the advantage of having microscopes of two different companies, of S&H and SDL/Zeiss, on hand in Tübingen (Stickl, 1951).

When Otto Stickl died on September 27, 1951, his Deputy, Professor Richard-Ernst Bader (1912-1996) took over the Institute and the sourcing of the electron microscope. Virus research and diagnostics required the instrument urgently and besides: to own an electron microscope and an ultracentrifuge was a "must" in those times in microbiology, contributing to prominence and esteem. In 1950, Bosch joined S&H and SDL/Zeiss on the microscopy market. The economic miracle that occurred in the early Federal Republic, now allowed funding even for expensive large equipment. Stiff competition arose. On November 12, 1953, S&H offered the ÜM 100D at a price of 80.000 DM (Fig. 29b). On December 15, 1953, S&H invited Bader to visit Berlin to compare the ÜM 100 models E and F and to prepare test images (Ruska, 1950, 1952; Siemens & Halske, 1953). When Bader was not too satisfied with the test results, he was offered another visit to Berlin sponsored by S&H (Siemens & Halske, 1954).

In view of the offers for the S&H ÜM 100E and the Bosch microscope, Bader sought professional advice. Professor Werner Schäfer (1912-2000) from the Max Planck Virus Research Group in Tübingen characterized both instruments as efficient, recommendable, but with different advantages and disadvantages. Next, in January 1954, Bader turned to Professor Otto Scherzer at the Technical College in Darmstadt (Bader, 1954a). Scherzer (1909-1982), an early member of the Brüche group in the AEG-Research Institute in Berlin (April 1932-December 1935), rejoined Brüche after World War II at the Süddeutschen Laboratorien in Mosbach (August 1946-April 1947) (Schulze, 2001; Typke, 2022, this volume). Otto Scherzer and also Professor Gottfried Möllenstedt (1912-1997; working at the Applied Physics Institute in Tübingen since 1950) had been involved in the construction of the EM 8 and were aware of the satisfactory experiences with the EM 8 in the virology group around Butenandt. Scherzer informed Bader that he himself had tested both the S&H and the Bosch microscope, and found both to be very useful ("beide sehr brauchbar gefunden"). He judged the Bosch instrument to be technically more versatile but probably more demanding in use. With Bosch being a newcomer on the market, this might turn into a disadvantage though the eager Dr. Rühle at Bosch would guarantee a close and reliable support. In summary: For the technically less well experienced staff in a life science EM laboratory both instruments, S&H and Bosch, would suit equally well. But Scherzer was astonished that the electrostatic microscope of SDL/Zeiss had not been considered a candidate instrument. Its smaller objective aperture would guarantee much higher contrast, especially with biological objects. In conclusion Scherzer pointed to the satisfying experiences accumulated by Möllenstedt in the Physics Institute and by the virologists (Scherzer, 1954). Bader followed Scherzer's advice and decided to compare all three instruments by taking test images (Bader, 1954b).

Bader finally chose the Bosch instrument, a choice that suited also the two other microscopy users in Tübingen, Professor Möllenstedt at the Physics Institute and the Butenandt virus group. As they already used the S&H ÜM and the SDL/Zeiss EM 8, respectively, the new Bosch instrument appeared to be of special interest. Bader himself and the colleagues from the other two microscopy groups preferred to have all three different microscopes available in Tübingen. Moreover: the after-sale service for the Siemens ÜM was reported to be poor (Bader, 1954a). The decision fell in March 1954 (Bader, 1954c). The Bosch microscope was installed between November 1954 and January 1955. Its 3 nm resolution was monitored and on February 8, 1955, Bader invited the young microbiologist cand. rer. nat. Werner Schreil to work for a week with the instrument (Bader, 1955).

Schreil was enthusiastic about electron microscopy. In Heidelberg, in the laboratory of Gustav-A. Kausche (1901-1960), he became familiar with the S&H ÜM and also with the EM 8 during working visits at Zeiss in Mosbach and Oberkochen. Summarizing his experiences with the Bosch microscope Schreil questioned the complicated handling of the camera and, criticizing the difficulty of aligning the condenser (see Fig. 19), he joked: "Dr. Rühle, the father of the instrument, can probably align the microscope even with his eyes closed - but for other microscopists the alignment might turn-out to be a bit demanding". Presumably not completely wrong, as Rühle had dealt with the centering topic intensely - by lecturing at the DGE-Meeting 1949 in Mosbach and by a specific publication (Rühle, 1949b). With the S&H instrument, in contrast, the condenser could be aligned readily as there are fewer possibilities for adjustment. And the ZEISS EM 8 - equipped with the Steigerwald fern-focus-cathode (Steigerwald, 1949) - operated even without any condenser. On the other hand, Schreil praised the Bosch instrument because of its brightness and versatility and judged it to be superior to the Zeiss EM 8/II. With regard to the future of the Bosch instrument, Schreil stated that Bosch was apparently unwilling to invest another million DM in its further development and marketing, thereby placing it aside as occurred with the Philips-Siemens-Zeiss instruments. With all three companies unable to benefit from microscope production, only RCA would make profits with its instruments (Schreil, 1955). This disparity might have changed when S&H introduced the Elmiskop I in 1954 (Ruska, 1954), and the Elmiskop IA in 1964 with more than 1500 instruments sold globally, and when Philips introduced its 200 and 300 series.

The team around Bader at the Hygiene-Institut was basically interested in the physiology of bacteria, e.g. to study the expression and properties of bacterial flagella. The Bosch microscope was mainly used by the physician Diether Jacherts; he examined flagella formation of *Pseudomonas aeruginosa* (Fig. 30) (Jacherts, 1959) and afterwards, in his habilitation treatise, described the action of lethal factors on staphylococci (Jacherts, 1960). In the presence of cell-free extracts of chicken chorio-allantoic membranes the cocci suffered growth inhibition. After shadow-cast preparation they showed gross morphological changes, swelling due to osmotic damage, and even rupture of the limiting cell wall (Fig. 31).

Yet apparently the Bosch instrument, designed by a physicist, required skills and understanding that was grossly uncommon in the life sciences team at the Hygiene institute. The microscope became increasingly unpop-

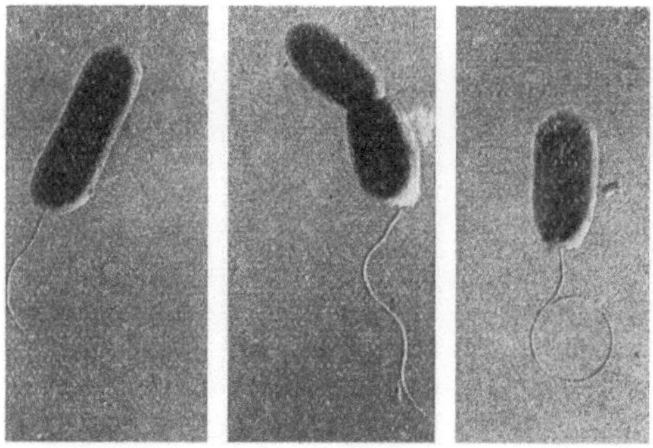

a b c

Abb. 1. a) Normales Bakterium zwischen zwei Zellteilungen. b) Teilungsform. c) Tochterbakterium kurz nach der Zellteilung. Elektronenoptische Aufnahmen. Vergrößerung: 7500 fach. Gerät: Bosch.

Figure 30 Studies on the formation of bacterial flagella. Samples were collected every 30 sec from a synchronized culture of *Pseudomonas aeruginosa* and rapidly cooled down to 0 °C. After fixation with OsO_4 and shadowing with palladium the specimens were studied using the Tübingen Bosch microscope. a) Normal elongated appearance of the cell body with flagellum. b) A dividing bacterium reveals the daughter cell still devoid of a flagellum. c) Shortly after cell division all cells reveal a flagellum. Dividing cell never showed two flagella, pointing to the very rapid formation of the motility organ (Jacherts, 1959) (*with permission of ELSEVIER*).

ular and accordingly, no further fine structural publications were generated using the Bosch microscope (UAT, 2021). The microscope was shut down in the 1960ties (Hans-Joachim Gerth, personal communication, 2021). Nevertheless it still exists today: in summer 1979, Professor Norbert Rieder from the Institute of Zoology of Karlsruhe University (now KIT) learned about the planned dismantlement and informed his colleague Dipl.-Ing. Winfried Send from the Laboratory of Electron Microscopy (LEM) at the Physics Institute of KIT. Being enthusiastic about electron microscopy, they decided to save this rare instrument. With a truck they went to Tübingen, dismantled the microscope and transported the parts to a cellar room of the LEM in Karlsruhe. Send is planning to reconstruct the instrument and then to exhibit it (Winfried Send, personal communication, 2021).

What about the two other early groups interested in electron microscopy in Tübingen, the Möllenstedt group at the university Institute for Applied Physics and the Working Group on Virus Research of the former Kaiser Wilhelm Institute (KWI) of Biochemistry in Berlin. Professor Gottfried Möllenstedt (1912-1997) had studied physics under Professor Walther Kossel (1888-1956) in Danzig. Focusing on geometric electron optics he did his doctoral degree in December 1940, his habilitation treatise in Jan-

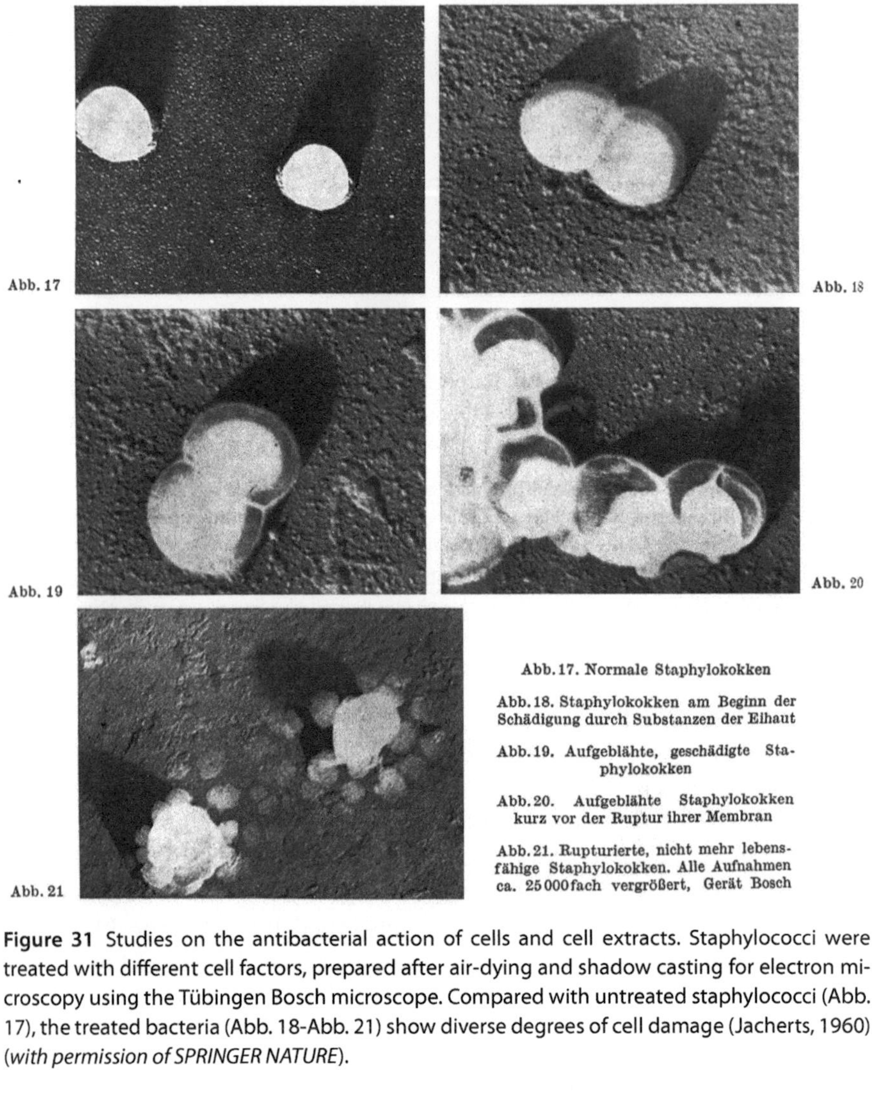

Figure 31 Studies on the antibacterial action of cells and cell extracts. Staphylococci were treated with different cell factors, prepared after air-dying and shadow casting for electron microscopy using the Tübingen Bosch microscope. Compared with untreated staphylococci (Abb. 17), the treated bacteria (Abb. 18-Abb. 21) show diverse degrees of cell damage (Jacherts, 1960) (*with permission of SPRINGER NATURE*).

uary 1945, and served then as an assistant professor at Kossel's Institute. In February 1945, the institute left for the West by ship and after an odyssey via Kiel and Schmalkalden, Möllenstedt was briefly detained by US Forces in Heidenheim. In 1947, Möllenstedt joined Ernst Brüche's Süddeutsche Laboratorien in Mosbach, helped to develop the SDL/AEG electron microscope and studied interference phenomena of electrons. In 1950, he became Dozent at the natural science faculty of the University of Tübingen and in 1960 full professor and Director of the Institute for Applied

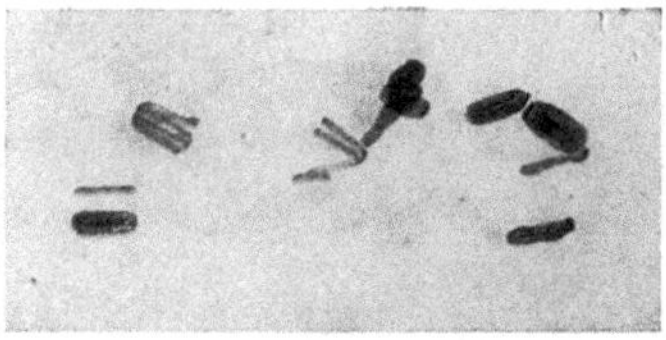

Abb. 1 u. 1a.* Polyedervirus-Partikel von *P. dispar* in Wasser. Bündelförmige Virus-aggregate in Auflösung zu Viruseinzel-stäbchen.

(a)

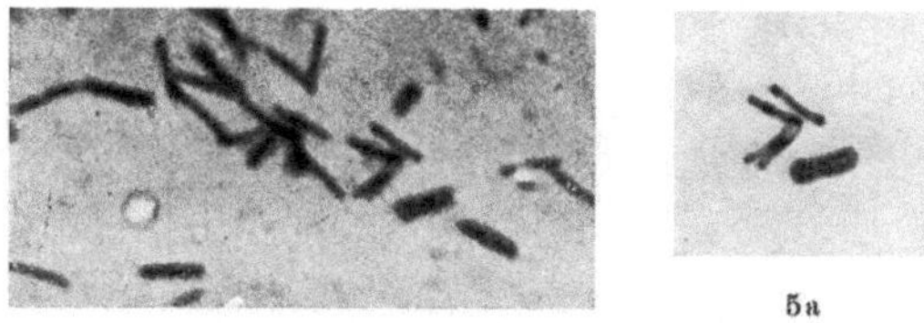

5a

5

Abb. 5 u. 5a. Polyedervirus-Partikel von *B. mori* in Wasser. Doppel- und Einzelstäbchen.

(b)

Figure 32 Baculoviruses (nuclear polyhedrosis viruses) are DNA containing agents attacking insects, e.g. the moths of *Lymantria dispar* and Bombyx *mori*: The rod-shaped virions and crystalline virus aggregates were isolated and characterized by Gernot Bergold at the KWI for Biology and Biochemisty in Tübingen 1943-1949. (a) Micrographs taken in Mosbach using the SDL-AEG microscope, (b) taken in Stuttgart using the Bosch microscope (Bergold, 1948) (*with kind permission of Walter de Gruyter publishing house, Berlin*).

Physics. The Möllenstedt–Düker biprism allowed him to study the wave optics of electrons. His institute focused on holography and electron- and ion beam-lithography (Lehmann & Lichte, 2002; Möllenstedt & Düker, 1956).

The virus group around the 1939 Nobel prize winner in Chemistry Adolf Butenandt (1903-1995), comprising the biochemist Gerhard Schramm (1910-1969), the biologist Hans Friedrich-Freksa (1906-1973), the veterinarian Werner Schäfer (1912-2000) and the insect virologist Gernot Bergold (1911-2003) by 1948, was basically familiar with and thus highly interested in electron microscopy, e.g., Friedrich-Freksa and Schramm were co-authors in the 1941 immune-electron microscopy paper (von Ardenne et al., 1941). In autumn 1943, when the bombing raids started to damage Berlin, Butenandt, Schramm and Friedrich-Freksa, moved to Tübingen, as Freksa was familiar with the small University town based on his previous studies. The "Arbeitsstätte für Virusforschung der Kaiser-Wilhelm-Institute für Biochemie und Biologie" was initially accommodated at different places in town (Munk, 1995). By order of the Allied Military Government the Kaiser Wilhelm Society was renamed the Max Planck Society in 1949. In 1950, the virus group moved to its own new building in Tübingen, Melanchtonstrasse 36, installed as the Max Planck Institute for Virus Research in 1954, and later moved to a new campus in the Spemannstrasse in 1959.

As early as 1948 Bergold had published on insect-pathogenic viruses (Fig. 32). More than 100 electron micrographs were taken by Brüche and

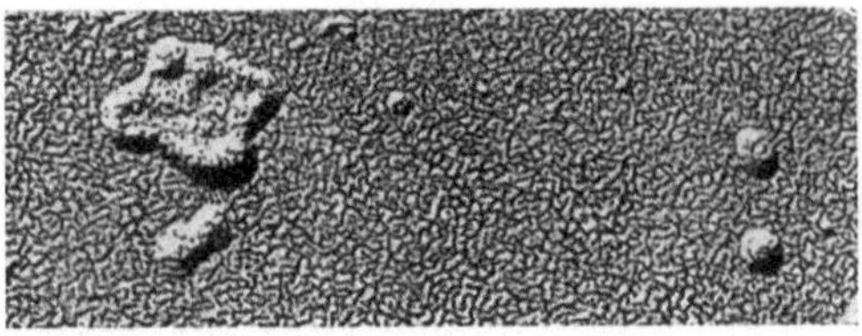

Abb. 1. Klassisches Geflügelpestvirus (Einzelteilchen und Aggregate). Goldbedampfung. Vergr. 25 000-fach. Aufgenommen mit dem elektrostatischen Mikroskop der S.D.L.

(a)

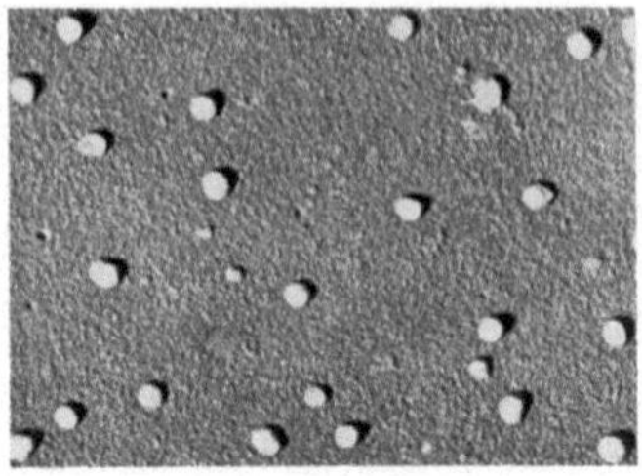

Abb. 3. K.P.-Virus aus 0,09-*m*. NaCl, 0,01-*m*. Phosphatpuffer-Lösung. Fixiert mit Osmiumsäure. Palladiumbedampfung im Winkel von 25°. Elektronenoptische Vergrößerung 5000 ×, Gesamtvergrößerung 30 000 ×.

(b)

Figure 33 Fowl plague virus ('Klassische Geflügelpest'), a highly pathogenic influenza virus, was purified by ultracentrifugation and characterized by Werner Schäfer and colleagues at the KWI for Biochemistry in Tübingen 1948-1955. Viruses appeared flattened after air-drying and shadow-casting with heavy metal at an angle of 25°. (a) Virus aggregates (left) and single virions (right) taken in Mosbach using the SDL-AEG microscope (Schäfer & Schramm, 1949) and (b) of osmium-vapor fixed viruses in Tübingen using a Zeiss EM 8 (Schäfer et al., 1952) (*with kind permission of Walter de Gruyter publishing house, Berlin*).

Seeliger at SDL in Mosbach and by Rühle and Zehender at Bosch in Stuttgart (Bergold, 1948).

The acquisition of the Zeiss EM 8 microscope facilitated a number of fundamental studies on fowl plague virus (Schäfer & Schramm, 1949, 1950; Schäfer et al., 1952, 1954) (Fig. 33) and tobacco mosaic virus (Schramm & Wiedemann, 1951; Schramm et al., 1955a, 1955b) (Fig. 34).

The gifted technical assistant Georg Berger (1913-1999) took care of the instrument and helped the academic users. From 1965 on the Electron Microscopy Laboratory was run diligently by Hermann Frank (1924-2015). Between 1968 and 1985, Frank together with the Virus Department of Werner Schäfer and a number of motivated young scientists - including the authors of this review - focused on essential structure-function-aspects in retrovirology.

19. Giving up the commercial electron microscope

We do not know who in the Bosch hierarchy decided to develop the internal tool into a commercial instrument. Rudolf Rühle's achievements in ESP/EAP 9 predestined him to guide production as well as promotion of the instrument. He did this with full personal commitment. Nevertheless, the microscope did not fit into the portfolio of the automotive company.

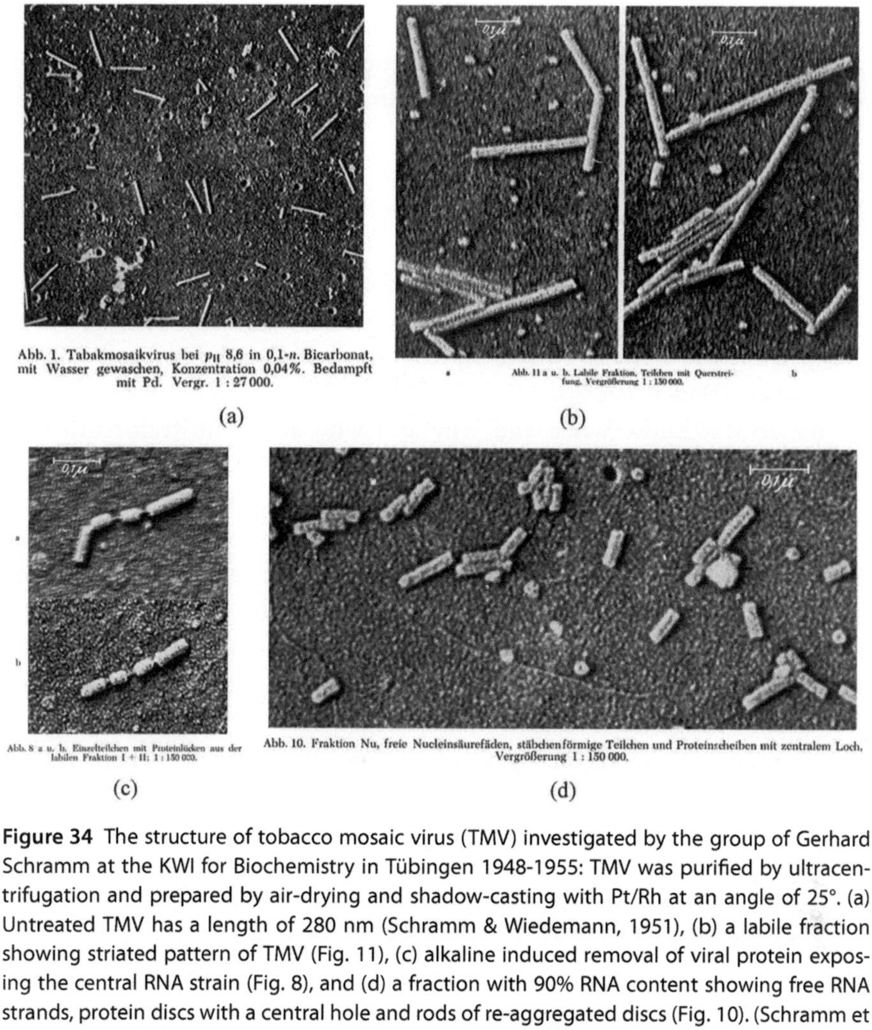

Figure 34 The structure of tobacco mosaic virus (TMV) investigated by the group of Gerhard Schramm at the KWI for Biochemistry in Tübingen 1948-1955: TMV was purified by ultracentrifugation and prepared by air-drying and shadow-casting with Pt/Rh at an angle of 25°. (a) Untreated TMV has a length of 280 nm (Schramm & Wiedemann, 1951), (b) a labile fraction showing striated pattern of TMV (Fig. 11), (c) alkaline induced removal of viral protein exposing the central RNA strain (Fig. 8), and (d) a fraction with 90% RNA content showing free RNA strands, protein discs with a central hole and rods of re-aggregated discs (Fig. 10). (Schramm et al., 1955b) (*with kind permission of Walter de Gruyter publishing house, Berlin*).

Bosch produced only three electron microscopes (Seeger, 1994). According to the Hohenheim-DFG expertise and the opinion of the united experts in Tübingen (Bader, 1954c; Maiwald, 1952) the Bosch product was considered absolutely equivalent to the Zeiss EM 8 and the S&H ÜM 100. Its usefulness was emphasized by Otto Scherzer, and Werner Schreil in letters to Professor Bader (Scherzer, 1954; Schreil, 1955). But Bosch was a newcomer on the market while the instruments of S&H and AEG/SDL/Zeiss were known in German science by the end of the 1930s. The Hohenheim

Bosch microscope was ordered by the Deutsche Forschungsgemeinschaft late in November 1952, the Bosch instrument in Tübingen in March 1954 by Professor Bader. Apparently, Bosch produced its microscope "on demand" and it was produced only during a brief period despite the efforts of Rudolf Rühle to make it known and to keep it functional. Also Otto Scherzer assessed the Bosch instrument to be comparable to the S&H ÜM and described the unquestionable ability and cordiality of Rudolf Rühle ("die unzweifelhafte Fähigkeit und Nettigkeit von Herrn Dr. Rühle") as an additional advantage and a guarantee for a functional microscope (Scherzer, 1954), while the Siemens Microscope Service was then known to be nearly nonexistent (Bader, 1954a).

We do not know when and why and who at Bosch decided to stop the project of a commercial microscope. Anyway, Rudolf Rühle accepted this decision probably with quiet relief (Manfred Rühle, personal communication, 2021). Yet, in-house electron microscopy remained essential in Bosch's R&D and this in-house microscope might represent the missing third unit of the series of three mentioned by Seeger (Seeger, 1994). The commercial Bosch group electron microscope was an ambitious, though unfortunately only short-lived experiment. In the authoritative book on the history of the Robert Bosch (Bähr & Erker, 2015) the instrument is not even mentioned and according to a contemporary witness in EAP (G.Z.) any knowledge of the commercial Bosch instrument was already lost by 1954 (Dr. G. Zielasek, personal communication, 2021).

20. A list of Rudolf Rühle's patents

Patentschrift Nr.720870, Klasse 48b, Gruppe 11 03. Dr. rer. nat. Walter Dorn in Stuttgart, Dr. phil. Karl Reger in Stuttgart-Sillenbuch und Dipl.-Ing. Rudolf Rühle in Stuttgart-Zuffenhausen als Erfinder genannt. Robert Bosch G.m.b.H. in Stuttgart: "Verfahren zur Herstellung von Spiegeln". Patentiert im Deutschen Reich vom 25. Februar 1937 an. Patenterteilung bekannt gemacht am 16. April 1942.

Patentschrift Nr. 753 937, Klasse 48b Gruppe 11 03. Dr.-Ing. Rudolf Rühle, Stuttgart ist als Erfinder genannt worden. Robert Boch G.m.b.H., Stuttgart: "Schleusenanlage zum Einbringen von z.B. durch Metallbedampfen zu metallisierenden Gegenständen". Patentiert im Deutschen Reich vom 4. Juli 1942 an.

Patentschrift Nr. 764 927, Klasse 48b, Gruppe 11 03. Dipl.-Ing. Rudolf Rühle als der Erfinder genannt. Robert Bosch G.m.b.H., Stuttgart: "Verfahren zur Verdampfung im Vakuum". Patentiert im Deutschen Reich vom 23. Februar 1939 an. Patenterteilung bekanntgemacht am 2. Mai 1951.

Patentschrift Nr. 878 883, Klasse 48b, Gruppe 11 03. Dipl.-Ing. Rudolf Rühle, Stuttgart-Zuffenhausen. Robert Bosch G.m.b.H., Stuttgart: "Vorrichtung zur Regelung der Verdampfung von Stoffen im Vakuum". Patentiert im Gebiet der Bundesrepublik Deutschland vom 1. Juli 1942 an. Patentanmeldung bekannt gemacht am 18. September 1952. Patenterteilung bekannt gemacht am 23. April 1953.

Patentschrift Nr. 882 174, Klasse 48 b, Gruppe 11 10. Dr.-Ing. Rudolf Rühle, Stuttgart-Zuffenhausen ist als der Erfinder genannt worden. Robert Bosch G.m.b.H.: "Verfahren zum Verdampfen von Stoffen im Vakuum mittels Elektronenstrahlen". Patentiert im Gebiet der Bundesrepublik Deutschland vom 7. Oktober 1942 an. Patentanmeldung bekannt gemacht am 21. Mai 1953.

Patentschrift Nr. 891 590, Klasse 21g, Gruppe 37 10. Dr.-Ing Rudolf Rühle, Stuttgart-Zuffenhausen ist als Erfinder genannt worden. Robert Bosch G.m.b.H., Stuttgart. "Verfahren zum Zuordnen von im Elektronenmikroskop aufgenommenen Stereobildern". Patentiert im Gebiet der Bunderepublik Deutschland vom 12. April 1944 an. Patentanmeldung bekanntgemacht am 18. Dezember 1952. Patenterteilung bekannt gemacht am 20. August 1953.

Patentschrift Nr. 959306. Dr.-Ing. Rudolf Rühle, Stuttgart-Zuffenhausen ist als Erfinder genannt worden. Robert Bosch G.m.b.H., Stuttgart: "Vorrichtung zum Halten und Verschwenken von Objekten in Elektronenmikroskopen oder Beugungsgeräten". Patentiert im Gebiet der Bundesrepublik Deutschland vom 13. April 1943 an. Patenterteilung bekannt gemacht am 17. August 1944

Patentschrift Nr. 972910, Klasse 21 g, Gruppe 37 01. Rudolf Rühle, Stuttgart-Zuffenhausen als Erfinder genannt worden. Für Robert Bosch G.m.b.H., Stuttgart: "Magnetisches Elektronenmikroskop mit veränderbarer Vergrößerung". Patentiert im Gebiet der Bundesrepublik Deutschland vom 16. April 1950 an.

Patentschrift Nr 1 034 285. Dr.-Ing. Rudolf Rühle, Stuttgart-Zuffenhausen, Dr.-Ing. Ernst Zehender, Stuttgart-Vaihingen und Dr.-Ing. Paul Schweitzer, Stuttgart sind als Erfinder genannt. Patentiert für Robert Bosch G.m.b.H., Stuttgart: "Elektronenstrahlröhre mit kalter Kathode". Anmeldetag 26. März 1954.

Patentschrift Nr. 1 061 917. Dr.-Ing. Rudolf Rühle, Stuttgart-Zuffenhausen, ist als Erfinder genannt worden. Patentiert für Robert Bosch G.m.b.H., Stuttgart: "Verfahren und Einrichtung zum Erzeugen von Durchbrüchen in Körpern mittels Korpuskularstahlen". Anmeldetag: 21. November 1959.

Acknowledgments

The authors are particularly grateful to Prof. Manfred Rühle for his patience and diligence in providing us with information, documents, and photos on his father Rudolf Rühle: the "Festschrift" donated to RR in 1971 on his retirement and the collection of RRs VAP-reports handed over to RR in 1982 on occasion of his 75th birthday. We also would like to thank the team at the University Hohenheim Archive, in particular Denisa Kohan, M.A., for the efficient supply of documents on the history of the Bosch microscope and the persons around in Hohenheim, and we have to thank here also Dr. Anna Krupp, Institut für Biologie Hohenheim for a number of Bosch-EM-relevant papers. Sincere thanks are also due to Drs. Anastasia Antipova and Susanne Rieß-Stumm at the Archives of the University Tübingen, for specific information and documents on the early microscopy scene in Tübingen and to Dr. Norbert Becker at the Archives of University Stuttgart supplying information on the studies and achievements of Rudolf Rühle at Stuttgart's Technical College. We would like to thank also Dr. Dietrich Kuhlgatz, Robert Bosch Corporate Communications, Stuttgart, for supplying information on the Bosch microscope and Rudolf Rühle's many activities at Bosch.

Susann Wolff at the Bundesarchiv in Berlin permitted access to the NSDAP membership data of Bodo von Borries and Dr. Astrid Küntzel at the Landesarchiv Nordrhein-Westfalen in Duisburg provided details of his denazification trial. Petra Lewandowski and Sabine Pakusa-Lerch at the Historical Archive of the Stiftung Deutsches Technik Museum in Berlin helped us with documents on Ernst Brüche and the AEG microscope. Photos were provided readily also by the Archives of the Max Planck Society in Berlin, Albrecht H. Kossel, Tübingen, and Wolfgang Gloede, Stockelsdorf. Sincere thanks are given to them all.

Special thanks are due to Prof. Dr. Hans-Joachim Gerth, Tübingen for remembering the early days of virology and of the Bosch microscope in Tübingen und to Dipl.-Ing. Winfried Send, KIT, Karlsruhe for his care for the abandoned Tübingen instrument and sharing information. We are obliged also to Walter Pietrusziak, Deutsche Forschungsgemeinschaft in Bonn for information on early acquisitions of electron microscopes in Germany and Dr. Gotthold Zielasek, Karlsruhe for his impressions on Rudolf Rühle and the quiet end of the commercial Bosch electron microscope.

The Librarians at the Robert Koch Institute in Berlin, Jens Erling, and Markus Schneller at the Max Planck Institute in Tübingen helped us generously to retrieve historic reading, and last, but not least, our sincere thanks are due to Brigitte Sailer, MPI for Biology Tübingen for preparing the final figures. Moreover, and not at least, the authors are deeply indebted to Brigitta Gelderblom and to Anne Schwarz for their abundance of patience.

References

Abbreviations for the Archives approached in this study:
Siemens Corporate Archiv = SAA
Uni Hohenheim Archiv = UAH
Uni Tübingen Archiv = UAT
Uni Stuttgart Archiv = UAST

AEG (1943). AEG Patent Office: Letter to S&H, 12.1.1943: "Regarding the level of licence costs". SAA Nachlass Flir, Sign 11/Lg 177.

Anderson, T. F., & Stanley, W. M. (1941). A study by means of the electron microscope of the reaction between tobacco mosaic virus and its antiserum. *Journal of Biological Chemistry, 139*, 339–344.

Bader, R.-E. (1954a). Letter Bader to Scherzer, 12.01.1954: "Request for advice to select between S&H ÜM and the BOSCH-EM". UAT 376/1.

Bader, R.-E. (1954b). Letter Bader to Scherzer, 29.01.1954: "Answer to Scherzer's advice of 18.01.1954. UAT 376/1.

Bader, R.-E. (1954c). Letter Bader to Klingler at S&H, 23.03.1954: "Why the decision pro Bosch?". UAT 376/63.

Bader, R. E. (1955). Letter Bader to Schreil, 10.01.1955: "Invitation to work at Bosch-EM". UAT 376/1.

Bähr, J., & Erker, P. (2015). *Bosch: History of a global enterprise*. Munich: C.H. Beck (translated by J.A. Underwood).

Barkleit, G. (2006). *Manfred von Ardenne: Selbstverwirklichung im Jahrhundert der Diktaturen*. Berlin: Duncker & Humblot.

Bergold, G. (1948). Bündelförmige Ordnung in Polyederviren. *Zeitschrift für Naturforschung, 3b*, 25–26.

Boersch, H. (1939). Das Schatten-Mikroskop, ein neues Elektronen-Übermikroskop. *Naturwissenschaften, 27*, 418.

Bosch (1947). Certificate for Dr. Rentschler of November 10, 947. UAH personnel files W. Rentschler.

Bosch (1950). *Bosch-Elektronenmikroskop*. Technische Daten.

Bosch (2019). *Automobiltechnik von Bosch: ANALOG, DIGITAL, VERNETZT*. Robert Bosch GmbH.

Brüche, E. (1940). Beilage zum Jahrbuch der AEG Forschung 7: Berichtigung vom 23.12.1940.

Brüche, E. (1943). Zum Entstehen des Elektronenmikroskops. *Physikalische Zeitschrift, 44*, 176–180.

Brüche, E. (1944). Bemerkungen zu vorstehender Zuschrift. *VDI-Zeitschrift, 88*, 687 (comment on: von Borries, B. & Ruska, E., 1944a).

Brüche, E. (1949). Neues vom Elektronenmikroskop. *Umschau in Wissenschaft und Technik, 13*.

Brüche, E., & Haagen, E. (1939). Ein neues, einfaches Übermikroskop und seine Anwendung in der Bakteriologie. *Naturwissenschaften, 27*, 809–811.

Brüche, E., & Johannson, H. (1932). Elektronenoptik und Elektronenmikroskop. *Naturwissenschaften, 20*, 353–358.

Brüche, E., & Scherzer, O. (1934). *Geometrische Elektronenoptik: Grundlagen und Anwendungen*. Berlin: Julius Springer.

Busch, H. (1927). Über die Wirkungsweise der Konzentrierungsspule bei der Braunschen Röhre. *Archiv für Elektrotechnik, 18*, 583–584.

De Broglie, L-V. (1925). *Recherches sur la théorie des quanta* (Thesis). Paris: Masson; Published at *Annales de Physique (Paris), 3* (1924), 22–128.

Deutsche Forschungsgemeinschaft (1952). Bestellung an Firma Robert Bosch GmbH. UAH, Signatur 3_08b. 6(7).

Driest, E., & Müller, H. O. (1935). Elektronenmikroskopische Aufnahmen (Elektronenmikrogramme) von Chitinobjekten. *Zeitschrift für Wissenschaftliche Mikroskopie, 52*, 53–57.

Emde, F., & Rühle, R. (1934). Die Debyeschen halbkonvergenten Reihen für die Zylinderfunktionen in Komplexen. *Jahresbericht der Deutschen Mathematiker-Vereinigung, 43*, 251–270.

Erbe (1949). Letter Erbe to Stickl, 12.04.1949: "Lack of money for ÜM". UAT 376/62.

Fischer, E., & Hamann, K. (1957). Elektronenmikroskopische Untersuchungen an Anstrichfilmen. *Farbe und Lack, 63*, 209–222.

Fischer, E. W., & Rentschler, W. (1957). Zur Verbesserung der Bodenstruktur durch synthetische Hochpolymere. *Zeitschrift für Pflanzenernährung, Düngung und Bodenkunde, 76*, 232–243.

Flir, D. (1940a). Internal note, 04.10.1940: "Differences Dr. Ruska vs. Prof. Ramsauer". SAA Nachlass Flir, Sign. 11/Lg 602.

Flir, D. (1940b). Summary of a meeting on S&H's relation to von Ardenne, 30.11.1940. SAA Nachlass Flir, Sign. 11/Lg 602.

Flir, D. (1940c). Internal note to Dr. von Buol, 22.12.1940: "Virology at KWI?". SAA Nachlass Flir, Sign. 11/Lg 602.

Flir, D. (1941). Note to Dr. v. Buol, 11. August 1941: "Low quality prints". SAA Nachlass Flir, Sign. 11/602.

Freundlich, M. M. (1963). Origin of the electron microscope. *Science, 142*, 185–188.

Gabor, D. (1957). Die Entwicklungsgeschichte des Elektronenmikroskops. *Elektrotechnische Zeitschrift, 78*, 522–530.

Gabor, D. (1974). The history of the electron microscope, from ideas to achievements. In *Proc. 4th internat. conf. electron microscopy, Canberra* (pp. 6–12).

Gaudilliere, J.-P. (2013). Changing the scale: Slides and electron microscopy at the virus laboratory of the Pasteur Institute. *History and Philosophy of the Life Sciences, 35*, 395–414.

Gelderblom, H. R. (2019). Bodo von Borries (1905-1956): Herforder und Friederizianer – Ein Pionier der Elektronenmikroskopie. In *Historisches Jahrbuch 2020 für den Kreis Herford: Band 27*. Bielefeld: Verlag für Regionalgeschichte (pp. 154–199).

Gelderblom, H. R., & Krüger, D. H. (2014). Helmut Ruska (1908–1973): His role in the evolution of electron microscopy in the life sciences, and especially virology. In *Advances in Imaging and Electron Physics: Vol. 182* (pp. 1–94).

Giuntini, J., Lépine, P., Nicolle, P., & Croissant, O. (1947). Images électroniques de quelques bactériophages et détermination de leur taille. *Annals de l'Institut Pasteur, 73*, 579–581.

Glasers Annalen (1941). Entwicklung und Ergebnisse der Elektronenmikroskopie. Comment by the Editor of 24.03.1941. *Zeitschrift für Verkehrstechnik und Maschinenbau, 10*.

Haguenau, F., Hawkes, P. W., Hutchison, J. L., Satiat-Jeunemaitre, B., Simon, G. T., & Williams, D. B. (2003). Key events in the history of electron microscopy. *Microscopy and Microanalysis, 9*(9), 96–138.

Jacherts, D. (1959). Untersuchungen über die Bildung von Geißeln von Pseudomonas aeruginosa in synchronisierten Kulturen. *Zentralblatt für Bakteriologie, II. Abteilung, 113*, 111–113.

Jacherts, D. (1960). Beobachtungen über die gegenseitige Beeinflussung von Zellen eines Makroorganismus und pathogenen Mikroorganismen an einem Infektionsmodell. *Zeitschrift für Hygiene und Infektions-Krankheiten, 147*, 169–200.

Jakob, A. (1941). Das elektrostatische Übermikroskop und seine Leistungsfähigkeit. *Zeitschrift für das gesamte Krankenhauswesen, 1941*, 8–11.

Jakob, A., & Mahl, H. (1940). Strukturdarstellung bei Bakterien, insbesondere die Kapseldarstellung bei Anaerobiern mit dem elektrostatischen Elektronen-Übermikroskop. *Archiv für Experimentelle Zellforschung, 24*, 87–104.

Karrenberg, F. H. (2019). *In the light of the electron microscope, in the shadow of the Nobel prize.* Norderstedt, Germany: Books on Demand.

Knoll, M., & Ruska, E. (1932a). Beitrag zur geometrischen Elektronenoptik (I, II). *Annalen der Physik (Leipzig), 404*, 607–661.
Knoll, M., & Ruska, E. (1932b). Das Elektronenmikroskop. *Zeitschrift für Physik, 78*, 318–339.
Krause, F. (1937). Das magnetische Elektronenmikroskop und seine Anwendung in der Biologie. *Naturwissenschaften, 25*, 817–825.
Krüger, D. H., Schneck, P., & Gelderblom, H. R. (2000). Helmut Ruska and the visualisation of viruses. *The Lancet, 355*, 1713–1717.
Kuhlgatz, D. (2021). Information on patents of R. Rühle and W. Rentschler from the Bosch patent office. Personal communication.
Kultminister (1939). Letter Ministry to Rektor, 13.06.1939: "Approval of the Tübingen ÜM". UAT 117/740.
Lehmann, M., & Lichte, H. (2002). Tutorial on off-axis electron holography. *Microscopy and Microanalysis, 8*, 447–466.
Levaditi, C., & Bonét-Maury, P. (1942). Les ultravirus: considerés a travers le microscope électronique. *La Presse Medicale*, 24.02.1942. (pp. 203–207).
Lieth, H., & Meyer, G. F. (1957). Über den Bau der Pigmentgranula bei den Myxomyceten. *Naturwissenschaften, 44*, 449–450.
Lin, Q. (1995). *Zur Frühgeschichte des Elektronenmikroskops*. Stuttgart: GNT-Verlag.
Löffler, S., & Rühle, R. (1942). Vergleich von Elektrolytkondensatorfolien unter Mitverwendung von elektronenmikroskopischen Untersuchungen. VAP Report no. 262, 27.02.1942. Courtesy Manfred Rühle, 2021.
Mahl, H. (1939a). Diatomeenaufnahmen mit dem elektrostatischen Übermikroskop. *Naturwissenschaften, 27*, 417.
Mahl, H. (1939b). Über das elektrostatische Elektronenmikroskop hoher Auflösung. *Zeitschrift für Technische Physik, 20*, 316–317.
Mahl, H. (1940a). Das elektrostatische Elektronen-Übermikroskop. *Jahrbuch der AEG-Forschung*, 7, 43–56.
Mahl, H. (1940b). Ein plastisches Abdruckverfahren zur übermikroskopischen Untersuchung von Metalloberflächen. *Metallwirtschaft, 19*, 488–491.
Maiwald, K. (1952). Letter Rektor to DFG, 15.11.1952: "Sincere thanks for funding of the Bosch microscope". UAH, Sign 3 08b.6(7).
Marton, L. (1934). Electron microscopy of biological objects. *Nature, 133*, 911.
Marton, L. (1968). *Early history of the electron microscope*. San Francisco Press.
Meyer, G. F. (1957). Elektronenmikroskopische Untersuchungen an den Apathyschen Neurofibrillen von *Hirudo medicinalis*. *Zeitschrift für Zellforschung, 45*, 538–542.
Möllenstedt, G., & Düker, H. (1956). Beobachtungen und Messungen an Biprisma-Interferenzen mit Elektronenwellen. *Zeitschrift für Physik, 145*, 377–397.
Müller, F. (2008). Scientific-technological system as a family business: the Ruskas, Bodo von Borries and the electron microscope. In A. Presas I Puig (Ed.), *Who is making science? Scientists as makers of technical-scientific structures and administrators of science policy* (pp. 21–33). Max Planck Institute for the History of Science. Preprint 361.
Müller, F. (2009). The birth of a modern instrument and its development during World War II. Electron microscopy in Germany from the 1930s to 1945. In A. Maas, & H. Hooijmaijers (Eds.), *Scientific research in World War II: What scientists did in the war* (pp. 121–146). London: Rouledge.
Müller, F. (2019). Weak and strong knowledge in industrial research: The rise of the "third physicist". In M. Epple, A. Imhausen, & F. Müller (Eds.), *Weak knowledge: Forms, functions, and dynamics* (pp. 231–261). Frankfurt/Main: Campus Verlag GmbH.
Müller, F. (2021). *Jenseits des Lichts: Siemens, AEG und die Anfänge der Elektronenmikroskopie in Deutschland. Reihe: Deutsches Museum. Abhandlungen und Berichte – Neue Folge: Band 35.* Wallstein Verlag, Göttingen.

Munk, K. (1995). *Virologie in Deutschland: die Entwicklung eines Fachgebietes.* S. Karger Freiburg/Basel.

Nobel Prize in Physics (1986). NobelPrize.org. Nobel Media AB 2021. Sun. 4 Apr 2021.

Oatley, C. W. (1982). The early history of the scanning electron microscope. *Journal of Applied Physics, 53*, R1–R13.

Raettig, H.-J. (1954). Die morphologische Wechselwirkung zwischen Phagen und Bakterien und die Lysosome als neue Formelemente. *Zentralblatt für Bakteriologie, I. Abteilung, Originale, 161*, 260–268.

Ramsauer, C. (1940). Letter to Dr. Trettin at Siemens: Limits to the "difficult" Ernst Ruska, 29.06.1940. SAA 11/ 602.

Ramsauer, C. (Ed.). (1941). *Zehn Jahre Elektronenmikroskopie: Ein Selbstbericht des AEG-Forschungs-Instituts* [Ten years of EM. A self portrait of the AEG Research Institute]. Berlin: Springer.

Ramsauer, C. (Ed.). (1943). *Elektronenmikroskopie: Bericht über Arbeiten des AEG-Forschungsinstituts 1930-1942.* Berlin: Springer.

Rang, O. (1992). Zur Elektronenmikroskopie – kurz nach 1945 – Erinnerungen eines Zeitzeugen. *Elektronenmikroskopie, 5*, 25–32.

Reichsforschungsrat (1942). Letter to Uni Tübingen, 14.03.1942: "ÜM is a war-deciding acquisition". UAT, Sign. 17/740.

Rektor Uni Tübingen (1939). Letter Rektor to S&H, 14.07.1939: "Order of the ÜM for two faculties". UAT, Sign. 117/740.

Rentschler, W. (1938). Elektronenmikroskop. Bosch-VAP Report Nr. 158, 29.09.1938. Copy courtesy Prof. Manfred Rühle, 2021.

Rentschler, W. (1953). Letter Rentschler to DFG, Bad Godesberg: Erster Jahresbericht über das der Landwirtschaftlichen Hochschule Hohenheim von der Deutschen Forschungsgemeinschaft zur Verfügung gestellte Elektronenmikroskop (First Annual Report on the Bosch EM in Hohenheim), 18.12.1953. UAH, Sign 3_08b.6 (7).

Rentschler, W. (1957). Rentschler to DFG, Bad Godesberg: 4. Jahresbericht über das der Landwirtschaftlichen Hochschule Hohenheim von der Deutschen Forschungsgemeinschaft zur Verfügung gestellte Elektronenmikroskop (4th Annual Report on the Bosch EM in Hohenheim), 16.10.1957. UAH, Sign 3_08b.6 (7).

Rentschler, W. (1963). Letter Rentschler to Rektor Uni Hohenheim, 11.07.1963: "Take over of the Bosch TEM". UAH, Sign 3_08b.6 (7).

Rüdenberg, R. (applied for on May 31, 1931a). Anordnung zur Beeinflussung des Verlaufs der Elektronenstrahlen durch elektrisch geladene Feldblenden. Patent No. 895635.

Rüdenberg, R. (applied for on May 31, 1931b). Anordnung zur vergrößernden Abbildung von Gegenständen mittels Elektronenstrahlen und mittels den Gang der Elektronenstrahlen beeinflussender elektrostatischer oder elektromagnetischer Felder. Patent No. 889660.

Rüdenberg, R. (applied for on May 31, 1931c). Anordnung zum vergrößernden Abbilden von Gegenständen mittels Elektronenstrahlen. Patent No. 906737.

Rüdenberg, R. (1932). Elektronenmikroskop. *Naturwissenschaften, 20*, 522.

Rühle, R. (1944). Fortschritte in den Untersuchungsmethoden am Elektronenmikroskop. (Dissertation Stuttgart Technische Hochschule).

Rühle, R. (1949a). *Das Elektronenmikroskop.* Stuttgart: Curt E. Schwab.

Rühle, R. (1949b). Zur Justierung des Strahlsystemes am Elektronenmikroskop. *Optik, 5*, 463–468.

Rühle, R. (1949c). Raumbilder am Elektronenmikroskop. *Optik, 5*, 534–543.

Rühle, R. (1950a). Übersichts- und Feinstrahlbeugungsbilder an einem Elektronenmikroskop mit Hilfsprojektiven. *Optik, 7*, 267–278.

Rühle, R. (1950b). Über Gesetzmäßigkeiten in Texturaufnahmen von Elektronenbeugungsbildern. *Optik, 7*, 279–283.

Rühle, R. (1971). Die zentrale Forschung in der Bosch-Gruppe. *BOSCH Technische Berichte*, *3*, 3–9.

Rühle, R., & Zehender, E. (1953). Die Beurteilung von Objekten im Elektronenmikroskop, die während der Beobachtung geschwenkt werden können. *Optik*, *10*, 111–115.

Ruska, E. (1934). Über Fortschritte im Bau und in der Leistung des magnetischen Elektronenmikroskops. *Zeitschrift für Physik*, *87*, 580–602.

Ruska, E. (1940). Letter to Rektor University Tübingen, 31.01.1940: "War-induced delays in the ÜM production". UAT, Sign. 117/ /740.

Ruska, E. (1943). Letter to Flir, 04.03.1943: "S&H patents in the fields of electron microscopy". SAA Nachlass Flier 1939-1944. Sign. 11/Lg 177.

Ruska, E. (1946). Letter E. Ruska to Stickl, 06.03.1946: "Offer ÜM 100B, personal letter". UAT, Sign. 376/62.

Ruska, E. (1949). Letter E. Ruska to Stickl, 09.11.1949: "Offer and note on first postwar ÜM". UAT, Sign.376/62.

Ruska, E. (1950). Über neue magnetische Durchstrahlungs-Elektronenmikroskope im Strahlspannungsbereich von 40...220 kV. *Kolloid-Zeitschrift*, *116*, 102–119.

Ruska, E. (1952). Über die Technik der Elektronenmikroskopie und das neue Siemens-Übermikroskop für 100 kV. *Siemens-Zeitschrift*, *26*, 6–15.

Ruska, E. (1954). Ein hochauflösendes 100-kV-Elektronenmikroskop mit Kleinfelddurchstrahlung. In *Proceedings of the third international conference on electron microscopy* (pp. 673–693). London: Royal Microscopical Society.

Ruska, E. (1979). Die frühe Entwicklung der Elektronenlinsen und der Elektronenmikroskopie. In *Acta Historica Leopoldina: Vol. 12*.

Ruska, E. (1980). *The early development of electron lenses and electron microscopy*. Stuttgart: Hirzel. Translated by T. Mulvey.

Ruska, H. (1939). Übermikroskopische Darstellung organischer Struktur (vom Größenbereich der Zelle bis zum Ultravirus). *Archiv für Experimentelle Zellforschung*, *22*, 673–680.

Ruska, H., von Borries, B., & Ruska, E. (1939). Die bedeutung des Übermikroskopes für die Virusforschung. *Archiv für die Gesamte Virusforschung*, *1*, 155–169.

Schäfer, W., & Schramm, G. (1949). Die Reindarstellung des Virus der klassischen Geflügelpest. *Zeitschrift für Naturforschung*, *4b*, 123–124.

Schäfer, W., & Schramm, G. (1950). Über die Isolierung und Charakterisierung des Virus der klassischen Geflügelpest. *Zeitschrift für Naturforschung*, *5b*, 91–102.

Schäfer, W., Munk, K., & Armbruster, O. (1952). Eigenschaften tierischer Virusarten untersucht an den Geflügelpestviren als Modell. III. Weitere Untersuchungen über die physikochemischen und morphologischen Eigenschaften der Geflügelpestviren. *Zeitschrift für Naturforschung*, *7b*, 29–33.

Schäfer, W., Zillig, W., & Munk, K. (1954). Isolierung und Charakterisierung hämagglutinierender, nicht-infektiöser Einheiten bei klassischer Geflügelpest. Ein Beitrag zur Kenntnis der, inkompletten Formen'. *Zeitschrift für Naturforschung*, *9b*, 329–340.

Scherzer, O. (1954). Letter to Bader, 18.01.1954: "A comparison of S&H, EM 8, BOSCH-EM". UAT, Sign. 376/1.

Schimmel, G. (1996). The history of the German society of electron microscopy. In *Advances in Imaging and Electron Physics: Vol. 96* (pp. 149–170).

Schimmel, G. (1998). Das Physikalische Laboratorium Mosbach. Erinnerungen an Prof. Dr. Ernst Brüche. *Elektronenmikroskopie*, *16*, 26–32.

Schramm, G., & Wiedemann, M. (1951). Größenverteilung des Tabakmosaikvirus in der Ultrazentrifuge und im Elektronenmikroskop. *Zeitschrift für Naturforschung*, *6b*, 379–383.

Schramm, G., Schumacher, G., & Zillig, W. (1955a). An infectious nucleoprotein from Tobacco Mosaic Virus. *Nature*, *157*, 549–550.

Schramm, G., Schumacher, G., & Zillig, W. (1955b). Über die Struktur des Tabakmosaikvirus. III. Mitt.: Der Zerfall in alkalischer Lösung. *Zeitschrift für Naturforschung, 10b*, 481–492.
Schreil, W. (1955). Letter to Bader, 05.04.1955: "Comparison of 3 EM instruments". UAT, Sign. 376/1.
Schulze, D. (2001). Ernst Brüche zum 100. Geburtstag. *Elektronenmikroskopie, 20*, 32–35.
Schütz (1942). Letter Staatsekretär Conti to Rektor Uni Tübingen, 20.04.1942: "Questions regarding the purpose of the ÜM". UAT, Sign. 117/740.
Seeger, A. (1994). Elektronenmikroskopie in Stuttgart. *Elektronenmikroskopie, 9*, 16–23.
Seiler, H., Schreiber, P., Wieser, P., & Kühme, H. (1981). Professor Walter Rentschler 70 Jahre. *Physikalische Blätter, 37*, 72–73.
Siebeck, R. (1936). Gutachten vom 02.10.1936 zu möglichen Fortschritten für die ärztliche Wissenschaft und Praxis durch das Elektronenmikroskop. In *Ernst Ruska (1979). Die frühe Entwicklung der Elektronenlinsen und der Elektronenmikroskopie. Acta Historica Leopoldina: Vol. 12* (pp. 123–124) (Anhang F).
Siebeck, R. (1941). *Das Übermikroskop als Forschungsmittel.* Vorträge, gehalten am 18. April 1940 anlässlich der Eröffnung des Laboratoriums für Übermikroskopie der Siemens & Halske AG, Berlin. Walter de Gruyter & Co, Berlin.
Siemens & Halske (1939). Letter to Rektor Uni Tübingen, 19.07.1939: "Acceptance of the order of ÜM". UAT, Sign. 117/740.
Siemens & Halske (1942). Letter S&H to Uni Tübingen, 27.05.1942: "Delivery note for the Tübingen ÜM". UAT, Sign. 117/740.
Siemens & Halske (1947). Letter S&H to Stickl, 07.01.1947: "Request for raw materials". UAT, Sign. 376/62.
Siemens & Halske (1953). Letter S&H to Prof. Bader, 15.12.1953: "Invitation to Berlin to compare ÜM E and F". UAT, Sign. 376/62.
Siemens & Halske (1954). Letter Klingler S&H to Prof. Bader, 10.02.1954: "Another invitation of Bader to S&H". UAT, Sign. 376/62.
Siemens & Halske (1955). Letter S&H to Ernst Ruska, 28.01.1955: "Bosch EM and licence fees". SAA, personnel file Ernst Ruska.
Steigerwald, K.-H. (1949). Ein neuartiges Strahlerzeugungs-System für Elektronenmikroskope. *Optik, 5*, 469–478.
Stickl, O. (accessed August 19, 2021). CV at website http://www.leo-bw.
Stickl, O. (1939a). Letter to Rektor, 13.07.1939: "Information to Med. Faculty on the S&H ÜM". UAT, Sign. 117/740.
Stickl, O. (1939b). Letter Stickl to Rektor, 08.07.1939: "S&H offer for the ÜM". UAT, Sign. 117/740.
Stickl, O. (1942). Letter Stickl to Dr. Schütz, 15.06.1942: "ÜM is of strategic importance". UAT, Sign. 117/740.
Stickl, O. (1946a). Letter to Rektor Uni Tübingen, 06.03.1946: "ÜM as a war reparation removed to France". UAT, Sign. 376/62.
Stickl, O. (1946b). Letter Stickl to S&H, 21.03.1946: "Order for the new type UM 100B". UAT, Sign. 376/62.
Stickl, O. (1950a). Letter Stickl to Dekan, 07.01.1950: "Request for funding a S&H ÜM, 120.000 DM". UAT, Sign. 376/60.
Stickl, O. (1950b). Letter Stickl to Dekan, 19.10.1950: "New request for funding an ÜM, 120.000 DM". UAT, Sign. 376/ 60.
Stickl, O. (1951). Letter Stickl to Erbe, 02.06.1952: "Reasons for "own" EM and for two different instruments in Tübingen". UAT, Sign. 376/62.
Typke, D. (2022). Otto Scherzer and his contributions to electron microscopy. In *Advances in Imaging and Electron Physics: Vol. 221*. Chapter 12 (this volume).

UAH archive (2021): Personnel files, Walter Rentschler: Among other items: "Verzeichnis der wissenschaftlichen Veröffentlichungen des Instituts" in: "Aufgaben des Instituts für Physik in Lehre und Forschung 1967".

UAT archive (2021): Cumulative list of publications of the Hygiene-Institut, UAT 376/21.

van Gorkom, J. (2018). The early electron microscopes: A critical study. In *Advances in Imaging and Electron Physics: Vol. 205* (pp. 1–137).

VAP reports (1934–1944). A selection of unpublished Bosch internal papers (## 26, 30, 47, 102, 122, 123, 152, 157, 159, 160, 161, 228, 249, 327) in private memorial book "BOSCH Versuchsberichte 1934-44" dedicated to Rudolf Rühle on occasion of his 75th birthday in 1982. Courtesy Manfred Rühle, 2021.

VDI (1944). Zur Entwicklung des Elektronen-Übermikroskops. *VDI-Zeitschrift, 88*, 687.

von Ardenne, M. (1938). Das Elektronen-Rastermikroskop. Theoretische Grundlagen. *Zeitschrift für Physik, 109*, 553–572;
von Ardenne, M. Praktische Ausführung. *Zeitschrift für Technische Physik, 19* (1938), 407–416.

von Ardenne, M. (1940a). Über ein Universal-Elektronenmikroskop für Hellfeld, Dunkelfeldfeld und Stereobetrieb. *Zeitschrift für Physik, 115*, 339–386.

von Ardenne, M. (1940b). *Elektronen-Übermikroskopie, Physik-Technik-Ergebnisse*. Berlin: Springer Verlag.

von Ardenne, M. (1944). Über ein neues Universal-Elektronenmikroskop mit Hochleistungsmagnet-Objektiv und herabgesetzter thermischer Objektbelastung. *Kolloid-Zeitschrift, 108*, 195–208.

von Ardenne, M. (1985). On the history of scanning electron microscopy, of the electron microprobe and of the early contributions to transmission electron microscopy. In *The beginning of electron microscopy. Adv. electron. electron phys.: Vol. Suppl 16* (pp. 1–21). Reprinted in this volume.

von Ardenne, M. (1997). *Erinnerungen, fortgeschrieben*. Düsseldorf: Droste Verlag.

von Ardenne, M., Friedrich-Freksa, H., & Schramm, G. (1941). Elektronenmikroskopische Untersuchungen zur Präzipitinreaktion von Tabak-Mosaikvirus mit Kaninchenantiserum. *Archiv für die Gesamte Virusforschung, 2*, 80–86.

von Borries, B. & Ruska, E. (1932). Magnetische Sammellinse kurzer Feldlänge. Deutsche Patentschrift Nr. 680284. Patent applied for on 17.03.1932, granted on 03.08.1939.

von Borries, B. (1939a). Bodo von Borries' membership in the NSDAP. Bundesarchiv Berlin, Sign. BArch, R 9361-I/326.

von Borries, B. (1939b). Note of 25.03.1939: "On his discussion with "Leibarzt des Führers" Prof. Karl Brandt". SAA Nachlass Flir, Sign. 11/ Lg 177.

von Borries, B. (1940). Letter to Dir. von Buol, 03.12.1940: "KWI virology institute". SAA Nachlass Flir 1938-1941, Sign. 11/Lg 602.

von Borries, B. (1943). Letter to Dir. Leifer, 04.01.1943: "Capacity of the EM production". SAA Nachlass Flir 1939-1944, Sign. 11/ Lg 177.

von Borries, B. (1944). Letter to S&H 19.04.1944: "Items for an answer to Prof. Rostock". SAA Nachlass Flir 1939-1944, Sign. 11 Lg 177.

von Borries, B. (1946). Documents on Bodo von Borries' denazification process, 06.08.1946. Landesarchiv NRW, Duisburg: Sign. NW 1064-V. Nr. 560.

von Borries, B. (1949). *Die Übermikroskopie. Einführung, Untersuchung ihrer Grenzen und Übersicht über ihre Ergebnisse*. Berlin: Saenger. 416 pages, 70 tables.

von Borries, B., & Ruska, E. (1938). Vorläufige Mitteilung über Fortschritte im Bau und in der Leistung des Übermikroskops. *Wissenschaftliche Veröffentlichungen aus den Siemens-Werken, 17*, 99–106.

von Borries, B., & Ruska, E. (1944a). Neue Beiträge zur Entwicklungsgeschichte des Elektronenmikroskops und des Übermikroskops. *Physikalische Zeitschrift, 45*, 314–326.

von Borries, B., & Ruska, E. (1944b). Zur Entwicklung des Elektronen-Übermikroskops mit elektrostatischen Linsen. *VDI-Zeitschrift, 88*, 686–687.

von Borries, B., Ruska, E., & Ruska, H. (1938a). Übermikroskopische Bakterienaufnahmen. *Wissenschaftliche Veröffentlichungen aus den Siemens-Werken*, *17*, 107–111.
von Borries, B., Ruska, E., & Ruska, H. (1938b). Bakterien und Virus in übermikroskopischer Abbildung. *Klinische Wochenschrift*, *17*, 921–925.
von Borries, H. (1991). Bodo von Borries: Pioneer of electron microscopy. In *Advances in Imaging and Electron Physics: Vol. 81* (pp. 127–176).
von Buol, H. (1942). Letter Flir to von Buol, 20.11.1942: "S&H—AEG negotiations on patents". SAA Nachlass Flir, 1939-1944, Sig. 11/ Lg 177.
von Buol, H. (1943). Letter to Prof. Ramsauer, spring 1943: "The E. Ruska-E. Brüche disputes". SAA Nachlass Flir, 1939-1944, Sign. 11/Lg 177.
Wikipedia (2021a). https://de.wikipedia.org/wiki/Otto_Stickl.
Wikipedia (2021b). https://de.wikipedia.org/wiki/Friedrich_Heinrich_Lüschen.
Wolpers, C. (1991). Electron microscopy in Berlin 1928-1945. In *Advances in Electronics and Electron Physics: Vol. 81* (pp. 211–227).

CHAPTER TWELVE

Otto Scherzer and his contributions to electron microscopy☆

Dieter Typke
Formerly Munich, Germany
e-mail address: harald.rose@physik.tu-darmstadt.de

Contents

1. Introduction 377
2. Hans Busch, "the father of electron optics" 378
3. Otto Scherzer 380
   3.1 Munich 1927-1931 380
   3.2 Berlin and Munich, 1932-1935 381
   3.3 Darmstadt 1935-1940, Scherzer's theorem 382
   3.4 The time during the Third Reich and the struggle against Deutsche Physik 386
   3.5 Peenemünde 1940-1945: radar investigations in war time 387
   3.6 Mosbach, Fort Monmouth, and again Mosbach 1946-1949; the start of aberration correction, "Scherzer Focus" 388
   3.7 Darmstadt, 1949-1982 390
   3.8 Scherzer's personality, other activities 400
   3.9 Scherzer anecdotes 401
Acknowledgments 403
References 403

1. Introduction

Otto Scherzer, born on March 9, 1909, was one of the leading personalities in the early development of electron optics and the electron microscope. Throughout his life, he made important contributions to many aspects of electron microscopy. He was involved almost from the beginning in the development of electrostatic microscopes during his short time, 1932-1934, at the *Zentrales Forschungslaboratorium der AEG* (Central

☆ Soon after completing this article, Dieter Typke fell victim to a malignant growth, too advanced for successful treatment. Professor Harald Rose has readily agreed to see the article through the press as Dieter Typke wished.

Advances in Imaging and Electron Physics, Volume 221
ISSN 1076-5670
https://doi.org/10.1016/bs.aiep.2022.03.012

Research Laboratory of the *Allgemeine Elektricitätsgesellschaft;* AEG: General Electricity Company). There, he was involved in the calculation and design of electron lenses, and he coauthored, with Ernst Brüche, the first comprehensive book on geometrical electron optics. After being appointed professor of theoretical physics at the *Technische Hochschule*[1] Darmstadt in 1936, a landmark contribution was his proof that the spherical and chromatic aberrations of "round" (i.e. axially symmetric) electron lenses cannot be avoided. Because of its basic importance, this proof has been named "The Scherzer Theorem". Overcoming the limitations of these aberrations by calculating, designing and testing aberration correctors remained a major goal of his work until his untimely death in November 1982 at age 73. His friends and former students were very sadly surprised when they heard about his sudden demise, since he had always enjoyed excellent health. If he had lived another 15 years, he would have seen his dreams of corrected microscopes and atomic resolution come true.

2. Hans Busch, "the father of electron optics"

When talking about the "Beginnings of Electron Microscopy", one should start with Hans Busch. He was the first to calculate the electron paths and imaging properties of axially symmetric electro-magnetic fields, in 1926. This approach became the theoretical basis for the development of electron lenses and the electron microscope. As there will be no separate article about Hans Busch in this volume, I will add some remarks about his life and achievements. See Fig. 1.

Born in 1884, Hans Busch studied physics in Strasbourg, Berlin and Göttingen (1904-1907). From 1907 until 1913, he worked as scientific assistant at the Institute for Applied Electronics in Göttingen, where, in 1911, he got his PhD with a thesis entitled *Stabilität, Labilität und Pendelungen in der Elektrotechnik* (Stability, Lability and Oscillations in Electrical Engineering). From 1913 to 1920, he was a scientific collaborator at the *Radiotechnische Versuchsanstalt für Marine und Heer* (Radiotechnical Laboratory of Navy and Army) in Göttingen. During that time, he also spent two years (1914-1916)

[1] Most of the *Technische Hochschulen* had been renamed *Technische Universitäten* by the end of the 20th century. They already had university character (including the right of promotions and *Habilitationen*, in contrast to "high schools" in the US). Promotions different from *Dr. phil.* (*doctor philosophiae*; Doctor of Philosophy, PhD) were and still are common such as *Dr. rer. nat.* (*doctor rerum naturalium;* Doctor of Natural Sciences) or *Dr. ing.* (*Doctor Ingenieur;* Engineering Doctor).

Figure 1 Hans Busch.

right at the Front in WWI, as an artillery officer. After the war, in 1920, he made his *Habilitation* with the thesis *Über die Erwärmung von Drähten in verdünnten Gasen durch den elektrischen Strom* (On warming of wires in dilute gases by an electric current), and he became a private lecturer at the University at Göttingen. He soon moved to the University at Jena, where he was appointed *Extraordinarius Professor* of the Physics Institute in 1922. There, he developed the basics of electron optics and electron lenses (H. Busch, 1926). The idea behind that was a new method to determine e/m_e, the specific charge of the electron (H. Busch, 1922). The experiments were carried out by his PhD student, Fritz Wolf. The method turned out to be suitable for a high-precision measurement (about 0.1 percent statistical error; F. Wolf, 1927).

In 1927 he became technical director of the telecommunications cable factory Oberspree of AEG in Berlin, and in 1930 he was appointed professor of electrical engineering at the *Technische Hochschule, Darmstadt.* There he stayed, with interruptions due to WWII, until his retirement. Even after his retirement, he regularly attended the Physics Colloquium which took place in Scherzer's *Institut für theoretische Physik.*

During WWII he and his group were involved in research and development of data-transfer methods for the *Heeresversuchsanstalt Peenemünde* (the Army Research Center in Peenemünde).

On February 16, 1949, Hans Busch was a cofounder of the *Deutsche Gesellschaft für Elektronenmikroskopie* (DGE; German Electron Microscopy

Society). On the same day he became the first honorary member of the society, he was additionally honored by the title "The Father of Electron Optics". He died on February 16, 1973 in Darmstadt, only a few days before his 89th birthday.

3. Otto Scherzer

3.1 Munich 1927-1931

Born in Passau, Otto Scherzer went to school in Passau and Kempten (Bavaria). After his *Abitur* (examination before leaving for university) in 1927, he studied electrical engineering at the *Technische Hochschule München*. After finishing his bachelor's degree, in autumn 1929, he began study of physics at the *Ludwig-Maximilians-Universität München* (LMU). At the LMU, the famous theoretical physicist Arnold Sommerfeld became his PhD supervisor. At the time, it was not yet required to get a *Diplom* (master's degree) before starting a PhD thesis. After two years, Scherzer finished his thesis entitled *Über die Ausstrahlung bei der Bremsung von Protonen und schnellen Elektronen* (On the radiation at the braking of protons and fast electrons), which was published in *Annalen der Physik* (Scherzer, 1932). On December 16, 1931 he passed his examination with distinction (*summa cum laude*). See Fig. 2.

At the time the examination (called the *Rigorosum*) included various subjects. The following is an anecdote from this examination. One of the examiners was a professor of electrical engineering and his question was "Which part of a high-voltage transfer line is the most important one?". Scherzer's answer was (maybe even knowing what was wanted, since the examiner specialized in insulators): "The wire.", The professor, of course, expected the answer to be "The insulator." As expected, Scherzer's answer provoked the professor, and an intensive discussion ensued.

The brilliance of the young Scherzer in mathematics and physics was astonishing. We may assume that he had already dealt intensively with these subjects while still at school, though there is no information about that. Arnold Sommerfeld's institute certainly provided a vivid and stimulating atmosphere. It was only a few years after the theory of quantum mechanics had been established, and Arnold Sommerfeld was, along with Werner Heisenberg, Max Born and Pascual Jordan in Göttingen, and Erwin Schrödinger in Zürich (later Berlin), one of the central figures in this development, with connections all over the world. His continuously updated book *Atombau und Spektrallinien* (Atomic Structure and Spectral Lines) was

ANNALEN DER PHYSIK

5. FOLGE, BAND 13, HEFT 2, APRIL 1932

Über die Ausstrahlung bei der Bremsung von Protonen und schnellen Elektronen[1])

Von Otto Scherzer

(Mit 3 Figuren)

Übersicht: Die Rechnungen werden im Anschluß an eine Arbeit von A. **Sommerfeld**[2]) durchgeführt. § 1 erbringt den Nachweis der Vollständigkeit des dort verwendeten Orthogonalsystems unter der Voraussetzung, daß die polaren Eigenfunktionen ein vollständiges System bilden. § 2 behandelt den Einfluß der Masse und des Vorzeichens der Ladung auf die Ausstrahlung. In § 3 wird die Intensität der bei der Bremsung von Protonen am nackten Kern zu erwartenden Röntgenstrahlung berechnet. § 4 enthält die Berechnung der Richtungsverteilung der gebremsten Elektronen unter Vernachlässigung der Relativitätskorrektion. Am Schlusse der Arbeit werden die Ergebnisse einer relativistischen Behandlung des Problems mitgeteilt.

---

§ 1. Nachweis der Vollständigkeit des den Rechnungen zugrunde liegenden Orthogonalsystems

Figure 2 Title and abstract of Scherzer's PhD thesis.

world-famous, and the first edition of the supplementary volume on wave mechanics appeared in 1929, the year in which Scherzer started his PhD thesis. Arnold Sommerfeld was also an ingenious and beloved academic teacher who really cared about the young people around him. He used to tell his students: "You should not care whether the integral converges, you should just calculate it!" At the institute, Scherzer shared a room with Hans Bethe (later famous as a theoretical physicist and Nobel Prize winner), who had finished his PhD in 1928 and was now working on his Encyclopedia of Physics article on "Quantum Mechanics of One- and Two-Electron Problems". Another PhD student, E.G. Ramberg from the USA (later at RCA for 30 years and a co-author of one of the first English-language books on the electron microscope (Zworykin et al., 1945), joined the lab one year after Scherzer and became a good friend.

3.2 Berlin and Munich, 1932-1935

Due to the economic difficulties at the time, Sommerfeld was not able to offer Scherzer a position at his institute and therefore recommended him to

Figure 3 Otto Scherzer with O. Hilsch, at the *Physikertagung* 1935.

Carl Ramsauer who headed the *AEG Zentrale Forschungsinstitut* in Berlin. Scherzer became an assistant of Ramsauer and worked in the Physics Laboratory, which was headed by Ernst Brüche. He went back to Sommerfeld's institute to complete his *Habilitation* and became an assistant of Sommerfeld in 1934. He was employed by AEG until the end of 1935. See Fig. 3.

At the *AEG Forschungsinstitut* Scherzer immediately became involved in electron optics, mainly from a theoretical point of view. Brüche and colleagues were experimenting with gas-concentrated electron beams that could easily be visualized and manipulated by external fields (Brüche & Johannson, 1932). Scherzer contributed a theoretical treatment of gas concentration (Scherzer, 1933a). Another effort was calculation of the potential distribution and the trajectories of an electrostatic three-electrode einzel lens. The result turned out to be in very good agreement with experimental data (Johannson & Scherzer, 1933). He then started his work on image aberrations (Scherzer, 1933b), and together with Ernst Brüche wrote the first comprehensive book on electron optics *Geometrische Elektronenoptik* (Brüche & Scherzer, 1934). See Figs. 4 and 5.

3.3 Darmstadt 1935-1940, Scherzer's theorem

In the summer of 1935, Scherzer was appointed lecturer of theoretical physics at the *Technische Hochschule Darmstadt*, replacing Hans Bärwald who

Figure 4 Otto Scherzer with Ernst Brüche; in the background Scherzer's wife and an unknown Scientist; about 1970.

Figure 5 Otto Scherzer, 1973.

Über einige Fehler von Elektronenlinsen.

Von **O. Scherzer** in Darmstadt.

Mit 3 Abbildungen. (Eingegangen am 4. Juni 1936.)

Unmöglichkeit des Achromaten. Die Bildfehler dritter Ordnung. Unvermeidbarkeit der sphärischen Aberration.

Figure 6 On some Aberrations of Electron Lenses.

had been evicted in 1933. On April 30, 1936 he was appointed *Ordinarius Professor* and Director of the *Institut für theoretische Physik*. At the time, he was the youngest physics professor in Germany. There, he continued with his electron-optical calculations (Scherzer, 1936a). Using the "electron trajectory method", he calculated all third-order aberrations occurring in arbitrary axially symmetric electro-magnetic imaging systems (Scherzer, 1937), represented by lengthy integral formulas.

Probably Scherzer's most important contribution to the field of electron microscopy was his proof, in 1936, that the spherical and chromatic aberrations of round (i.e. axially symmetric) lenses cannot be avoided. See Fig. 6. For this proof, in the case of the spherical aberration, he changed the integrand of the aberration integral, enforcing the condition that a real image is formed, by using partial integrations in such a way that the formula only contained quadratic terms with the same sign. The spherical aberration constant, C_s, therefore always being positive, means that electrons that travel farther away from the optical axis always return to the axis in front of the paraxial ones, meaning that focusing by the outer parts of a lens is too strong. For chromatic aberration, he showed that the lens action always becomes stronger if the accelerating voltage of the electrons is decreased. This means that electrons starting from one point but with different velocities cannot all be focused in the same plane. Both statements together have been termed "The Scherzer Theorem" (Scherzer, 1936b). Scherzer's theorem is also valid in the relativistically correct case, as has been shown by H. Rose (Preikszas & Rose, 1995; Rose, 1967). According to a personal comment by Scherzer to Harald Rose, he developed his theorem during "enforced idleness", an air-raid-protection exercise he was ordered to participate in, which lasted an entire night.

In the publication, Scherzer added a footnote referring to an article by W. Glaser (1933), in which Glaser stated that round lenses free of spherical aberration are possible. Scherzer states that this contradicts his own results and he adds: *Zusatz bei der Korrektur: Wie mir Herr Glaser freundlicherweise mitteilt, ist dort der Ausdruck "sphärische Aberration" nicht in der üblichen Bedeutung, sondern in einem übertragenen Sinne zu verstehen* (Addition at proof stage: As Mr. Glaser kindly informs me, the term 'spherical aberration' has not to be understood in its usual meaning but in a metaphorical sense). Even after several years, Glaser was not convinced that Scherzer's proof was correct (Glaser, 1940). Glaser had recognized that, for purely magnetic lenses, the integrand of the spherical aberration integral could be set to zero, from which a condition for the magnetic field distribution resulted. Solving the differential equation for the magnetic induction, $B(z)$, on the optic axis, he obtained a distribution for which the spherical aberration should vanish. The problem was that this field did not form a real image, the condition that Scherzer had enforced for his proof.[2]

It is interesting that the invention of holography by Dennis Gabor in 1947 was triggered by Scherzer's theorem. In an article on the occasion of Scherzer's 60th birthday, Gabor wrote "Holography owes its birth to Professor O. Scherzer. It was his brilliant proof, in 1936, of the theorem that the spherical aberration can never change sign which induced me, in 1947, to look for a way around this fundamental difficulty" (Gabor, 1969).

Having established that the spherical and chromatic aberrations cannot be avoided for the usual round lenses, Scherzer started thinking about various possibilities to improve the situation, the final goal being to get to atomic resolution. As the wavelength of electrons is 30- to 50-times smaller than atomic distances for the accelerating voltages being used, atomic resolution should in principle be achievable, the main obstacle being a spherical aberration constant, C_s, of the order of millimeters. C_s was thus not particularly large; it was just of the order of magnitude of the lens dimensions. The problem was the enormous factor in sizes, about 7 orders of magnitude, between manageable instrument dimensions and the atomic distances that one would like to observe. Reducing the lens dimensions clearly was not a viable option. He considered the option of minimizing the aberration constants by means of suitable field distributions. He himself calculated an electrostatic einzel lens with minimum spherical aberration (Scherzer, 1936a). Later, in the 1950s, his student W. Tretner calculated field distributions that minimized C_s (Tretner, 1959). However, these could be considered theoretical games rather than viable options. Another possibility was to correct C_s in two sections. This was investigated in the 1960s by a *Diplom* student, Müller, in a geometric-optical treatment. Wave-optical

[2] This has been shown later by R. Rebsch and for electrostatic lenses, by A. Recknagel.

calculations were carried out by D. Typke (Scherzer & Typke, 1967). It turned out that the resolution limit could in fact be improved by a factor of 2 to 3, however the strange point-spread function, with four rather strong lobes, would make image interpretation more difficult. The proposal has never been realized.

In conclusion, fully correcting the aberrations would be the best way for attaining the goal of atomic resolution. Because of the political situation and World War II, it took several years until he was able to follow up on this task.

3.4 The time during the Third Reich and the struggle against Deutsche Physik

In a lecture on *Physik im totalitären Staat*, presented 1965 within a lecture series *Deutsches Geistesleben im Nationalsozialismus* (German cultural life during national socialism), Scherzer described the situation in the case of physics during that time (Scherzer, 1965b). From the very beginning of the Third Reich, the *National-Sozialistische Deutsche Arbeiterpartei* (NSDAP) (National-Socialist German Workers Party) tried to get intellectuals to join the Party. In 1934, a few weeks after his *Habilitation*, Scherzer was ordered, together with many other lecturers, to participate in a 10-weeks *Dozentenlager* (lecturer's camp). When arriving at the camp, he found to his astonishment a nameplate on his cabinet with the words *SA-Mann Scherzer*, indicating that he was considered a member of the *Sturmabteilung* (SA; Storming Division), which during the Weimar republic had been the paramilitary organization of the party. In fact, all the other participants also had such nameplates. He had thus, *nolens volens* (involuntarily), become a member of the SA.

According to a Wikipedia article, Scherzer enlisted in the SA on October 1, 1933, and four years later in the NSDAP. He also became a member of the *Deutsche Dozentenbund* (NSDDB; the lecturers' union).[3]

In his 1965 lecture, Scherzer described the situation. Even though physics should be far from politics, physics teachers came under political pressure during National Socialism. Nazi-leaning professors had created the so-called *Deutsche Physik* in which contributions by Jewish colleagues, in particular relating to relativity theory, but also to quantum theory, had been

[3] The 1933 date of enlistment in the SA seems to be in contradiction with the 1934 date given in the previous section. Most likely, the Party internally dated the enlistment back to the previous year.

eliminated. Even two Nobel prize winners, Philipp Lenard and Johannes Stark, adhered to *Deutsche Physik*, and in 1936 Lenard wrote a four-volume series of physics textbooks under this heading. The textbook series was distributed to all high school and university teachers and urgently recommended to be used. As this did not work, daily newspapers were used to exert pressure. The official government organ *Völkischer Beobachter* even attacked famous physicists like Sommerfeld, Planck and Heisenberg. Scherzer himself, when preparing a talk to German engineers, was asked by the rector of the *Hochschule* to please not mention Einstein. When Scherzer answered "If I mean Einstein, I have to say Einstein" the rector begged him to do it as discreetly as possible.

In this situation, Scherzer and his colleague Wolfgang Finkelnburg who was head of the *Dozentenbund* (NSDDB) in Darmstadt decided to try an active defense by means of a discussion between (normal) physicists and "German" physicists. The disputation, later called the *Münchner Religionsgespräch*, took place in November 1940 in the *Münchner Ärztehaus*. Six people participated on each side: Wolfgang Finkelnburg, Otto Heckmann, Georg Joos, Hans Kopfermann, Carl-Friedrich von Weizsäcker and Otto Scherzer on the physicists' side, but no known names on the side of *Deutsche Physik*. The chief negotiator was a medical professor. In addition, there were several *Gutachter* (censors), all of them party people. The disputation started like a court trial, the "German" physicists acting like plaintiffs, accusing the other side of obscuring the true character of physics by faith in theories and bondage to Jews. However, after some intense discussion the "German" physicists' side was not able to come up with any reasonable arguments. It was disastrous for "*Deutsche Physik*", and the chief negotiator became really upset about the development. Finally, as a result of the disputation, they all arrived at a decision by consensus that quantum theory and the theory of relativity are well-established and indispensable parts of physics. They agreed on a truce that reduced the party's pressure on the teachers of physics.

During the late 1930s, Scherzer worked on various theoretical physical problems such as the quadratic Doppler effect, electrons in the radiation field and the twin paradox, resulting in publications that appeared in *Annalen der Physik* (Scherzer, 1938a, 1938b, 1939a, 1939b).

3.5 Peenemünde 1940-1945: radar investigations in war time

Scherzer spent the years of WWII at the *Nachrichtenmittel-Versuchskommando der Kriegsmarine* (communications research headquarters of the Navy) near

Sphärische und chromatische Korrektur von Elektronen-Linsen.

Von O. Scherzer, z. Zt. USA.

(Aus den Süddeutschen Laboratorien in Mosbach.)

(Mit 7 Textabbildungen.)

Figure 7 Paper in *Optik* on spherical and chromatic correction of electron lenses.

Kiel, where he was head of research on radar. At the end of WWII, Scherzer was under American imprisonment for 1 year.

After the war, in a denazification lawsuit, he was first rated as a follower, but in a follow-up suit he was exonerated. This was certainly due to his critical stand towards the party but also to his and his colleagues' efforts against *Deutsche Physik*.

3.6 Mosbach, Fort Monmouth, and again Mosbach 1946-1949; the start of aberration correction, "Scherzer Focus"

Since Scherzer had lost his position as a professor in 1945, he worked for two years as a scientific consultant at the *Süddeutsche Laboratorien in Mosbach*. This lab was headed by his friend Ernst Brüche; it was the continuation of the *AEG Forschungsinstitut*. They worked on design and manufacture of electron microscopes. There, he dealt intensively with methods of spherical and chromatic correction, the results being communicated in a first comprehensive paper (Scherzer, 1947). The conditions when deriving his theorem had been that the electric and magnetic field distributions were (1) axially symmetric, (2) free of charges within the region of the beam and (3) constant in time. (A fourth condition was that no electron mirrors were used). In the paper, he showed that both the chromatic and spherical aberrations can be corrected by giving up any one of these conditions, and he made design proposals. See Fig. 7.

Already in 1939, Scherzer had determined the theoretical resolution limit of an uncorrected electron microscope to be $d \approx 100\lambda$, where λ is the wavelength of the electrons (Scherzer, 1939a, 1939b).

In 1947 Scherzer worked for about one year at the Signal Corps Engineering Laboratories Fort Monmouth, N.J. USA. There, he dealt with the resolution limit of the electron microscope (Scherzer, 1949). Considering the imaging of either (1) two holes in an opaque foil, (2) two small absorbing objects or (3) two atoms, he evaluated the minimum distance for which these objects can still be distinguished when defocus and spherical aberra-

The Theoretical Resolution Limit of the Electron Microscope

O. SCHERZER
Signal Corps Engineering Laboratories, Fort Monmouth, New Jersey
(Received June 14, 1948)

The resolving power of the electron microscope and the contrast in the image are calculated for different conditions of focusing, illumination and aperture. These conditions can change the limit of resolution by a factor of about 3. The contrast in the image of an atom is appreciably increased by defocusing and spherical aberration. Nevertheless, the contrast improves when the numerical value of the aberration constant is diminished. The effect of different methods of spherical correction is discussed briefly.

Figure 8 Title and abstract of the paper in J. Appl. Phys. **20** (1948) in which the Scherzer focus is derived.

tion of the lens are taken into consideration. As atoms are essentially phase objects which cause a 90° phase shift of the scattered electron waves, they would not be visible in focus for an ideal lens, while some contrast could be obtained by defocusing the lens. In the case of the electron microscope, Scherzer showed that one can get optimum contrast and resolution using a small underfocus by which, in combination with the spherical aberration, about 90° phase shift of the lens is achieved over an extended range of scattering angles, which, together with the 90° phase shift of the scattered electrons, produces phase contrast. This focus has been termed "Scherzer focus". The wave-optical treatment of image formation was a precursor of contrast transfer theory which is the basis of image reconstruction methods developed years later. See Fig. 8.

Back in Germany, he continued to work in Mosbach. He proposed a device to correct for the axial astigmatism that is caused by small deviations from axial symmetry of the imaging lenses. He called the device a "stigmator", meaning sharpener. The device was constructed and successfully tested by Otto Rang (1949).

Pursuing the aim of improving the resolution of the microscope, he followed his most promising idea, to give up axial symmetry. He calculated and proposed the first spherically corrected electrostatic objective lens system. This system consisted of a round objective lens and a lengthy corrector system that was composed of a first cylinder lens, a round lens and a second cylinder lens and, additionally, three octopole elements, two of them positioned in the locations of astigmatic intermediate images. It was built by Scherzer's PhD student R. Seeliger and tested for a period of about 5 years. Seeliger could show that the correcting system produced a negative spherical aberration, which could be adjusted to compensate the spherical

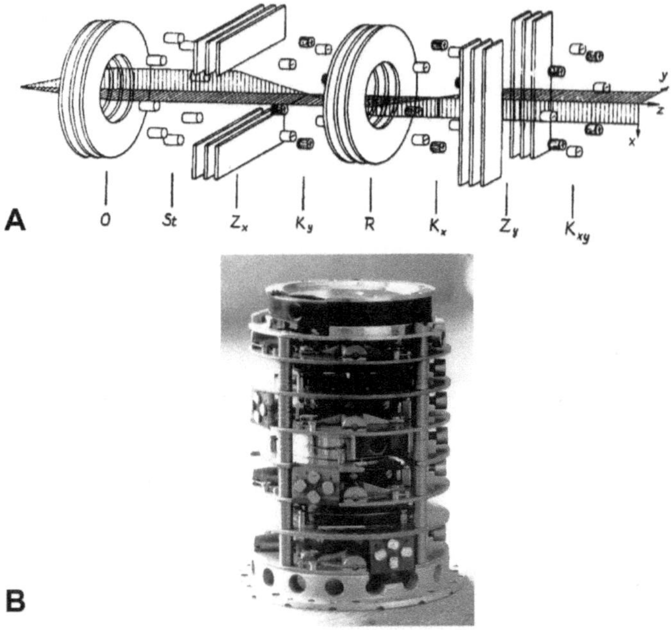

Figure 9 Spherically corrected electrostatic objective-lens system according to Scherzer. A Electron-optical arrangement; B experimental device (Seeliger, 1951).

aberration of the front lens. But due to mechanical and electromagnetic instabilities, the resolution could not be improved (Seeliger, 1951). See Fig. 9.

Several years later, G. Möllenstedt again experimented with this device. He used a large illumination angle of 2×10^{-2} rad and thus produced, due to the spherical aberration, a strongly blurred image. After compensating the spherical aberration by means of the corrector octopoles, the resolution was improved by a factor of 7, accompanied by a striking increase of contrast (Möllenstedt, 1956).

3.7 Darmstadt, 1949-1982

3.7.1 Professor of theoretical physics

At the beginning of 1949, Scherzer was again appointed Professor of Theoretical Physics at the *Technische Hochschule Darmstadt*, at first as an *Extraordinarius* (special appointment). He had accepted the appointment under the condition that he would get a mechanical workshop in his institute, to be able to pursue physical experiments. As head of the workshop, he persuaded an excellent mechanical engineer, Bruno Bastian, to move with

Figure 10 Otto Scherzer lecturing on thermodynamics in Darmstadt.

him from Mosbach to Darmstadt. Bastian had worked at the experimental physics institute in Danzig before the war. After the war he had, from scratch, built an electrostatic electron microscope at Mosbach. He was extremely valuable when later, in the seventies, the experiments on spherical and chromatic correction within the "Darmstadt project" were carried out.

After Scherzer had received an offer to move to Cologne University, his position was converted in 1954 to Ordinarius Professor (a permanent position), and he was named Head of the *Institut für theoretische Physik*. At the time, he was the only theoretical physics professor. He gave the basic theoretical-physics lectures, cycling through electrodynamics, optics, quantum mechanics and thermodynamics in four semesters. These lectures were accompanied by exercises in solving theoretical physical problems, which were corrected by advanced students, and finally discussed by Scherzer himself in a weekly exercise hour. See Figs. 10 and 11. In addition, Scherzer gave, in every semester, a lecture on a special theoretical physics topic such as the quantum theory of chemical bonding, the Dirac equation, the theory of nuclear forces, and electron optics. Students liked his lectures, and he attracted good students to work on theoretical problems. He was not an eloquent speaker, but what he said was well-considered and substantial.

Figure 11 Discussion after the lecture.

He had the ability to explain complicated matters in a comprehensible way. According to his own wide interest in theoretical physics, his *Diplom* and doctoral students worked on various subjects, ranging from electron optics to problems in quantum theory, nuclear physics and solid-state physics. Several of his *Diplom* and PhD students later became professors. However, he was rather critical and not very supportive in helping them with their career. Most of them succeeded at other universities. There was a saying that Scherzer's fugitive students were his best.

Scherzer was open to new, unconventional ideas. An example of his open-mindedness was that he accepted a paper as PhD thesis by an external scientist, Hieronymus Zygan, in which the author assumed the quantization of space. From his assumptions, Zygan had, among other results, derived Sommerfeld's fine-structure constant, a basic quantity in atomic theory, to an astonishing accuracy. Scherzer was at first skeptical, but when he checked all the mathematical derivations, he did not find an error. Although Scherzer was not convinced by Zygan's basic assumptions, he decided the paper was worth publishing. Scherzer's theoretical physicist colleagues at the time (in the late 1960s) disagreed with his decision, but he did it anyway, with mathematics professor D. Laugwitz as the first referee.

3.7.2 Work on spherical and chromatic correction continues, 1955-1982

To find out why the first corrector of spherical aberration (which had been investigated by Seeliger) had failed to improve the resolution, Scherzer tasked a PhD student, W.E. Meyer, with investigating the reasons for the failure. Seeliger had already found that the corrector system produced a large fifth-order spherical aberration, which he termed *Restfehler* (residual aberration), with the aberration constant C_5 between 11 m and 35 m, depending on the azimuth angle. In a first study, Meyer investigated this aberration, later termed the "Combination Aberration". He found that it was produced by the action of the third-order aberrations of the single elements in combination with their distances. By rearranging and changing some of the elements, he was able to devise a system for which this error was strongly reduced. The theoretical resolution limit could thereby be reduced by a factor of 2 from 2 Å to 1 Å.

In a more extended study (Meyer, 1961, his PhD thesis), W. E. Meyer carefully investigated various perturbations that should be considered and eliminated for success. He identified two groups of parasitic perturbations. Aberrations of the first group were defined as alignment aberrations; they are produced by static imperfections and misalignments of the objective lens and the constituent elements of the corrector. The second group were time-dependent disturbances such as mechanical instabilities or electromagnetic interfering fields causing an image blur.

Scherzer liked to compare the task of identifying and then eliminating one error after the other to a fakir's board equipped with hundreds of nails. The fakir senses which nail momentarily hurts most and hits it with a hammer, and then the next; in this way he goes on and on until he feels comfortable (or more likely not).

The purely electrostatic corrector for spherical aberration was not pursued further. The next generation was to be combined with a magnetic objective lens, and it would use electric and magnetic elements, which would facilitate correction of spherical as well as chromatic aberration. In the first effort, Scherzer's student, Harald Rose, received his PhD with the thesis *Allgemeine Eigenschaften unrunder Elektronenlinsen mit gerader optischer Achse* (General properties of unround electron lenses with a straight

Figure 12 Dieter Typke and Harald Rose with Walter Heimann, ca. 1970.

optical axis; Rose, 1965). Scherzer and Rose had intense discussions, one point being an appropriate nomenclature. In an article on the occasion of E. Brüche's 65th birthday, Scherzer proposed a specific nomenclature for the Gaussian imaging properties of this kind of system (Scherzer, 1965a). In 1970, Rose received his *Habilitation* with a thesis on *Korrektur elektronenmikroskopischer Objektive* (published as *Elektronenoptische Aplanate*; Rose, 1970, 1971). The proposed system provided spherical and chromatic correction, and in addition the third-order coma was completely eliminated, resulting in a rather large field of view.

In 1970, D. Typke (the author of this article), finished his PhD with a thesis entitled *Sphärische und chromatische Korrektur von Elektronenlinsen mit Hilfe von Influenzladungen* (Spherical and Chromatic Correction of Electron Lenses by means of Induced Charges; Typke, 1972 and 1976). Experiments in this direction have never been carried out. See Figs. 12 and 15.

3.7.3 The Darmstadt project

H. Rose's *Aplanator* was the system to be built and tested within the "Darmstadt project". Already in 1970, B. Bastian, the head of the mechanical workshop, started designing and manufacturing some of its elements. One of these components, an electric-magnetic octopole element was then tested by D. Typke and a *Diplom* student, K. Spengler (Bastian et al., 1971).

After Typke had left the institute, Scherzer hired a scientist, H. Koops, who had earned his PhD in Möllenstedt's institute in Tübingen.

The scope, progress and limitations of the Darmstadt project have been extensively described by Marko and Rose (2010). So only an overview will be given in the following. The project was funded by the *Deutsche Forschungsgemeinschaft* (German Research Foundation) starting in 1972, and, after some experimental success, was continued until Scherzer's sudden death in 1982. Quite a few *Diplom* and PhD students were involved in the project, and of course the skills and experience of B. Bastian, and his successor as head of the mechanical workshop, W. Bernhardt, were also important for progress.

According to Rose, the *Aplanator* design consisted of a round lens that is free of third-order azimuthal coma, followed by a symmetric quadrupole quintuplet. As such, it should correct the third- and fifth-order spherical and first-degree chromatic aberration. The theoretical resolution was expected to be 0.4 Å. As the base of the test setup, a modified Zeiss EM 10 was used.

The experiments were carried out in several steps, starting with a simplified setup for which chromatic correction could be demonstrated (Koops et al., 1977). In the following steps, simultaneous correction of chromatic and spherical aberrations could be demonstrated for a half-Aplanator as well as the full-Aplanator corrector (Bernhard, 1980; Hely, 1982a, 1982b). In spite of the correction capability, only a resolution of 10-15 Å was achieved at 40 kV accelerating voltage. Reasons may have been insufficient stability of the base setup and mechanical and magnetic disturbances at the location in the *Technische Hochschule*, but also the complexity of aligning the multiple elements of the corrector (Fig. 13A, B). At about the same time, another group, V. Beck and A.V. Crewe in Chicago, were also experimenting on spherical correction (Beck, 1979; Crewe, 1980); these experiments were not successful, either.

3.7.4 Theoretical papers

From the beginning, Scherzer accompanied the Darmstadt project with theoretical considerations.

The beam damage of objects as a limit for high-resolution electron microscopy (Scherzer, 1970): In this paper, Scherzer investigated, in a comprehensive way, atomic imaging with a focus on light atoms, taking into account nearly all imaginable imaging conditions, using protons as well as

Figure 13 Full-Aplanator system of the Darmstadt Project. A System cut in half; B experimental setup.

electrons, and energies from the low voltage up to the MeV range; bright-field as well as dark-field TEM, image reconstruction, and STEM imaging. A lower limit to beam damage is given by knock-on events, which of course depend on the bonding energy and the atomic number. For imaging carbon atoms in biological specimens or other sensitive objects of more than 100 Å thickness, he states with some confidence that only bright-field imaging in a chromatically and spherically corrected microscope, possibly combined with image reconstruction, will be viable.

Conditions for imaging carbon atoms in biomolecules (Scherzer, 1971): In a second paper, he particularly investigated imaging of carbon atoms in biological specimens using a corrected microscope. In the introduction, he explains that a strong incentive for improving the electron microscope would be to visualize carbon atoms in biomolecules. This would enable determination of the sequence of nucleotide bases in DNA. He then thoroughly discussed the conditions to be met for this goal. He concluded that for thin specimens up to 30 Å in thickness, either a corrected microscope, a high-voltage (6 MV) microscope, or even protons (at 100 kV) may be used. For thicker specimens, a spherically and chromatically corrected microscope (with 0.4-Å nominal resolution) should provide the best conditions to minimize beam damage. He then goes on to discuss stabilization of electric and magnetic fields and means to avoid mechanical perturbations. He concludes that, though difficult problems are to be solved, there are no insurmountable obstacles.

On the theory of imaging single atoms in thick specimens (Scherzer, 1973): When specimens are rather thick, the structural noise of unresolved specimen layers will limit the recognizability of single carbon atoms if parallel illumination of the specimen is used. Replacing parallel by conical illumination, these disturbances can be greatly reduced. This paper extensively and thoroughly discusses the questions related to this approach.

A spherically corrected foil lens for phase microscopy with electrons (Scherzer, 1980): By 1980, the Darmstadt project had not succeeded in improving the resolution, even though the correction of chromatic as well as spherical aberration could be demonstrated. Scherzer then proposed to again try a foil lens system. At the beginning of this paper, he recalls a discussion with V.K. Zworykin in 1947: "*Zworykin told me: You Germans are strange people. You know that there will be a fence somewhere along the way over which you have to get at some point, and immediately you start jumping. I answered: At least we will be trained once we get to the fence. But he growled: oh no, you will be tired.*" After some consideration of the present situation, Scherzer states: "*Now we are in front of the fence, and we have to jump. C_3 must be reduced by several orders of magnitude to obtain significant progress. Once again there is the question: What is the best way to achieve this?*"

The correcting system that he proposed uses a thin foil with a central hole. In an ingenious way, he calculated an electric field distribution in such a way that the field strength, and thus the induced charges as well, are minimized at the edges of the hole. To keep the object space free of an electric field, charges are only induced on the surface of the foil away from the object space. In principle, a quadratic increase of the charge density on the foil surface compensates for the influence of the spherical aberration. At 60 kV accelerating voltage, a nominal resolution limit of 0.3 Å should be achievable. At the end, he discussed various settings of the parameters to optimize the imaging conditions for various types of specimen.

Phase tomography in a corrected TEM (Scherzer, 1982a): In this paper, Scherzer refers to a statement by W. Hoppe (1975) that in tomographic data recording, the dose required to significantly image a detail within the 3D volume is about the same as that required when the detail is isolated. This statement has been proven by R. Hegerl (Hegerl & Hoppe, 1976) and is, of course, very important for the technique of tomography. Scherzer applies the same consideration to data recording in the corrected

Figure 14 Otto Scherzer (about 1980).

TEM. In a previous paper on thick specimens, he had already proposed to use conical illumination to sufficiently reduce the structural noise from layers other than the focused one. Now, he proposed to reduce the illumination to only a small part of the cone, and to record multiple images at different illumination angles around the cone. From the images, any layer within the volume could be reconstructed with low interference from the other layers. The 3D image elements would be cigar-shaped, but at 60 kV accelerating voltage and 0.4 Å nominal (lateral) resolution, it should be possible to recognize single atoms if the specimen is not too densely packed (Scherzer, 1978).

How not to correct an electron lens: In his last published note, Scherzer responds to a proposal for spherical correction, pointing out that the proposal is not suitable for its purpose. (Scherzer, 1982b). See Fig. 14.

3.7.5 Aftermath of the Darmstadt project

After Scherzer's demise, the project was abruptly nearly ended. The last person working on his *Diplom* thesis, M. Haider, tried to proceed with the experiments. Due to insufficient support, he had soon to give up. He finished his *Diplom* in Rose's group, and then became a PhD student at the

EMBL (European Molecular Biology Laboratory) in Heidelberg under H. Rose's supervision. After two years, the experimental setup of the Darmstadt Project was brought to the *Institut für Angewandte Physik* in Tübingen. Years later, the Darmstadt corrector was obtained by M. Haider's CEOS (Corrected Electron Optical Systems) company and cut in half to show its interior design (Fig. 13A).

3.7.6 After 1982: persistence and breakthrough in spherical correction

It took quite some determination by H. Rose and M. Haider to persist in pursuing the goal of practical correction, particularly under the circumstances at the end of the 1980s. A panel of "experts" advised the NSF (American National Science Foundation) to cease funding correction projects. The DFG would no longer fund the German project. On the other hand, at about the same time CCD cameras suitable for TEMs had become available, and personal computers had become fast enough to accomplish automatic control procedures. In addition, Rose had devised a simpler design consisting of two hexapoles in combination with a round-lens doublet (Rose, 1981). Together with K. Urban from the Ernst Ruska Centre in Jülich, Rose and Haider succeeded in obtaining a grant from the Volkswagen foundation. The development of the new setup started in 1992 at the EMBL. At an intermediate step in 1994, they could show that the correction worked, and in 1996 they succeeded in improving the resolution from 2.4 to ~1.2 Å. In 1997, B. Kabius from K. Urban's lab obtained the first atomic-resolution images of gallium arsenide (Kabius et al., 1998). After this breakthrough, multipole corrected systems of various kinds have been built for TEMs and also STEMs, one of the most advanced being the TEAM corrector (Rose, 1990). As of 2019, about 1000 corrected systems had been delivered worldwide, about 95% of them equipped with a Rose-type corrector built by CEOS or with a slightly modified version of the corrector by the Japanese company JEOL (K. Urban, personal communication).

3.7.7 Breakthrough in biological structure research

Biological structure determination using the EM took a course different from what Scherzer had imagined, not aiming at the highest possible resolution when recording the data. In a similar way to x-ray protein-structure analysis it was sufficient to obtain 3D data to about 3 Å resolution, and then to use primary and secondary structure data to fit the amino acid chain into the 3D map. This made it possible to get atomic-resolution structures. By further improving the resolution to 2.5 or 2 Å, the precision of positioning single atoms within the structure can be greatly improved, e.g. at 2.5 Å resolution, positioning could be achieved to about ±0.4 Å accuracy.

In comparison to the case of x-ray crystallography, image data are used in 3-D EM. About thirty years after Scherzer's passing, structure determination at atomic resolution by electron microscopy has been achieved for various types of specimen. It started with two-dimensional crystalline protein arrays such as glucose-embedded bacteriorhodopsin (Henderson et al., 1990), and went on with "single particles" with high internal symmetry in frozen-hydrated preparations like the envelopes of a class of bacteriophages or helically arranged structures. Finally, atomic resolution could be obtained even for ice-embedded asymmetric single particles. The breakthrough in this last case came with the availability of solid-state direct-electron detectors in 2012, which made it possible to record more than 100 images per second. In this way, the electron dose per image was reduced to extremely small values and beam-induced movement could be largely compensated afterwards by means of correlation techniques, thus improving the resolution to 3 or 2.5 Å. For more information about these developments see, e.g., the Nobel lectures of R. Henderson (2018), J. Frank (1988) and J. Dubochet (2018). Since 2010, aberration corrected TEMs have also been introduced to biological structure research thus enabling even higher resolution to be attained. A recent result is the determination of the (rather stable) structure of apoferritin to 1.25 Å resolution (Yip et al., 2020).

3.8 Scherzer's personality, other activities

Scherzer was a self-confident and strong-willed personality. He would not accept anything unless he himself had carefully thought it over. Instead of searching the literature for solutions to some specific problem, he preferred to solve the problem himself. He was well aware that he was a first-class physicist, and in fact he had a broad knowledge of both theoretical and experimental physics. He was also an excellent mathematician. When, in 1935, A. Sommerfeld wrote a letter of recommendation for Scherzer, he added that "*Scherzer is a highly skilled theoretical physicist, who does not know any mathematical problems*". To Scherzer this praise seemed to be ambivalent, as it could be understood in a very different way.

Scherzer was not an easygoing person; he was critical of others and demanding of his collaborators, but he was always fair. He was a perfectionist to the point of pedantry. For example, he set a high value on small language details. Even the way the institute's name was written highlights this personal trait. He insisted that the adjective "*theoretische*" was to be written in lower case, to emphasize that physics was the main subject of his research and teaching. Another example of his attention to language details is the word "*stigmator*" that he created for his proposed tool to correct the (axial) astigmatism of electron lenses. In contrast to that, people in light optics had termed an objective lens for which the (off-axial) astigmatism had been corrected as "anastigmat."

For many years Scherzer was the host of the physics colloquium. He was famous for his probing questions and for his ability to clarify matters that seemed to be unclear. He also had various activities outside his institute. For instance, he was, within the *Technische Hochschule Darmstadt*, pivotal in establishing the Institute of Nuclear Physics. He was twice the Dean of the Faculty of Mathematics and Physics. Outside the *Hochschule*, he was one of the founding fathers of the *Helmholtzzentrum für Schwerionenforschung* (GSI; Helmholz center for heavy-ion research, near Darmstadt), which has become a worldwide large and unique accelerator site for heavy-ion research.

He received numerous tributes for his scientific accomplishments. Something very special was the celebratory renewal of his dissertation in 1981 at Munich University, where he gave a lecture on *Bildfehlerkorrektur mit unrunden Elektronenlinsen* (correction of image aberrations by means of unround electron lenses). In 1975 he was made an honorary member of the German Electron Microscopical Society, and in 1983 he was posthumously awarded the "Distinguished Scientist Award" of the Electron Microscopy Society of America (EMSA). See Fig. 15.

Scherzer was married to Elisabeth Sindel in February, 1934. His wife must have been a very patient person. For example, Scherzer reportedly proofread the Brüche/Scherzer book during their honeymoon. Elisabeth and Otto Scherzer had 5 daughters together.

3.9 Scherzer anecdotes

A strong, opinionated character like Scherzer is often the subject of anecdotes. In fact, there were many "Otto" anecdotes around. Below I provide just a few of them.

a. "Scherzer Focus": At an international workshop, during the talk of one of the participants, Scherzer was not satisfied with the sharpness of the projector. He got up, went to the projector and adjusted the focus. As he returned to his seat, Albert Crewe said with a low voice but such that everybody could hear: this is now Scherzer Focus.

Figure 15 Rest during an excursion of the institute about 1965.

b. At a physics colloquium, H. Lichte was speaking about electron holography. In his talk, he often used the term "Scherzer defocus." After the talk Scherzer, sitting in the first row, asked: "Mr. Lichte, could you please explain to us what the Scherzer defocus is?"

c. When a *Diplom* student remarked to him that he had to do a lot of work, Scherzer answered: "You see, for this reason it is called *Diplom* work and not *Diplom* pleasure."

d. In a talk with H. Rose and E. Plies he expressed his skeptical view of the Ω-filter: "When the electrons come out on the other side, they will be totally dizzy; do you really think that this will work?" Rose's answer was: "The electrons behave just like dancers; in order not to get dizzy they first turn left and then right just as electrons are doing in the Ω-filter".

e. For many years Scherzer drove a Volkswagen Beetle. Once, when driving with his colleagues in his car on a two-lane highway, he passed a big truck loaded with logs. When he completed passing the truck, one of his colleagues, who was obviously scared, asked him what he would have done if a car had come along in the second lane. His answer: "I would have pondered what would have had the least risk, and probably would have moved under the logs." On another trip with his car, the motor failed at some point. He opened the hood and fiddled around for a short while. Then he started the motor, and it worked again. To his colleagues, it appeared a bit like magic, but he knew that the distributor contacts were a weak element in those days, and he fixed the bad contacts by some fiddling.

f. When some students came late to his lecture, he once remarked: "Academic freedom means that one may come or may not come to a lecture, but it does not mean that one may be late." After that, a latecomer only dared enter the small physics lecture hall through the back door as silently as possible, taking a seat in the most distant row.

g. Before the violation of parity was discovered, the theoretical physicist Friedrich Hund was in Darmstadt at a colloquium. In the discussion after the talk, Scherzer (in favor of parity violation) offered a bet to Hund (in favor of preservation of parity). But Hund did not want to bet. Even when Scherzer offered to only bet for a bar of chocolate, Hund did not accept. Today, we know that Scherzer would have won the bet.

Acknowledgments

I would like to thank Peter Hawkes for inviting me to write this article of my physics teacher and mentor and for information exchange during its progress. Apart from my own memory of the time in Scherzer's institute, I mainly used articles by Beck (1983), Marko and Rose (2010), and Rose (2009a, 2009b) and Typke et al. (2010) as sources. See too Rose (1983) [HR]. I also appreciate the personal correspondence with Erich Plies, Harald Rose, Max Haider and Knut Urban, and I am very thankful to Mike Marko for copy-editing my draft, converting German into native English.

References

Bastian, B., Spengler, K., & Typke, D. (1971). Ein elektrisch-magnetisches Oktopol-Element zur sphärischen und chromatischen Korrektur von Elektronenlinsen. *Optik*, *33*, 591–596.

Beck, F. (1983). Otto Scherzer: Wegbereiter der Elektronenoptik. *Physikalische Blätter*, *39*, 50.

Beck, V. D. (1979). A hexapole spherical aberration corrector. *Optik*, *53*, 241–255.

Bernhard, W. (1980). Erprobung eines sphärisch und chromatisch korrigierten Elektronenmikroskops. *Optik*, *57*, 73–94.

Brüche, E., & Johannson, H. (1932). Einige neue Kathodenuntersuchungen mit dem elektrischen Elektronenmikroskop. *Physikalische Zeitschrift*, *33*, 898–899.

Brüche, E., & Scherzer, O. (1934). *Geometrische Elektronenoptik*. Berlin and New York: Springer-Verlag.

Busch, H. (1922). Eine neue Methode zur e/m-Bestimmung. *Physikalische Zeitschrift*, *23*, 438–441.

Busch, H. (1926). Berechnung der Bahn von Kathodenstrahlen im axialsymmetrischen elektromagnetischen Feld. *Annalen der Physik*, *81*, 974–993.
Crewe, A. V. (1980). Studies on sextupole correctors. *Optik*, *57*, 313–327.
Dubochet, J. (2018). On the development of electron cryo-microscopy (Nobel lecture). *Angewandte Chemie. International Edition in English*, *57*, 10842–10846.
Frank, J. (1988). Single-particle reconstruction of biological molecules—story in a sample (Nobel lecture). *Angewandte Chemie. International Edition in English*, *57*, 10826–10841.
Gabor, D. (1969). The outlook for holography. *Optik*, *28*, 437–441.
Glaser, W. (1933). Zur geometrischen Elektronenoptik des axialsymmetrischen Feldes. *Zeitschrift für Physik*, *81*, 647–686.
Glaser, W. (1940). Über ein von sphärischer Aberration freies Magnetfeld. *Zeitschrift für Physik*, *116*, 19–33.
Hegerl, R., & Hoppe, W. (1976). Influence of electron noise on three-dimensional image reconstruction. *Zeitschrift für Naturforschung A*, *31*, 1717–1721.
Hely, H. (1982a). Technologische Voraussetzungen für die Verbesserung der Korrektur von Elektronenlinsen. *Optik*, *60*, 307–326.
Hely, H. (1982b). Messungen an einem verbesserten korrigierten Elektronenmikroskop. *Optik*, *60*, 353–370.
Henderson, R. (2018). From electron crystallography to single particle CryoEM (Nobel lecture). *Angewandte Chemie. International Edition in English*, *57*, 10804–10825.
Henderson, R., Baldwin, J. M., Ceska, T. A., Zemlin, F., Beckmann, E., & Downing, K. H. (1990). Model for the structure of bacteriorhodopsin based on high-resolution electron cryo-microscopy. *Journal of Molecular Biology*, *213*, 899–929.
Hoppe, (1975). Private communication between Hoppe and Scherzer.
Johannson, H., & Scherzer, O. (1933). Über die elektrische Elektronensammellinse. *Zeitschrift für Physik*, *80*, 183–192.
Kabius, B., Urban, K., Haider, M., Uhlemann, S., Schwan, E., & Rose, H. (1998). First application of a spherical-aberration corrected microscope. Material science. *Microscopy and Microanalysis*, *4*(Suppt 2: Proceedings), 384–385.
Koops, H., Kuck, G., & Scherzer, O. (1977). Erprobung eines elektronenoptischen Achromators. *Optik*, *48*, 225–236.
Marko, M., & Rose, H. (2010). The contributions of Otto Scherzer (1909–1982) to the development of the electron microscope. *Microscopy and Microanalysis*, *16*, 366–374.
Meyer, W. E. (1961). Das Auflösungsvermögen sphärisch korrigierter elektrostatischer Elektronenmikroskope. *Optik*, *18*, 69–91.
Möllenstedt, G. (1956). Elektronenmikroskopische Bilder mit einem nach O. Scherzer sphärisch korrigierten Objektiv. *Optik*, *13*, 209–215.
Preikszas, D., & Rose, H. (1995). Procedures for minimizing the aberrations of electromagnetic compound lenses. *Optik*, *100*, 179–187.
Rang, O. (1949). Der elektrostatische Stigmator, ein Korrektiv für astigmatische Elektronenlinsen. *Optik*, *5*, 518–530.
Rose, H. (1965). *Allgemeine Eigenschaften unrunder Elektronenlinsen mit gerader optischer Achse* (Dissertation). Darmstadt.
Rose, H. (1967). Über die Korrigierbarkeit von Linsen für schnelle Elektronen. *Optik*, *26*, 289–298.
Rose, H. (1970). *Korrektur elektronenmikroskopischer Objektive* (Habilitationsschrift). Darmstadt.
Rose, H. (1971). Elektronenoptische Aplanate. *Optik*, *34*, 285–311.
Rose, H. (1981). Correction of aperture aberrations in magnetic systems with threefold symmetry. *Nuclear Instruments and Methods*, *187*, 187–199.
Rose, H. (1983). Otto Scherzer gestorben. *Optik*, *63*, 185–188.
Rose, H. (1990). Outline of a spherically corrected semi-aplanatic medium-voltage TEM. *Optik*, *85*, 19–24.

Rose, H. (2009a). Historical aspects of aberration correction. *Journal of Electron Microscopy, 58*, 87–97.
Rose, H. (2009b). In remembrance of Otto Scherzer, the eminent pioneer of electron optics. *Microscopy and Microanalysis, 15*(Suppl 2), 1452–1453.
Scherzer, O. (1932). Über die Ausstrahlung bei der Bremsung von Protonen und schnellen Elektronen. *Annalen der Physik, 405*, 137–160.
Scherzer, O. (1933a). Zur Theorie der Gaskonzentration von Elektronenstrahlen. *Zeitschrift für Physik, 82*, 697.
Scherzer, O. (1933b). Zur Theorie der elektronenoptischen Linsenfehler. *Zeitschrift für Physik, 80*, 193–202.
Scherzer, O. (1936a). Die elektrostatische Einzellinse geringster sphärischer Aberration. *Zeitschrift für Physik, 101*, 23;
Also see Busch, H., & Brüche, E. (Eds.), *Beiträge zur Elektronenoptik* (pp. 14–18). Leipzig: J.A. Barth, (1937).
Scherzer, O. (1936b). Über einige Fehler von Elektronenlinsen. *Zeitschrift für Physik, 101*, 593–603.
Scherzer, O. (1937). Die Berechnung der Bildfehler dritter Ordnung nach der Bahnmethode. In H. Busch, & E. Brüche (Eds.), *Beiträge zur Elektronenoptik* (pp. 33–41). Leipzig: J.A. Barth.
Scherzer, O. (1938a). Die Messbarkeit des quadratischen Dopplereffekts. *Annalen der Physik (Leipzig), 424*, 242–244.
Scherzer, O. (1938b). Die imaginäre Einheit in der Diracgleichung. *Annalen der Physik (Leipzig), 425*, 593–595.
Scherzer, O. (1939a). Das Elektron im Strahlungsfeld. *Annalen der Physik (Leipzig), 426*, 585–602.
Scherzer, O. (1939b). Das Elektron im Strahlungsfeld II. *Annalen der Physik (Leipzig), 427*, 665–670.
Scherzer, O. (1947). Sphärische und chromatische Korrektur von Elektronenlinsen. *Optik, 2*, 114–132.
Scherzer, O. (1949). The theoretical resolution limit of the electron microscope. *Journal of Applied Physics, 20*, 20–29.
Scherzer, O. (1965a). Vorschläge zur Terminologie unrunder Elektronenlinsen. *Optik, 22*, 314–318.
Scherzer, O. (1965b). Physik im totalitären Staat. In A. Flitner (Ed.), *Deutsches Geistesleben im Nationalsozialismus, Tübingen, 1965* (pp. 47–58);
English translation in: Hentschel, K., & Hentschel, A. M. (Eds.), *Physics and national socialism: An anthology of primary sources*. Basel: Birkhäuser, (1996).
Scherzer, O. (1970). Die Strahlenschädigung der Objekte als Grenze für die hochauflösende Elektronenmikroskopie. *Berichte der Bunsengesellschaft für Physikalische Chemie, 74*, 1154–1167.
Scherzer, O. (1971). Voraussetzungen für die Abbildung von Kohlenstoff-Atomen in Biomolekülen. *Optik, 33*, 501–516.
Scherzer, O. (1973). Zur Theorie der Abbildung einzelner Atome in dicken Objekten. *Optik, 38*, 387–415.
Scherzer, O. (1978). *Limitations for the resolving power of electron microscopes*In *Proceedings ICEM-9: Vol. 3*. San Francisco, CA: San Francisco Press (pp. 123–129).
Scherzer, O. (1980). Eine sphärisch korrigierte Folien-Linse für die Phasenmikroskopie mit Elektronen. *Optik, 56*(2), 133–147.
Scherzer, O. (1982a). Phase tomography in the corrected electron microscope. *Ultramicroscopy, 9*, 9–16.
Scherzer, O. (1982b). How not to correct an electron lens. *Ultramicroscopy, 9*, 285–286.
Scherzer, O., & Typke, D. (1967). Die Auflösungsgrenze eines in zwei Schnitten sphärisch korrigierten Objektivs. *Optik, 26*, 564–573.

Seeliger, R. (1951). Die sphärische Korrektur von Elektronenlinsen mittels nicht rotationssymmetrischer Abbildungselemente. *Optik*, *8*, 311–317.

Tretner, W. (1959). Existenzbereiche rotationssymmetrischer Elektronenlinsen. *Optik*, *16*, 155–184.

Typke, D. (1972 and 1976). Sphärische und chromatische Korrektur von Elektronenlinsen mit Hilfe von Influenzladungen, I. *Optik*, *34*, 573–594;

Typke, D. Sphärische und chromatische Korrektur von Elektronenlinsen mit Hilfe von Influenzladungen, II. *Optik*, *36* (1972 and 1976), 124–138;

Typke, D. Sphärische und chromatische Korrektur von Elektronenlinsen mit Hilfe von Influenzladungen, Erratum. *Optik*, *44* (1972 and 1976), 509.

Typke, D., Rose, H., Plies, E., & Haider, M. (2010). Some personal memories of Otto Scherzer. *Microscopy Today*, *18*, 82–84.

Wolf, F. (1927). Eine Präzisionsmessung von e/m_0 nach der Methode von H. Busch. *Annalen der Physik*, *83*, 849.

Yip, K. M., Fischer, N., Paknia, E., Chari, A., & Stark, H. (2020). Atomic-resolution protein structure determination by cryo-EM. *Nature*, *587*, 157–161.

Zworykin, V. K., Morton, G. A., Ramberg, E. G., Hillier, J., & Vance, A. W. (1945). *Electron optics and the electron microscope*. New York: John Wiley and Sons.

CHAPTER THIRTEEN

1950-1960: a decade from the viewpoint of an applications laboratory

Cilly Weichan and Maren Heinzerling (Afterword)*
Formerly Berlin, Germany
*Corresponding author. e-mail address: hawkes@cemes.fr

Contents

1. Introduction 407
2. Experience with testing a new microscope 409
   2.1 The importance of a higher electron-optical magnification 409
   2.2 Astigmatism and testing the resolving power 411
   2.3 Contrast and aperture problems 413
   2.4 Calibration and high-resolution test specimens 415
3. A microscope with higher resolving power 416
   3.1 Contamination and an anticontamination device 418
   3.2 Contrast improvement and focusing aids 420
   3.3 I take up the aperture problem again 422
   3.4 Extended application of microscopical techniques 423
   3.5 One end point: 0.4-nm high resolution 423
4. Development and improvement of preparation methods 424
   4.1 Investigation of rubber and filling material 424
   4.2 Further applications: modifications of the replica technique 428
   4.3 Shadowing and special techniques 432
5. Teaching duties in an applications laboratory for electron microscopy 435
6. Final remarks 438
7. Afterword by the late Dipl.-Ing. Maren Heinzerling (Deutscher Akademikerinnenbund e.V.) 439
References 442

1. Introduction

Toward the end of the 1940s the basic work in developing the electron microscope was already complete, and a variety of preparation techniques had been successfully tested. In a second development phase, starting about the middle of the 1940s, the most urgent problem became

Advances in Imaging and Electron Physics, Volume 221
ISSN 1076-5670
https://doi.org/10.1016/bs.aiep.2022.03.013

the improvement of the high resolving power, which in turn required new preparation methods. The producers of electron microscopes soon recognized that the manufacture of a high-quality microscope and the use of suitable specimens were not yet a guarantee for the production of high-resolution pictures. Another requirement of similar importance was the knowledge of how to use the equipment best. For this reason, it was not long before special laboratories for teaching instrument manipulation and for checking newly developed preparation techniques were created.

At the end of the 1940s, I came by chance in touch with electron microscopy after finishing my studies in plant physiology and doing some research work in a food control section. Thanks to my practical experience in physics and chemistry, I was engaged by the Siemens Company for their applications laboratory in order to prepare specimens. My knowledge in biology was not yet needed at that time because of the lack of convenient preparation methods, so at first only a few biologists showed any interest in electron-microscopic investigations. But a short time later, due to the new technique of ultramicrotomy, many applications in the field of biology and medicine became possible. This was of consequence for me as I was capable of participating, without any trouble in communication, from the beginning of such work.

Help was given in the preparation of specimens and in the use of the microscope. The members of the applications laboratory had themselves to learn how to attain the best optical performance before teaching the beginners what to do. Ever since 1956, when I started guiding the applications laboratory, I saw my main duty in this teaching, and I felt myself obliged to pass on to our learners as much knowledge as possible. Years of practical microscopy were necessary to find out the optimum working conditions for each microscopical technique and the best way of imaging various specimens. Based on this experience, instruction manuals could be written in order to get the best out of the microscope.

Looking back on about 30 years' work in an applications laboratory, the years between 1950 and 1960 seem to be the most significant, as it was mainly in this period that the development of the electron microscope to a high-performance instrument took place, and it was a good period, too, for the introduction of new preparation methods. As well as reminiscences of those times concerning the microscope equipment, some techniques of preparation and the duties of a teacher will be reported.

2. Experience with testing a new microscope

2.1 The importance of a higher electron-optical magnification

The new microscope designed after the war was technically advanced compared with the instruments built before 1945. However, it still took a long time to attain the high performance of modern microscopes and to develop all the auxiliary accessories which are today a matter of course. There was no longer any doubt that improving the optical quality rendered visible structures much smaller than had formerly been imagined possible.

Using the earlier type of Siemens electron microscope, the best point-to-point resolution of 2 nm could be obtained when the instrument was used at its ultimate magnification of about 40,000×. It was evident that for imaging still finer specimen structures, a higher electron-optical magnification was needed. Only at an original magnification of at least 80,000× did it seem possible to image clearly particles in the range between 1 and 2 nm without being misled by the statistical granularity of the photographic emulsion.

As a consequence of this consideration, the earlier design with two magnifying lenses was given up in favor of three-stage magnification. In the development phase of the ÜM 100 (Ruska, 1950) toward the end of the 1940s, nobody in the construction team favored the idea of changing the magnification by altering the objective-lens current, bearing in mind the difficulties of constructing an objective stage for accurate axial movement. To gain higher magnification, a projector drum was built containing four polepiece systems for four fixed steps in magnification. The image remained in focus when exchanging the polepieces.

As expected, working at higher magnification caused some brightness problems. At low magnification, a weak beam current gave enough light for image observation, but doubling the final magnification caused a reduction of the brightness to a quarter of its original value, so that a beam density four times higher was necessary for the same illumination conditions. To prevent loss of light, the beam had to be precisely aligned along the axis of the objective by means of a mechanical condenser adjustment for displacement and tilt. Some loss in brightness owing to bending of the cathode was mastered by the utilization of prealigned filaments. As already detected in the early days of electron microscopy, a hairpin-shaped tungsten filament suffered from slight deformation during its first heating up, but afterward remained stable. Making use of this observation we worked only with filaments preheated during fabrication and thereafter centered to the bore of the control electrode before insertion in the gun.

The better adjustment possibilities helped, but did not solve all brightness problems. The microscope was still equipped with a single-condenser system, which meant that with no exceptions a large specimen field was inevitably irradiated. In order to get sufficient light for high magnification on the screen the beam current had to be increased, thus generating a high thermal load on the specimen, which frequently drifted or even became damaged.

As members of an applications laboratory we devoted our efforts to the improvement of the preparation. Nevertheless, the microscopical work remained difficult, though we learned how to prepare stable specimens. It normally took a long time before the specimen drift stopped, and it was therefore not surprising that the number of high-quality photos taken at high magnification was minimal and the number of failures immense. Not only did the specimen movement due to overheating prevent successful high-resolution work, but even more embarrassing were the electrical and mechanical instabilities. The best pictures were mostly produced during the night hours because of the high demands for electrical and mechanical stability. Only rarely could high-resolution pictures be produced during the normal working hours of the factory, when vibrations would shake the column and mains instabilities and external magnetic stray fields would damage the image quality. It was a hard time of learning by doing, but in this way we soon became familiar with all the requirements for the production of high-resolution pictures. Disappointments from the first years' work bore good fruit later on, as many future improvements were the result of this experience.

At first, however, for best adaptation, we sat hour by hour in the darkness of the microscope room trying to reduce the intensity for better preservation of the specimen. Exposure times below 10 sec were only a wish-dream even when using the most sensitive photographic emulsions. A voltage constancy of 3.5×10^{-5} was needed in order to get a resolution value of 1 nm. The stability of the high tension and the lens currents was still insufficient for the use of long exposure times.

In 1951 the required performance was achieved by the introduction of an electronic circuit. By improving the insulation of the gun system, even very small discharges in the accelerating system were cut off. In addition, the objective-lens current was stabilized and the current adjustment improved. Mechanical faults of the stage movement system responsible for the specimen drift were corrected. We came to realize that a transverse movement of only 2×10^{-6} mm would damage the picture quality because this value corresponded to the dimension of the interval to be imaged and degraded the resolving power in the direction of the fault.

2.2 Astigmatism and testing the resolving power

Even with better electrical and mechanical stability the recording of high-quality pictures remained problematic. In consequence of the improvement in image quality, other defects, negligible in the initial period because of the dominant influence of other faults, now became important.

It was now the axial astigmatism that bothered us most, even when near the point of least confusion so that the astigmatism recognizable as directional structure was nearly invisible. In the long run, the loss of fine detail due to this fault was intolerable.

Carbon films with small and smoothly bordered holes of different diameters were found to be the most suitable specimens for checking the astigmatism. In the first years of my work, a corrector system was not yet available, and we therefore made use of the most primitive aids to prevent loss of image quality due to astigmatism. Different lens groups from the factory production were combined to determine the best combination with regard to coincidence of the magnetic axes and low basic astigmatism thanks to lens field homogeneity. Alternatively, the correction was effected in an unscientific way by the introduction of a contaminated aperture, using its stray field as stigmator. Contamination of this aperture situated near the specimen still caused the most serious instability factor at that time (von Borries, 1949). Although we knew about the risks of such working procedures, we succeeded in producing our first real high-resolution pictures by compensating the astigmatism in orientation and strength by a slight movement of a contaminated aperture. Such an eccentrically irradiated aperture could be used at best for a few moments as corrector because of the instability of the field.

In January, 1953, we were able to demonstrate for the first time a resolving power of 1 nm on a natural specimen containing fine granules. Techniques for the production of artificial high-resolution test specimens needed for checking the optical quality of each microscope in the test bay of the factory had not yet been worked out. To check the resolving power we always looked for natural samples containing very fine particles not too poor in contrast. This was a tiring and by no means rational method (see Fig. 1).

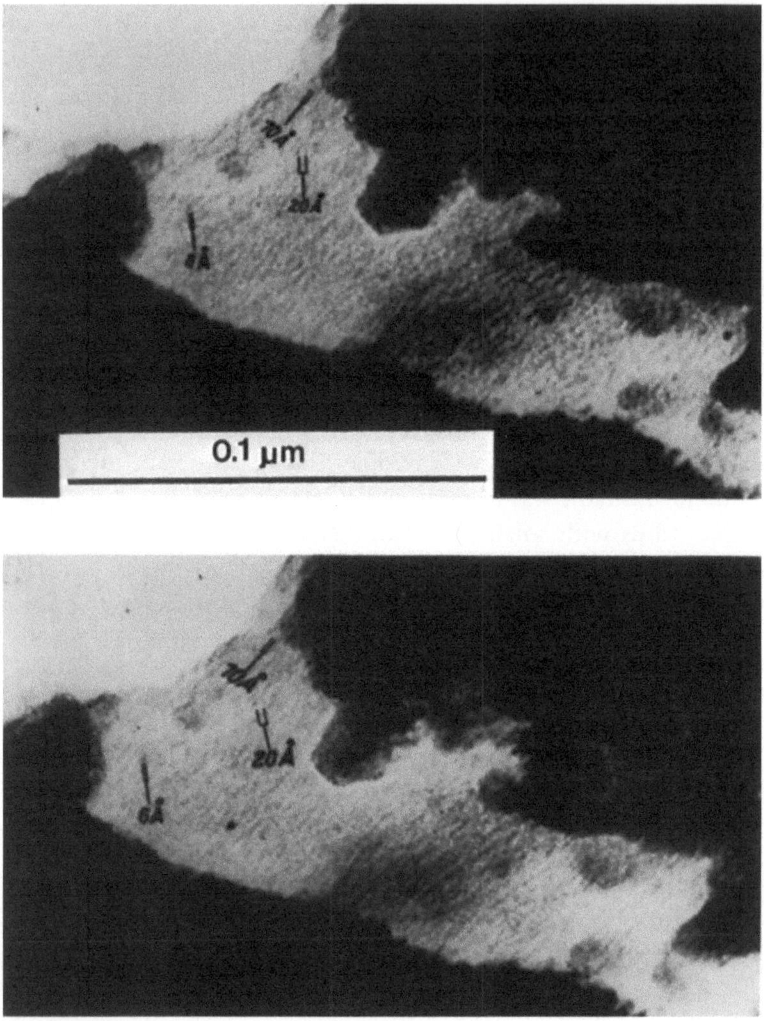

Figure 1 Early high-resolution test—peat coated with beryllium: electron-optical magnification, 100,000:1; acceleration voltage, 80 kV; beam density in the specimen, $j = 1.7\ \mathrm{A/cm^2}$; illumination aperture, $\alpha_s = 4 \times 10^{-3}$; objective aperture, $a_0 = 1 \times 10^{-2}$; astigmatic focal difference, $\Delta f = 0.1\ \mu\mathrm{m}$.

In order to establish a resolution of 1 nm, crystalline precipitates of peat were imaged at 100,000× magnification. This magnification was sufficiently high to make the finest particles visible, avoiding irritations by statistical grain aggregations. In addition, the resolution value was checked by imaging the same specimen area at the same focal setting at least twice.

After the publication of these images some called the credibility of the electron-optical magnification into question, and, with that, the reliability of the indicated resolution value. Such doubts from the outside generated further research work in our group on the accuracy of the calibration. At the same time we started to occupy ourselves with the development of new preparation methods in order to produce systematically artificial high-resolution specimens. It was difficult and time-consuming work which required considerable skill.

2.3 Contrast and aperture problems

The publication of pictures showing finer structures than had been seen hitherto attracted the interest of potential buyers. Many now came visiting our applications laboratory with their own specimens to be investigated. Visitors from foreign countries also came for the first time. We took pains to prepare their material as well as possible, invoking all known techniques. The specimens frequently consisted of light elements, and the contrast was low. In order to improve the visible contrast in such a case either the beam voltage was reduced to the lowest value of 40 kV or an objective aperture with a hole of small diameter was used. The experimental work at low voltage was found inconvenient as the brightness dropped, and in spite of a remarkable gain in contrast it became difficult to observe and focus the image properly due to the loss in light. Furthermore, it was disadvantageous that the specimen was heated up, absorbing more energy. The use of higher voltages was preferable, and under these conditions the loss in contrast was compensated by the insertion of a small aperture.

In this stage of the experiments we were disturbed by the rapid contamination of the objective aperture. Due to the reduction of the hole diameter, disturbing electrostatic fields increasingly influenced the beam. Today, in an era of ultrahigh vacuum and contamination-free apertures, the everyday problems of our first years sound strange, but 30 years ago conditions were less favorable. The apertures used in our microscope were platinum-iridium disks 1 mm thick with one central bore. Heat resulting from the electron bombardment was quickly dispersed over the large body so that the temperature around the hole remained low due to the good conductivity of the metal. It was this temperature reduction that accelerated the building of contamination layers. We were confronted with the fact that the rate of growth of contamination is inversely proportional to the temperature. To make the microscopical work easier the original aperture holder for only one aperture was now replaced by a provisional device which permitted the introduction of two apertures.

All apertures had to be used several times for financial reasons. The price was high because of the cost of the mechanical boring, and it was not economical to throw them away after only one use. New methods of cleaning were first developed, but were not much appreciated by our customers, as they had to manipulate hazardous chemicals and to glow the apertures. During the glowing process, careful control of the temperature was necessary to avoid melting of the alloy during the heating procedure.

We next searched for another metal more suitable for cleaning, and, additionally, we tried to change the construction model. Molybdenum was found to be a suitable substitute for platinum-iridium, as the fabrication of smoothly bordered round holes was technically possible. Cleaning by glowing under vacuum could be done at a temperature sufficiently high to burn away all beam-influencing deposits. The problem was, however, that users of Mo apertures had to possess a vacuum evaporator to perform this cleaning treatment. For more universal use, a modification of the construction seemed to be more promising. The outer dimensions of the new aperture model remained unchanged for further use of the existing holder. Only the bore channel was widened by thinning the metal around the hole. This complicated manual work had its natural limitations. Only in the most favorable cases could the material be thinned down to the point where the thickness of the walls around the bore channel was 70 μm. A special 15-μm Pt-Ir aperture was fabricated using this construction and was introduced to gain higher contrast. In this way the contamination was retarded as the thinned metal was heated. Before the new design was introduced, small apertures remained free of contamination not longer than a few minutes.

The contamination of the contrast aperture was not the sole source of problems of this kind. The contamination of the specimen also became increasingly a problem. During exceptionally long exposure times, the contours of the specimens became blurred so that a correct setting in focus became more difficult from minute to minute. The outside shapes of particles no longer appeared distinct and made measurements for calibration senseless.

The amount of residual gas in microscope systems was still high at the beginning of the 1950s because the greased airlock for the insertion of the specimens was located near the sample, and there were also many greased connections at the gun system and the vacuum valve. The three-stage mercury diffusion pump reached at best a pressure of 10^{-4} Torr. The rubber mixtures used for the gaskets needed a long time for degassing. Despite the presence of so many contamination sources, the problem of contaminated specimens was not yet limiting in the lifetime of the ÜM 100, probably because of the elevated temperature in the large irradiated specimen area.

2.4 Calibration and high-resolution test specimens

A well-defined high-resolution value must be based on a knowledge of the electron-optical magnification. In the case of our microscope the calibration conditions were favorable in that the dimensions used were measured first in a light microscope. Adjustments in the test field were made with a specimen holder containing a hole exactly 70 μm in diameter which was imaged on a circular marking of the intermediate screen. Attached controls referred to light-optical measurements of the visual aperture in the projector and its imaging on the final screen. For casual checking the same specimen holder was found suitable, covered with a flat film pierced with holes. Starting from the light-optical magnification, some characteristic holes could always be detected and the distance between them measured in two perpendicular directions. The test specimens contained structural details usable for the whole range, from the light microscope up to the highest magnifications. All magnification steps were documented by photographs (Elbers & Pieters, 1964).

For a while, replicas from fine diffraction gratings were tested for calibration purposes, but these were soon abandoned because they frequently suffered from contraction caused by the electron bombardment. In addition, they did not contain lines narrow enough to permit the highest magnification values to be checked. Many years had still to pass until the imaging of lattice planes with well-determined spacings was generally introduced as a quick and convenient calibration method (Dowell, 1964).

After 1952, our applications group was engaged in the preparation of artificial test specimens. We wanted to replace specimens having an accidental arrangement, as found in some natural samples, with specimens produced systematically, valuable for a high-resolution test. The next step was to improve the specimens in such a way that at any time and with reasonable effort a resolution of 1 nm could be obtained, not only by specialists but even by personnel with little training.

To check the optical power required specimens that show big changes in the local arrangement of mass thickness in the resolution range but are still of sufficient contrast for the detection of these changes in a highly magnified image. The easiest method for the preparation of such specimens was the evaporation of a heavy metal onto a very thin supporting film, but even a heavy metal could not be imaged with visible contrast if the evaporation layer was too thin. Our intention was to produce extremely thin and fine granular specimens suitable for quick and accurate focusing. In order to do so, focusing aids were needed. Perforated films thickened by carbon evaporation, as successfully tested for astigmatism control, were also convenient for the high-resolution test because the diffraction fringes on the border of the holes were clearly detectable. A second but very thin carbon film covering the holes in such fortified perforated film was prepared as support for the evaporated metal. Only specimen areas located on this thin film spread over a hole were used for testing purposes (Weichan, 1970).

Different evaporation materials were investigated to find the best. In doing this work we could profit from our experience in shadowing replica foils. Light substances such as silicon monoxide were avoided because of insufficient contrast. Beryllium was excluded because of its high toxicity. Evaporated gold and palladium particles melted at the intensities necessary for higher magnification work, thus forming bigger granules. An evaporated film of platinum was found to be best, as, due to the lower vapor pressure, no coarsening of the structure took place. Finally, a platinum-iridium alloy (percentage ratio, 80:20) was used for testing the resolving power in the range between 0.5 and 1 nm. This evaporated film consisted of a large number of small, isolated aggregates of the required dimension. The distance between the centers of two small particles defined the resolution. Several pairs of particles located in different orientations were used to demonstrate the lack of astigmatism. Two successive pictures at the same focal setting were always taken in order to find the identical particles separated on each plate. For the calibration, negatives were never used, but rather we used enlargements of at least four times to prevent measurement errors. From 1954 on this kind of point-to-point separation served as a quality document for each individual microscope.

3. A microscope with higher resolving power

Our increased knowledge of the microscope's capacity allowed us to define our demands for the next generation of microscopes to be developed. An objective stigmator was needed because the high resolution was mainly limited by the axial astigmatism. The illumination system had to be improved to get sufficient image brightness for the work even at much

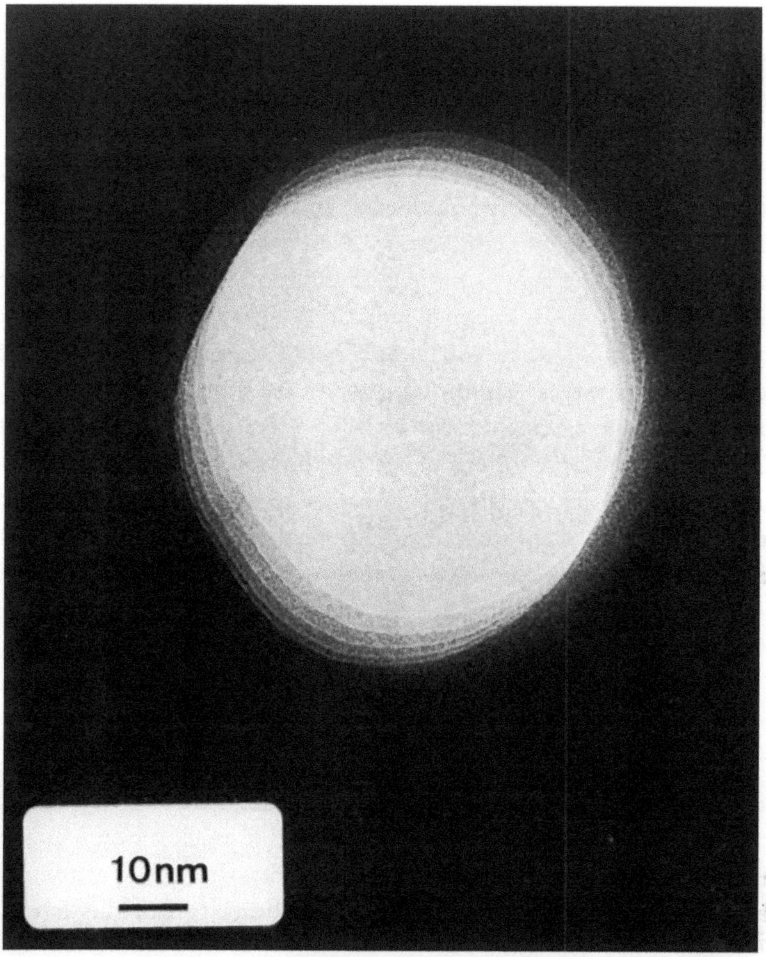

Figure 2 Five-times-exposed micrograph of a hole in a carbon film over a time period of four times 5 min showing specimen drift, focus stability, and the deposit of contamination layers.

higher magnifications, but without the risk of damaging the specimen by temperature influences. Our other wishes were again concerned with better mechanical and electrical stability and greater convenience of manipulation (see Fig. 2).

Most of these demands were fulfilled in the summer of 1954, when an improved high-resolution microscope, the Elmiskop I, was brought on the market. To use the high resolving power, the maximum magnification of the new instrument was raised once more, up to 160,000× for 100 kV of acceleration voltage, to ensure that even the smallest specimen details would not be confused with image noise. A mechanical stigmator as corrector for the objective lens now made it possible to oppose the residual astigmatism by setting up a cylindrical lens field in any desired orientation and strength. The residual astigmatism of the objective lens had been strikingly reduced by advances in fabrication and by care taken in assembling the polepiece system. Aperture slides, now equipped with three instead of two apertures, permitted the exchange of contaminated apertures without interruption of work and, at the same time, a choice of different hole diameters according to contrast conditions. By the incorporation of a double-condenser system, a critical overheating of the specimen could be prevented even for the work at high intensity, as diminishing the beam spot diameter lowered the temperature in the irradiated area.

The electrical supply system of the microscope was so far modified that the acceleration voltage remained constant to within 10^{-5}/min and the objective-lens current did not vary by more than 5×10^{-6}. These stability values offered a chance to use longer exposure times and made it possible to give up incidental focusing and to take focus series in fine steps (Ruska & Wolff, 1956).

3.1 Contamination and an anticontamination device

Not experienced in the use of a double-condenser system, we tried to work first under the most promising conditions, that is, at the highest electron-optical magnification in the light of the smallest beam spot, 3 μm in diameter, and we were fascinated by the idea that the specimen under these conditions would not suffer from overheating. But this fascination did not last long. We soon learned that the use of a fine beam created severe contamination problems unknown until then. Lowering the temperature on the irradiated field increased remarkably the speed of contaminant growth, as the contamination increased at a rate inversely proportional to the temperature. Advantage could not be taken of the higher electrical and mechanical stability as the specimen in the light of the smallest spot was covered within minutes by a layer consisting of cracked carbon-hydrogen vapor. Due to the increasing thickness of the contamination, the specimen border was no longer sharp and the focal setting became undefined. We learned by studying the literature that the contamination problem had already bothered the scientists in the early days of developing the cathode-ray oscillograph, and that, in addition to the temperature and the amount of residual gases, the beam density was one of the most important factors.

There was little chance of reducing the intensity. For the highest magnifications a certain brightness was necessary for the observation and for setting the focus. We experimented temporarily with lighter fluorescent screens but stopped this work soon because of the coarser granularity of the fluorescent phosphor. The only reasonable way promising success was the removal of gaseous material from the neighborhood of the specimen. We tried to improve the vacuum in the microscope by cleaning carefully all parts of the column, by heating procedures, by the insertion of cold traps in the pumping system, by the introduction of new chemical cleaning procedures, and by the construction of auxiliary tools that allowed the manipulation of the specimen holder. The specimen chamber was closed upward and downward with fixed apertures. But none of these attempts succeeded in lowering the deposition of contaminant layers. Microscopical work using the smallest beam spot promised success only if the contaminating vapors could be removed from the vicinity of the specimen. To overcome this problem the working mode was altered. The beam spot was diminished, but only to a diameter between 5 and 10 μm, so that the illuminated field, larger than at a fully excited fine-focus condenser, was still smaller than at large-field illumination when using one condenser system alone. The consequence of this working method was a higher temperature than in an area illuminated with the smallest spot but still a much lower temperature than in an irradiated large field, so that destruction by temperature could be mastered. Otherwise the temperature was sufficiently elevated to retard the rate of contamination.

Working in the light of a medium-size beam, the problem of contamination could be temporarily neglected for many applications. However, for imaging beam-sensitive material the smallest beam spot had to be used to prevent thermal destruction. Such specimens could be preserved only at the lowest possible temperature, and that meant a higher rate of contaminant deposition.

A new problem arose with the introduction of ultramicrotomy, not yet fully controlled by some users in the early days. Residues of the chemicals used for polymerization caused an enrichment of gassing vapors on the specimen, thus providing an additional source for the formation of contamination layers.

A prototype of an anticontamination device was constructed in 1954 but was not at first found satisfactory. It was a matter of course that the vacuum in the vicinity of the specimen had to be improved in order to reduce the amount of residual gas. A cooled cartridge was built for investigations at low temperature, but served as a cryopump used mainly to keep the specimen free of contamination. The temperature of this special cartridge, cooled by liquid nitrogen, was controllable and could be kept in the most favorable range of approximately — 80°C. We noticed soon that a cooled specimen container was not an ideal anticontamination device for various reasons. The temperature balance always took between 15 and 30 min, and during this time the specimen drifted. When removed from the microscope for the exchange of specimens, the cartridge was immediately frosted and had to be dried carefully before every new insertion. In addition, water vapors forming ice deposits at apertures caused a deterioration of image quality due to astigmatism. Such disappointing observations showed that an effective and practicable anticontamination device should not include the cooling of the specimen in observation.

In 1955, systems were developed that enclosed the isolated specimen in a cooled chamber. All the vapors forming contaminants were frozen on the cold walls of this equipment. When the specimen was illuminated with the smallest spot, weak contamination layers were still formed because the vapors could be completely removed only under ultrahigh-vacuum conditions. In general, however, the contamination problem was mastered (Weichan, 1965). Occasionally, protracted irradiation of the same specimen field caused some problems, at high electron-optical magnification, for instance, when a series of pictures was taken at higher intensity, but these inconveniences belonged to a later development phase. About the year 1955 lengthy observations were required only in the case of weak contrast, as focusing became unsafe and therefore time consuming.

3.2 Contrast improvement and focusing aids

On another occasion we had to return to the contrast problem. In 1954 the relatively young technique of ultramicrotomy had already reached a high standard in the leading laboratories of the Unites States and Sweden. There was an enormous demand for high-resolution pictures of tissues. Ultramicrotomy made extensive work in histology possible as ribbons of thin sections below 0.1 μm thickness were now available (Sjöstrand, 1954). Staining methods were still in the development phase, however, so it followed that the contrast in the sections was mostly weak. In order to improve the image contrast, well-known techniques were once more applied, such as the use of the lowest acceleration voltage, the choice of a small illumination angle, and the selection of a small bore in the contrast aperture. Of course the image contrast at 40 kV was much better than at the 80 kV normally used, but the image quality became poorer. Working in the higher voltage range required the use of the smallest possible objective aperture to secure increased contrast. Conversely, the high resolution suffered through the use of too small aperture bores. In addition, the anticontamination device retarded but could not completely prevent the contamination of the contrast aperture under such extreme conditions. Focusing aids were needed for correct focus setting.

Focusing aids had been known since the beginning of electron-microscopical work. Randomly attached dirt particles, such as carbon black, served as focusing aids because of the good visibility of fringes. In cases of very weak contrast a small quantity of higher contrast substances was added to the specimen. Latex balls were employed early, not only for focusing but also because their diameter was so constant that they gave a good indication of the image magnification. Unfortunately, it was found disadvantageous that the latex particles, with an average diameter of 0.3 μm, were not situated in the same plane as the specimen and were in focus at another current, thus provoking an additional uncertainty factor. Because of their size, latex balls were suitable at best for work in the medium range of magnification. For a high-magnification check, they were too large. In the early days of ultramicrotomy we gratefully accepted the advice of American colleagues to use gold particles as a focusing aid, as these had been successfully tested since 1951 (Flewitt & Tymms, 1956). If the preparation instructions were carried out carefully and the chemical work properly done, a dispersion from a golden chloride and a sodium citrate solution could be prepared consisting of regular particles in the 1- to 2-nm diameter range. A droplet of such dispersion placed on the specimen surface was removed after it remained there for about 5 min, and in that short time only a few of the strongly contrasted gold particles, statistically distributed, were sedimented. At least one particle was found in the neighborhood of an area to be imaged and was there used as a focusing aid. The fringes were clearly visible and the objective current needed to focus the focusing aid was just the same as that needed to focus the low-contrast specimen. Even this method was not found to be ideal, however, as particles not belonging to the specimen had to be added, and not infrequently these were deposited on structural details to be studied. A much better way of solving focusing problems was to use double films, as successfully tested before for high-resolution checks. By observation of the fringes appearing at the border of the holes in the thick support film, the focus point could be seen easily. The second film stretched over the holes could be delicate but was still stable enough to carry a specimen. Supporting films as generally used could not be prepared too thin because they had to cover much larger openings in the specimen grids and to resist heavier mechanical and thermal stresses (Sjöstrand, 1956).

Today nobody bothers about artificial focus aids in thin-section work because the granularity of carbon caused by phase effects has been found to be a most suitable focusing aid when working at higher electron-optical magnification (Thon, 1967).

3.3 I take up the aperture problem again

In 1954, we occupied ourselves once more with modification of the objective apertures. We could now make use of multihole apertures because in the meantime the stage movement space had been extended. Multihole apertures in the form of small sieves containing either 19 or 37 holes with diameters of 50, 30, or 20 μm were arranged in circles around a central opening. The distance between the individual holes was chosen in such a way that the fine beam only irradiated the border of the aperture located in the center but not the adjacent holes, which remained free of contamination. Doubtless, the use of multihole apertures helped to simplify the microscopical work, but even this was not yet the final and best solution. The material thickness around the hole channel was still too great to stop the conductivity over the metal. These areas still remained at a low temperature, thus encouraging contamination. The depth of the mechanically drilled channels was still 30 μm. Another production method was required, and in 1959 we began to replace the platinum-iridium disks by metallic foil apertures. The disks still remained in use but only as a matrix for the production of foils. A thin copper or silver layer was evaporated onto the front side of these apertures, there forming a replica. The thin film was removed by floating, then picked up and fixed between two rings for insertion into the ordinary aperture holder. Using this method, the depth of the channel could be reduced below 10 μm, at best even below 7 μm. Thinner replicas were too fragile and broke during the mounting procedure. The thin foil grew quite hot at the border of the hole, and now the thermal conductivity became insufficient. It now took much longer for contamination layers to form. Even metallic foil apertures with small holes could be used for a much longer time. We succeeded in doing our microscopical work utilizing 37-hole metallic foil apertures for a full year without any necessity for exchange.

3.4 Extended application of microscopical techniques

Though the main point of our work in the applications laboratory was the use of high resolution, we never neglected other possible applications. To these belonged all kinds of diffraction studies, such as diffraction without imaging lenses, the simultaneous diffraction of a known and an unknown specimen, and diffraction with grazing incidence on conducting materials. Diffraction studies using the imaging lenses became increasingly important because the fine diffraction method enabled the precise coordination of image area and diffraction down to a smallest specimen area of only 2 μm, and was therefore chiefly essential for technological investigations.

An extensive array of accessories was developed in response to the demands of the users. Some users wished, for example, to observe alterations in thin etched metal during a heat treatment and to record this process at the same time with the aid of a movie camera. Customers called for goniometer devices for correct adjustment of their sections because of the difficulty of cutting precisely in the transverse or longitudinal direction. In X-ray microanalysis, another very important technique was being developed to obtain additional information about the chemical composition of a specimen.

To do routine work, even untrained personnel could be quickly taught because of the superiority of the high-resolution microscope. Until the mid-1960s, most buyers decided to purchase the high-resolution microscope, though we had offered since 1950 a medium-class instrument for routine investigations (Krisch & Weichan, 1971; Weichan, 1971). Two reasons probably lie behind this decision: The 1950s and early 1960s were a good time for new investments in research institutes because funds were readily available, and a big electron microscope was regarded as a prestige instrument. On the other hand, it was well known that following the purchase of an electron microscope no more money could be expected for a long time, and therefore a maximum of technical possibilities, not yet for present but for future work, was bought on the same occasion.

3.5 One end point: 0.4-nm high resolution

The extensive work in our and many other continually growing electron microscope laboratories all over the world created an increased interest in this field of science. The results were always considered spectacular and indeed they still are with the present-day results in high resolution. In 1962, our development laboratory could announce a new top resolution due to further instrumental improvement. The use of a pointed filament made it possible to get sufficient brightness on the screen at a final magnification of 200,000× at 80 kV.

The first tests using image intensifiers ended successfully (Herrmann et al., 1972), offering hopes for the imaging of beam-sensitive substances of technological importance such as polymers. The vacuum of the microscope was improved by the installation of more powerful pumps. Electromagnetic beam-adjusting systems were tested, and the recording system was changed to automatic operation. The lens-cooling device became temperature controlled, giving supplementary stabilization to minimize disturbances due to drift.

On the occasion of the Fifth International Congress for Electron Microscopy in Philadelphia in 1962, twin pictures with a point-to-point resolution of 0.4 nm were demonstrated for the first time. Advances in the high-voltage system as introduced by Wolff (Wolff et al., 1962) should be considered as mainly responsible for this gain in optical quality. The measuring resistor, situated until this time in the oil of the high-voltage tank, was removed and placed in a separate vessel in a protecting gas atmosphere. As a result, the tension remained highly constant over a long period (see Fig. 3).

Temporary advances in preparation techniques were now compensated by a rise in the quality of the microscope itself.

4. Development and improvement of preparation methods

4.1 Investigation of rubber and filling material

Since the beginning of electron microscopy, technologists were interested in investigating rubber mixtures in order to study the properties of the filling material and to detect how different production methods were influencing the structure. Filling material had already been frequently observed, but a convenient method of imaging the whole rubber system did not yet exist. The known preparation techniques were not appropriate to make specimens thin enough for transmission. Various methods for thinning were tested although it was realized that the value of the information provided by maltreated samples was limited. Small rubber pieces, swollen with organic solvents, were gently passed over a specimen grid. Under the influence of this mechanical stress, a system of fibers and islets containing filling-powder particles was formed, but such specimens hardly showed the original arrangement (Endter, 1952). Improved methods of imaging larger areas not

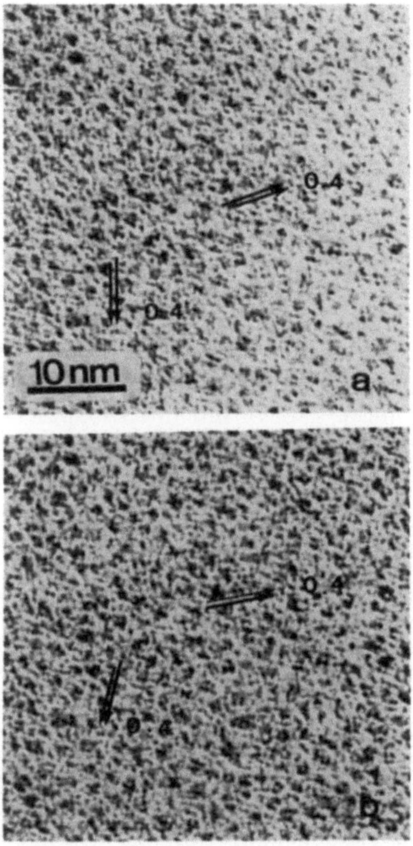

Figure 3 Point-to-point resolution test (distances shown in nanometers) (1962). Two successive micrographs (a) and (b) of an evaporated Pt-Ir alloy on a perforated carbon film: electron-optical magnification, 200,000:1; acceleration voltage, 80 kV.

yet damaged by the manipulation were needed. A promising method was found in a modification of the freeze-sectioning technique at the time when I started my work in electron microscopy in 1950 (Kruse, 1951). Using a Leitz sliding microtome equipped with a freezing stage, we succeeded, though with much waste, in the production of transparent sections. For this, the surface of a little rubber piece was moistened with benzene before sectioning to increase its thickness by swelling. The idea was that the sections, normally too thick for transmission, would be thinned afterward by the evaporation of the solvent. The preparation was extremely difficult, and only the developer was well acquainted with this method! Not only was the choice of glue to fix the rubber piece for freezing problematic, but even more so was the handling of the sections. These adhered to a support only if they were extended fixed by means of preparation needles at the moment the benzene evaporated. Despite these complications we were soon capable of producing rubber sections. In transparent areas of these sections the shape and the size of filling particles was clearly visible. In addition to dispersed particles, aggregates were imaged, and in this way interrelations could be derived between the technical properties of a rubber mixture and its structural composition (see Fig. 4).

In the early days the main point of interest was the development of adequate preparation methods usable for controlling industrial products. As members of a microscope-producing company we tried to find new fields of application to promote sales of our microscopes. Thus the first experiments on rubber led to further studies concerning its behavior during straining. Clearly, straining of organic specimens under the beam is critical, as even after a short time carbonization takes place. To analyze the strain process an ordinary specimen cartridge was provided with a straining mechanism activated from the outside by means of a drive (Weichan, 1955). Two flat planes pressed together to a narrow slit held the specimen, but this construction did not stand the test in practical work (Fisher, 1959). Symmetrical fastening of the specimen using a conductive lacquer could not be done easily, and quite often the rubber section slid off during the straining procedure. Even if the investigation in the microscope took place within seconds, the specimen elasticity had gone. For this reason, we stopped this way of straining after a short while and later exerted a well-defined strain before the insertion of the specimen in the microscope. Nevertheless, this strange method allowed us to establish that the rubber system was loosened by weak straining, while the filling particles kept their arrangement (Grasenick, 1961).

Better preparation methods for powders made possible an effective investigation of diverse filling substances. Powders were among the first specimens ever studied by means of electron microscopy because such samples could not be viewed under a light microscope. But even in 1950 the preparation technique was far from ideal. For technical reasons it was urgently necessary to record more precisely the shape and the dimensions of numerous particles located in different areas of the specimen.

For statistical investigations single powder particles must be imaged distinctly. Such samples unfortunately tend to form agglomerates, from which it follows that measurements of clumped particles could not be carried out exactly. As the simplest preparation method, dusting was used but was

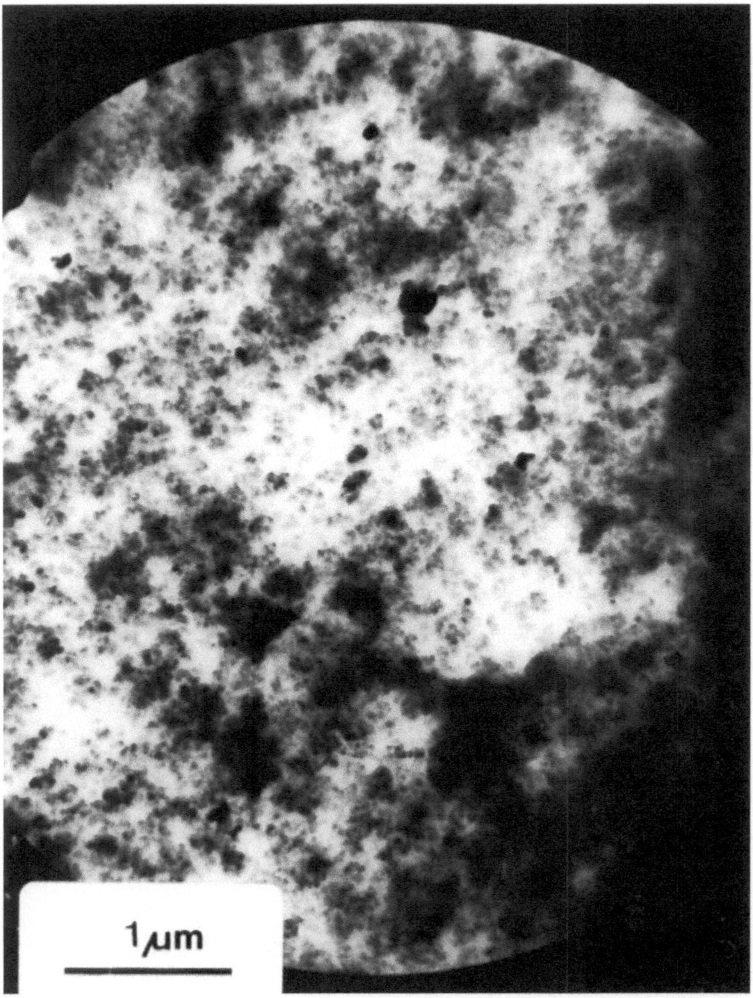

Figure 4 Rubber section cut on a sliding microtome equipped with a freeze stage (1951).

rarely found convenient. Particles dispersed in liquids by ultrasonic treatment again formed aggregates when drying on the support. A preparation method in which single particles were deposited by using thermal or electrical fields was found troublesome because it was nearly impossible to clean the containers thoroughly. Mixing collodion with the sample in order to prepare support and powder simultaneously proved to be suitable only in the case of fine and uniform powders, as the support broke under the load of heavier and more extended particles. Finally, it was thanks to auxiliary equipment assembled in the laboratory of a customer who had to struggle with the same preparation difficulties that a satisfactory solution to this problem was found (Kehler et al., 1954–1955). In the first years of electron microscopy especially, it happened quite often that users who had to solve special problems extended the applicability by developing their own new preparation techniques. After many fruitless efforts, a fine dispersion of powders was now obtainable by atomizing a suspension in the sonic field of a 3-MHz ultrasonic quartz oscillator.

For calibration we needed pictures of particles with a sharply determined outline. The use of higher electron-optical magnifications was advisable to render visible the shape of each individual particle. The stereoscopic method for three-dimensional studies was increasingly applied to electron-microscopical technology in order to prevent errors in interpretation (Helmcke, 1954). In this way it was frequently found that clumps in rubber vulcanizates really consisted of single particles located in different planes. All had appeared in focus on a single picture. Coming back to the rubber problem, our investigations confirmed that there is an intimate relation between the formation of clumps and the breaking stress and between the size of the particles and the material's resistance against breakage. Rubber technologists acknowledged these results but it was some years before they honored them. At the beginning of the 1950s, most industrial plants were not yet willing to bear the high expense of an electron-microscopical laboratory, being convinced that a foreman in the factory could judge the suitability of a filling material because of his practical experience in this field.

4.2 Further applications: modifications of the replica technique

The improved optical capacity of the electron microscope justified hopes that still finer structures could be imaged by using higher electron-optical magnifications. Besides the enlargement of magnification range, the microscopes were supplied with many new accessories which had to be tested by us before they were handed over to the customers. Thus the members of the applications laboratory were the first users, and often their practical tests stimulated the development of additional techniques. It happened not infrequently that a suitable preparation method had first to be worked out before the testing of newly developed accessories was even possible. We were fascinated by the idea of a final magnification of about 100,000×. But at first the specimens required for such a high magnification were not available and the microscope performance was not yet good enough. In addition, we had to learn how to use our own microscope best. Instruction manuals were not yet available at this time, and all advice concerning the manipulation of the microscopes had to be worked out first by us in the applications laboratory; such a situation caused many disappointments. Over the years we experienced the following situation again and again: Interested people came full of hope with their own specimens, which were either too thick, too fragile, or too low in contrast, but always they expected images at highest magnification. Not infrequently, the disappointing result was a fuzzy image which often appeared out of focus in one direction due to specimen drift.

While the physicists and engineers were busy developing the electron microscope still further, my colleagues and I worked in the applications laboratory on perfecting the preparation techniques.

Ultramicrotomy was still in the experimental stage and the surface scanning microscope was not yet in practical use at this time. This meant that thin sections of biological material were not yet available, and all other samples could be investigated only if they were transparent or could be thinned by mechanical or chemical means. Such drastic treatment caused the structural elements to become detached from their connections, so that although some information about structural details could perhaps be acquired, only in the most optimistic case could speculations be made about the original structure.

Because of these restrictions the first years of postwar microscopy were the most successful period for surface replica work. On the basis of early publications (Mahl, 1940), the replica technique was modified in many ways for indirect imaging. At the beginning only smooth surfaces were replicated correctly. For the preparation to be true to nature, the replica material had to be free of structure and flexible enough to image even the slightest unevenness of the specimen surface. The replica method was used by us preferably for the investigation of technically important metals and fibers. Depending on the special problem to be solved, either single-stage or double-stage replicas were prepared. In the case of metallurgical applications, single-stage oxide replicas were most suitable because they replicated cavities as well as elevations, thus forming a true copy of the surface (Hunger & Seeliger, 1947). The plastic replica method was more versatile. It could be applied to a variety of materials and was therefore favored, though such replicas were difficult to handle due to the delicacy of the thin film. Furthermore, the whole procedure took too much time to do a sequence of investigations, as was necessary for repeated imaging of the same surface after various treatments. We were confronted with the need to get information about rough fracture surfaces rapidly and we therefore searched for more suitable matrix substances. Such a material was found in acetobutyrate foils with the trade name Triafol, frequently used as an insulating material in electricity. The surface of a piece of this foil was wetted with acetone and pressed onto the surface of interest. Under the influence of the solvent the foil took up the contours of the specimen and could easily be peeled off after drying a few minutes later. As the coating material, silicon monoxide was found best because of the high stability of the thin evaporated layer when dissolving the replica matrix. The number of unbroken replica films which showed very fine structural details in the 30-μm range was high (Pfisterer, 1953). The Triafol-silicon monoxide technique was time saving, as the whole procedure from replication to imaging could be carried out in a few hours. This was extremely important for us, as many visitors came to the applications laboratory for only 1 day for information and needed their results quickly.

The accuracy of the replication was considered to be a most important question. This was the reason we occupied ourselves more intensely with the problem of artificial structures caused by varying affinity of the matrix to the specimen surface (Weichan, 1951). The presence of empty regions in a replica led us to suspect that the replication was imperfect. We found it worthwhile to take more than one replica from the same specimen area or, even better, to image a surface using various replica techniques (Heidenreich & Peck, 1943). Especially in the case of rough fracture structures the comparison of replicas obtained by the use of diverse techniques was helpful for interpretation.

Experiments invoking the imaging of wet surfaces by means of gelatin had little success. Due to a poor affinity for structural details the topography appeared too flat. In addition, the heat treatment needed to dissolve the matrix frequently caused breakage of the thin evaporated layer (Weichan, 1952). In order to enhance the contrast and to accentuate the surface topography, the coating material was deposited at a small angle. At a known angle quantitative tests were possible by measuring the length of the shadow. The use of the stereoscopical method helped to avoid misinterpretations about the formation of deep cavities. These, appearing as high elevations in plastic replicas, were frequently bent under the load of metal deposited by the evaporation only from one side. In order to obtain a replica true to nature, this mode of shadowing was given up, and with this the accentuation

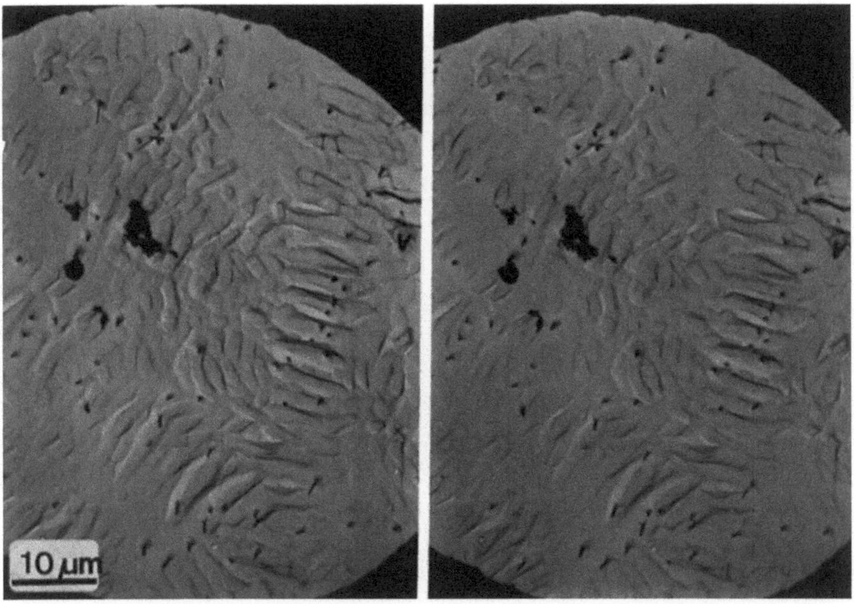

Figure 5 Triafol replica from the surface of a beryllium crystal after etching. Conical evaporation in conjunction with the stereo method permits study of the orientation of the deep cavities.

of topography. We later used chiefly conical shadowing to get a uniform deposit all around the fine structure of the replica (see Fig. 5).

The improvement of the optical quality of the electron microscope enabled the imaging of even the tiniest surface irregularities. To make these visible the most suitable angle of shadowing had to be carefully found. Larger shadows, which might hide small asperities, resulted from too low a shadow angle. There was another problem with the use of a very small shadow angle: under these conditions the evaporated material frequently formed artifacts at sharp corners with the appearance of little caps which looked like turrets (Helwig & König, 1950a). Luckily the investigation of most specimens did not require a small shadow angle because they contained large features. Silicon monoxide as light element was shown to be a suitable shadowing material for technological testing normally done at lower magnifications in the medium-resolution range. Silicon was no longer suitable for imaging structural details of molecular dimensions because of its poor scattering power and had to be replaced by heavier metals.

4.3 Shadowing and special techniques

Biological specimens which could be prepared as thin and transparent layers were among the first specimens ever investigated in an electron microscope, but for this they needed to be of sufficiently high contrast for observation. To increase the contrast, most such specimens were shadowcast. With the improvement of the resolving power it became desirable to image much smaller structures than had been detected hitherto. This made it necessary to seek appropriate materials which did not form aggregates under the influence of the beam, thus covering the finest specimen details. Tungsten oxide, which forms extremely fine granular layers, proved to be such a material (Fig. 6). The only obstacle was the fact that tungsten oxide was not commercially available and had to be prepared for every need by heating tungsten wires in air (König et al., 1951). For many years this was the only way of revealing the finest structures; the negative-staining method was later introduced as a substitute method.

Nearly all heavy metals tended to condense and formed granules coarser than the specimen fine structure. Under these circumstances the high resolving power could not be exploited. This was a compelling reason to change the preparation technique. Instead of evaporating, a metal cathode was atomized in an oxygen atmosphere by glow discharge, thus forming much smoother deposits (Helwig & König, 1950b). This method, first used only for platinum to obtain very fine granular layers with high contrast, was soon modified and expanded with a view to its possible application in other domains (Helwig & König, 1951). Layers of cracking products could be produced by glow discharge in a benzene atmosphere, usable both as supporting films because of their high-temperature resistance and mechanical stability and for the stabilization of sensitive specimens.

These cracking layers proved to be most valuable for the preparation of replicas taken from very small samples which, due to their density, could not be imaged before. The specimen was enclosed completely within a stable layer which still remained permeable to chemical solvents. In this way, many small crystals were imaged, as after the substance had dissolved replicas remained showing the inner structure of each individual particle and the construction of the whole assembly (Grasenick, 1956). This special kind of carbon replica method was used in our laboratory for studying crystals which were difficult to dissolve and for the investigation of mixtures. Dissolving only one component in such a case made it possible to observe the composition of a mixture (see Fig. 7).

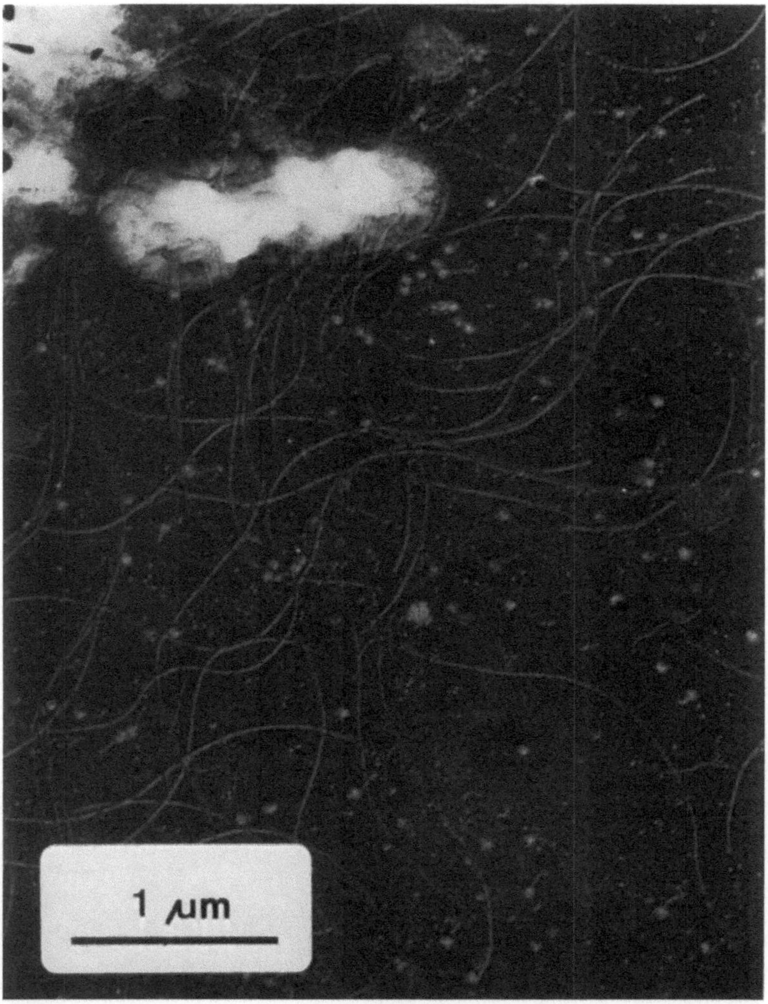

Figure 6 Bacteriophage images taken with tungsten oxide evaporation to better show the specimen fine structure.

Other experiments were started in order to detect very small quantities of metals by using electrolysis (Wiesenberger, 1955). Such experiments were extremely difficult and required great manual skill because the electrolysis took place on a tiny field of the specimen support. By intense electron bombardment the supporting film was transformed into conductive carbon while the metal surface around the holes became insulated by a thick deposit of lacquer. On the conductive film used as electrode, all the

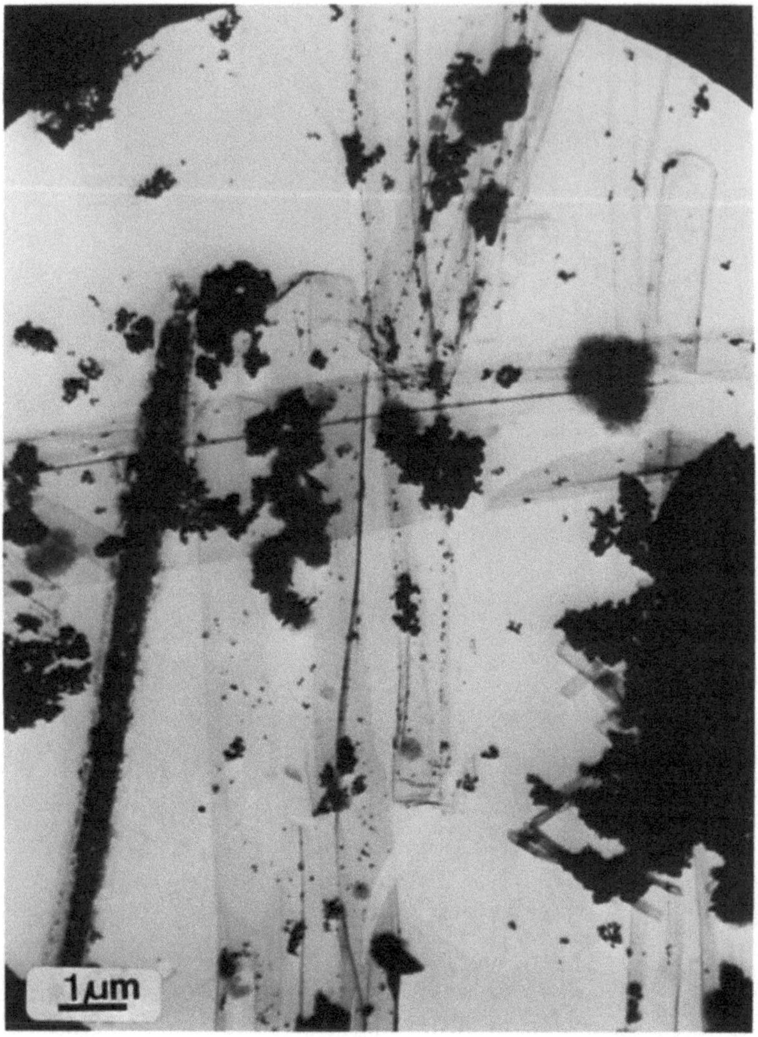

Figure 7 Palladium-asbestos carbon envelope replica; the asbestos fibers were dissolved with fluorine acid; the palladium remained undissolved.

metal was deposited. This method permitted direct observation of all the electrolytically precipitated metal and could be applied for imaging substances in quantities below 10^{-9} g (Wiesenberger, 1957).

Other microchemical experiments for checking whether a total reaction product could be imaged on the supporting film were also successful. The specimen holders were prepared as described for electrolysis. Reaction products resulted from two solvents diffusing through the film from both sides. The reaction product deposited in this way showed the total deposit and could be investigated altogether in the electron microscope. It was found surprising that on a tiny surface of only 500 μm^2 a chemical reaction could still take place and that even a local limitation was possible (Wiesenberger, 1960).

This selected choice of applications does not mean that only the preparation techniques mentioned here were used by us, but, as they were either developed or improved in our laboratory, their description belongs to the remarkable sequence of events of the period from 1950 to 1960.

5. Teaching duties in an applications laboratory for electron microscopy

Looking back over 30 years' activity in an applications laboratory, I remember not only the technical highlights but also the pleasure of observing excellent pictures frequently not obtained before preparatory problems, which at first seemed insoluble, had been overcome. But with the same pleasure I think back on many hundreds of microscopists who visited our laboratory to be introduced to the manipulation of the microscope they had bought. It was truly a special teaching duty. First we ourselves had to learn how to use our own instrument best before it was possible to hand this knowledge over to others. A continual adaptation was necessary to instruct the well-known scientist together with a young technician at the beginning of her career or an expert from industry alongside a worker in fundamental research. After the mid-1950s the majority of our microscopes were exported and as a consequence some knowledge of foreign languages was required for this instruction. Rudimentary linguistic skills dating back to our school days had to be polished up, and luckily in most cases we succeeded in making ourselves understood. If not, some manipulations were taught by means of sign language.

We needed much steadiness, patience, and empathy to provide instruction not only about operation but even more important, to teach correct vision. When starting electron microscopy it was hard for nearly everyone to detect astigmatism well enough to judge the value of a correction procedure. Only if the microscopist saw the unequal fringes was he capable of compensating it. Without exception all our trainees had to repeat this special kind of correction again and again because of the dominating influence on the image quality of astigmatism. It is therefore not astonishing that the pain of learning to detect astigmatism resulted in amusing drawings and remarks in our visitor's books. Here are some samples in the languages mainly used for the teaching:

> *Speak to us of aperture, alignment, Angstroms, Fresnel fringes*
> *And how we can astigmatism cure—on this the image hinges.*

> *Je remercie vivement les personnes dans le laboratoire de m'avoir délivré de la peine de l'astigmatisme, qui contaminait mes rêves ....*

> *Da sitzt er nun der arme Tropf und dreht an dem verkehrten Knopf.*

We also had to fight against the prejudice that a high-resolution microscope would only yield good results in the hands of qualified physicists. On the other hand, it was necessary to explain that excellent pictures did not automatically appear. In a great many cases, the specimens which the customers brought with them were not suitable for microscopical work at high magnification. Sometimes we had to be very diplomatic to convince beginners that for the solution of their special problems a medium magnification would be more effective than the highest possible magnification. Concerning problems which required the use of high magnification to gain the best possible resolution, we had to make the beginners understand that the attainment of the highest optical performance always demanded a greater effort. Learners had to be taught that single photographic snapshots, taken by chance, were only of value for superficial information but not for exact research work. For users accustomed to light microscopy, the image with the strongest contrast was expected to be the best, and it was a new experience for them to accept the point of minimum contrast as a correct focus setting (Gansler & Nemetschek, 1958).

For the training, we compiled collections of micrographs in order to demonstrate how optical faults can spoil image quality and to teach at the same time the basis of electron-optical imaging. An important duty was the making of two different kinds of manuals, one with an extensive description of construction, function, and mode of working, and the other in the style of a cookbook, describing the manipulation in brief form step by step. Instructions resulting from our own experiences prevented users from wasting too much time with adjustments, sometimes not necessary for their special investigations. Everyone understood that a high-performance electron microscope could not be operated by turning a single knob. But on the other hand a great many users thought that the numerous controls available must always be employed, as a result of which they condemned the microscope as being too complicated. One of our main duties was therefore to demonstrate how easy manipulation was for most routine applications.

Some errors were typical of beginners. Consider, for instance, the lifetime of the filament, frequently criticized as too short. The reason for this complaint was that, without necessity, the tungsten wire was overheated, higher than required for saturation, so as "to be sure," as the user told us. Another typical mistake was the destruction of the specimen by thermal overload: when inserting it in the beam the user had forgotten to put in first a protecting condenser aperture. Our discussions of all conceivable errors were important to teach how to avoid errors such as these.

Though most of the visitors were trainees, another group was even more important from the commercial point of view, namely people interested in checking whether the microscope was suitable for their specialized research work. They visited the applications laboratories of the different companies with a catalogue of questions and with specimens more or less usable. Anyone lucky enough to present his own microscope convincingly and to produce a number of high-quality pictures could hope to get an order.

On the basis of such exploratory visits, the development of additional accessories, needed for technological studies, was stimulated. Steadily we tried to adapt the mode of working to the customer's demands. Thus we experimented for about a year trying to improve the contrast of thin-section pictures until, in 1955, the first Elmiskop I was ordered for the Rockefeller Institute in New York "because of such amazing micrographs." My obligation was to supervise the assembly of this microscope and to keep the members of the laboratory informed about this installation. It was not only the first delivery of one of our microscopes to America but also the first time that the training course took place in the customer's institute and not in the applications laboratory of the plant. Some years later when the microscopes were sold worldwide, it happened more often that for economic reasons the course was given in the customer's laboratory. It was understandable that the customers regretted such decisions as they would have preferred to travel to us in Berlin. Instead we visited them, being capable of informing more people at the same time in this way. For me this meant traveling all over the world. It was a fascinating adventure, even if I had little chance to see the beauty of the countries I visited. In most cases it was more important to perform the astigmatism correction than to see the sights.

Whenever time was available between the nearly continuous training courses, we did our own microscopy to get as many instructive pictures as possible from all fields of application. Information from different projects converged in our laboratory, thus contacts could be arranged between scientists active in similar fields of application, stimulating in this way a fruitful collaboration. We became a central information exchange and even years after they had participated in our training courses customers contacted us asking for help in critical preparation or microscope situations. We were proud that we could help, usually from our own experience, or by contacting other more qualified specialists.

The national and international meetings of electron microscopists offered excellent opportunities for advanced training not only by receiving new information in the lectures but also by way of dialogues between the former teacher and the learner. I remember that I stayed most of the time in the lobby and had only little chance to go to the lecture halls and hear about new developments in the field. Instead of this I talked about problems and tried to find ways of solving them. The majority of the users during the first 15 years of my teaching profession in electron microscopy were biologists and physicians. They put their questions without fear, expecting that I as a trained biologist could understand them better than a physicist.

6. Final remarks

The microscope family grew, first slowly but then more rapidly mainly as a consequence of progress in the techniques of preparing biological material. We counted, in Europe, before 1948, 16 laboratories doing research work in electron microscopy. Ten years later, in only 7 months, 18 new laboratories were established. The improvement of the preparation techniques enabled the user to make his choice realistically and not merely on the basis of the successful imaging of some selected samples. Now each specimen of interest could be investigated. Toward the end of the 1950s, research work by chance ended, mainly thanks the development of ultramicrotomy and to the introduction of such new techniques as negative staining and freeze etching. The morphological fine structure of all tissues of interest was already known. Electron microscopy began to lose its position as an independent discipline and became allied to other disciplines in biology, for instance, to cytochemistry and cytophysics.

The number of biologists and physicians among the electron microscopists rose permanently so that by around 1960 nearly 70% of the users came from these research fields. The high resolution of the electron microscope was increasingly used for fundamental research work in physics, and its technical applications were extended by the development of analytical accessories.

Figure 8 A funny drawing from a visitor's book.

Today electron microscopy is a very valuable technique as it relates to biology, chemistry, and physics. It is therefore understandable that electron microscopy itself does not play a unique starring role as it did in the beginning when we were "only electron microscopists." Nevertheless, numerous scientists still feel strongly about this marvellous research instrument, as expressed by an amusing drawing found in one of our visitor's books (Fig. 8).

7. Afterword by the late Dipl.-Ing. Maren Heinzerling[1] (Deutscher Akademikerinnenbund e.V.)

Cilly Weichan (Fig. 9) was an astonishing woman of many different talents und interests. Born on 16 January 1922, in her youth she was affected by the difficulties of a vanquished Germany. From 1940 to 1944 she studied biology, chemistry, and physics at the Friedrich-Wilhelms-Universität

[1] 1938–2021.

Figure 9 Dr. rer.nat. Cilly Weichan.

Berlin, since 1949 named Humboldt-Universität zu Berlin. During her doctorate in January 1945, her professor Dr. Kurt Noack told her to leave Berlin; for young women especially it was too dangerous to stay. She fled to the countryside and returned one year later, in order to finish her doctorate in 1948. In 1951 her institute was disbanded und Cilly found herself unemployed. Half a year later she came in contact with Siemens through a friend. An applications laboratory for electron microscopy had to be assembled. She explained that physics had always been her main interest, so she was accepted for the department. This task determined her subsequent life.

She worked together with the later Nobel prize winner Prof. Dr. Ernst Ruska, participated in all developments, was the leader of the customer advisory service all over the world and became a "scientific employee of

Siemens", which meant she belonged to the management. At that time Siemens was still a very conservative company: Beside other restrictions, managers of the higher level had to present their fiancé to the board, had to ask for permission to marry or to get divorced. Nevertheless Cilly decided not to marry when she became pregnant. She wanted to keep her independence. The parents of her son Andreas (born in 1962) lived in two different cities but took both care of the child – quite a rare situation in the 1960s. As a result of this arrangement, little Andreas often traveled from Berlin to Munich and back. But Siemens had to accept Cilly's decision, because she was too important for the development of the electron microscopes and for the treatment of clients. She was not only extremely diplomatic and competent, but also spoke four languages, English, French, Italian, and of course German.

In 1981 Siemens ceased production of electron microscopes. Consequently Cilly's career also ended; she was 59 years old. But she had a lot of other interests. She was interested in art and history and traveled several times to Italy. She also was one of those women who revived the Berlin group of the Deutscher Akademikerinnenbund (German University Women's Association) and she continuously participated in their cultural life. I remember her lecture in 2009 "Mein Leben im Hause Siemens – Berufliches und Authentisches" told with many anecdotes. She had really enjoyed her professional life und therefore she spoke in a very relaxed way about all the appreciation she had received. But she was also aware that after the war there might not have been much male competition, men did not survive the war. But that would change and our female scientists should be prepared. In 1991 she was appointed to the "Kuratorium der Technischen Fachhochschule" (Board of the Technical College) as a representative of the women of Berlin, a position she held until 2003. She focused on questions of women's emancipation. Her advice and her motto: Work as hard as men but always remain a lady.

Announcing her death her son Andreas wrote: "She had an unconventional and happy life". It seems to me she had also had a very exciting, respected und successful life.

Berlin, 11 March 2021

We also draw attention to a long chapter by Ulla Ruschhaupt (later Diallo-Ruschhaupt) on 'Cilly Weichan, Als Frau unter Männern im Management eines großen Konzerns' in H. Wüst, U. Ruschhaupt, and

S. Ehrhardt (Eds), Profile – Frauengeschichte(n) der Technische Fachhochschule, Berlin (2004), pp. 254–265.

References

Dowell, W. C. T. (1964). *Optik, 21*, 26.

Elbers, P. F., & Pieters, J. (1964). *Journal of Ultrastructure Research, 11*, 25.

Endter, F. (1952). *Kautschuk und Gummi, 5*, 17.

Fisher, R. M. (1959). *Review of Scientific Instruments, 30*, 925.

Flewitt, T. H., & Tymms, P. (1956). *Nature (London), 177*, 98.

Gansler, H., & Nemetschek, T. (1958). *Zeitschrift für Naturforschung. B: Anorganische Chemie, organische Chemie, Biochemie, Biophysik, Biologie, 13B*, 190.

Grasenick, F. (1956). *Radex Rundschau, 4/5*, 226.

Grasenick, F. (1961). *Technische Rundschau, 21*, 3.

Heidenreich, R. D., & Peck, V. G. (1943). *Journal of Applied Physics, 14*, 23.

Helmcke, J. G. (1954). *Optik, 11*, 201.

Helwig, G., & König, H. (1950a). *Optik, 6*, 111.

Helwig, G., & König, H. (1950b). *Optik*, 7, 294.

Helwig, G., & König, H. (1951). *Zeitschrift für Physik, 129*, 491.

Herrmann, K.-H., Krahl, D., & Rindfleisch, V. (1972). *Siemens-Forschungs- und Entwicklungsberichte, 1*, 167.

Hunger, J., & Seeliger, R. (1947). *Metallforschung, 2*, 65.

Kehler, H., Koch, A., & Tesser, K. (1954–1955). *Zeitschrift für Wissenschaftliche Mikroskopie, 62*, 521.

König, H., Knoch, M., & Winkler, A. (1951). *Naturwissenschaften, 38*, 241.

Krisch, B., & Weichan, C. (1971). In *Annual Meeting Electron Microscopy Society of America* (pp. 50–51).

Kruse, J. (1951). *Kolloid-Zeitschrift, 122*(2), 65.

Mahl, H. (1940). *Zeitschrift für Technische Physik, 21*, 1.

Pfisterer, H. (1953). *Naturwissenschaften, 40*, 199.

Ruska, E. (1950). *Kolloid-Zeitschrift, 116*, 102.

Ruska, E., & Wolff, O. (1956). *Zeitschrift für Wissenschaftliche Mikroskopie, 62*, 465.

Sjöstrand, F. S. (1954). *Zeitschrift für Wissenschaftliche Mikroskopie, 62*, 65.

Sjöstrand, F. S. (1956). *Experimental Cell Research, 10*, 657.

Thon, F. (1967). *Physikalische Blätter, 23*, 450.

von Borries, B. (1949). *Die Ubermikroskopie*. Berlin: Verlag Dr. W. Saenger.

Weichan, C. (1951). *Berichte der Deutschen Botanischen Gesellschaft, 64*, 29.

Weichan, C. (1952). *Physikalische Verhandlungen, 3*, 104.

Weichan, C. (1955). *Zeitschrift für Wissenschaftliche Mikroskopie, 62*, 147.

Weichan, C. (1965). *Prague, 1964*, 131.

Weichan, C. (1970). *Methodensammlung fur Elektronenmikroskopie, vol. 5*. Stuttgart: Wiss. Verlagsges. (p. 3.1).

Weichan, C. (1971). *Elektrische Ausrüstung, 6*, 15.

Wiesenberger, E. (1955). *Zeitschrift für Wissenschaftliche Mikroskopie, 62*, 163.

Wiesenberger, E. (1957). *Mikrochimica Acta, 3/4*, 527.

Wiesenberger, E. (1960). *Berlin, 1958, 1*, 769.

Wolff, O., Engel, A., & Koppen, G. (1962). *Philadelphia, 1962, 1*, Article E13.

CHAPTER FOURTEEN

From the cathode-ray oscillograph to the high-resolution electron microscope

Otto Wolff[a] **and Peter Hawkes (Afterword)**[b,*]
[a]Formerly Auf dem Sand, Muehltal, Germany
[b]CEMES-CNRS, Toulouse, France
*Corresponding author. e-mail address: hawkes@cemes.fr

Contents

1. Introduction 443
2. Developments up to the end of World War II 444
   2.1 Working toward Braun's tube 444
   2.2 From Braun's tube to the cathode-ray oscillograph 445
   2.3 Further development of the cathode-ray oscillograph into a technical instrument 447
   2.4 The recording oscillograph 451
3. Period of transition 453
4. Improvements in the electron microscope after 1945 457
   4.1 The task ahead 457
   4.2 Improvements in the electron optics 458
   4.3 Stability of the operating parameters 458
   4.4 Replacement of the previous control circuits 460
   4.5 Test equipment 460
   4.6 A new high-voltage system 461
   4.7 Improvement of resolution 464
5. Concluding remarks 465
6. Afterword by Peter Hawkes 467
References 467

1. Introduction

The history of the electron microscope has been described and published by several authors, for example, after periods of 24 and 40 years. The article by D. Gabor (1957) is of particular interest because the author not only traced the technical development of the electron microscope, but

Advances in Imaging and Electron Physics, Volume 221
ISSN 1076-5670
https://doi.org/10.1016/bs.aiep.2022.03.014

also brought in the human details that are often the accompaniment of epoch-making inventions. In his book "Die frühe Entwicklung der Elektronenlinsen und der Elektronenmikroskopie" ("The Early Development of Electron Lenses and Electron Microscopy", Ruska, 1980), E. Ruska provided a detailed illustration of the development of the entire field including optical microscopy. In the above-mentioned book one also finds a sketch (Fig. 5) that proves the fact of imaging with two lenses in a simplified beam path.

The electron microscope owes its emergence to a problem associated with high voltage that arose in the 1920s. At that time high-voltage engineers found that they needed a cathode-ray oscillograph with which they could investigate, in high-voltage plants, faults caused by switching processes, for instance, or lightning strikes close to or on overhead lines. The electron microscope came about as a continuation of the cathode-ray oscillograph as a result of the wide-ranging investigations and improvements that were undertaken when it was discovered that the "concentration coil," from an electron-optical standpoint, was a true lens and could thus be used for imaging.

If one goes further back into history, one can see the cathode-ray oscillograph as a final stage in the development of Braun's tube, while the study of gas discharges in evacuated glass tubes began even earlier. I would like to trace the course of developments from this point onward and point out problems associated with the electron microscope that had to be solved. I believe that I may be accredited some judgment on this subject in the light of the fact that I developed a technically improved cathode-ray oscillograph in my younger years, while serving as an assistant at the electrotechnical institute of the technical college in Aachen (headed by W. Rogowski), and, many years later, was engaged in development in the field of electron microscopy at an industrial concern.

2. Developments up to the end of World War II

2.1 Working toward Braun's tube

Geissler's tube was a closed glass tube that was evacuated to reduced pressure and had at least two electrodes to which a direct voltage could be applied. If the voltage was high enough, an electric discharge, i.e., an electric glow, occurred. This phenomenon was investigated in 1858 by J. Plücker, who observed that the glow could be deflected by a magnetic field. W. Hittorf observed the same thing and described it in 1869. He set a tube with a flush glass base upon the face of a magnet and was able to determine that, if the force of the magnet was strong enough, the glow contracted to a small spot. Today one would say that the electron beams were focused by the magnetic field, on the base of the glass tube for example. One had to be careful, otherwise the glass would melt. In the course of time it was realized that the light phenomena were caused by electrically charged particles of finite mass, the invisible electron beams. Evidence of this was provided by E. Wiechert in 1898 and 1899.[1]

In 1897 F. Braun announced a tube—later named Braun's tube after him—that was used for studying the behavior of varying currents with time in electric circuits. Braun's tube was employed in many different ways for investigations in the field of electrical engineering because it soon became clear that it could be a valuable measuring instrument, in particular when working with higher frequencies. It was limited in its voltage and current values, however, if one looked for ease when working with it.

From the beginning of the twentieth century onward, attempts were made to devise an easily handled measuring device from this experimental laboratory apparatus. As industrial developments advanced there grew a need for being able to perform measurements on overhead lines at high voltages to investigate faults.

2.2 From Braun's tube to the cathode-ray oscillograph

At the turn of the century Braun's tubes were all made of glass and thus were very fragile. Furthermore, it was no easy matter to adapt them to differing operational requirements. The result of an investigation shown on the fluorescent screen had to be perceived externally. It was known that cathode rays darken photographic emulsion, so the first task was that of introducing the photographic material into the apparatus so that it could be exposed directly to the electron beam.

A. Dufour, a lecturer at the Sorbonne in Paris, was, around 1910, probably the first person to start turning Braun's tube into a so-called cathode-ray oscillograph. He reported on his work in 1914 and again in 1920. He presented the results in 1923 in a work entitled "Oscillographe Cathodique pour l'Étude des Basses, Moyennes et Hautes Fréquences" ("Cathode-Ray Oscillograph for the Study of Low, Medium and High Frequencies"), published by Etienne Chiron of Paris.

[1] And by J.J. Thomson in *The Electrician* **39** (1897) 104–109 and *Phil. Mag.* **48** (1899) 547–567 (PWH).

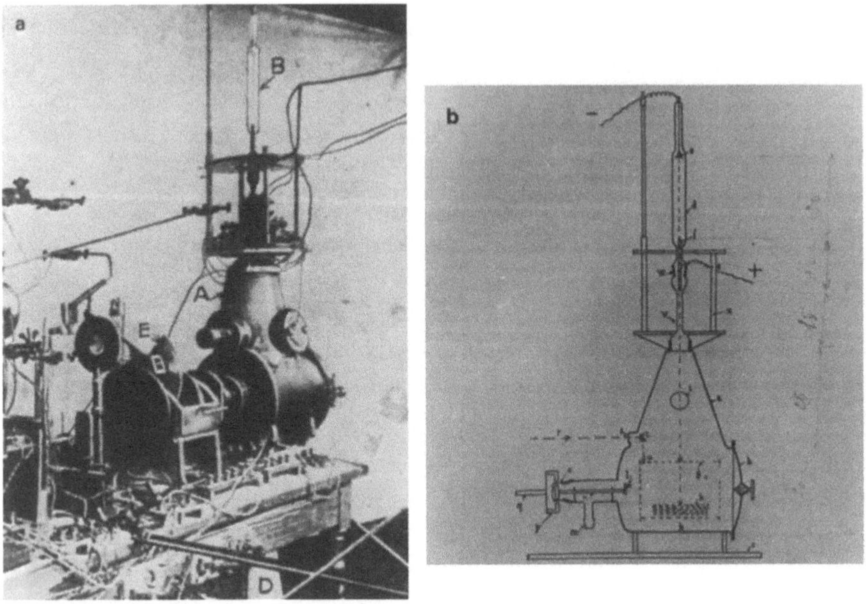

Figure 1 Dufour's cathode-ray oscillograph. (a) View of the instrument. (b) Cross-section.

Dufour tried to convert Braun's tube into an instrument that was technically more useful and more versatile. But his "oscillographe cathodique" remained inadequate and rather elaborate in use.

What Dufour did manage to achieve was a metallic housing holding the devices for photographic recording. Fig. 1a is a view of the cathode-ray oscillograph; in Fig. 1b, a schematic cross section, the metal housing is marked *a* and *b* is the metal cover forming the airtight sealing of the chamber. Above this opening there was a drum from which photographic material could be drawn (the photographic device is denoted by the letters *g*, *h*, *l*, *n*, and *q*). The torsion rod *n* moved the drum by means of a small crank (1) and thus provided the time axis for the recording. At the opposite end of the rod there was a magnet at right angles to the axis that was sealed off but driven by the turning device *q* and thus recorded the time axis. For determining the timing of a recording there were time markers, which were produced on the recording by *s*, *t*, and *u*. The time interval between the markers was set according to the process that was recorded. In Fig. 1a, below the letter E, one can see the exterior of the mechanical construction for the movement of the photographic material. The window can also be seen below the letter A (Fig. 1b, *s*).

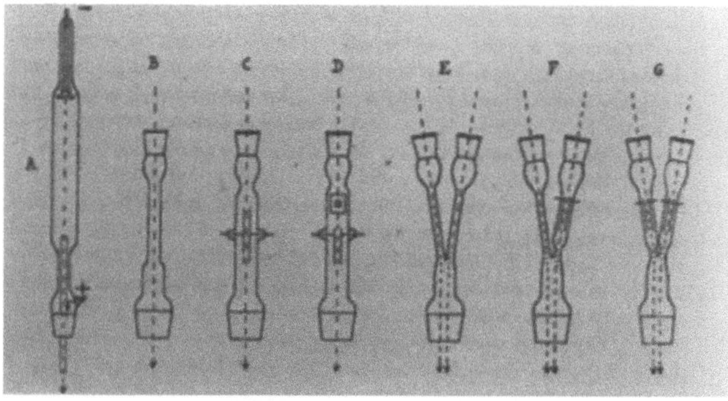

Figure 2 Dufour's cathode-ray oscillograph: parts for various testing methods.

The remaining glass components of the cathode-ray oscillograph were mounted on the metal housing with a conical ground-in joint. These consisted of two glass parts, the discharge tube being located in the upper part (Fig. 2, A). This contained the cathode (—) at the top and the anode below, the latter in the form of a tube with matching diameter and a connection to the exterior (+). The discharge tube was set upon a glass spacer (Fig. 2, B), again by means of a conical ground-in joint. The part B was provided with a deflection coil for recording magnetic fields, for example. Instead of the spacer B, parts C–G could also be incorporated; C and D were for displaying voltages, and parts E–G were to enable a second discharge tube to be attached. In this way two synchronous events could be recorded, although with considerable limitations.

Dufour's cathode-ray oscillograph pointed the way for the further development of Braun's tube that was necessary. The next step in this development commenced in the early 1920s, primarily at technical colleges in Germany.

2.3 Further development of the cathode-ray oscillograph into a technical instrument

As the economy grew from the outset of the twentieth century onward, there was an increasing need for electrical energy. Power stations were built and high-voltage lines spread out from them to distribute the energy they produced. As the power stations increased in number, the distribution lines became more and more subject to atmospheric disturbance or suffered from lightning strikes in the vicinity of the lines. The surges that were thus produced on the lines caused frequent damage to circuits and transformers in the switching stations. Consequently it became necessary to devise suitable measuring devices for examining effects of this kind.

Because of the low mass of its particles the cathode-ray oscillograph seemed to be a suitable instrument for investigating very fast phenomena. But the Dufour apparatus was not yet up to the demands made of a technical measuring instrument. A more powerful instrument was called for, a high-grade cathode-ray oscillograph with accessories enabling high voltages to be handled, and with recording speeds up to the speed of light.

A spate of development set in. The major part of this activity was at the electrotechnical institutes of the technical colleges in Aachen and Berlin (W. Rogowski, A. Matthias), which constantly published the results of their ongoing work. Of the many publications only the first are mentioned, concerning the recording of surges. There is a more extensive list of references in the book by E. Ruska (1980).

From April 15, 1924, onward I was an assistant in the Electrotechnical Institute in Aachen, under W. Rogowski, and I am thus in a position to report on the work that was carried out there, inasmuch as it had to do with the further development of the cathode-ray oscillograph.

We knew of the publications by the institute in Berlin, where, in the course of further development of the cathode-ray oscillograph, the first electron microscope emerged in 1931, but there was no scientific exchange between the two institutes.

Around 1923/1924 in Aachen, E. Flegler began work on the technical design of a cathode-ray oscillograph that was largely made of metal; he was later assisted in this by R. Tamm. This work was a valuable pointer, but the laboratory-model cathode-ray oscillograph that Flegler and Tamm developed was not yet up to the demands of use in power stations.

In 1926 I was assigned the task of developing an automatically recording cathode-ray oscillograph that was to be suitable for use in power stations and switching stations. It soon became clear to me that I would need a cathode-ray oscillograph of my own for dealing with some of the problems confronting me. An airtight, basic case was already available, and also a deflection chamber plus other standard institute items, with the result that my cathode-ray oscillograph was operational as early as the beginning of 1927 (Fig. 3). In 1927 the new Physico-electrotechnical Institute in Aachen was finished, which fortunately had a flat roof. Here there was sufficient space for laying a 450-m, dual high-voltage line that could be fed with 75 kV; i.e., it could safely sustain surges of 150 kV.

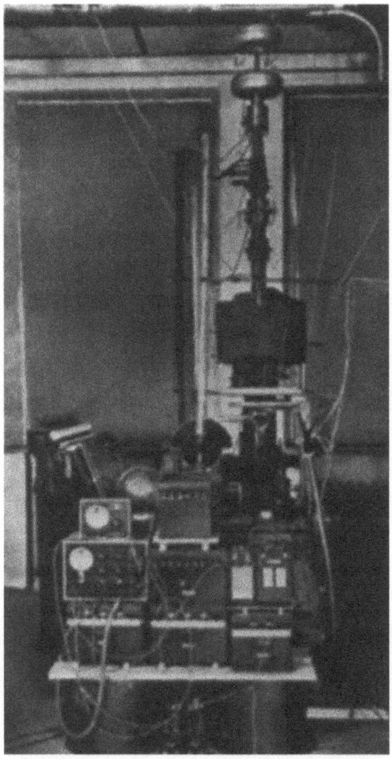

Figure 3 Laboratory model of the automatically recording cathode-ray oscillograph.

At the beginning of 1928 I was able to move into the new building with the operational cathode-ray oscillograph, which differed markedly from former ones. It had been usual to incorporate the deflecting plates so that the voltage terminals on the deflection tube were diametrically opposed. I retained this constructional feature for the time sweep because the time-base box was behind the instrument and the two connections were thus parallel and equal in length. The switch box contained capacitors and switches for setting the speed of the recording beam to match the application. The vertical deflecting plates were arranged next to one another on the right, however, so that the two leads could be routed closer together and parallel.

The deflection tube, containing the deflecting plates for the time base and the measuring voltages in twin form, was about 50 cm high (16/15 cm outer and inner diameter). Directly over the end of the deflection tube was the main focusing coil, which—as was usual in Aachen—was sheathed in iron. The electrical beam cutoff was located in this tube. Following the suggestion of the Swiss Berger, this was a so-called quadruple cutoff

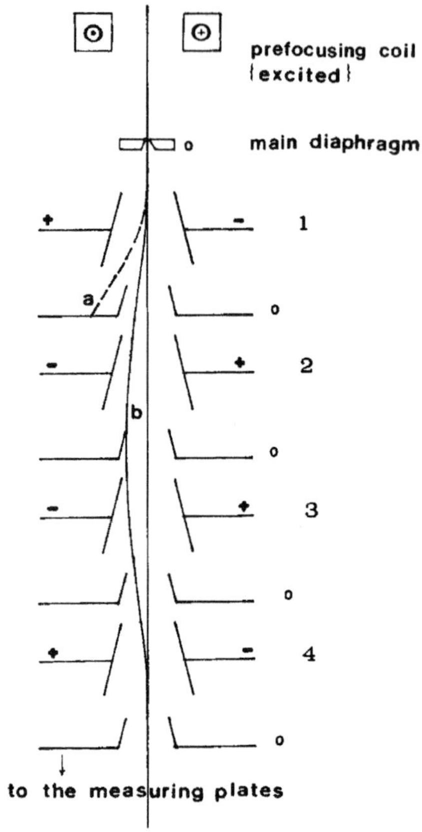

Figure 4 Quadruple cutoff: (1–4) four deflecting plates (polarity, +/—); 0 indicates the beam cutoff. All deflecting plates were equally excited (drawn only for plate 1). (a) Beam in cutoff position (dashed line); (b) beam, just unblocked, is still weakly curved, but leaves the cutoff perfectly axially.

(Fig. 4). A cutoff of this kind offered the advantage of also being able to effectively block any scattered beams. Even with a certain partial, residual voltage, the beam entered the deflection tube perfectly axially. On the blocking-chamber tube there was a set of deflecting coils for centering the beam if necessary. The remaining parts, proceeding upward, were the anode diaphragm, the centering coils, the shielded prefocusing coil, the initial diaphragm for limiting the beam, and the discharge tube with the high-voltage shield grids. With the vacuum equipment there was gas discharge in the discharge tube and a high vacuum in the remaining parts of the cathode-ray oscillograph. The vacua were separated by the anode diaphragm, measuring 0.3 mm at the most.

The cathode-ray oscillograph was fitted with an internal photographic device. The magazine was opened and closed with a rotary magnet on the left-hand outside face of the basic case. The process could be triggered with another magnet underneath the case.

By the end of 1928 three tasks had been completed on this cathode-ray oscillograph that were necessary for investigations on overhead lines carrying high voltage: (1) a time base for the cathode-ray oscillograph (Rogowski & Wolff, 1929), (2) the voltage division in the cathode-ray oscillograph (Rogowski, Wolff, & Klemperer, 1930), and (3) the automatic recording of spontaneous events with the cathode-ray oscillograph (Rogowski, Wolff, & Schaffer, 1930). In the second and third tasks I was assisted by H. Klemperer and H. Schäffer.

A minor episode is worth mentioning. During the early days of 1929 I was tinkering with the lens currents before the day's work began. Suddenly I saw on the screen an image of the fine anode diaphragm with a diameter of 60 mm. The jagged edge of the diaphragm could be easily made out. It was an image of the 0.3-mm diaphragm magnified 200×. The prefocusing coil was excited slightly more than it should have been and the main focusing coil was disconnected. Fig. 5a shows that this was not an image but a projection of the emission spot of the cathode ray transmitted by the prefocusing coil to a point above the anode diaphragm. Using different settings I managed to display the emission region with a diameter of between 15 and 20 mm. H. Viehmann, an acquaintance who was then working for his doctorate and who died some years ago, saw the pictures and suggested that I record them. But there was no more film left. From our knowledge of Busch we presumed that the effect we had observed had to do with "electron lenses." But each of us had a defined scientific program to get on with, and what the future had in store was quite an open matter. So we kept what we had seen to ourselves.

2.4 The recording oscillograph

For the planned investigations in a switching station, a recording oscillograph (Fig. 6) was built that was suitable for practical use. In its physical form this was largely the same as what has already been described, but the recording technique was altered. For the purposes of "spontaneous" operation it was necessary to change to exterior recording, because only in this way could the exposed film be withdrawn and replaced by another without interrupting the operating vacuum.

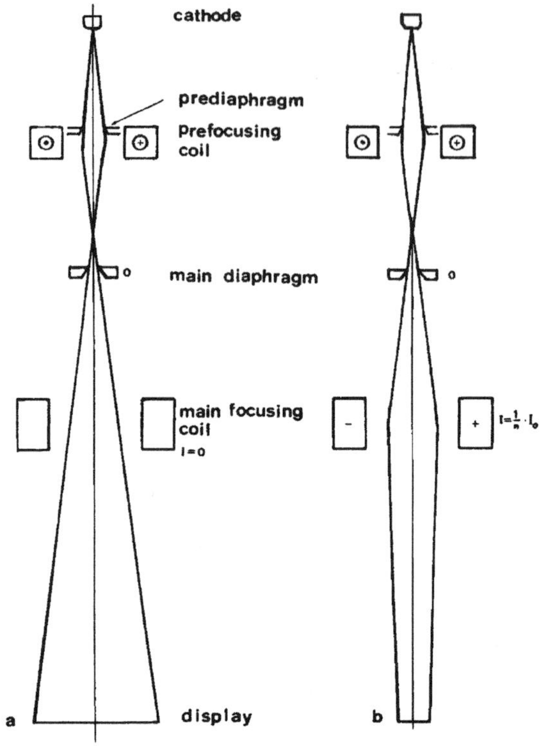

Figure 5 Ray diagram for (a) and (b): prefocusing coil excited in such a way that the emission spot of the cathode was transmitted to a point above the main diaphragm, (a) Main coil not excited; the border of the diaphragm is imaged on the final screen by projection, (b) Main coil excited such that the diameter of the image was approximately 15 mm. This was an image of the emission region. Now the light spot's diameter was reduced to a fourth compared with (a), but its intensity became 16 times stronger (see Ruska, 1980, p. 106).

This oscillograph also called for an additional control device. As already mentioned, there were two different vacua in the cathode-ray oscillograph: the cathode-ray vacuum in the discharge space from the cathode to the anode diaphragm with a minimal pressure of 10^{-2}–10^{-3} mm Hg, and the high vacuum in the deflection part. For this latter part of the cathode-ray oscillograph, because of its considerable volume, powerful diffusion pumps were provided. On the pump for the discharge space was the air-inlet valve that produced the working pressure for the cathode ray. To automatically stabilize the cathode-ray current once it had been set, I devised a "vacuum regulator," together with my colleague Beyerle, which kept the set beam current constant. The regulator controlled the valve by means of a setting

Figure 6 The recording oscillograph for practical use.

actuator. Beyerle needed a control device of this kind for the demonstration of a cathode-ray oscillograph with a very large fluorescent screen. Later the entire cathode-ray oscillograph could be set into operation by a single switch procedure with the aid of a timing relay, while observing a running-in time for the backing pump (Rogowski et al., 1929).

Aided by E. Flegler and H. Klemperer, J. Röhrig of the Rheinische Elektrizitäts- und Kleinbahnen AG undertook to transport the equipment to the Heinsberg power station, some 40 km from Aachen, and to install it there (Flegler et al., 1931). Initially, with assistance from the institute in Aachen, randomly switched events were examined. The work at Heinsberg extended into 1930. I myself could no longer participate from the beginning of 1930 onward because I left the institute on April 30, 1930, after my doctorate examination (February 28, 1930), and my work at Aachen was thus at an end. For financial reasons I was not able at that time to continue working in this field, which was soon to lead to the building of the first electron microscope.

3. Period of transition

The work of the Berlin institute must be described in more detail at this junction because its research presented science with a new tool: the electron microscope. From 1928 onward a group of undergraduates and postgraduates, headed by M. Knoll, attempted to turn the cathode-ray oscillograph into a usable measuring instrument. E. Ruska, as part of

Figure 7 Ruska's first electron microscope (copy), 1931.

his studies, was entrusted with investigating the concentration of electron beams in a coaxial magnetic field, and subsequently occupied himself with the bunching of electrons in an electrostatic field in his thesis. On the basis of Busch's theory (Busch, 1926) concerning the concentrating effect of a magnetic field on electrons, he made an experimental examination of the latter's imaging equation. For the main part he was able to verify it, inasmuch as the coil was arranged approximately in the middle between the anode diaphragm and the fluorescent screen.

An improved experimental setup that he built in 1931 was already fitted with two short coils, one behind the other, and proved, with its at that time modest magnification factor of 17.4, that metal grids permeated by electrons could be shown with good definition. Knoll and Ruska realized that the overall magnification would be increased if one considerably enlarged the spaces between the two lenses and between them and the fluorescent screen, or if one used lenses with a shorter focal length. Their instrument, which they presented at the Cranz colloquium in Berlin on June 3, 1931, was named the electron microscope. (Fig. 7 shows a copy of the first electron microscope.)

The initial results naturally sparked a lot of discussion about the optical possibilities of the new technique. From calculations published by de Broglie in 1925 it was known that the wavelength of electrons was some four to five orders of magnitude shorter than that of visible light, but at that time physicists did not foresee any practical applications. Parallel to the development work at the technical college in Berlin, a research laboratory (C. Ramsauer) headed by E. Brüche and established by AEG in Berlin in 1928 was conducting systematic studies in electron optics. A number of physicists worked on the development of electrostatic electron lenses, on the development of electron-optical instruments such as Braun's tubes, image converters, and mass spectrographs, and on the imaging of thermally or photo-electrically emitting cathodes. As early as 1934 the research group produced a first omnibus volume on the subject of electron optics (Brüche & Scherzer, 1934). The future later proved that electromagnetic lenses would be of great practical value.

To increase the electron-optical magnification, the focal lengths of the lenses had to be reduced. E. Ruska had already suggested encapsulating the current coil in magnetically soft iron and had provided an iron-free ring slot alone in the interior of the coil. It thus had a strong field of short axial expansion. In 1932 B. von Borries and E. Ruska suggested the use of conical polepieces with iron-encapsulated coils for achieving focal lengths that were as short as possible. This was the decisive step forward, the way to the genuine electron supermicroscope, because it was only through the use of lenses with short focal lengths that magnifications greater than those of photo-optics could be realized without a large number of projection stages and excessively long microscope tubes. An electron microscope with improved polepieces was built by E. Ruska in 1933 and achieved a magnification factor of 12,000. (Fig. 8 shows a copy of the instrument of 1933.)

At that time it was already possible to envisage what the electron microscope promised for the future, but there was a general wariness in the face of investing the money that was required, even though at the beginning of 1936 a Ruska microscope that had been further developed by students exhibited a never before achieved resolution of less than 45 nm in an image of etched silver foil.

For a long time the mistaken opinion was held that in an instrument of this kind any specimen that was of interest would be destroyed by the vacuum and the intensive electron bombardment. Impulses from elsewhere (Marton, 1934) and evidence on the part of respected medical men (Siebeck; see Ruska, 1980) finally dispelled the nightmare, at the end of

Figure 8 Ruska's second electron microscope (copy), 1933; magnifications greater than those of photo-optics.

1936, that biological objects could not be investigated. And at last there came the support from industry that had been awaited in vain. E. Ruska and B. von Borries were able to start their work at Siemens und Halske in Berlin at the beginning of 1937 and produced a first laboratory model, on the basis of their experience, by the end of the same year. This was developed into the first commercial, series-produced electron microscope, the ÜM, an instrument with three polepiece lenses and maximal magnification of 30,000×. The point-resolution capability at that time was 7 nm. This ultramicroscope, which was used for operational checks in the chemical industry, lasted through the war years and was in full-time use until 1959, when it took up its place of honor in the Deutsches Museum in Munich. (Fig. 9 shows a ÜM from the first production batch at the beginning of 1940.)

Most of the instruments that were manufactured during World War II were either destroyed or requisitioned. It was time for a new beginning. There were no longer any doubts about the value of electron-optical investigations. On all sides there was hope that development and production could be started up again so that, on the basis of what had already been

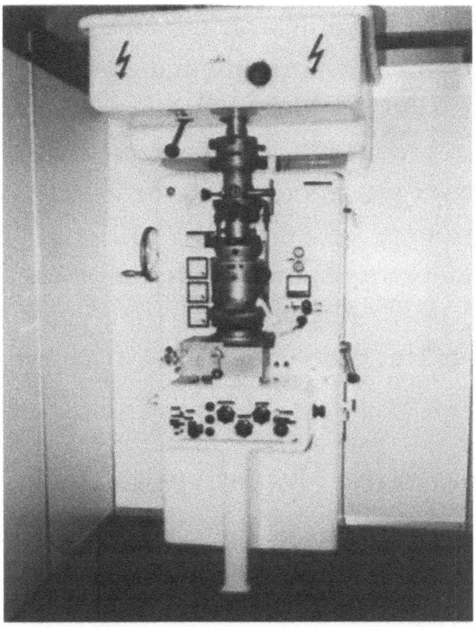

Figure 9 First commercial, series-produced electron microscope, the ÜM 100, produced by Siemens in 1939.

learned, the electron-optical magnification could be further increased and, if possible, greater resolution achieved.

On October 16, 1945, I joined the research group of E. Ruska to work on the further development of the electron microscope. I had already been at Siemens und Halske since May of 1930, but for the previous 15 years I had been working on other applications in the central laboratory in Berlin, for example, the supervision and control of railway traffic; the development of sensitive telegraph relays; and finally the measurement of very weak magnetic fields, the penetration of electromagnetic waves from air into water, and—in cooperation with M. von Ardenne—the development of glass Braun's tubes. With the end of World War II these tasks also came to an end.

4. Improvements in the electron microscope after 1945

4.1 The task ahead

The date of my joining the department for the development of electron microscopes was favorable inasmuch as the first postwar model, the ÜM 100a, was being prepared. As a result of my experience in developing cathode-ray oscillographs, I was able to familiarize myself with the work very quickly. To begin with, I gave the new instrument a good looking over.

The high-voltage system came from Siemens and was derived from a system that had been produced for X-ray equipment. The current source for the magnetic lenses was an accumulator battery of 60 V. The setting resistors were wire wound.

To be able to achieve an advance in resolution capability, magnification of some 10^5 was planned. According to experience obtained up to that time, magnification of this order could not be achieved in two stages with the given dimensions.

4.2 Improvements in the electron optics

Earlier there had been plans to fit the objective polepiece with two slits. A polepiece of this kind had been built, but magnetic saturation effects led to its failure. I suggested a different design (Fig. 10). The iron body of the objective was slanted inward—the virtually axially parallel vacuum channels that can be seen in the figure had no magnetic role—so that its axial "iron length" was at least 30 mm and the entire magnetic flux could pass through without saturation effects. In the central polepiece the slit of the objective lens can be distinguished above and a second slit, an intermediate lens slit, below. With this optical arrangement magnification by a factor of 350 was achieved up to the projective. Using a projective with a small focal length it was possible to obtain overall magnification by $10^5\times$ with limited design height. A distance of 2×10^{-6} mm was thus displayed on the screen by a length of 0.2 mm, but only if the electron microscope remained stable in operation. This meant that the path of 0.2 mm was imaged without any blurring. It was obvious that very high demands had to be made as regards operational stability.

4.3 Stability of the operating parameters

The high voltage was derived from the ac mains by means of transformers and rectifiers. A subsequent regulating device leveled out fluctuations originating in the mains and any other fluctuations caused by the components that were used. The user noticed little or nothing of these measures as long as everything worked properly. The stability achieved here depended on many factors. A value of 2×10^{-6} was easily lost. This limit value did not have to be maintained for any particularly long period of time of course. But it did have to remain constant long enough for observation, focusing,

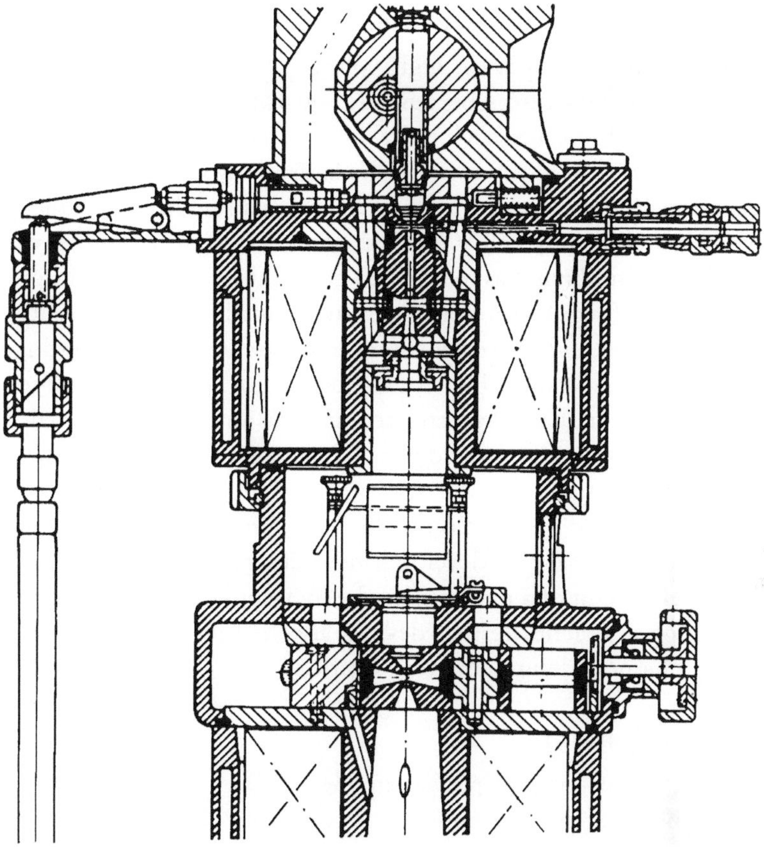

Figure 10 Objective; intermediate lens adjusted together with the objective to increase the magnification.

and recording. Focusing the image calls for high stability of the lens current. For this reason a battery was initially selected as the current source. The resistors required for setting the current were not to alter in resistance as a result of inadmissible current fluctuations.

Mechanical effects also had to be taken into consideration. The controls for moving the specimen in the object stage had to work without any play, for instance. Other possible sources of interference were reported on in detail in a description of the ÜM 100a (Ruska, 1950).

Seen from the standpoint of the technology of the time, the ÜM 100a was a very capable, state-of-the-art instrument. Nevertheless it was clear that an electron microscope that was even better optically could be developed using technical possibilities that were already at hand.

4.4 Replacement of the previous control circuits

Our wish to have an improved high-voltage system developed was turned down by the manufacturers because the quantities would have been too small. I myself did not like having the lens current supplied by batteries either, so I started work of my own on the basis of my experience in Aachen. The fastest way to replace the battery was with a stabilized rectifier power supply for approximately 250 V dc on the lens coils. In this manner the current in the lens windings was reduced to practically one-quarter its previous value. In terms of regulation this meant substantially smaller currents across the setting contacts and the replacement of the previously used setting resistors with wound resistances by decade resistors for coarse and fine settings. The decade resistors had large contact surfaces and thus very small transfer resistances. There was an additional advantage in that the new setting controls led to no significant difficulties in design and production.

The high-voltage system could not be altered at the first attempt. It was absolutely necessary to develop a new design. But it was clear to me that the development of a new and markedly improved high-voltage system alone was not sufficient. Testing devices also had to be developed so that the quality of a new design could be verified.

4.5 Test equipment

For testing the lens-current power supplies it was possible to use available devices from other departments and to combine them into a test setup. These were conventional and commercial components. Fig. 11a shows the test equipment and Fig. 11b its circuit diagram.

Measuring equipment was also built in our laboratory for testing the high-voltage stability. Fig. 12a illustrates the equipment and Fig. 12b again shows the circuit diagram. In the foreground of the illustration a high-voltage system can be seen, the stability of which was to be tested. In the background is the cylindrical precision resistor, in a corner of the room and well clear of the walls to protect it against interference fields. The design of the precision resistor is shown in Fig. 13. A resistance coil with 5.5 high-voltage resistors for each winding was suspended around a central insulating column on horizontal insulating arms. A total of $29 \times 5.5 \approx 160$ resistors were aligned to form a bleeder chain of 1600 MΩ and a divided residual voltage of 100 V was measured in a bridge circuit. For the rating the recorder indicated zero.

As Fig. 13 also shows, a cylinder of laminated paper was set over the voltage-divider chain. Laminated paper is a poor conductor and so a ho-

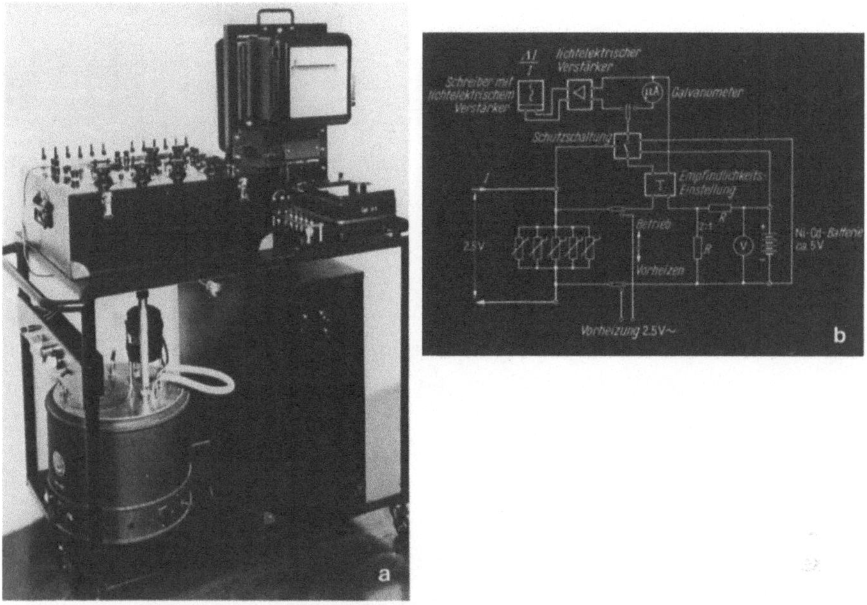

Figure 11 Equipment for testing the lens-current power supplies: (a) view of the test equipment; (b) its circuit diagram.

mogeneous electric field was created in the interior which, with a height of 100 cm between the baseplate and the upper shielding hood, resulted in an axial field strength of 1 kV/cm. Corona effects within were thus suppressed. No interference could penetrate from the exterior because the precision resistor was set up sufficiently far from the walls.

4.6 A new high-voltage system

Fig. 14 shows views of the new high-voltage system that was constructed in line with my data. Fortunately there were only a few small initial difficulties to be overcome. The view from the right-hand side shows the precision resistor partially surrounded by a shielding panel. Only about half of the high-impedance resistors can be seen, the other half being behind the support plate. Twenty resistors were sufficient because the entire high-voltage system was immersed in a tank filled with transformer oil.

The work on the new high-voltage system lasted into 1954. In conjunction with improvements to the microscope itself, a new model, the Elmiskop I, was produced. At the International Congress for Electron Microscopy in London in 1954, it was possible to present test recordings of

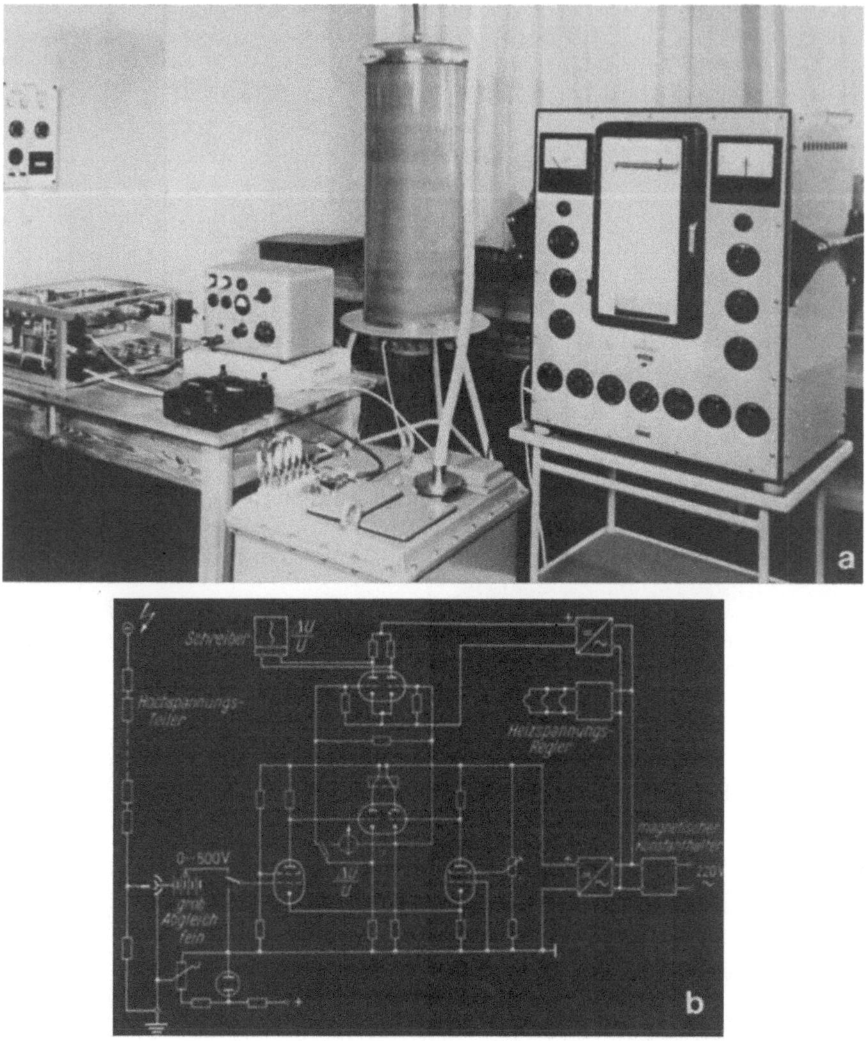

Figure 12 Equipment for testing the high-voltage stability: (a) view of the equipment; (b) its circuit diagram.

the resolution, triple recordings with point resolution of 0.9 nm and dual recordings with point resolution of 0.8 nm (Ruska & Wolff, 1956). The Elmiskop I proved a success and was manufactured unaltered for 6 years. But further improvement of the resolution capability was nevertheless still a goal.

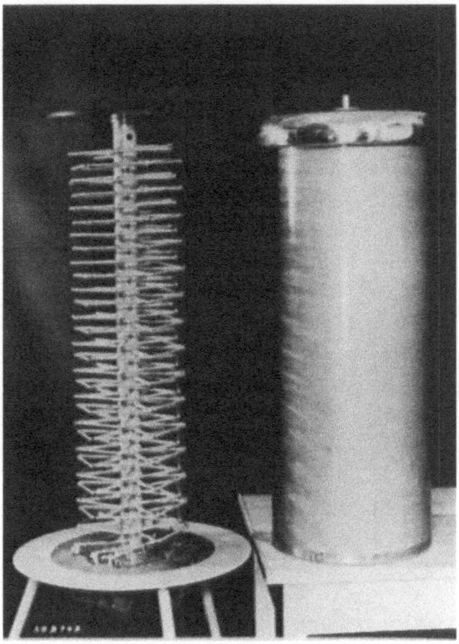

Figure 13 Design of the precision resistor.

Figure 14 Views of the new high-voltage system with the internal precision resistor, 1954.

Figure 15 External high-voltage resistor, since 1962.

4.7 Improvement of resolution

At the end of the 1950s we observed occasional irregularities in the high voltage. If the temperature in the transformer oil was measured after a longish period of operation, a rise of about 5°C could be detected. This temperature increase was too high because of the temperature coefficients of the resistance material (carbon resistors). To achieve even greater resolution than previously, the high-voltage precision resistor had to be taken out of the oil. Earlier on I had had a flat precision resistor built that was filled with air for insulation and later with sulfur hexafluoride. On each side of the support plate were arranged about 66 resistors. The results were satisfactory and we decided to introduce an external precision resistor of this kind for the improved Elmiskop I, the Elmiskop IA. The arrangement of the 132 resistors can be seen in Figs. 13 and 15. Whereas with the built-in precision resistor the high voltage of 100 kV was distributed among 20 resistors (i.e., 5 kV per resistor), there was now only 750 V per resistor. Fig. 16 shows the result of a test with the test equipment previously described. Within a few minutes the high voltage was stable at 10^{-6} of its rating. On a test recording of the resolution (Fig. 17) separate points can be clearly made out down to a separation of 0.4 nm (4 Å). This dual recording was in 1962—according to a report, the world resolution record of the

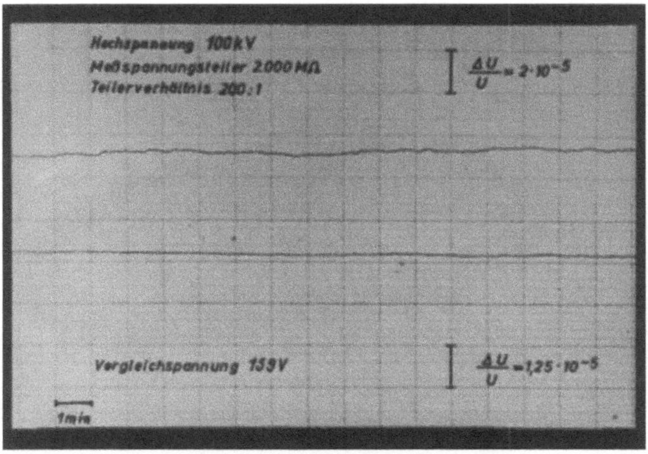

Figure 16 Results of a test with the high-voltage test equipment.

time at the International Congress for Electron Microscopy in Philadelphia (Wolff et al., 1962).

5. Concluding remarks

My observations have been extended over a period of almost 30 years because I see this as a kind of constant further development. With the Elmiskop I, which was produced in large numbers and virtually unaltered for many years, a certain final point was reached.

This statement was proved true even for the electron microscope production of other factories. Glancing now back on the numerous attempts we made concerning "improvements" in construction and the selection of components, we were, compared with the situation in 1962, unable to obtain further remarkable progress. A point-to-point resolution of about 0.4 nm seemed to us a final point, considering normal efforts. It became evident that for further improvement to reach values in the 0.1-nm range, that is, in the range of atomic dimensions, enormous additional labor would be necessary. For my own consolation I remember relatively early findings of B. von Borries. He estimated toward the end of the 1930s that approximately 90% of all electron-microscopical images were taken at a resolution level much below the optical limit of the electron microscope.

In its first 30 years of development the electron microscope confirmed its importance as a valuable tool in science and technology. In this time pe-

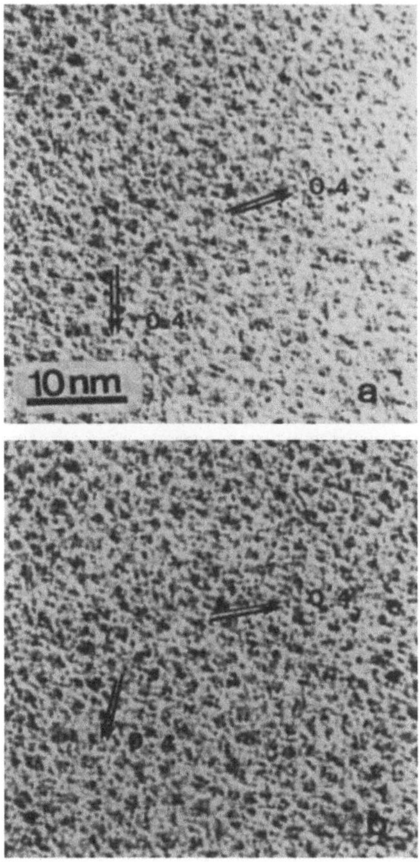

Figure 17 High-resolution test, dual recording (a) and (b), Elmiskop I, 1962. Pt–Ir alloy: electron-optical magnification, 200,000:1; acceleration voltage, 80 kV.

riod also, varieties of simpler microscopes for practical use were developed in addition to the high-performance instruments. Even relatively early, in the 1930s, the first scanning microscope was designed and built as a laboratory model.

About 70 years elapsed between the first experiments with gas discharges (1858) and the realization of a high-quality cathode-ray oscillograph. Reviewing, we must confess that none of the numerous scientists had concluded that the converging cathode-ray lens could be used for imaging purposes. Already the high-performance cathode-ray oscillograph from 1925, with its two stages, was an imaging instrument with an enlargement factor of 1:1. It took then only another 6 years until, in 1931, the first electron microscope as a magnifying instrument was realized.

We need to thank many scientists, engineers, and constructors in research institutes and factories for their important development work in the first 30 years of electron microscopy.

6. Afterword by Peter Hawkes[2]

Otto Wolff was born on 2 November 1898 in Kaiserslautern (Rheinpfalz). He joined Siemens & Halske in Berlin in May 1930 as Labor-Ingenieur; on 5 November 1953, he became a Wissenschaftliche Mitarbeiter of the firm. He retired on 1 January, 1966. He was, however, co-author of a later publication in the *Siemens Zeitschrift* (Asmus et al., 1968) and *Siemens Review* (Asmus et al., 1969).

References

Braun, F. (1897). *Annalen der Physik und Chemie, 60*, 552.
Brüche, E., & Scherzer, O. (1934). *Geometrische Elektronenoptik*. Berlin and New York: Springer-Verlag.
Busch, H. (1926). *Annalen der Physik (Leipzig) [4], 81*, 974.
Dufour, A. (1914). *Comptes Rendus Hebdomadaires des Seances de l Academie des Sciences, 158*, 1339.
Dufour, A. (1920). *Journal de Physique Et Le Radium [6], 1*, 147.
Flegler, E., Wolff, O., Röhrig, J., & Klemperer, H. (1931). *Elektrotechnische Zeitschrift, 52*, 1.
Gabor, D. (1957). *ETZ. Elektrotechnische Zeitschrift. Ausgabe A, 78*, 522.
Hittorf, W. (1869). *Annalen der Physik und Chemie, 16*. 1 and 197.
Marton, L. (1934). *Physical Review, 46*, 527.
Plücker, J. (1858). *Annalen der Physik und Chemie, 103*. 88 and 151.
Rogowski, W., & Wolff, O. (1929). *Archiv für Elektrotechnik (Berlin), 21*, 645.
Rogowski, W., Beyerle, K., & Wolff, O. (1929). *Archiv für Elektrotechnik (Berlin), 22*, 507.
Rogowski, W., Wolff, O., & Klemperer, H. (1930). *Archiv für Elektrotechnik (Berlin), 23*, 509.
Rogowski, W., Wolff, O., & Schaffer, H. (1930). *Archiv für Elektrotechnik (Berlin), 23*, 707.
Ruska, E. (1950). *Kolloid-Zeitschrift, 116*, 102.
Ruska, E. (1980). *The Early Development of Electron Lenses and Electron Microscopy*. Stuttgart: Hirzel Verlag.
Ruska, E., & Wolff, O. (1956). *Zeitschrift fur wissenschaftliche Mikroskopie und mikroskopische Technik, 62*, 465.
Wiechert, E. (1898). *Nachrichten von der Gesellschaft der Wissenschaften zu Göttingen*, 260–293.
Wiechert, E. (1899). *Annalen der Physik und Chemie, 69*, 739.
Wolff, O., Engel, A., & Koppen, G. (1962). *Philadelphia, 1962, 1*, Article E13.

[2] I am very grateful to Dr Claudia Salchow at the Siemens Archives, who provided the following information.

Afterword references

Asmus, A., Herrmann, K.-H., & Wolff, O. (1968). ELMISKOP 101, ein neues Hochleistungs-Elektronenmikroskop. *Siemens Zeitschrift*, *42*, 609–619.

Asmus, A., Herrmann, K.-H., & Wolff, O. (1969). ELMISKOP 101, a new high-power electron microscope. *Siemens Review*, *36*, 57–67.

CHAPTER FIFTEEN

Reminiscences☆

R.W.G. Wyckoff[a] and Peter Hawkes (Afterword)[b,*]
[a]Formerly Duval Corporation, Tucson, AZ, United States
[b]CEMES-CNRS, Toulouse, France
*Corresponding author. e-mail address: hawkes@cemes.fr

Contents

1. Afterword by Peter Hawkes 473
References 473

These are casual memories of the electron microscope by one of the few persons still living in the United States who has been concerned with the instrument since the early days of its development. It deals only with those aspects of its history which involved me personally.

As we all know, electron microscopy grew out of Louis de Broglie's recognition that electrons have a wavelength which, at easily attained accelerating voltages, is vastly shorter than that of light, and out of Sir George Thomson's and C. J. Davisson's demonstration of electron diffraction. I am of the generation of these men and their work and have known them personally. My first exposure to the subject goes back to a visit, probably in 1927, to Davisson in the old Bell Laboratories in West Street, New York City. He had built a small projection-type microscope and was getting an image 17× magnified of a metal screen. I recall feeling with him that this experimental arrangement did not seem to have a very promising future, and in fact he did not continue work along this line. He has been dead for some years now, and so has Sir George, whom I knew rather better, but de Broglie is still alive. He recently resigned as a Permanent Secretary of the French Académie des Sciences, where I saw him whenever I went to its weekly meetings.

The electron microscope is in a certain sense the child of the oscilloscope, which was being developed in the middle 1920s. It is a consequence of Busch's demonstration (1926) of the conditions under which a suitably

☆ Dr. Wyckoff planned to expand this brief contribution into a much fuller chapter for this book, but his health has sadly prevented this.

Advances in Imaging and Electron Physics, Volume 221
ISSN 1076-5670
https://doi.org/10.1016/bs.aiep.2022.03.015

shaped magnetic field could focus a beam of electrons. Knoll and Ruska in Berlin used this information a few years later to construct their first electron microscope having magnetic lenses. Steadily refined in succeeding years by von Borries and Ruska, it was, just before the beginning of World War II, in a state to be produced commercially by Siemens. While this was going on L. Marton built in Belgium a primitive microscope and in 1934 produced the first micrographs of a biological specimen.

In 1939 RCA brought Marton to this country to build a microscope and this he did at Camden. Professor Burton at Toronto also had become interested and just before the war he constructed a functioning microscope with the aid of graduate students, notably A. Prebus and J. Hillier.

When Marton's microscope was operational at Camden I heard about it from George Morton, who some years before had spent a short time in my laboratory at the Rockefeller Institute and who was then part of the group beginning to develop television. At the time I was with Lederle, a branch of American Cyanamid, and had the opportunity to take Mr. William Bell, Cyanamid's president, to see the microscope. It was not commercially available but it illustrated the potentialities of electron microscopy sufficiently well so that when a bit later RCA began manufacturing microscopes, Cyanamid bought the first. This instrument, produced under the supervision of James Hillier, who had come from Toronto, followed the general lines of the Siemens microscope except for a more mobile column and a novel, improved high-voltage supply. Siemens used a conventional low-frequency source which involved an oil-immersed transformer, rectifiers, and large capacities to reduce unacceptable ripple. Vance at RCA designed a high-frequency supply utilizing only radio-type vacuum tubes and other electronic components. This was the EMB model of which RCA built a number of examples during the war years. The #1 instrument was installed by Cyanamid in its Stamford laboratories and, being engaged at Lederle on war problems of a very different nature, I did not have the opportunity to use it; nevertheless the interest fostered by my instrumentality in its purchase found later expression.

When at the University of Michigan in Ann Arbor during the middle war years, I heard that a practically unused RCA electron microscope was in the basement of the bacteriology building and thought it might profitably be used in connection with current studies of influenza. It proved to be #2, the next one to have been produced after that purchased by Cyanamid. T. F. Anderson, working in the RCA laboratory in Camden, had made photographs of influenza virus but its particles could show only as small featureless circles. I was anxious to know their shapes and discussed the matter with Robley Williams of the Astronomy Department, who was aluminizing the department's telescope and vacuum-coating mirrors for the Navy. We developed the notion that the height of a particle could be ascertained by evaporating a layer of metal from one side and measuring the length of the "shadow" cast on the supporting substrate. On trial such "shadows" were obtained using particles of dust (Williams & Wyckoff, 1944). I then decided to examine a suspension of virus particles. The result caught me completely by surprise, for we had not foreseen the three-dimensional effect of "shadowing" nor was it evident in the pictures of dust. The value of metal shadowing was so apparent from these first virus pictures that I decided to concentrate for the time being on its applications. Various metals were tested and a variety of microscopic objects shadowed (Mosley & Wyckoff, 1946; Price et al., 1945; Thomassen et al., 1945; Williams & Wyckoff, 1945a, 1945b, 1945c, 1945d).

At this time RCA was finishing up the design of its improved EMU model microscope. The #2 EMB microscope in the bacteriology building was not in good condition and through my previous relations with RCA I was able to arrange the sale to the university at a low price of the excellent EMB that had been the laboratory instrument at RCA. Much improved micrographs were made after its installation in Ann Arbor.

At the outset of the war I had been engaged at Lederle in the manufacture for our Army of vaccine against epidemic typhus fever and the study of the *Rickettsia* spp. which cause it and Rocky Mountain spotted fever. After Norman Topping from the National Institutes of Health saw, more or less by accident, some of my shadowed micrographs, I was given a commission in the Public Health Service to further in Bethesda the application of the electron microscope.

Some time after my laboratory there was in full operation, Perry Smith, the engineer at RCA in charge of developing microscopes, suggested that I become their consultant. The NIH agreed and I continued in this capacity for several years until in 1952 I went to London for 2 years as Science Attaché.

At the end of the war Philips in Holland entered the electron microscope field, starting out with a model which Le Poole had developed in Delft during the war years. At about the time it began to market this instrument in the United States, Jack Woods, sales manager for North American Philips, asked me to become its consultant. For a number of years I consulted for both companies. During the stay in London my Philips con-

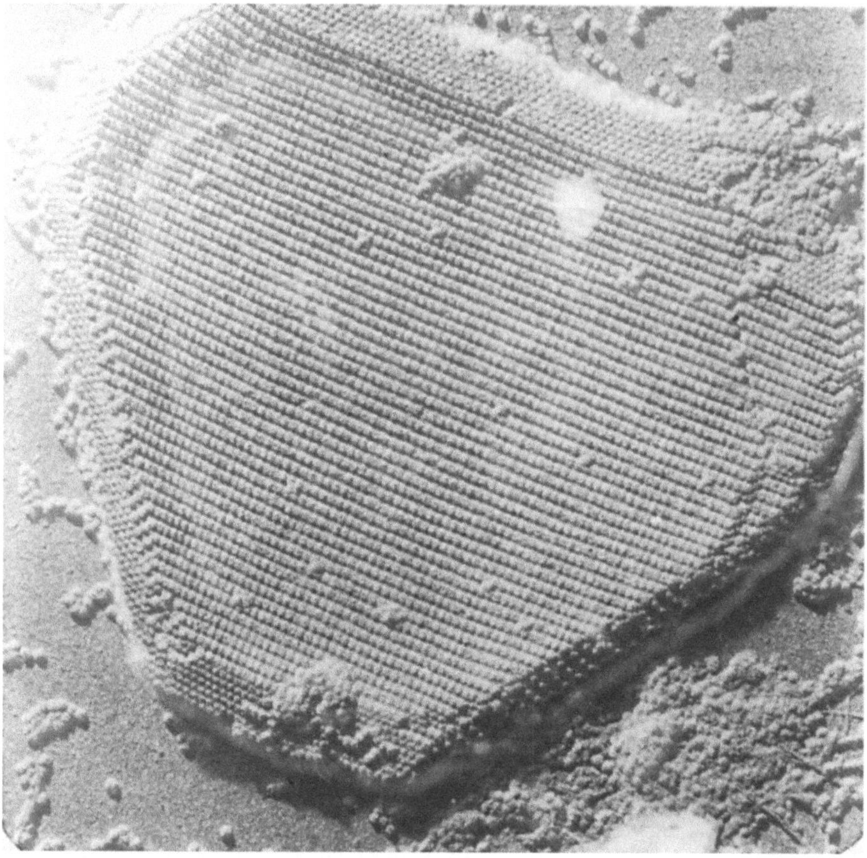

Figure 1 Micrograph of crystalline molecular arrangement of southern bean mosaic virus protein; ×82,000. [From L. W. Labaw and R. W. G. Wyckoff, *Nature (London)* **176**, 455 (1955).]

nection continued on an informal basis—I was loaned an instrument which was installed in my embassy office. After my return this association with Philips flourished until a few years ago when its instrument development became concentrated in Holland.

In the years following the end of the war, the use of electron microscopes spread so widely that societies dealing with it became necessary. We formed the Electron Microscopy Society of America, of which I was an early president. During my London years we tried to set up an international union but were not accepted by the ICSU.

At the NIH my electron microscopy was concerned mainly with viruses and with collagen in its association with hard tissues. Shortly after peace was established I met Pierre Lépine, who was bringing electron microscopy into his Virus Laboratory at the Pasteur Institute in Paris. We developed a collaboration, aided by Odile Croissant, which continued until his recent retirement. Being throughout my scientific life primarily concerned with the crystalline state, a high point of the Bethesda years was making the first electron micrographs showing the regular three-dimensional arrangement of the macromolecular particles in a crystal (Fig. 1). Another was the introduction, in collaboration with David Scott, of the microscope as an essential tool in the study of the microstructure of teeth. After retiring from the NIH in 1959, I was able to use the experience thus gained in investigating at the University of Arizona the ultrastructure of vertebrate fossils. Now, after more than 20 years at the university, I am engaged in applying my experience with electron microscopes and a still longer involvement with X-ray diffraction in studying the ore minerals that concern the Duval Corporation.

1. Afterword by Peter Hawkes

Ray Walter Graystone Wyckoff was born in Geneva (NY) on 9 August 1897. He obtained his B.S. degree at Hobart College in Geneva (NY) in 1916 and his PhD at Cornell in 1919. He first concentrated on x-ray crystallography but after moving to the University of Michigan, he also turned to electron microscopy. He was elected Foreign Member of the Royal Society in 1951. He died on 3 November 1994 in Tucson (AZ).

References

Williams, R. C., & Wyckoff, R. W. G. (1944). Thickness of electron microscopic objects. *Journal of Applied Physics, 15*, 712.

Early papers, published through 1945, are listed below

Mosley, V. M., & Wyckoff, R. W. G. (1946). Electron micrography of the virus of influenza. *Nature (London), 157*, 263.

Price, W. C., Williams, R. C., & Wyckoff, R. W. G. (1945). The electron micrography of crystalline plant viruses. *Science, 102*, 277.

Thomassen, L., Williams, R. C., & Wyckoff, R. W. G. (1945). Surface replicas for electron microscopy. *Review of Scientific Instruments, 16*, 155.

Williams, R. C., & Wyckoff, R. W. G. (1945a). Electron shadow-micrography of virus particles. *Proceedings of the Society for Experimental Biology and Medicine, 58*, 265.

Williams, R. C., & Wyckoff, R. W. G. (1945b). Shadowed electron micrographs of bacteria. *Proceedings of the Society for Experimental Biology and Medicine, 59*, 265.

Williams, R. C., & Wyckoff, R. W. G. (1945c). Electron shadow micrography of the tobacco mosaic virus protein. *Science, 101*, 594.

Williams, R. C., & Wyckoff, R. W. G. (1945d). Electron shadow-micrographs of haemocyanin molecules. *Nature (London), 156*, 68.

CHAPTER SIXTEEN

Complementary accounts of the history of electron microscopy

Peter Hawkes[a,b]

[a]Laboratoire d'Optique Electronique du Centre, National de la Recherche Scientifique, Toulouse, France
[b]Now CEMES-CNRS, Toulouse, France
e-mail address: hawkes@cemes.fr

Contents

1. Appendix 488
Acknowledgments 495
2. Afterword 496
Acknowledgments 498
References 498

By far the most detailed exploration of the development of the first electron microscopes is to be found in Ernst Ruska's historical volume (Ruska, 1979, 1980), but this was by no means the first account of the subject. A book of informal recollections was published by Ladislaus ("Bill") Marton in 1968 containing many anecdotes of the 1930s, when Marton made his first observations of biological specimens in Brussels with his homemade instruments. Ruska and Dennis Gabor wrote long historical papers in 1957, which are a valuable source of references to early work in many countries. In addition to these better-known publications, there is a host of shorter papers, some on points of detail, others concerned with developments in a particular country or firm or domain. It is with these that we shall be concerned in the remainder of this article.

We commence with Germany, where both the magnetic and electrostatic microscopes were born in the early 1930s. The book of Ruska's already mentioned (Ruska, 1979, 1980) covers this period very thoroughly so far as magnetic systems are concerned, while Ruska (1957) covers both electrostatic and magnetic microscopes in Germany and elsewhere (see also Ruska, 1956, 1974, 1984; Giesbrecht, 1981; Catalogue, 1980). Postwar developments are intimately associated with the Siemens company, and these may be traced through their house journal as well as the regular published

Advances in Imaging and Electron Physics, Volume 221
ISSN 1076-5670
https://doi.org/10.1016/bs.aiep.2022.03.016

descriptions. Unconventional instruments, such as those using permanent magnets, are mentioned by Riecke (1982).

A good idea of the development of the electrostatic instrument may be obtained from the volumes prepared by the Allgemeine Elektrizitäts-Gesellschaft (AEG), the first recording 10 years of electron microscopy (Ramsauer, 1941, 1942, 1943). The *Jahrbücher* of the AEG likewise contained regular and very detailed progress reports (see in particular Vol. 7, 1940). The appendix (by Brüche, 1941) to the book by Brüche and Recknagel (1941) sets out the principal publications meticulously, since the authors felt that an earlier review by von Borries and Ruska (1940) did not do justice to the AEG work. See also Kedesdy and Matthias (1939), Matthias (1942), Brüche (1943), Küpfmüller (1944), von Borries and Ruska (1944), and Brüche and Scherzer (1947); for other relevant papers, by Brüche especially, see Cosslett (1950) and Marton et al. (1950).

The Suddeutsche Laboratorien and later Carl Zeiss (Oberkochen) took up the AEG instrument after the war (Rang & Schluge, 1951; Mahl et al., 1957) but eventually turned over to magnetic lenses. Electrostatic microscopes were also produced for some years by Carl Zeiss (Jena) but they too ceased production in favor of their EF device (e.g., Recknagel, 1952–1953; Hahn, 1955a, 1955b; Guyenot, 1955; Fickler & Guyenot, 1963; Soa, 1965; Görlich, 1966).

A vexed question is the patenting of the electron microscope. The first patents were applied for by Reinhold Rüdenberg on May 31, 1931, and four further patent applications were filed between June, 1931, and March, 1932. German patents were not granted until 1953/1954, but patents were accepted in France in 1932, in Britain in 1933, in the United States in 1936/1973, in Switzerland in 1934, and in Austria in 1934. Exact details of all these are given by Freundlich (1963), from whom we have taken the foregoing details. A note concerning these was published in 1932 by Rüdenberg, who added some further details later (Rüdenberg, 1943). A vivid account of the way the first patent applications came to be made is given by Mulvey (1962, 1973).

A further element in the discussion of Rüdenberg's patent application is the exchange of letters between the late Max Knoll and the late Max Steenbeck, a colleague of Rüdenberg at that time, referred to by Ruska in his historical volume (English edition, p. 31; German edition, p. 35). We reproduce these two letters in full in the Appendix in the hope that anyone else with recollections of those times will now make them public. Fig. 1 is a photostat of part of the two letters. See also Borasky (1982) and Afzelius (1981).

On a more personal note, Ernst Brüche has written a brief log of his *Forschungsreise* (Brüche, 1966–1967; see also Brüche, 1957). Manfred von Ardenne has published his memoirs (1972) as well as a historical article (1978), incorporated into his contribution in volume 220. Festschriften too add informal details: that for Brüche's eightieth birthday (Brüche, 1980) even summarizes his career in verse! Bethge (1982) has also contributed to the history of the subject. The accounts by Gabor (1957, 1974) are likewise full of interesting comments.

A reasonably complete list of textbooks, which are yet another useful source of early information and references, is to be found in Glaser (1956) and is not reproduced here; Glaser (1952) is likewise a rich source.

In the same decade, interest in the electron microscope sprang up rapidly in several other countries, taking root in Belgium, England, and Canada in particular and also in Russia and Japan. The Belgian work was stimulated by Ladislaus Marton, who built a series of instruments in Brussels and examined a variety of biological materials, surpassing the light microscope with his high-voltage (90-kV) instrument. For additional information, see the essay by Süsskind (this volume) as well as Marton's own writings (1946, 1960, 1965, 1968). Developments in Great Britain are examined in the following works, as well as in this volume and volume 220 (Bradbury, 1968; Challice, 1982; Cosslett, 1971a, 1971b, 1979, 1980, 1981a, 1981b, 1982; Hawkes, 1979; Horne & Wildy, 1979; Jervis, 1971–1972; Manton, 1978; Mulvey, 1962, 1964, 1967, 1971, 1972, 1973, 1974; Oatley, 1982; Oatley et al., 1965; Preston, 1983). See too Klemperer (1939), Myers (1939), Langrish et al. (1972), and Georghiou et al. (1985).

The beginnings of electron microscopy in Russia are much less easy to trace, though there may well be historical articles of which I am unaware. The only connected account that I have seen is that of Nyrikov and Kuschnier [Kushnir] (1956), which contains no references to published work. They state that "It was in 1937 when some laboratories began to construct electron microscopes for their own use. Among them the work of Lebedev A. A. was the most successful. In 1946, the first commercial electron microscope was produced in our country...." Ruska (1957) established (presumably from personal contact, since no references are listed) that

> In Russland kam die Entwicklung von Elektronenmikroskopen 1940 im Staatlichen Optischen Institut [GOI] in Petersburg in Gang, das 1946 die ersten fünf Geräte herstellte ... Seit

> 1949 wird ein magnetisches 50-keV-Gerät Typ УМ-3 serienmässig hergestellt. Auch ein magnetisches 100-keV-Gerät Typ УЭМ-100 für 25 000 fache und ein elektrostatisches 40-keV-Gerät Typ ЭСМ-40 für 6 000 fache elektronische Vergrösserung wurden 1950 entwickelt.

In fact, however, the early GOI instrument is fully described in the literature. Thus Ruska's opening remarks are essentially confirmed in a short paper of 1944 by Vertsner, in which further details of the GOI instrument are given:

> Work on the construction ... of an electron microscope was begun in 1940 in the State Optical Institute in the Laboratory of A.A. Lebedev. [The fact that a long review article on electron optics and its applications had appeared the previous year (Zernov, 1939) may be no coincidence.] Toward the end of the year a prototype had been completed with a magnification of 10 000×.

Subsequent improvements indicated that a resolving power of 50 Å had been reached. The article shows a photograph of the instrument and several micrographs, taken at magnifications ranging from 6000 to 25,000×. Further accounts by Lebedev (1946) and Vertsner (1945, 1946, 1947) followed soon after, and in 1947 Kushnir published two long survey papers, the first (Kushnir, 1947a) on electron microscopy and the second (1947b) on the construction and applications of electron microscopes. The latter contains descriptions and photographs of the RCA EMU 2 and of one of the Ruska and von Borries instruments, and provides an extended account of the GOI microscope with a line diagram. See also Vertsner (1980).

As well as this work on the transmission microscope, some attention was paid to emission microscopy by Sushkin as early as 1937 (Sushkin, 1937, 1941) and subsequently by Artsimovich (1944) and by Spivak[1] and Lukatskaya (1948), working at relatively high pressures; see too Milyutin (1949). Meanwhile, electron optics was attracting the attention of Russian theoreticians. Very general expressions for the motion of electrons in electric and magnetic fields were derived by Grinberg (1942, 1943), and these are used as a starting point in modern texts on electron optics (e.g., Kel'man & Yavor, 1959, 1963, 1968). Other theoretical studies followed,

[1] It is amusing to note that Spivak's earliest work was concerned with modeling the images formed by lightning (*ceraunographie*, as Flammarion styled the phenomenon) by means of electrons: "The positive space charge focusses the electron beam thus playing a rôle of optical arrangement ('electric lens'). . . . With the increase of the current intensity, the proper magnetic field of the lightning played the rôle of a 'magnetic lens,' which focusses the electron beam. The peculiar electron optics in the discharge at the atmospheric pressure is just the cause of the appearance of the beautiful images" (Spiwak & Kardash, 1945, 1946).

17. Okt. 1960

======= 5592/576

Herrn
Prof. Dr. Steenbeck
Physikalisches Institut
der Universität Jena
J e n a

Institut für
Magnetohydrodynamik
Deutsche Akademie
der Wissenschaften
J e n a

Lieber Herr Steenbeck !

Leider ist es mir noch nicht möglich gewesen, Ihnen zu Ihrer glücklichen Rückkehr zu gratulieren und über unsere Erlebnisse in den verschiedenen Ländern Meinungen auszutauschen. Dagegen habe ich Herrn von Engel mehrere Male seit dem Krieg getroffen, das letzte Mal in Princeton.

Ich möchte Sie heute bitten, wenn es Ihnen möglich ist, mir eine Frage in der leidigen Elektronenmikroskop-Angelegenheit zu beantworten. Von verschiedenen Seiten wird mir geschrieben, daß Rüdenberg in Amerika sich als den Erfinder des Elektronenmikroskops hinstellt aufgrund seiner Patentanmeldung vom 31. Mai 1931. Wie Sie wissen, hielt ich nicht nur einige Tage später (4. Juni 1931) einen Vortrag im Kranz-Kolloquium über das Elektronenmikroskop (wobei Rüdenberg in der ersten Reihe saß und in der Diskussion kein Wort sprach), sondern wir hatten auch unser Elektronenmikroskop schon seit 1 1/2 bis 2 Jahren in Betrieb. Sie hatten uns, wenn ich mich recht erinnere, im April oder Mai 1931 besucht, und zwar auf Veranlassung von Rüdenberg, um Einzelheiten unseres Instrumentes kennenzulernen, worüber Sie ihm dann berichtet haben. Finden Sie es nicht sehr merkwürdig von Rüdenberg, daß er aufgrund dieser Sachlage, die ja ganz allgemein bekannt ist und durch meine Doktoranden als Zeugen belegt werden kann, sich ständig als alleinigen Erfinder des Elektronenmikroskops ausgibt? Schließlich hat er doch auch später, wie aus seinen Publikationen zu entnehmen ist, nicht ein einziges optisches Experiment gemacht, geschweige ein Elektronenmikroskop gebaut.

Neuerdings hat Herr Rüdenberg wiederum einen Vorstoß in dieser Richtung gemacht, den ich nicht unwidersprochen lassen kann. Der Herausgeber der Encyclopedia of Chemistry schreibt mir:

Figure 1 Excerpts from the Knoll–Steenbeck correspondence. A full transcript and an English translation are given in the Appendix. (Above) Letter from Knoll to Steenbeck; (overleaf) Steenbeck to Knoll. The publishers of the "Encyclopedia of Chemistry" no longer have any record of the Rüdenberg manuscript mentioned.

-2-

"Professor Rüdenberg hat uns einen langen Aufsatz geschickt, in dem er behauptet, Alleinerfinder zu sein. Ihr Name wird so am Rande erwähnt. Sie hätten zwar über Fokussierungsbedingungen an Elektronenstrahlen gearbeitet, aber niemals an eine Bilderzeugung gedacht. Dies hätte er allein zuerst getan und in einem französischen Patent im Jahre 1932, mit der Priorität von 1931, veröffentlicht. . .etc."

Ich möchte Sie nun bitten, mir verstehen zu helfen, wie Rüdenberg solche Äusserungen machen kann. Ich habe nun seit vielen Jahren nicht dazu gesagt, weil Prioritätsstreitigkeiten immer für beide Teile unangenehm sind, aber ich kann mir auf die Dauer solche Entstellungen tatsächlicher Vorgänge nicht gefallen lassen und Herr Ruska ist der gleichen Meinung.

Ich hoffe, daß es Ihnen und Ihrer Familie gut geht und daß Sie doch einmal durch München kommen, um uns zu besuchen und sich über alte Zeiten zu unterhalten.

Mit besten Grüßen
Ihr

PS: Vor zwei Wochen hatten wir einen schweren Verlust zu beklagen. Prof. M. Schön ist unerwartet nach einer Operation gestorben.

Figure 1 (*continued*)

which can be traced with the aid of later monographs without undue difficulty. However, Soviet interest in electron optics was really born not in Leningrad, where Grinberg was also working, but in the Ukrainian city of Kiev. Between 1935 and 1939, several papers were published in Ukrainian by Dyachenko and Sakharov (Dyachenko, 1935; Dyachenko and Sakharov, 1935, 1937a, 1937b, 1938a, 1938b, 1939) on electrostatic lens properties, as well as studies of immersion lenses by Savchenko (1938, 1939); the early work of Sushkin already mentioned likewise appeared in Ukrainian.

DEUTSCHE AKADEMIE DER WISSENSCHAFTEN ZU BERLIN

FORSCHUNGSGEMEINSCHAFT DER NATURWISSENSCHAFTLICHEN, TECHNISCHEN UND MEDIZINISCHEN INSTITUTE

Institut für Magnetohydrodynamik
Der Direktor

JENA, den [illegible]. November 1960
FRÖBELSTIEG 3
Fernsprecher: 4741/42, 4409

Prof.Stb/Kö.-

Herrn
Professor Dr.Ing. Max K n o l l
Direktor des Instituts für Technische Elektronik
der TH München

M ü n c h e n 2
Arcisstrasse 21

Lieber Herr Knoll !

Es war für mich wirklich eine große Freude, von Ihnen nach rund 20 Jahren wieder einmal zu hören. Uns alle hat der große Knall ziemlich durch die Welt gewirbelt und aus diesem Chaos hat sich ja in zum Teil erst sehr langer Zeit ein neuer Zustand sedimentiert. Ich freue mich, daß Sie dabei offensichtlich an einen Platz gekommen sind, der mir für Sie wie geschaffen erscheint.

Zu Ihrer Frage in der "leidigen Elektronenmikroskop-Angelegenheit" will ich gern aus der Erinnerung so genau antworten, wie ich es heute noch kann. Dabei kann ich mich nicht an die einzelnen Daten erinnern. Da aber die Patentanmeldung von Rüdenberg und Ihr Vortrag im Kranz-Kolloquium zeitlich genau festliegen, müssen sich die Vorgänge in Ihrem Zeitablauf ja genau genug einordnen lassen.

Ich besuchte Sie im Institut vor dem Zeitpunkt, an dem Rüdenberg seine Patentanmeldung über das Elektronenmikroskop machte. Ich sah damals bei Ihnen die ersten Abbildungen mit einer Luftspule und ich erinnere mich, daß Sie mir berichteten, daß Sie experimentell nachprüften, wie weit die Linsenformel von Busch richtig sei. Sie hatten eine Bestätigung der Linsenformel gefunden. Mir war dabei das Interessanteste, daß damit die Vorkonzentrierungsspule von Rogowski sozusagen rechenmäßig dimensionierbar geworden war. Gerade diese Vorkonzentrierungsspule von Rogowski gab damals Anlaß zu Streitigkeiten. – Ich sah bei Ihnen weiter Abbildungen der Anodenblende, auch einer Anodenblende mit mehreren Löchern. Ich erinnere mich weiter, daß Sie von Elektronenstrahl durchsetzte feine Drahtnetze mit einer verhältnismäßig kurzbrennweitigen Spule vergrößert

[illegible] - V/18/63

-2-

Figure 1 (*continued*)

\- 4 -

Ihre Kosten geht. Daß Sie auf dem Weg waren, der unbedingt zum Elektronenmikroskop führen mußte, daß Sie auf diesem Weg mit sicheren Resultaten auf Grund solider Rechnungen und Experimente voranschritten, so daß dann durch Sie und Ihren Schüler Ruska schließlich der große reale Erfolg erreicht wurde, macht den ganzen Vorgang so tragisch. Mir hat in diesem Zusammenhang immer leid getan, daß hier die saubere systematische wissenschaftliche Arbeit ins Hintertreffen geriet gegenüber den nach dem Patentrecht möglichen fast börsenmäßigen Spekulationen, die ja oft genug zu Wegelagerermethoden führen. Daß Ihr persönlicher Anteil an der Entwicklung der Elektronenmikroskopie dann in dem sich später entwickelnden Riesengezänk zurücktrat, hat mir sachlich leid getan, obwohl mich die Tatsache, daß Sie von niemandem angegriffen wurden, immer gefreut hat.
Soweit ich weiß, bekamen Sie mit einigen anderen Elektronen-Optikern einmal von der Physikalischen Gesellschaft einen Preis; ich fand, daß dies das mindeste sei, was die Wissenschaftler Ihnen schuldig waren. Die ersten mehrlinsigen elektronen-optischen Vergrösserungen haben sicher Sie gemacht.

Ich hoffe, daß dieses aus meiner Erinnerung konstruierte Bild - gesehen durch die Trübung einer sehr bewegten dazwischenliegenden Zeit - im wesentlichen richtig ist und ich hoffe, daß es Ihnen nützt. In der vorliegenden Form ist es völlig unbeeinflußt und ich halte es für möglich, daß es etliche Bildfehler enthält. Wenn Sie glauben, daß mich meine Erinnerung in wesentlichen Zügen trügt, dann schreiben Sie mir bitte. Vielleicht wird dann bei mir wieder einiges bewußt, was zur Zeit untergegangen ist.

Das, lieber Herr Knoll, war für heute alles, was ich dazu sagen kann. - Ich hoffe, daß es Ihnen menschlich, gesundheitlich und arbeitsmäßig gut geht und ich hoffe, daß Sie wieder etwas wohler aussehen als damals, als wir uns das letztemal sahen. Ich weiß nicht mehr, wann und wo es war; ich weiß nur noch, daß es während des Krieges war und daß ich erschrocken darüber war, wie angegriffen und elend Sie damals aussahen. Wenn wir uns einmal wieder begegnen, hoffe ich, daß dieses Erinnerungsbild durch ein besseres Gegenwartsbild ersetzt werden kann.

Mit den besten Grüßen
bin ich Ihr
M. Steenbeck
(Prof.Dr.Max Steenbeck)

Figure 1 (*continued*)

A number of books appeared in Russian in the 1940s and 1950s, from which some further details may be gleaned: Sushkin (1949); Vainrib and Milyutin (1951); Lebedev (1954); Kel'man (1955, 1968); Kas'yankov (1956); Kel'man and Yavor (1959); and Strashkevich (1959); to which we may add Luk'yanovich (1960) and Zinchenko (1961).

Of the other countries in Europe, the most active in electron microscopy have been Holland, Switzerland, France, and Czechoslovakia, with local efforts in Sweden, Finland, and Italy. Of Holland we need say little, since one of the founders of the subject in that country is a contributor to this volume,[2] and the development of the Philips series of electron microscopes may be traced through the descriptions published in the *Philips Technical Review*. The abstracts volume of the twenty-fifth anniversary meeting of the Dutch electron microscope society contains several historical accounts (de Bruijn, 1977; van Dorsten, 1977; Elbers, 1977; van Iterson, 1977; Jan, 1977; Le Poole, 1977; Mellema, 1977; Spit, 1977; van Zuylen, 1977), some of which were reprinted in *Ultramicroscopy* (**3**, 1978, 111–124). See also Le Poole (1966–1967). France too is well represented with reviews by the two scientists who were the driving forces behind French microscope development, leading to the commercial models produced by CSF and OPL (for which see Fert & Selme, 1956) and the first very high-voltage instruments.

The history of electron microscope development in Switzerland is closely associated with the firm of Trüb-Täuber and with the work of the late Giovanni Induni and of L. Wegmann. Several descriptions of the Trüb-Täuber instruments and of their unusual cold-cathode gun and reduced-distortion arrangement were published (Extermann, 1948; Gribi et al., 1959; Induni, 1944, 1945a, 1945b, 1946, 1947a, 1947b, 1949, 1951, 1954, 1955, 1956; Kellenberger, 1949a, 1949b; Mühlethaler, 1948; Wegmann 1950a, 1950b, 1950c, 1951a, 1951b, 1952, 1953a, 1953b, 1954; Wegmann & Gribi, 1957; Wegmann & Studer, 1950). A collection of Induni's papers, including many patent applications, has been deposited in the Library of the Eidgenossische Technische Hochschule in Zürich.[3]

[2] A second contribution, by Dr. A. C. van Dorsten, will appear in a later volume of *Advances in Electronics and Electron Physics*. Dr van Dorsten's health did not allow him to complete this [PWH].

[3] I am indebted to the librarian for much helpful correspondence concerning these documents and to Dr. L. Wegmann for telling me of their existence and for supplying a virtually complete list of publications by Induni and himself and various collaborators.

The situation in Sweden is more obscure and I can add little to the account given by Ruska (1957); an essay by Siegbahn (1939) on the Research Institute for Experimental Physics of the Academy of Sciences contains a description and a photograph of their electron microscope, and later papers by Bergqvist (1946a, 1946b) give further details of this horizontal instrument, which was to have been marketed by the firm of Georg Schönander in Stockholm. Ruska adds that this plan was abandoned. Some early results obtained with the microscope were described in *Nature* in 1944 by Ingelman and Siegbahn. See too the articles by Fernández-Morán and Cosslett in volume 220. Delong tells us that one such instrument was imported into Czechoslovakia.

The history of electron microscope development in Finland is essentially the story of Alvar Wilska's quest for high performance at low accelerating voltage. An account of his first microscopes is to be found in his recollections (Wilska, 1964a), and further details are given in Wilska (1958, 1961, 1962, 1964b, 1965). As in many other countries, however, introductions to this new instrument were written much earlier (e.g., Erkama, 1941).

In Italy, electron microscopy began with the arrival in 1943 in the Istituto Superiore di Sanità in Rome of a Siemens instrument, ordered in 1939. It remained there until October of the same year, when "d'ordine del Comando militare germanieo, il microscopio veniva retirato per essere 'posto al sicuro' con la promessa di restituircelo dopo 'l'immancabile vittoria.' Noi non ponemmo molta fiducia in questa promessa . . . " (Trabacchi, 1947, 1948), and so a microscope was built to replace it, despite the difficulty of obtaining parts. By June, 1944, half of the instrument was ready but power cuts added to the problem of completing the other half. It was not until 1946, therefore, that the microscope could be assembled and tested (Fig. 2). For further details see Bocciarelli and Trabacchi (1946, 1948) and Trabacchi (1948). Before leaving this wartime work in Italy, we mention that an impressive number of results were obtained during the short visit of the Siemens instrument; see Babudieri and Bocciarelli (1943), Babudieri et al. (1944), Cavalli (1946), and Strampelli and Posarelli (1946) [none of which is recorded in Cosslett (1950) or Marton et al. (1950); q.v. for other early papers by some of these authors].

The Istituto Superiore di Sanità celebrates its fiftieth anniversary in 1985, and on this occasion an exhibition charting the beginnings of electron microscopy will be held and a complete list of publications emanating from this institute will be available (cf. Donelli et al., 1982, 1983).

a

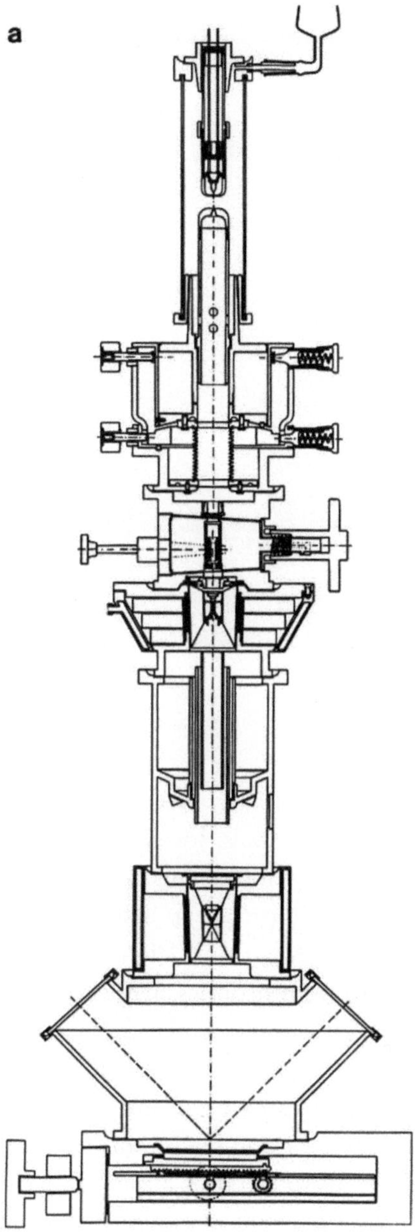

Figure 2 The Italian electron microscope, (a) Diagram, (b) General appearance.

Figure 2 (*continued*)

I have found no mention of early electron microscope construction in Spain. A full list of dates of arrivals, locations, and models of the various instruments imported is to be found in González-Santander (1968)—we note that the first electron microscope in Spain, an RCA 2A, was installed in the Instituto de Optica in Madrid in 1946.

Two early papers on electron optics may also be mentioned: in 1950, Baltá Elías and Gómez García published a theoretical study of an electrostatic cylindrical lens, and a year later Gómez García (1951) published a study of the optical properties of electromagnetic cylindrical lenses. See too Ortiz Fornaguera (1948). Long before this, however, Brú (1936, 1942) had published introductions to electron optics in Spanish.

In Eastern Europe, the most active country was Czechoslovakia, as the article by Professor Delong (volume 220) shows; the effect of the arrival of one of the imported instruments mentioned there, the Trüb-Täuber, is described by Dohnálek and Rozsíval (1948). Some interest was shown in electron optics in Poland in the 1930s (Majewski, 1936[4]) but I know of no other early work from that country. Papers introducing the instrument were also published in Hungarian (Renner, 1939, 1940) and Serbo-Croatian (Vouk, 1941–1942) and no doubt in many other Eastern European languages.

Turning now to the American continent, we list a few papers that may be of interest in conjunction with the articles in this volume and volume 220 by C. E. Hall and H. Fernández-Morán. By far the fullest account of the Canadian work is that compiled by Franklin et al. (1978), supplemented by Kohl et al. (1978), but the after-dinner lecture delivered by Watson in 1974 (Watson, 1974) should not be missed. The textbook of Burton and Kohl (1942) also records some of this work. See too Prebus (1942), Wyckoff (1954) and Calbick (1944).

Less well known is the instrument built by P. A. Anderson and K. Fitzsimmons in the Physics Department of Washington State College (now University). A photograph of it, dated May, 1937, is reproduced in Cohen and Steever (1971), the only reference I have seen to this instrument, and neither Anderson nor Fitzsimmons figures in the bibliographies of C. Marton et al. (1950) and Cosslett (1950), both rich sources of recondite papers.

The RCA contribution is easily traced through the *RCA Scientific Instrument News*, but there have been some smaller American commercial ventures (e.g., the Forgflo "Paragon," 1970; Rempfer et al., 1972). Brüche (1957) states that the Farrand Optical Co. of New York produced an electrostatic microscope but I have found only one published reference to it (Fleming et al., 1948); we note in passing that Walter Glaser spent some time with the Farrand Co. in 1955, where he produced a number of (unpublished) reports on the correction of spherical aberration by means of quadrupoles and octopoles. The better known GEC instrument is described by Bachman and Ramo (1943).

The Electron Microscopy Society of America *Bulletin* contains occasional "Reflections": the origins of the EMSA have been retraced by Reisner (1981a, 1981b; 1982). T. G. Rochow (1983) describes some "significant early days in industrial electron microscopy."

A list of the publications of the late Vladimir Zworykin is to be found in the tribute to the latter by Abramson (1981).

[4] Not in Cosslett (1950) or Marton et al. (1950); cited in the Polish edition of the textbook by Paszkowski (1960, 1965) but not in the English translation (Paszkowski, 1968).

That Japan has been extremely active in the field since the late 1930s needs no comment here; the contributions by Kanaya and Hibi give abundant evidence of this. An early paper by Tadano (1953) may perhaps be added to the already long list given by Kanaya. This is an account of "Progress of Electron Microscopes in Hitachi," with photographs of many early Hitachi instruments, including the first, as well as many technical details. The bibliography of this article is particularly useful, since many early Japanese papers are listed, with titles (in English), including of course their first paper on the whole instrument (Tadano & Siraga, 1942). Hitachi (1959) has also published a collection of their published papers. By far the finest pictorial record of the various Japanese electron microscopes is, however, that edited by Sugata (1968). The albums with the general title "The World through the Electron Microscope" produced by JEOL (1959–1973) are very evocative. See too the surveys by Higashi (1967) and Tani (1956) and the more personal account by Hashimoto (1983).

Of the remainder of Asia, I can say little. A horizontal electron microscope was built at the University of Calcutta in the 1940s and a detailed description is to be found in Das Gupta et al. (1948); see also De et al. (1953) and Das Gupta and De (1968). Books on electron optics and electron microscopy were available in Chinese and Korean as early as 1959 (Seman) and 1963 (Kim and Choi), respectively, and a sprinkling of papers on electron optics appeared in the Chinese scientific journals from at least 1956 onward (e.g., Chiang, 1956; Syui, 1956). A paper by Kim appeared in the *Korean Journal of Mathematics and Physics* in 1962; for anyone wishing to pursue this further, the Russian *Referativnyi Zhurnal* may well prove the least inaccessible source of information.

I am very conscious of the lack of completeness of this account, but I hope it may provide a few pointers for anyone working on the history of the first half-century of electron microscopy. Two collections of accounts of the neighboring domain of electron diffraction should also not be missed (Dobson et al., 1978; Goodman, 1981); see too an article by Molière (1977).

1. Appendix

Letter from Prof. Dr. Ing. Max Knoll, in Munich, dated 17 October 1960, to Prof. Dr. Steenbeck, Physikalisches Institut der Universität Jena and Institut für Magnetohydrodynamik, Deutsche Akademie der Wissenschaften, Jena (translation follows).

Lieber Herr Steenbeck!

Leider ist es mir noch nicht möglich gewesen, Ihnen zu Ihrer glücklichen Rückkehr zu gratulieren und über unsere Erlebnisse in den verschiedenen Ländern Meinungen auszutauschen. Dagegen habe ich Herrn von Engel mehrere Male seit dem Krieg getroffen, das letzte Mal in Princeton.

Ich möchte Sie heute bitten, wenn es Ihnen möglich ist, mir eine Frage in der leidigen Elektronenmikroskop-Angelegenheit zu beantworten. Von verschiedenen Seiten wird mir geschrieben, daß Rüdenberg in Amerika sich als den Erfinder des Elektronenmikroskops hinstellt aufgrund seiner Patentanmeldung vom 31. Mai 1931. Wie Sie wissen, hielt ich nicht nur einige Tage später (4. Juni 1931) einen Vortrag im Kranz-Kolloquium über das Elektronenmikroskop[1] (wobei Rüdenberg in der ersten Reihe saß und in der Diskussion kein Wort sprach), sondern wir hatten auch unser Elektronenmikroskop schon seit 1 1/2 bis 2 Jahren[2] in Betrieb. Sie hatten uns, wenn ich mich recht erinnere, im April oder Mai 1931 besucht, und zwar auf Veranlassung von Rüdenberg, um Einzelheiten unseres Instrumentes kennenzulernen, worüber Sie ihm dann berichtet haben. Finden Sie es nicht sehr merkwürdig von Rüdenberg, daß er aufgrund dieser Sachlage, die ja ganz allgemein bekannt ist und durch meine Doktoranden als Zeugen belegt werden kann, sich ständig als alleinigen Erfinder des Elektronenmikroskops ausgibt? Schließlich hat er doch auch später, wie aus seinen Publikationen zu entnehmen ist, nicht ein einziges optisches Experiment gemacht, geschweige ein Elektronenmikroskop gebaut.

Neuerdings hat Herr Rüdenberg wiederum einen Vorstoß in dieser Richtung gemacht, den ich nicht unwidersprochen lassen kann. Der Herausgeber der Encyclopedia of Chemistry schreibt mir:

"Professor Rüdenberg hat uns einen langen Aufsatz geschickt, in dem er behauptet, Alleinerfinder zu sein. Ihr Name wird so am Rande erwähnt. Sie hätten zwar über Fokussierungsbedingungen an Elektronenstrahlen gearbeitet, aber niemals an eine Bilderzeugung gedact. Dies hätte er allein zuerst getan und in einem französischen Patent im Jahre 1932, mit der Priorität von 1931, veröffentlicht … etc."

Ich möchte Sie nun bitten, mir verstehen zu helfen, wie Rüdenberg solche Äusserungen machen kann. Ich habe nun seit vielen Jahren nicht [sic] dazu gesagt, weil Prioritätsstreitigkeiten immer für beide Teile unangenehm sind, aber ich kann mir auf die Dauer solche Entstellungen tatsächlicher Vorgänge nicht gefallen lassen und Herr Ruska ist der gleichen Meinung.

Ich hoffe, daß es Ihnen und Ihrer Familie gut geht und daß Sie doch einmal durch München kommen, um uns zu besuchen und sich über alte Zeiten zu unterhalten.

Mit besten Grüßen

Ihr

[Knoll]

PS: Vor zwei Wochen hatten wir einen schweren Verlust zu beklagen. Prof. M. Schön ist unerwartet nach einer Operation gestorben.

[1] Nicht richtig, das Vortragsthema "der Kathodenstrahloszillograph" war (E. Ruska).

[2] Zwei Jahre im Betrieb war nur mein einstufig vergrössemdes Gerät, das zweistufige Gerät war erst seit März 1931 in Betrieb (E. Ruska).

Dear Herr Steenbeck,

Unfortunately I have not yet been able to congratulate you on your happy return and to compare notes on our experiences in different countries. On the other hand, I have encountered Herr Von Engel several times since the war, the most recent being in Princeton.

I shall be most grateful if it is possible for you to reply to a question concerning the distressing electron-microscope affair. I have received letters from several places to the effect that in America, Rüdenberg is representing himself as the discoverer of the electron microscope on the grounds of his patent application of 31 May 1931. As you know, not only did I give a lecture in the Kranz-Colloquium on the electron microscope[1] a few days later (4 June 1931) (at which Rüdenberg sat in the first row and said not a word during the discussion) but we had also had our electron microscope in operation for $1\frac{1}{2}$ to 2 years.[2] If I remember correctly, you visited us in April or May 1931, in fact, at Rüdenberg's instigation, in order to become acquainted with the details of our instrument, which you subsequently reported to him. Do you not find it very remarkable that Rüdenberg should set himself up as the sole inventor of the electron microscope on the basis of this situation, which is quite generally known and can be proved by my Ph.D. students? Finally, as can be seen from his publications, he later performed not a single optical experiment, let alone built an electron microscope.

Recently, Herr Rüdenberg has taken yet another step forward in this direction, which I cannot leave uncontradicted. The editor of the Encyclopedia of Chemistry writes to me as follows:

"Professor Rüdenberg has sent us a long contribution in which he claims to be the only discoverer. Your name is mentioned only in passing. You had certainly worked on the focusing conditions for electron rays but never thought of forming an image. He alone was the first to do this, and this was published in a French patent in 1932, with priority dating from 1931 ... etc".

I shall therefore be grateful if you will help me to understand how Rüdenberg can make such statements. I have said nothing about this for many years now, as wrangles over priority are always unpleasant for both parties but I cannot tolerate permanently such misrepresentations of actual facts and Herr Ruska is of the same opinion.

I hope that all is well with you and your family and that you will come again to Munich to visit us and talk over old times.

Mit besten Grüssen
Ihr
[KNOLL]

P.S. Two weeks ago, we suffered a severe loss. Prof. M. Schön died unexpectedly after an operation.

Letter from Prof. Dr. Max Steenbeck, Director of the Institut für Magnetohydrodynamik (Deutsche Akademie der Wissenschaften zu Berlin), Jena, dated 8 November 1960, to Prof. Dr. Ing. Max Knoll, Director of

[1] This is not correct: the theme was "The cathode-ray oscillograph" (E. Ruska).

[2] Only Ruska's one-stage magnifying instrument had been in Operation for two years; the two-stage instrument first came into Operation in March, 1931 (E. Ruska).

the Institut für Technische Elektronik der TH München (translation follows).

Lieber Herr Knoll!

Es war für mich wirklich eine große Freude, von Ihnen nach rund 20 Jahren wieder einmal zu hören. Uns alle hat der große Knall ziemlich durch die Welt gewirbelt und aus diesem Chaos hat sich ja in zum Teil erst sehr langer Zeit ein neuer Zustand sedimentiert.

Ich freue mich, daß Sie dabei offensichtlich an einen Platz gekommen sind, der mir für Sie wie geschaffen erscheint.

Zu Ihrer Frage in der "leidigen Elektronenmikroskop-Angelegenheit" will ich gern aus der Erinnerung so genau antworten, wie ich es heute noch kann. Dabei kann ich mich nicht an die einzelnen Daten erinnern. Da aber die Patentanmeldung von Rüdenberg und Ihr Vortrag in Kranz-Kolloquium zeitlich genau festliegen, müssen sich die Vorgänge in Ihrem Zeitablauf ja genau genug einordnen lassen.

Ich besuchte Sie im Institut vor dem Zeitpunkt, an dem Rüdenberg seine Patentanmeldung über das Elektronenmikroskop machte. Ich sah damals bei Ihnen die ersten Abbildungen mit einer Luftspule und ich erinnere mich, daß Sie mir berichteten, daß Sie experimentell nachprüften,wie weit die Linsenformel von Busch richtig sei. Sie hatten eine Bestätigung der Linsenformel gefunden. Mir war dabei das Interessanteste, daß damit die Vorkonzentrierungsspule von Rogowski sozusagen rechenmäßig dimensionierbar geworden war. Gerade diese Vorkonzentrierungsspule von Rogowski gab damals Anlaß zu Streitigkeiten. - Ich sah bei Ihnen weiter Abbildungen der Anodenblende, auch einer Anodenblende mit mehreren Löchern. Ich erinnere mich weiter, daß Sie von Elektronenstrahl durchsetzte feine Drahtnetze mit einer verhältnismäßig kurzbrennweitigen Spule vergrößert abbildeten. Ich erinnere mich aber nicht, ob es sich um reine Luftspulen, oder schon um eine eisenarmierte Spule handelte; wenn das letztere der Fall gewesen sein sollte, hätte ich alle Details vergessen. Ich erinnere mich an die bei der Abbildung des Netzes deutlich sichtbar werdenden Vorzeichnungen [sic], wobei ich nicht mehr weiß, ob sie kissen-oder tonnenförmig waren (obwohl man etwas derartiges heute natürlich ohnehin wissen müßte). Ich glaube mich weiter daran zu erinnern, daß ich bereits bei diesem Besuch Versuche gesehen habe, auch mit elektrischen Feldern Reflektion und Sammellinsen zu erproben; dabei waren die elektrostatischen Elektroden durch geladene feine ebene oder gewölbte Drahtnetze hergestellt. Die Resultate dieser Abbildungen waren erheblich schlechter als die magnetischen Abbildungen, bedingt und verständlich durch die Mikrostruktur des von Drahtnetzen erzeugten elektrischen Feldes.

Mein Eindruck von diesen Versuchen war der, daß die Behandlung des Elektronenstrahles analog zur Optik effensichtlich auch experimentell durchaus gerechtfertigt sei, aber ich selbst hatte damals noch nicht begriffen, daß auf diesem Wege eine praktisch brauchbare Vergrößerung zustande kommen könne, die das optisch mögliche Gebiet wesentlich ergänzen und erweitern könnte. Ich halte es für möglich, daß Sie diese Aussichten damals schon sahen und ich halte es auch für möglich, daß Sie sogar davon gesprochen haben könnten. Ich selbst hielt das Gesehene für eine durchaus solide und interessante, für die Abbildungsschärfe von Oszillografen auch praktisch wichtige grundsätzliche Arbeit; aber an eine praktische Verwertbarkeit als Elektronenmikroskop habe ich jedenfalls zu dieser Zeit nicht geglaubt und hätte eine solche Erwartung wahrscheinlich als hemmungslose

Phantasie abgelehnt; sicher auch aus der gefühlsmäßigen Erwartung, daß Raumladungen im Elektronenstrahl eine scharfe Abbildung ohnehin ausschließen müßten.

Beeindruckt hatten mich aber die experimentellen Vorführungen der optischen Gesetze, vor allem im Hinblick auf die Anwendung für den Kathodenstrahl-Oszillografen, mit dem wir um diese Zeit bei Rüdenberg ernsthafte Schwierigkeiten hatten. Ich glaube, um diese Zeit versuchte Dr. Seitz bei uns mit einem von Rogowski gekauften Kathodenstrahl-Oszillografen zu arbeiten und hatte immer nur sehr mäßige Erfolge in der Fleckschärfe.

Ich berichtete in diesem Zusammenhang Rüdenberg sehr bald nach meinem Besuch von dem, was ich bei Ihnen gesehen hatte, in der eben angegebenen Richtung für den Kathodenstrahl-Oszillografen, und dabei habe ich sicher von der überzeugenden Demonstrierung der Busch'schen Linsenformel gesprochen. Rüdenberg, der durch keine experimentellen Erfahrungen gehemmt diese Zusammenhänge hörte, wird dann für sich die Kombination gemacht haben: Wenn Magnetspulen und elektrostatische Felder nachgewiesenermaßen wie Linsen arbeiten, dann kann man vermutlich die bekannten optischen Geräte mit Elektronenstrahlen nachmachen. So hat er diesen Gedanken zum Patent angemeldet, ohne daß ich davon etwas wußte. Wenn er mit mir gesprochen hätte, hätte ich die Sache damals ohnehin für Unfug gehalten. Die Anmeldung von Rüdenberg ist sicher erst nach meinem Besuch bei Ihnen erfolgt und sicher erst durch meinen Bericht über das Gesehene ausgelöst worden. Die Schlußfolgerung, daß nach den bei Ihnen durchgeführten Experimenten ein Elektronenmikroskop möglich oder doch zumindest denkbar sei, hat Rüdenberg in seiner bekannten generalisierenden Art des Denkens aber ebenso sicher unabhängig selbst gezogen. Ich halte es daher für denkbar, daß Herr Rüdenberg sich subjektiv mit allem Recht als den Erfinder des Elektronenmikroskops ansieht, obwohl er für die Realisierung dieses Gerätes weder ernsthafte Rechnungen noch Experimente gemacht oder angeregt hat. - Ihr Vortrag in dem Kranz-Kolloquium erfolgte ziemlich bald nach meinem ersten Besuch bei Ihnen und gefiel mir durch die Geschlossenheit und Klarheit, obwohl ich den sachlichen Inhalt im wesentlichen von meinem Besuch her schon kannte; insofern war ich von dem Vortrag nicht mehr so überrascht, wie die übrigen Zuhörer. Was mir an Ihren Arbeiten so imponierte, war die sorgfältige Systematik und das Fehlen von noch nicht bewiesenen Behauptungen. - Daß Abbildungen durch Elektronen Dinge aufzeigen konnten, die grundsätzlich über das optisch Mögliche hinausgingen, zeigte mir als erstes die Veröffentlichung von Brüche in den "Naturwissenschaften" über die Emissionsfelder von Oxydkathoden. Erst von diesem Augenblick an erkannte ich die grundsätzlich neuen Möglichkeiten in der elektronen-optischen Abbildung. Ich betone aber nochmals, daß diese Feststellung nur für mich persönlich gilt und daß ich es für durchaus wahrscheinlich halte, daß Sie auf Grund Ihrer Arbeiten viel weiter sahen. Ich kann mir selbst jetzt nur das keineswegs befriedigende Zeugnis geben, eine damals sicher nicht fernliegende Möglichkeit überhaupt nicht gesehen zu haben. Anderenfalls hätte ich mich Rüdenberg gegenüber sicher anders verhalten, und ich kann nur bedauern, ohne jede Absicht eine Entwicklung ausgelöst zu haben, die so ausschließlich auf Ihre Kosten geht. Daß Sie auf dem Weg waren, der unbedingt zum Elektronenmikroskop führen mußte, daß Sie auf diesem Weg mit sicheren Resultaten auf Grund solider Rechnungen und Experimente voranschritten, so daß dann durch Sie und Ihren Schüler Ruska schließlich der große reale Erfolg erreicht wurde, macht den ganzen Vorgang so tragisch. Mir hat in diesem Zusammenhang immer leid getan, daß hier die saubere systematische wissenschaftliche Arbeit ins Hintertreffen geriet gegenüber den nach dem Patentrecht möglichen fast börsenmäßigen Spekulationen, die ja oft genug zu Wegelagerermethoden führen. Daß Ihr persönlicher Anteil an der Entwicklung der Elektronenmikroskopie dann in dem sich später entwickelnden Riesengezänk zurücktrat, hat mir sachlich leid getan, obwohl mich die Tatsache, daß Sie von niemandem angegriffen wurden, immer gefreut hat.

Soweit ich weiß, bekamen Sie mit einigen anderen Elektronen-Optikern einmal von der Physikalischen Gesellschaft einen Preis; ich fand, daß dies das mindeste sei, was die Wissenschaftler Ihnen schuldig waren. Die ersten mehrlinsigen elektronen-optischen Vergrösserungen haben sicher Sie gemacht.

Ich hoffe, daß dieses aus meiner Erinnerung konstruierte Bild - gesehen durch die Trübung einer sehr bewegten dazwischenliegenden Zeit - im wesentlichen richtig ist und ich hoffe, daß es Ihnen nützt. In der vorliegenden Form ist es völlig unbeeinflußt und ich halte es für möglich, daß es etliche Bildfehler enthält. Wenn Sie glauben, daß mich meine Erinnerung in wesentlichen Zügen trügt, dann schreiben Sie mir bitte. Vielleicht wird dann bei mir wieder einiges bewußt, was zur Zeit untergegangen ist.

Das, lieber Herr Knoll, war für heute alles, was ich dazu sagen kann. - Ich hoffe, daß es Ihnen menschlich, gesundheitlich und arbeitsmäßig gut geht und ich hoffe, daß Sie wieder etwas wohler aussehen als damals, als wir uns das letztemal sahen. Ich weiß nicht mehr, wann und wo es war; ich weiß nur noch, daß es während des Krieges war und daß ich erschrocken darüber war, wie angegriffen und elend Sie damals aussahen. Wenn wir uns einmal wieder begegnen, hoffe ich, daß dieses Erinnerungsbild durch ein besseres Gegenwartsbild ersetzt werden kann.

Mit den besten Grüßen
bin ich Ihr

[PROF. DR. MAX STEENBECK]

Dear Herr Knoll,

It really was a great pleasure for me to hear from you again after about 20 years. The great catastrophe has scattered us all over the world and from this chaos a new state of affairs has partly sedimented out, after this long interval. I am delighted that you have clearly attained a position for which you are, I believe, perfectly suited.

In answer to your question concerning the "distressing electron microscope affair," I shall reply from memory as accurately as I now can. I cannot remember the exact dates. As, however, the dates of the patent application by Rüdenberg and of your lecture in the Kranz-Colloquium are known exactly, it must be possible to arrange the incidents correctly in the period involved.

I visited you in the Institute before the date on which Rüdenberg applied for his patent on the electron microscope. On that occasion I saw there the first images obtained with an air-coil and I remember that you informed me that you had investigated experimentally the correctness of Busch's lens formula. You had found agreement with the lens formula. The most interesting thing for me was that it thereby became possible to calculate Rogowski's pre-concentrating coil with regard to dimensions, so to speak. It was precisely this pre-concentration coil of Rogowski that was then giving rise to argument. I also saw images of the anode aperture and of an anode aperture with several holes. I remember too that you formed a magnified image of a fine wire mesh, through which the electron beam passed, by means of a relatively short-focal-length coil. I do not however remember whether the coils were merely air-coils, or if there was already an iron-cased coil. If the latter was the case, I have forgotten all the details. I remember that distortion was clearly present in the image of the mesh but I do not know now whether it was pincushion- or barrel-shaped (although it must be possible to clear up such a point today). Furthermore, I think that I remember seeing, during this same visit, experiments to test the reflection and convergent lens action of electric fields; the electrostatic electrodes consisted of charged fine wire-mesh, plane or curved. The resulting images were distinctly worse than the magnetic images, being understandably limited by the microstructure of the electric field produced by the wire mesh.

My impression of these experiments was that the handling of electron beams in the same way as in optics was completely vindicated, clearly and experimentally, but I myself did not grasp at that time that a useful magnification could in practice be achieved, significantly extending and widening the domain accessible in (light) optics. I consider it possible that you did indeed then see this possibility and also consider it possible that you even spoke of it. I myself found what I saw to be a thoroughly solid and interesting fundamental study, of practical importance for sharp imaging in oscillographs; however, at that time at least, I never thought of its practical exploitation as an electron microscope, and should probably have dismissed such an idea as pure fantasy; particularly from the instinctive expectation that space charges in the electron beam would exclude sharp image formation, in any case.

I was particularly struck by the experimental demonstrations of the optical laws especially with a view to their use for cathode-ray oscillographs, with which we were having serious difficulties at that time in Rüdenberg's laboratory. I believe that at about that time Dr. Seitz (from our group) tried to work with a cathode-ray oscillograph bought from Rogowski, and had only very moderate success regarding the spot sharpness.

I reported to Rüdenberg very soon after my visit what I had seen in your Laboratory and the tendency thus created for cathode-ray oscillographs, and I certainly talked about the convincing demonstration of the Busch lens formula. Rüdenberg, hampered by the absence of experimental information and hearing of this relation, will have made for himself the following conjecture: if magnetic coils and electrostatic fields work as lenses as has been shown, then the known kinds of optical instruments can presumably be imitated with electron rays. He thus applied for a patent for this idea, without my knowing anything about it. If he had spoken to me about it, I would have found the suggestion nonsensical at that time anyway. Rüdenberg's [patent] application is certainly a consequence of my visit to you and was certainly stimulated by my report on what I had seen. The conclusion, that as a result of your experiments an electron microscope was possible or at least conceivable, was certainly drawn independently by Rüdenberg himself, thanks to his well-known habit of generalizing. It therefore seems to me conceivable that Herr Rüdenberg really regards himself as the legitimate discoverer of the electron microscope, even though he did not make or even suggest any serious calculations for the realization of such an instrument.

Your lecture in the Kranz-Colloquium followed fairly soon after my first visit to you and pleased me by its completeness and its clarity, even though I was already familiar with the contents from my visit; for the same reason, I was not quite so astonished by the lecture as the other members of the audience. What I found so impressive in your studies was the careful and methodical approach and the absence of any statements not yet proved. The fact that images of things could be produced by electrons, essentially surpassing the possibilities of (light) optics, was first shown to me by the publication of Brüche in "Naturwissenschaften" on the [imaging of] emission fields from oxide cathodes. From this very first moment, I recognized the fundamentally new possibilities in electron-optical imaging. I stress however, once again, that this remark is only true of me personally and I find it highly probable that you saw a great deal further thanks to your work: I myself can now give only the not at all satisfactory testimony that I definitely did not see a possibility which was surely within reach at the time. Otherwise, I should certainly have behaved differently with Rüdenberg and I can only deplore that I was quite unintentionally the cause of a development, which has been made exclusively at your expense. The fact that you were already on the path that would undoubtedly have led to the electron microscope, and that you were ahead on this path with firm results based on solid calculations and experiments, so that you and your pupil Ruska would finally achieve real and substantial success, makes the whole affair so tragic. In this connection, I have always regretted that here, healthy systematic scientific work takes second place to what almost amounts to financial speculations, possible in patent law, often leading to highwayman-like behaviour. The fact that your personal contribution to the development of electron microscopy has been diminished in the giant quarrels that later developed made me truly sorry although the knowledge that nobody criticized you has always been gratifying.

So far as I am aware, you were once awarded a prize by the [German] Physical Society, together with some other scientists in the field of electron optics; this was the very least that your fellow-scientists owed you. The first many-lens electron-optical magnifications were certainly obtained by you.

I hope that this image, reconstructed from my memory—seen through the cloudiness of such disturbed intervening years—is essentially correct, and I hope that it is useful for you. As it stands, it is completely free of outside influences and I believe it possible that it may contain a few aberrations [image-errors]. If you think that my memory has led me astray on any important point, please write to me. Perhaps it will then be possible for me to reestablish what at present is lost.

That, dear Herr Knoll, is all that I can say about this now. I hope that all goes well for you, personally and in your health and in your work, and I hope that you look better than when we last met. I no longer know when and where that was; I only know that I was alarmed to see how exhausted and wretched you then appeared. If we meet again, I hope that this image in my memory can be replaced by a better contemporary image.

Mit den besten Grüssen
bin ich Ihr

[Prof. Dr. Max Steenbeck]

Acknowledgments

I have incurred many debts of gratitude while writing this account. I wish to thank very warmly all those who have helped by sending or lending elusive documents and books, and, in particular, the librarian of the Zentralbibliothek of AEG-Telefunken in Berlin; Professor Borovskii, who enabled me to see a copy of Sushkin (1949); Professor Donelli of the Istituto Superiore di Sanità in Rome; Professor H. Hashimoto; Dr. Jan Mellema of the Nederlands Instituut voor Zuivelonderzoek, formerly of the University of Leiden, for information about microscopy in The Netherlands; Monsieur Michot of JEOL (Europe) for the loan of many documents; Mademoiselle Monique Pénochet, librarian of my own laboratory and her colleagues at the CNRS Library in Paris and at the British Lending Library in Boston Spa; Dr. George T. Simon of McMaster University; Professor E. Sugata for the gift of a copy of his historical album; Professor U. Valdrè of the University of Bologna; Dr. L. Wegmann and the librarian of the Wissenschafthistorische Sammlungen of the ETH-Bibliothek in Zürich; Dr. Stella Yakoblevna Yavor of the A. F. Ioffe Institute in Leningrad. My greatest single debt is again, as so often in the past, to the librarian and staff of the Cambridge Scientific Periodicals Library, who have been unfailingly helpful and patient in the face of requests for rarely consulted journals and imperfect, even wrong, references.

2. Afterword

Books and articles on the history of electron microscopy and biographical studies on individual pioneers of the subject continue to appear. Germany has attracted the most attention. The historian of science Falk Müller has written an account of the first years of the electron microscope (Müller, 2009). A much fuller account can be found in his book, *Jenseits des Lichts* (Müller, 2021). See too Müller (2006). The book by Lin Qing (1995) has the merit that a long chapter is devoted to work at the AEG Research Institute during the 1930s, complementing the chapter on the work of the 'Ruska-Gruppe'. The controversy that sprang up after the award of the Nobel Prize to Ernst Ruska is summarized and the references include letters to newspapers as well as articles in refereed journals. Individual chapters in *The Growth of Electron Microscopy* (Mulvey, 1996) are not listed here apart from the magisterial chapter by Alan Agar (1996) on commercial European electron microscopes.

The early years are also re-examined by John van Gorkom (2018a, 2018b), who enjoyed several interviews with Ruska. The Nobel Prize lecture is of course an essential document (Ruska, 1987a, 1987b, 1987c). We also list the books by Gloede (1986) and by Karrenberg (2019); the latter contains many unfamiliar illustrations of early Siemens instruments.

Several articles on the early history of the electron microscope were listed in my original article; to these should be added the paper by Grünewald (1976), which was overlooked there.

The second group of publications contains biographical articles. A full biography of Ernst Ruska has been written by Lambert and Mulvey (1996) as well as a light-hearted essay by Ruska's secretary, Lotte Lambert, for his 80th birthday (Lambert, 1986). See too Mulvey (1987, 1988) and Schulze (2007). Less has been written about Max Knoll (1897–1969); the fullest account that I have seen is the 'Laudatio' by Gottfred Möllenstedt (1968) when Knoll was given honorary membership of the German Electron Microscope Society (DGE). Ernst Ruska's brother Helmut played an important role in encouraging Siemens to produce a commercial model of the electron microscope. His biography has been written by Hans Gelderblom and Detlev Krüger (2014), cf. Gelderblom et al. (2014). The life and work of Bodo von Borries, the other major figure in the construction of the first electron microscopes with magnetic lenses, has been described by his widow Hedwig (von Borries, 1991) and a short book by his cousin Götz von Borries is also devoted to his work (von Borries, 2001). More recently Gelderblom (2020) has also re-examined his contribution to early electron microscope development. The life of Rudoph Rühle and his connection with the little-known instruments constructed by the Bosch company is described by Gelderblom and Schwarz (this volume). Rühle's book on the electron microscope contains several micrographs obtained with the Bosch microscope (Rühle, 1949). Manfred von Ardenne has written a later autobiography than that cited earlier (von Ardenne, 1997). Before leaving Germany, we mention Reinhold Rüdenberg, who applied for patents on the electron microscope shortly before Knoll described Ruska's work at the Cranz colloquium in June 1931. Rüdenberg wrote an account of the invention of the electron microscope emphasizing his own role, which was submitted to the editor of an encyclopedia of microscopy (not chemistry, as stated in Knoll's letter). The editor asked Knoll for his opinion of Rüdenberg's article, which led to the exchange of letters reproduced above. As a result the article was rejected; nevertheless, it is an element in the early history of the subject and it was therefore published posthumously in these *Advances* (Rüdenberg, 2010) followed by a long commentary by Rüdenberg's sons (Rudenberg & Rudenberg, 2010). Ruska had published an extremely meticulous analysis of the patent sequence in German and English (1984, 1986).

The beginnings of electron microscopy in Holland are described by Marian Fournier (2009) and in a long biography of Jan Le Poole by Mulvey and van de Laak-Tijsen (2001). See too van Delft and van Helvoort (2018).

The history of the French Cameca company has been written by de Chambost (2009, 2011). A little-known French scanning electron microscope has been retrieved from oblivion by Hawkes and McMullan (2004).

An account of microscopy in Italy during the fifty years from 1956 to 2006 has been edited by Quaglino et al. (2006).

The first decade of electron microscope development in Great Britain is explored in detail by Ruiz-Castell (2013). Gabor's brainwave that gave birth to holography is evoked by Mulvey (1995). A whole volume of these *Advances* has been devoted to Sir Charles Oatley and the scanning electron microscope, with contributions by many of his research students (Breton et al., 2004). To this we may add two articles by Smith (2009, 2013), an obituary of Dennis Macmillan (Hawkes, 2015) and an essay by Paden (2015) on the 50th anniversary of the launch of the Cambridge Instrument Company Stereoscan.

A book and a long review article add considerably to our knowledge of electron microscope development in the USA. Nicholas Rasmussen describes the effect of the arrival of the electron microscope on biological research with many references to unpublished material (Rasmussen, 1997). This was preceded by a short article concentrating on the first RCA instruments (Rasmussen, 1996) and by a discussion of epistemiology with the electron microscope (Rasmussen, 1993). The review by John Reisner (1989) describes the various instruments that were constructed in the USA in the first decades. The reasons why electron microscopes with magnetic lenses had more success than those with electrostatic lenses in the USA are examined by Kunkle (1994).

Electron microscopy in Australia and especially in Sydney is recorded in the collection of 50 'Moments' gathered by Kyle Ratinac (2008). The arrival of biological electron microscopy is surveyed by Rasmussen (1999).

The commercial exhibition committee of the eleventh International Congress on Electron Microscopy held in Kyoto in 1986 decided to publish *History of Electron Microscopes* as a commemoration of the Congress. The book that resulted (Fujita, 1986) has three sections: The beginnings of electron microscopes in Japan; Development of commercial electron microscopes with chapters on Akashi, Hitachi, Philips, JEOL, and Carl Zeiss (Oberkochen); and a collection of pages from 'Catalogues of electron microscopes and their related instruments'.

Acknowledgments

I am very grateful to Mr Götz von Borries (not a relative of Bodo von Borries or Götz the author) who found me a copy of the book by Götz von Borries and to Professor Simon Ringer who sent me a copy of "50 Moments".

References

Abramson, A. (1981). Pioneers of television. Vladimir Kosma Zworykin. *SMPTE Journal*, *90*, 579–590 (includes a list of Zworykin's publications).

Afzelius, B. A. (1981). Half a century of electron microscopy: The early years. *Ultrastructural Pathology*, *3*, 309–311.

Agar, A. W. (1996). The story of European commercial electron microscopes. *Advances in Imaging and Electron Physics*, *96*, 415–584.

Artsimovich, L. A. (1944). Electron optical properties of emission systems (in Russian). *Izvestiâ Akademii Nauk SSSR. Seriâ Fizičeskaâ, 8*, 313–329.

Babudieri, B., & Bocciarelli, D. (1943). Ricerche di microscopia elettronica. I. Studio morfologico di *Rickettsia prowazeki*. *Rendiconti dell'Istituto Superiore di Sanità, 6*, 298–304; Babudieri, B., & Bocciarelli, D. Ricerche di microscopia elettronica. II. Studio morfologico del genere *Spironema*. *Rendiconti dell'Istituto Superiore di Sanità, 6* (1943), 305–314.

Babudieri, B., Bietti, G., & Bocciarelli, D. (1944). La batteriolisi prodotta dal "Lysozym" lacrimale studiata al supermicroscopio elettronico. *Rendiconti dell'Istituto Superiore di Sanità*, 7, 604–615.

Bachman, C. H., & Ramo, S. (1943). Electrostatic electron microscopy. I. *Journal of Applied Physics, 14*, 8–18;
Bachman, C. H., & Ramo, S. Electrostatic electron microscopy. II. *Journal of Applied Physics, 14* (1943), 69–76;
Bachman, C. H., & Ramo, S. Electrostatic electron microscopy. III. *Journal of Applied Physics, 14* (1943), 155–160.

Baltá Elías, J., & Gómez García, J. A. (1950). Contribución al estudio de la lente electrostática cilindrica constituida por una rendida indefinida. *Anales de la Real Sociedad Española de Física y Química. Series A, 46*, 83–92.

Bergqvist, A. (1946a). Det svenska elektronenmikroskopet. *Tek. Tidskr., 76*, 649–655.

Bergqvist, A. (1946b). Ett Svenskt elektronmikroskop. *Industritidningen Norden, 74*. 323–327 and 331.

Bethge, H. (1982). Electron microscopy—beginnings and present. *Ultramicroscopy, 10*, 181–186.

Bocciarelli, D., & Trabacchi, G. C. (1946). Il microscopio elettronico dell'Istituto Superiore di Sanità. *Rendiconti dell'Istituto Superiore di Sanità, 9*, 762–768.

Bocciarelli, D., & Trabacchi, G. C. (1948). Modifica al microscopio elettronico dell'Istituto Superiore di Sanità. *Rendiconti dell'Istituto Superiore di Sanità, 11*, 794–800.

Borasky, R. (1982). The electron microscope: Priority of invention. *Microscope, 30*, 11–15.

Bradbury, S. (1968). *The microscope past and present*. Oxford: Pergamon.

Brú, L. (1936). El Microscopio de electrones. *Las Ciencas, 3*, 1–11.

Brú, L. (1942). *Introducción al Estudio de la Óptica Geométrica Electrónica*. I. Lentes electricas. *Publ. de la Universidad de la Laguna: Vol. 2*. 39 pp.

Brüche, E. (1941). Hinweis auf einige Abweichungen in der Darstellung dieses Buch von der Darstellung der Herren v. Borries und Ruska (1940) über Mikroskopie hoher Auflösung mit schnellen Elektronen. I. Zur Frage der Abbildungssysteme und der ersten Bilder. II. Zur Entwicklung des Übermikroskops. Appendix to Brüche and Recknagel (1941, pp. 437–441).

Brüche, E. (1943). Zum Entstehen des Elektronenmikroskops. *Physikalische Zeitschrift, 44*, 176–180.

Brüche, E. (1957). 25 Jahre Elektronenmikroskop. *Naturwissenschaften, 44*, 601–610.

Brüche, E. (1966–1967). Forschungsreise durch das Gebiet der Elektronenoptik. *Optik, 24*, 290–295.

Brüche, E. (1980). Papers for his eightieth birthday by various colleagues. *Physikalische Blätter, 36*(3), 57–64.

Brüche, E., & Recknagel, A. (1941). *Elektronengeräte, Prinzipien und Systematik*. Berlin and New York: Springer-Verlag.

Brüche, E., & Scherzer, O. (1947). Stand und Zukunft des Elektronenmikroskops. *Physikalische Blätter, 3*, 263–266.

Burton, E. F., & Kohl, W. H. (1942). *The electron microscope*. Princeton, New Jersey: Van Nostrand-Reinhold.

Calbick, C. J. (1944). Historical background of electron optics. *Journal of Applied Physics, 15*, 685–690.

Catalogue (1980). *Three centuries of microscopes and cell biology* [Mikroskope und Zellbiologie in drei Jahrhunderten]. Berlin: Catalogue of an exhibition at the Second International Congress on Cell Biology.

Cavalli, L. (1946). Contributo al problema del nucleo batterico. *Rendiconti dell'Istituto Superiore di Sanità, 9*, 578–596.

Challice, C. E. (1982). International microscopy meeting, London 1954. *Proceedings of the Royal Microscopical Society, 17*, 4–5.

Chiang, M.-Y. (1956). On the Petzval coefficient in electron optics (in Russian). *Acta Physica Sinica, 12*, 439–446.

Cohen, A. L., & Steever, R. G. E. (1971). An early American electron microscope. In *Proceedings of the 29th annual meeting electron microscopy society of America* (pp. 4–5).

Cosslett, V. E. (1950). *Bibliography of electron microscopy*. London: Arnold.

Cosslett, V. E. (1971a). The early history of the Electron Microscopy and Analysis Group. In W. C. Nixon (Ed.), *Electron microscopy and analysis* (pp. 1–5). London: Institute of Physics.

Cosslett, V. E. (1971b). The early history of the Electron Microscopy and Analysis Group. *Physics Bulletin, 22*, 339–341.

Cosslett, V. E. (1979). Electron microscopy: Retrospect and prospect. A personal history of the subject. *Proceedings of the Royal Microscopical Society, 14*, 101–107.

Cosslett, V. E. (1980). Early days in electron microscopy and microprobe analysis. *South African Journal of Science, 76*, 494–497.

Cosslett, V. E. (1981a). The development of electron microscopy and related techniques at the Cavendish Laboratory, Cambridge, 1946–1979. Part 1, 1946–1960. *Contemp. Phys., 22*, 3–36.

Cosslett, V. E. (1981b). The development of electron microscopy and related techniques at the Cavendish Laboratory, Cambridge, 1946–1979. Part 2, 1961–1979. *Contemp. Phys., 22*, 147–182.

Cosslett, V. E. (1982). Origins of electron microscopy in Britain. *Bulletin of Electron Microscopy Society of America, 12*(2), 11–15.

Das Gupta, N. N., & De, M. L. (1968). The electron microscope—its past and present. *Indian Journal of History of Science, 3*, 25–41.

Das Gupta, N. N., De, M. L., Bhattacharya, D. L., & Chaudhury, A. K. (1948). A new horizontal electron microscope. *Indian Journal of Physics, 22*, 497–513.

De, M. L., Guha, A., & Das Gupta, N. N. (1953). Phase contrast and electron microscopic studies on the nuclear apparatus of *Escherichia coli*. *Proceedings of the Royal Society of London. Series B, 141*, 199–203.

de Bruijn, W. C. (1977). Van glasbreek-kunst tot glasbreek-kunde. In *EM Conferentie Leiden*. Leiden: New Rhine Publishers (pp. 28–30).

Dobson, P. J., Pendry, J. B., & Humphreys, C. J. (Eds.). (1978). *Electron diffraction 1927–1977*. Bristol: Institute of Physics.

Dohnálek, R., & Rozsíval, M. (1948). Elektronová optika v technické praksi. *Chemický Obzor, 23*(4), 63–66.

Donelli, G., Merli, P. A., Pasquali Ronchetti, I., & Valdré, U. (1982). La microscopia elettronica in Italia: stato attuale e prospettive di sviluppo. *Annalli dell'Istituto Superiore di Sanità, 18*, 153–162 (to be published in English in Proceedings of the Royal Microscopical Society).

Donelli, G., Morace, A., Ardita, G., & Notargiacomo, S. (1983). Bibliografia analytica delle ricerche ultrastrutturali svolte nell' Istituto Superiore de Sanità: 1943–1982. *ISTISAN* Rep. No. 1983/46, Rome.

Dyachenko, V. (1935). The ring electron lens (in Ukrainian). *Ukraïns'ki Fizichni Zapiski, 3*, 53–65.

Dyachenko, V., & Sakharov, I. (1935). Calculation of the distribution of the field for a lens consisting of two and three plates with round apertures (in Ukrainian). *Ukraïns'ki Fizichni Zapiski, 4*, 23–32.
Dyachenko, V. E., & Sakharov, I. I. (1937a). Calculation of the trajectory of an electron in the field of an electric lens (in Ukrainian). *Ukraïns'ki Fizichni Zapiski, 6*, 37–48.
Dyachenko, V. E., & Sakharov, I. I. (1937b). Potential field of two long co-axial successively located cylinders (in Ukrainian). *Ukraïns'ki Fizichni Zapiski, 6*, 49–52.
Dyachenko, V. E., & Sakharov, I. I. (1938a). Potential fields of electron electrical lenses (in Ukrainian). *Ukraïns'ki Fizichni Zapiski*, 7, 3–21.
Dyachenko, V. E., & Sakharov, I. I. (1938b). Determination of the optic parameter in electric electronic lenses (in Ukrainian). *Ukraïns'ki Fizichni Zapiski*, 7, 175–191.
Dyachenko, V. E., & Sakharov, I. I. (1939). Distribution of potential on electric lens of two cylinders (in Ukrainian). *Ukraïns'ki Fizichni Zapiski*, 7, 299–303.
Elbers, P. F. (1977). Ontwikkeling van de ultramicrotomie. In *EM Conferentie Leiden*. Leiden: New Rhine Publishers (pp. 25–27).
Erkama, J. (1941). Elektronimikroskoopista ja sen käyttömahdollisuuksista. *Suomen Kemistilehti A, 14*(5–6), 47–53.
Extermann, R.-C. (1948). *Le Microscope Electronique*. Zurich, Trüb-Täuber, 6 pp.
Fert, C., & Selme, P. (1956). Le microscope électronique O.P.L.. *Bulletin de Microscopie Appliquée, 6*, 157–164.
Fickler, W., & Guyenot, E. (1963). Über die elektronenoptische Anlage EF. *Jenaer Jahrbuch*, (Pt. II), 345–375.
Fleming, G., Rempfer, R., Asherman, E., & Nolan, P. (1948). Recent electronmicrographs obtained with a new electrostatic electron microscope. *Journal of Applied Physics, 19*, 125.
Forgflo "Paragon" (1970). Personal communication and technical data sheet. The specification guarantees 2Å point-to-point resolution at 150 kV while the text states that "the Paragon Column *as is* is capable of better than 10 Å resolving power at 750 kV".
Franklin, U. M., Weatherly, G. C., & Simon, G. T. (1978). A history of the first North American electron microscope. *Toronto, 1978* Vol. 3 (pp. 4–17).
Freundlich, M. M. (1963). Origin of the electron microscope. *Science, 142*, 185–188.
Gabor, D. (1957). Die Entwicklungsgeschichte des Elektronenmikroskops. *ETZ. Elektrotechnische Zeitschrift. Ausgabe A, 78*, 522–543.
Gabor, D. (1974). The history of the electron microscope, from ideas to achievements. *Canberra, 1*, 6–12.
Georghiou, L., Metcalfe, J. S., Gibbons, M., Ray, T., & Evans, J. (1985). *Post-innovation performance*. London and New York: Macmillan.
Giesbrecht, P. (1981). Ernst Ruska and the reconstruction of the first electron microscope. *Proceedings of the Royal Microscopical Society, 16*, 48–49.
Glaser, W. (1952). *Grundlagen der Elektronenoptik*. Vienna: Springer-Verlag.
Glaser, W. (1956). Elektronen- und Ionenoptik. In S. Flügge (Ed.), *Handbuch der Physik: Vol. 33*. Berlin and New York: Springer-Verlag (pp. 123–395) (esp. pp. 389–390).
Gómez García, J. A. (1951). Contribución al estudio de una lente electrónica electromagnética cilíndrica. *Anales de la Real Sociedad Española de Física y Química. Series A, 47*, 15–28.
González-Santander, R. (1968). *Tecnicas de Microscopia Electrónica en Biologia*. Madrid: Aguilar.
Goodman, P. J. (Ed.). (1981). *Fifty years of electron diffraction*. Dordrecht, The Netherlands: Reidel.
Görlich, P. (1966). Die Entwicklung des wissenschaftlichen Gerätebaus seit Ernst Abbes Wirken. *VEB Carl Zeiss Jena Nachrichten,* Series 9, No. 8, 3–54.
Gribi, M., Thürkauf, M., Villiger, W., & Wegmann, L. A. (1959). Ein 70 kV-Elektronenmikroskop mit kalter Kathode und elektrostatischer Linse. *Optik, 16*, 65–86.

Grinberg (Grünberg), G. A. (1942). General theory of the focusing action of electrostatic and magnetostatic fields. I. *Doklady Akademii Nauk SSSR*, *37*, 172–178;
Grinberg (Grünberg), G. A. General theory of the focusing action of electrostatic and magnetostatic fields. II. *Doklady Akademii Nauk SSSR*, *37* (1942), 261–268.
Grinberg, G. A. (1943). General theory of the focusing action of electrostatic and magnetostatic fields. III. *Doklady Akademii Nauk SSSR*, *38*, 78–81; Parts I–III also published as *Zhurnal Tekhniceskoj Fiziki*, *13*, 361–388 and in Grinberg (1948).
Grinberg, G. A. (1948). *Selected problems in the mathematical theory of electric and magnetic phenomena* (in Russian). Moscow and Leningrad: Izd. Akad. Nauk SSSR.
Guyenot, E. (1955). Uber die Entwicklung eines fünfstufigen elektrostatischen Elektronenmikroskopes. *Jenaer Jahrbuch*, (Pt. I), 28–34.
Hahn, E. (1955a). Die elektronenoptischen Elemente des Elektronenmikroskopes Typ D. *Jenaer Jahrbuch*, (Pt. I), 35–53.
Hahn, E. (1955b). Ein neuer elektrostatischer Strahlkondensor und ein neues fünfstufiges Linsensystem. *Experimentelle Technik der Physik*, *3*, 58–65.
Hashimoto, H. (1983). Direct imaging of atomic processes in crystals: Some personal steps towards this goal. *Proceedings of the Royal Microscopical Society*, *18*, 298–308.
Hawkes, P. W. (1979). An account of electron optics in the Cavendish Laboratory and related developments. *Journal of Microscopy (Oxford)*, *117*, 39–53.
Higashi, N. (1967). A brief survey covering the development of the electron microscopy in Japan. *Journal of Electron Microscopy*, *16*, 2–4; see also *Tokyo, 1966* (pp. 1–2).
Hitachi (1959). *Publication on the Hitachi electron microscope.* This volume of 124 pages is a collection of publications from Hitachi ranging from Tadano's paper in the first volume of the *Hitachi Review* to an account by T. Komoda and S. Sakata of the "direct observation of K_2 Pt Cl_4 and K_2 Pt Cl_6 crystal lattices" which first appeared in Vol. 7 of the *Journal of Electronmicroscopy*.
Horne, R. W., & Wildy, P. (1979). An historical account of the development and applications of the negative staining technique to the electron microscopy of viruses. *Journal of Microscopy (Oxford)*, *117*, 103–122.
Induni, G. (1944). Uebermikroskopie in der Schweiz. *Neue Zürcher Zeitung*, *10 May*, 4.
Induni, G. (1945a). Das Schweizerische Uebermikroskop. *Neue Zürcher Zeitung*, *28 Feb.*, 4.
Induni, G. (1945b). Das Schweizerische Elektronenmikroskop. *Vierteljahresschrift der Naturforschenden Gesellschaft in Zürich*, *90*, 181–195.
Induni, G. (1946). Eine neue Methode der Elektronenbeugung. *Helvetica Physica Acta*, *19*, 231–233.
Induni, G. (1947a). Sur une source d'électrons pour microscopes électroniques. *Helvetica Physica Acta*, *20*, 463–467.
Induni, G. (1947b). Le microscope électronique suisse. *L'Indicateur Industriel*, *35*(45), 1–4.
Induni, G. (1949). Elektronenmikroskop und Elektronenbild. *Prisma*, *3*, 356–360.
Induni, G. (1951). Dall'oscillografo catodico al microscopio elettronico. *Elettrotecnica*, *38*, 2–9.
Induni, G. (1954). Absolute Spannungsmessung. *Scientia Electrica, Zeitschrift für moderne Probleme der theoretischen und angewandten Elektrotechnik*, *1*, 41–53.
Induni, G. (1955). La cathode froide et ses applications. In *Les Techniques Récentes en Microscopie Electronique et Corpusculaire* (pp. 189–197). Toulouse.
Induni, G. (1956). Il microscopio elettronico svizzero a catodo freddo. In *Atti del Congresso Scientifico, Vol. III (Sezione Elettronica): 3e Rassegna Internazionale Elettronica Nucleare.* Rome: CNRN.
Ingelman, B., & Siegbahn, K. (1944). Dextran and levan molecules studied with the electron microscope. *Nature (London)*, *154*, 237–238.
Jan, E. M. (1977). Enkele feiten uit het 25-jarig bestaan van de Nederlandse Vereniging voor Elektronenmikroskopie. In *EM Conferentie Leiden*. Leiden: New Rhine Publishers (pp. 7–9).

JEOL (1959–1973). *The world through the electron microscope*. This series of albums of electron micrographs and of JEOL instruments was published in three series, Metals, Chemistry, and Biology, between 1959 and 1973. The details are as follows. 1959: Metals; 1960: Chemistry; 1961: Biology; 1962: Metallurgy II; 1963: Chemistry II; 1964: Biology II; 1965: Metallurgy III; 1966: Chemistry III; 1967: Biology III; 1968: Metallurgy IV; 1969: Chemistry IV; 1970: Biology IV; 1971: Metallurgy V; 1972: Chemistry V; 1973: Biology V.

Jervis, P. (1971–1972). Innovation in electron-optical instruments—two British case histories. *Research Policy, 1*, 174–207.

Kas'yankov, P. P. (1956). *Teoriya Elektromagnitnikh Sistem c Krivolineinoi Os'yu*. Leningrad: Leningrad Univ. Press.

Kedesdy, H., & Matthias, A. (1939). *Entwicklungsarbeiten am elektromagnetischen Elektronenmikroskop*. Gesellschaft von Freunden der Technischen Hochschule Berlin zu Charlottenberg, Geschäftsjahr 1939, 3. Forschungsber. (pp. 97–104).

Kellenberger, E. (1949a). La microscopie électronique. *Journal Suisse D'horlogerie Et De Bijouterie, 3–4*, 3–16.

Kellenberger, E. (1949b). Elektronenmikroskop und Biologie. *Prisma, 3*, 7–10.

Kel'man, V. M. (1955). *Elektronnaya Optika*. Moscow: Nauka.

Kel'man, V. M. (1968). *Elektronnaya Optika* (2nd ed.). Moscow: Nauka.

Kel'man, V. M., & Yavor, S. Ya. (1959). *Elektronnaya Optika*. Moscow and Leningrad: Izd. Akad. Nauk SSSR.

Kel'man, V. M., & Yavor, S. Ya. (1963). *Elektronnaya Optika* (2nd ed.). Moscow and Leningrad: Izd. Akad. Nauk SSSR.

Kel'man, V. M., & Yavor, S. Ya. (1968). *Elektronnaya Optika* (3rd ed.). Leningrad: Nauka.

Kim (= Kim, Gil' Khon), K.-H. (1962). Study of electron lenses by the method of successive approximations (in Korean). *Suzhvakgva Mulli (The Korean Journal of Mathematics), 6*(3), 28–41.

Kim, K.-H., & Choi, S.-K. (1963). *Taejon Ipja Kwanghak* [Optics of Charged Particles]. Pyong-Yang: Academy of Sciences of Korea; listed in *Referativnyi Zhurnal (Fizika), 9*.

Klemperer, O. (1939). *Electron optics*. London and New York: Cambridge Univ. Press.

Kohl, W. H., Hall, C. E., Prebus, A. F., Hillier, J., & Watson, J. H. L. (1978). Recollections. *Toronto, 1978* Vol. 3 (pp. 18–25).

Küpfmüller, K. (1944). Zur Geschichte des Elektronenmikroskops. *Physikalische Zeitschrift, 45*, 47–51.

Kushnir, Yu. M. (1947a). Physico-technical principles of electron microscopy (in Russian). *Elektrichestvo, 5*, 3–16.

Kushnir, Yu. M. (1947b). Construction and applications of electron microscopes (in Russian). *Elektrichestvo*, 7, 17–31.

Langrish, J., Gibbons, M., Evans, W. G., & Jevons, F. R. (1972). *Wealth from knowledge*. London and New York: Macmillan.

Lebedev, A. A. (1946). The Soviet model of the electron microscope (in Russian). In *Obshchee Sobranie Akademii Nauk SSSR*. Moscow and Leningrad: Izd. Akad. Nauk SSSR (pp. 31–44).

Lebedev, A. A. (Ed.). (1954). *Elektronnaya Mikroskopiya*. Moscow: Gos. Izd. Tekh.-Teor. Literatury.

Le Poole, J. B. (1966–1967). Über die Entwicklung der Elektronenmikroskopie in Holland. *Optik, 24*, 375–382.

Le Poole, J. B. (1977). The early days of electron microscopy in Holland. In *EM Conferentie Leiden*. Leiden: New Rhine Publishers (pp. 12–13).

Luk'yanovich, V. M. (1960). *Elektronnaya Mikroskopia v Fiziko-khimicheskikh Issledovaniyakh*. Moscow: Izd. Akad. Nauk.

Mahl, H., Volkmann, H., & Weitsch, W. (1957). Uber ein neues Gebrauchs-Elektronenmikroskop. *Stockholm, 1956* (pp. 34–37).

Majewski, W. (1936). Podstawy fizyczne geometrycznej optyki elektronowej orazjej zastosowanie w technice i w nauce. *Wiadomosci I Prace-Panstwowego Institutu Telekomunikacyjnego Warszawa*, 7(3/4), 58–83.
Manton, I. (1978). Recollections in the history of electron microscopy. *Proceedings of the Royal Microscopical Society*, *13*, 45–57.
Marton, C., Sass, S., Swerdlow, M., Van Bronkhorst, A., & Meryman, H. (1950). *Bibliography of electron microscopy*. NBS Circ. 502. Washington, D.C.: U.S. Natl. Bur. Stand.
Marton, L. (1946). Electron microscopy. *Reports on Progress in Physics*, *10*, 204–252.
Marton, L. (1960). Note historique sur les premières années de la microscopie électronique. *Bulletin de la Classe Des Sciences. Académie Royale de Belgique*, *46*, 119–121.
Marton, L. L. (1965). Historical background of image formation. *Laboratory Investigation*, *6*, 739–746.
Marton, L. (1968). *Early history of the electron microscope*. San Francisco: San Francisco Press.
Matthias, A. (1942). Bemerkung zur Entstehung des Elektronenmikroskops. *Physikalische Zeitschrift*, *43*, 129–130.
Mellema, J. E. (1977). Buitenlandse ontwikkelingen van de elektronenmikroskopie. In *EM Conferentie Leiden*. Leiden: New Rhine Publishers (pp. 10–11).
Milyutin, V. I. (1949). The emission electron microscope (in Russian). *Uspehi Fizičeskih Nauk*, *38*, 377–409.
Molière, K. (1977). 50 Jahre Elektroneninterferenzen. *Naturwissenschaften*, *64*, 310–316.
Mühlethaler, K. (1948). Elekronenmikroskopie. *Forschung und Technik*, *585*.
Mulvey, T. (1962). Origins and historical development of the electron microscope. *British Journal of Applied Physics*, *13*, 197–207.
Mulvey, T. (1964). The development of the electron microscope, AEI's contribution. *AEI Engineering*, *Jan/Feb*, 3–11.
Mulvey, T. (1967). The history of the electron microscope. In S. Bradbury, & G. Turner (Eds.), *Historical aspects of microscopy* (pp. 201–227). Cambridge: Heffer; also published as *Proceedings of the Royal Microscopical Society*, *2*, 201–227.
Mulvey, T. (1971). The development of electron optics. In *Proceedings of the 29th annual meeting electron microscopy society of America* (pp. 2–3).
Mulvey, T. (1972). The electron microscope. *Laboratory Equipment Digest*, *March*, 73–99.
Mulvey, T. (1973). Forty years of electron microscopy. *Physics Bulletin*, *24*, 147–154.
Mulvey, T. (1974). Science and technology under the microscope. Inaugural lecture published by the University of Aston in Birmingham.
Myers, L. M. (1939). *Electron optics, theoretical and practical*. London: Chapman & Hall.
Nyrikov, V. G., & Kuschnier, U. M. (1956). Electron microscope and electron diffraction cameras developed in the U.S.S.R. *Tokyo, 1956* (pp. 30–34).
Oatley, C. W. (1982). The early history of the scanning electron microscope. *Journal of Applied Physics*, *53*, R1–R13.
Oatley, C. W., Nixon, W. C., & Pease, R. F. W. (1965). Scanning electron microscopy. *Advances in Electronics and Electron Physics*, *21*, 181–247.
Ortiz Fornaguera, R. (1948). Acerca de unas formulas de Picht en optica electronica. *Anales de la Real Sociedad Española de Física y Química. Series A*, *44*, 70–79.
Paszkowski, B. (1960). *Optyka Elektronowa* (1st ed.). Warsaw: Wydawnictwa Naukowo-Techniczne.
Paszkowski, B. (1965). *Optyka Elektronowa* (2nd rev. ed.). Warsaw: Wydawnictwa Naukowo-Techniczne.
Paszkowski, B. (1968). *Electron optics*. London/New York: Iliffe/Elsevier (this is a translation of Paszkowski, 1965).
Prebus, A. (1942). The electron microscope. *Ohio State University Engineering Experimental Station News*, *14*(Part 3), 6–32.
Preston, R. D. (1983). Early days in electron microscopy. *Proceedings of the Royal Microscopical Society*, *18*, 37–40.

Ramsauer, C. (Ed.). (1941). *Zehn Jahre Elektronenmikroskopie. Ein Selbstbericht des AEG Forschungs-Instituts*. Berlin: Springer-Verlag.
Ramsauer, C. (Ed.). (1942). *Elektronenmikroskopie* (2nd ed.). Berlin: Springer-Verlag.
Ramsauer, C. (Ed.). (1943). *Elektronenmikroskopie* (3rd ed.). Berlin: Springer-Verlag.
Rang, O., & Schluge, H. (1951). Aufbau des AEG-Zeiss-Elektronemikroskops EM 8. *AEG-Mitteilungen, 41*, 151–154. This issue (No. 7-8) contains several articles on the EM 8.
Recknagel, A. (1952–1953). Über ein elektrostatisches Eletronenmikroskop. *Wissenschaftliche Zeitschrift der Technischen Hochschule Dresden, 2*, 515–522.
Reisner, J. H. (1981a). An historical perspective on the origins of the EMSA. *Bulletin of Electron Microscopy Society of America, 11*(1), 13–15.
Reisner, J. H. (1981b). The formation of the EMSA. *Bulletin of Electron Microscopy Society of America, 11*(2), 16–19.
Reisner, J. H. (1982). The EMSA's first year. *Bulletin of Electron Microscopy Society of America, 12*(1), 11–14.
Rempfer, G., Connell, R., Mercer, L., & Louiselle, I. (1972). An electrostatic transmission electron microscope. *American Laboratory*, 39–46.
Renner, J. (1939). The electron microscope (in Hungarian). *Természettudományi Közlöny, 71*, Article 285294.
Renner, J. (1940). New applications of electron microscopy. *Természettudományi Közlöny, 72*, 48–50.
Riecke, W. D. (1982). Practical lens design. In P. W. Hawkes (Ed.), *Magnetic electron lenses*. Berlin and New York: Springer-Verlag (pp. 163–357).
Rochow, T. G. (1983). Significant early days in industrial electron microscopy. *Bulletin of Electron Microscopy Society of America, 13*(1), 10–18.
Rüdenberg, R. (1932). Elektronenmikroskop. *Naturwissenschaften, 20*, 522.
Rüdenberg, R. (1943). The early history of the electron microscope. *Journal of Applied Physics, 14*, 434–436.
Ruska, E. (1956). Geschichte des Elektronenmikroskops. *Tokyo, 1956* (pp. 19–29).
Ruska, E. (1957). 25 Jahre, Elektronenmikroskopie. *ETZ. Elektrotechnische Zeitschrift. Ausgabe A, 78*, 531–543.
Ruska, E. (1974). Zur Vor- und Frühgeschichte des Elektronenmikroskops. *Canberra, 1*, 1–5.
Ruska, E. (1979). Die Frühe Entwicklung der Elektronenlinsen und der Elektronenmikroskopie. *Acta Historica Leopoldina, 12*, 1–136.
Ruska, E. (1980). *The early development of electron lenses and electron microscopy*. Stuttgart: Hirzel (transl. into English by T. Mulvey).
Ruska, E. (1984). Die Entstehung des Elektronenmikroskops (Zusammenhang zwischen Reaiisierung und erster Patentanmeldung, Dokumente einer Erfindung). *Archiv der Geschichte der Naturwissenschaften*, (11/12), 525–551.
Savchenko, F. A. (1938). Experimental determination of the optical characteristics of electron lenses (in Ukrainian). *Ukraïns'ki Fizichni Zapiski, 7*, 147–157.
Savchenko, F. A. (1939). The dependence of certain optical parameters on the electric and geometric parameters of electric immersion lenses (in Russian). *Zhurnal Tekhniceskoj Fiziki, 9*, 2211–2219.
Seman, O. I. (1959). *Theoretical foundations of electron optics* (in Chinese). Peking.
Siegbahn, M. (1939). Vetenskapsakadamiens Forskningsinstitut för Experimentell Fysik. *Nordisk Familjeboks Manadskronika, 2*, 571–576.
Soa, E. A. (1965). Ein neues Emissionselektronenmikroskop. *Optik, 22*, 66–76 (several early papers published in the *Jenaer Jahrbuch* and *Experimentelle Technik der Physik* are listed here).
Spit, B. J. (1977). Twenty-fifth anniversary of the Dutch Society for Electron Microscopy. In *EM Conferentie Leiden*. Leiden: New Rhine Publishers (p. 6).

Spivak, G. V., & Lukatskaya, P. A. (1948). On a second limiting case in electron optics (in Russian). *Doklady Akademii Nauk SSSR, 60*, 375–378.
Spiwak, G., & Kardash, J. (1945). On the nature of the images formed by lightning. *Journal of Physics (Moscow), 9*, 447.
Spiwak, G., & Kardash, J. (1946). The image of objects produced by lightning and impulse discharge in atmosphere. *Journal of Physics (Moscow), 10*, 252–256.
Strampelli, B., & Posarelli, A. (1946). La struttura del vitreo studiato al microscopio elettronico. *Rendiconti dell'Istituto Superiore di Sanità, 9*, 769–778.
Strashkevich, A. M. (1959). *Elektronnaya Optika Elektrostaticheskikh Polei ne Obladayushchikh Ocevoi Simmetriei*. Moscow: Gos. Izd. Fiz. Mat. Literatury.
Sugata, E. (Ed.). (1968). *Electron microscopes of Japan 1936–1965, a historical survey*. Tokyo: Maruzen.
Sushkin, N. G. (1937). *Ukraïns'ki Fizichni Zapiski, 6*, 288; Cited, apparently incorrectly, by Lebedev (1954); not yet traced by PWH.
Sushkin, N. G. (1941). Study of the oxide cathode with the aid of the electron microscope (in Russian). *Ukraïns'ki Fizichni Zapiski, 9*(2), 152–156.
Sushkin, N. G. (1949). *Elektronnyi Mikroskop*. Moscow: Gos. Izd. Tekh.-Teor. Literatury.
Syui, C.-K. (1956). Determination of the aberration of third order of electrostatic cylindrical lenses by the point eikonal method (in Russian). *Acta Scientiarum Naturalium Universitatis Pekinensis, 2*, 457–465.
Tadano, B. (1953). Progress of electron microscopes in Hitachi. *Hitachi Review, 1*(4), 19–30.
Tadano, B., & Siraga, T. (1942). Development of the electron supermicroscope and its applications (in Japanese). *Hitachi Hyoron, 25*, 431–440.
Tani, Y. (1956). General survey of development and present state of electron microscope in Japan. *Tokyo, 1956* (pp. 50–57).
Trabacchi, G. C. (1947). Il microscopio elettronico. *Rendiconti dell'Istituto Superiore di Sanità, 10*, 1222–1235.
Trabacchi, G. C. (1948). Il microscopio elettronico. *Elettrotecnica, 35*, 313–319.
Vainrib, E. A., & Milyutin, V. I. (1951). *Elektronnaya Optika*. Moscow and Leningrad: Gos. Energ. Izd.. Translated into German by K. Keller as K. Keller as Wainrib, E.A., Miljutin, W.I. *Elektronenoptik*. Berlin: VEB Verlag Technik.
van Dorsten, A. C. (1977). De begintijd van de elektronenmicroscopie bij Philips, enige historische notities, and 25 jaar geschiedenis van de ontwikkeling van de preparaattafel in de Philips electronenmicroscopen. In *EM Conferentie Leiden*. Leiden: New Rhine Publishers (pp. 14–20 and 22–24).
van Iterson, W. (1977). Vijf en twintig jaar electronenmicroscopie in Nederland. In *EM Conferentie Leiden*. Leiden: New Rhine Publishers (pp. 3–5).
van Zuylen, P. (1977). Hoogspanningsmicroscopie in Nederland. In *EM Conferentie Leiden*. Leiden: New Rhine Publishers (p. 21).
Vertsner, V. N. (1944). The electron microscope of the State Optical Institute (in Russian). *Izvestiâ Akademii Nauk SSSR. Seriâ Fizičeskaâ, 8*, 232–234.
Vertsner, V. N. (1945). Electron microscopy (in Russian). *Zavodskaya Laboratoria, 11*, 543–554.
Vertsner, V. N. (1946). The electron microscope of the GOI (in Russian). *Optiko-Mekhanicheskaya Promyshlennost', 13*, 3–15.
Vertsner, V. N. (1947). The Soviet electron microscope (in Russian). *Zavodskaya Laboratoria, 13*, 1364–1375.
Vertsner, V. N. (1980). Obituary (1909–1980). *Optiko-Mekhanicheskaya Promyshlennost', 47*(7), 56. *Soviet Journal of Optical Technology (English Translation), 47*, 434–435.
von Ardenne, M. (1972). *Ein Glückliches Leben für Technik und Forschung*. Munich: Kindler [republished by Suhrkamp, Taschenbuch 310 (1976)].
von Ardenne, M. (1978). Zur Geschichte der Rasterelektronenmikrosonde. *Optik, 50*, Article 177188.

von Borries, B., & Ruska, E. (1940). Mikroskopie hoher Auflösung mit schnellen Elektronen. *Ergebnisse der Exakten Naturwissenschaften, 19*, 237–322.

von Borries, B., & Ruska, E. (1944). Neue Beiträge zur Entwicklungsgeschichte der Elektronenmikroskopie und der Übermikroskopie. *Physikalische Zeitschrift, 45*, 314–326; reprinted with minor changes in *Frequenz, 2*, 267–278.

Vouk, V. (1941–1942). Electron microscope (in Serbo-Croat). *Kemijski Vjestnik, 15/16*, 80–91.

Watson, J. H. L. (1974). Toronto "daze". *Bulletin of the Microscopy Society of Canada, 2*, 10–18.

Wegmann, L. (1950a). Der Aufbau des Trüb-Täuber Elektronenmikroskopes. *Optik, 7*, 263–266.

Wegmann, L. (1950b). Zur Bestimmung des Auflösungsvermögens durch Fresnelsche Beugung im Elektronenmikroskop. *Helvetica Physica Acta, 23*, 437–452.

Wegmann, L. (1950c). Zur Frage der Abbildung von Atomabständen im Elektronenmikroskop. *Helvetica Physica Acta, 23*, 867–868.

Wegmann, L. (1951a). *Helvetica Physica Acta, 24*, 63–71.

Wegmann, L. (1951b). Das schweizerische Elektronenmikroskop. Betrachtungen zu seiner praktischen Anwendung. *Chemische Rundschau, 16*, 3–8.

Wegmann, L. (1952). Ein neues schweizerisches Elektronenmikroskop. *Neue Zürcher Zeitung, 16 July*.

Wegmann, L. (1953a). Ein neues Trüb, Taäuber-Elektronenmikroskop. *Transactions on Instrumentation and Measurement Conference, Stockholm 1952*. Norrköping: Norrköpings Tidningars (pp. 73–83).

Wegmann, L. (1953b). Objektveränderungen und Beleuchtungssystem im Elektronenmikroskop. *Optik, 10*, 44–59.

Wegmann, L. (1954). Ein verzeichnungsarmes elektromagnetisches Doppelprojektiv für Elektronenmikroskope. *Optik, 11*, 153–170.

Wegmann, L., & Gribi, M. (1957). Richtstrahlwert der kalten Kathode. *Stockholm, 1956* (pp. 41–42).

Wegmann, L., & Studer, H. (1950). Das Elektronenmikroskop im Dienste der chemischen Forschung. *Chimia, 4*, 26–33.

Wilska, A. P. (1958). Brief account in Finnish of work by A. P. Wilska. *Suomen Kemistilehtii A, 31*, 61.

Wilska, A. P. (1961). Low voltage electron microscope. *Delft, 1960* Vol. 1 (pp. 105–109).

Wilska, A. P. (1962). A new low-voltage electron microscope. *Philadelphia, 1962* Vol. 1 Artic. D-6.

Wilska, A. P. (1964a). Recollections and experiences from fifteen years of constructing electron microscopes. *Annales Academiæ Scientiarum Fennicæ, Series A5, 106*(23), 1–15.

Wilska, A. P. (1964b). Low-voltage electron microscope: A 6 kV instrument. *Journal of the Royal Microscopical Society, 83*, 207–211.

Wilska, A. P. (1965). Expectations and limitations of low voltage electron microscopy. *Laboratory Investigation, 14*, 825–828.

Wyckoff, R. W. G. (1954). Entwicklungslinien in der amerikanischen Elektronenmikroskopie. *Zeitschrift für wissenschaftliche Mikroskopie und mikroskopische Technik, 62*, 87–91.

Zernov, D. (1939). Electron optics and its applications (in Russian). *Uspehi Fizičeskih Nauk, 21*, 162–210.

Zinchenko, N. S. (1961). *Kurs Lektsii po Elektronnoi Optike* (2nd ed.). Khar'kov: Izd. Khar'kovsk. Ordena Trudob. Krasn. Znameni Gos. Univ.

Afterword references

Breton, B. C., McMullan, D., & Smith, K. C. A. (Eds.). (2004). *Sir Charles Oatley and the scanning electron microscope. Advances in Imaging and Electron Physics: Vol. 133.*

de Chambost, E. (2009). *Histoire de Cameca (1954–2009)*. Limours: Emmanuel de Chambost.

de Chambost, E. (2011). A history of Cameca (1954–2009). In *Advances in Imaging and Electron Physics: Vol. 167* (pp. 1–119).

Fournier, M. (2009). Electron microscopy in Second World War Delft. In A. Maas, & H. Hooijmaijers (Eds.), *Scientific research in World War II: What scientists did in the war* (pp. 77–95). London and New York: Routledge.

Fujita, H. (Ed.). (1986). *History of electron microscopes 1986*.

Gelderblom, H. R. (2020). Bodo von Borries (1905–1956): Herforder und Friederiziana – ein Pionier der Elektronenmikroskopie. *Historisches Jahrbuch für den Kreis Herford, 27*, 154–199.

Gelderblom, H. R., & Krüger, D. H. (2014). Helmut Ruska (1908–1973): His role in the evolution of electron microscopy in the life sciences, and especially virology. In *Advances in Imaging and Electron Physics: Vol. 182* (pp. 1–94).

Gelderblom, H. R., & Schwarz, H. (2022). Rudolph Rühle and the BOSCH electron microscope, another early commercial instrument. In *Advances in Imaging and Electron Physics: Vol. 221*. Chapter 11 (this volume).

Gelderblom, H. R., Krüger, D. H., & Hawkes, P. W. (2014). Publications from the Düsseldorf University Institute for Biophysics and Electron Microscopy: (Institut für Biophysik und Elektronenmikroskopie der Universität Düsseldorf) 1958–1973. In *Advances in Imaging and Electron Physics: Vol. 182* (pp. 95–122).

Gloede, W. (1986). *Vom Lesestein zum Elektronenmikroskop*. Berlin: VEB Verlag Technik.

Grünewald, H. (1976). Zur Entwicklungsgeschichte des Elektronenmikroskops mit elektromagnetischen Linsen. *Technikgeschichte, 43*, 213–222.

Hawkes, P. W. (2015). Dennis McMullan (1923–2015). *In Focus, 38*, 116–118.

Hawkes, P. W., & McMullan, D. (2004). A forgotten French scanning electron microscope and a forgotten text on electron optics. *Proceedings of the Royal Microscopical Society, 39*, 285–290.

Karrenberg, F. H. (2019). *In the light of the electron microscope, in the shadow of the Nobel prize*. Norderstedt: BoD Books on Demand.

Kunkle, G. C. (1994). Technology in the seamless web: "Success" and "failure" in the history of the electron microscope. *Technology and Culture, 36*, 80–103.

Lambert, L. (1986). Some anecdotes around and with Colonel Ernst August Ruska. *Ultramicroscopy, 20*, 337–340.

Lambert, L., & Mulvey, T. (1996). Ernst Ruska (1906–1988), designer extraordinaire of the electron microscope: A memoir. In *Advances in Imaging and Electron Physics: Vol. 95* (pp. 2–62).

Möllenstedt, G. (1968). Laudatio zur Ernennung von Prof. em. Max Knoll zum Ehrenmitglied der Deutschen Gesellschaft für Elektronenmikroskopie anlässlich der Tagung der Gesellschaft in Marburg 1967. *Mikroskopie, 23*, 61–96.

Müller, F. (2006). Zwischen Bilderbuch und Messgerät: der elektronenmikroskopische Blick auf die Realstruktur von Festkörpern. In *Technikgeschichte nach dem Pictorial Turn?* (pp. 77–101). Mannheim: Gesellschaft für Technikgeschichte.

Müller, F. (2009). The birth of a modern instrument and its development during World War II. In A. Maas, & H. Hooijmaijers (Eds.), *Scientific research in World War II: What scientists did in the war* (pp. 121–146). London and New York: Routledge.

Müller, F. (2021). *Jenseits des Lichts*. Göttingen: Wallstein-Verlag.

Mulvey, T. (1987). Ernst Ruska and his electron microscope. *Proceedings of the Royal Microscopical Society, 22*, 103–109.

Mulvey, T. (1988). Ernst Ruska Hon FRMS (25 December 1906–27 May 1988). Father of the electron microscope. *Proceedings of the Royal Microscopical Society, 23*, 239–241.

Mulvey, T. (1995). Gabor's pessimistic 1942 view of electron microscopy and how he stumbled on the Nobel prize. In *Advances in Imaging and Electron Physics: Vol. 91* (pp. 259–283).

Mulvey, T. (Ed.). (1996). *The Growth of Electron Microscopy. Advances in Imaging and Electron Physics: Vol. 96*.

Mulvey, T., & van de Laak-Tijssen, D. J. J. (2001). Jan Bart Le Poole (1917–1993), pioneer of the electron microscope and particle optics. In *Advances in Imaging and Electron Physics: Vol. 115* (pp. 287–354).

Paden, R. S. (2015). The birth of the scanning electron microscope. The 50th anniversary of the launch of the Stereoscan. *In Focus, 38*, 70–79.

Qing, L. (1995). *Zur Frühgeschichte des Elektronenmikroskops*. Stuttgart: Verlag für Geschichte der Naturwissenschaften und der Technik.

Quaglino, D., Falcieri, E., Catalano, M., Diaspro, A., Montone, A., Mengucci, P., & Pellicciari, C. (Eds.). (2006). *1956–2006. Cinquanta Anni di Microscopia in Italia tra Storia, Progresso ed Innovazione*. Società Italiana de Scienze Microscopiche.

Rasmussen, N. (1993). Facts, artifacts, and mesosomes: Practicing epistemology with the electron microscope. *Studies in History and Philosophy of Science, 24*, 227–265.

Rasmussen, N. (1996). Making a machine instrumental: RCA and the wartime origins of biological electron microscopy in America, 1940–1945. *Studies in History and Philosophy of Science, 27*, 311–349.

Rasmussen, N. (1997). *Picture control. The electron microscope and the transformation of biology in America*. Stanford: Stanford University Press.

Rasmussen, N. (1999). What moves when technology migrates? "Software" and hardware in the transfer of biological electron microscopy to postwar Australia. *Technology and Culture, 40*, 47–73.

Ratinac, K. R. (Ed.). (2008). *50 great moments, celebrating the golden jubilee of the University of Sydney's Electron Microscope Unit*. Sydney: Sydney University Press.

Reisner, J. H. (1989). An early history of the electron microscope in the United States. In *Advances in Electronics and Electron Physics: Vol. 73* (pp. 133–231).

Rudenberg, H. G., & Rudenberg, P. G. (2010). Origin and background of the invention of the electron microscope: Commentary and expanded notes on memoir of Reinhold Rüdenberg. In *Advances in Imaging and Electron Physics: Vol. 160* (pp. 207–286).

Rüdenberg, R. (2010). Origin and background of the invention of the electron microscope. In *Advances in Imaging and Electron Physics: Vol. 160* (pp. 171–205).

Rühle, R. (1949). *Das Elektronenmikroskop*. Stuttgart: Curt E. Schwab.

Ruiz-Castell, P. (2013). Seeing the invisible: The introduction and development of electron microscopy in Britain, 1935–1945. *History of Science, 51*, 221–249.

Ruska, E. (1984). Die Entstehung des Elektronenmikroskops (Zusammenhang zwischen Reaiisierung und erster Patentanmeldung, Dokumente einer Erfindung). *Archiv der Geschichte der Naturwissenschaften, 11/12*, 525–551.

Ruska, E. (1986). The emergence of the electron microscope: Connection between realization and first patent application, documents of an invention. *Journal of Ultrastructure and Molecular Structure Research, 95*, 3–28.

Ruska, E. (1987a). The development of the electron microscope and of electron microscopy. *Angewandte Chemie, International Edition (English), 26*, 595–605.

Ruska, E. (1987b). The development of the electron microscope and of electron microscopy. *Reviews of Modern Physics, 59*, 627–638.

Ruska, E. (1987c). The development of the electron microscope and of electron microscopy. In *Les Prix Nobel/The Nobel prizes 1986*. Stockholm: Almqvist & Wiksell (pp. 58–83).

Schulze, D. (2007). Pioniere des Elektronenmikroskops [Ruska and von Borries]. *Praktische Metallographie, 44*, 557–565.

Smith, K. C. A. (2009). STEM at Cambridge University: Reminiscences and reflections from the 1950s and 1960s. In *Advances in Imaging and Electron Physics: Vol. 159* (pp. 387–406).
Smith, K. C. A. (2013). Electron microscopy at Cambridge University with Charles Oatley and Ellis Cosslett: Some reminiscences and recollections. In *Advances in Imaging and Electron Physics: Vol. 177* (pp. 189–277).
van Delft, D., & van Helvoort, T. (2018). *Beelden zonder Weerga*. Amsterdam: Prometheus.
van Gorkom, J. (2018a). The early electron microscopes: A critical study. In *Advances in Imaging and Electron Physics: Vol. 205* (pp. 1–137).
van Gorkom, J. (2018b). The early electron microscopes: Incubation. In *Advances in Imaging and Electron Physics: Vol. 208* (pp. 43–128).
von Ardenne, M. (1997). *Erinnerungen Fortgeschrieben*. Düsseldorf: Droste.
von Borries, G. (2001). *Bodo von Borries und das Elektronenmikroskop. Erfindung und Entwicklung*. Engelsbach: Fouqué.
von Borries, H. (1991). Bodo von Borries: Pioneer of electron microscopy. In *Advances in Electronics and Electron Physics: Vol. 81* (pp. 127–176).

Index

A

Aberrations, 52, 79, 84, 95, 96, 109, 241, 242, 255, 378, 386, 393
- alignment, 393
- astigmatic, 84, 114
- astigmatic spherical, 88, 89, 93
- lens, 52, 80, 84, 230

Abraded specimens, 262
Accelerating voltage, 77, 92, 131, 214, 216, 219, 226, 233, 237, 239, 248, 329, 356, 484
Achromatic lens, 109
AEG
- Forschungsinstitut Scherzer, 382
- instrument, 476
- microscope, 317, 368

Anastigmatic lens, 109
Anisotropic astigmatism, 124
Applications laboratory, 408, 410, 413, 423, 428–430, 435, 437
Applications laboratory for electron microscopy, 435, 440
Associated Electrical Industries (AEI), 249
Astigmatic
- aberrations, 84, 114
- spherical aberration, 88, 89, 93

Astigmatism, 79, 80, 86, 89, 91, 93, 98, 106, 119–121, 241, 261, 401, 411, 416, 420, 435
- coefficient, 115
- control, 416
- correction, 437

Asymmetric
- aberrations, 88, 91, 93
- lens, 75, 93

Atomic
- resolution, 221, 222, 239–241, 248, 378, 385, 386, 400
- resolution electron holography, 220

Attainable resolving power, 20
Attaining high resolution, 271
Axisymmetrical lenses, 219

B

Bacteriological specimen, 185
Beam
- current, 84, 127, 409, 410, 452
- cutoff electrical, 449
- deflection blanking, 270
- deflection direction, 122
- density, 409, 418
- electron, 15, 35, 37, 38, 40–42, 77, 79, 84, 110, 117, 122–125, 144, 178, 181, 183, 203–205, 222, 226, 247, 257, 322, 329, 346, 445, 454, 470
- electron diffraction, 203
- energy, 186
- path, 183, 444
- spot, 117, 418, 419
- spot diameter, 418
- voltage, 413

Berlin
- electron microscopy, 311, 313, 321
- institute, 453

Biological specimens, 39, 130, 182, 187, 228, 396, 432, 475
Biological specimens electron microscopy, 180
Biophysics, 184
Bipotential electrostatic lens, 131
Bosch
- brochure, 343
- choir, 325
- company, 324, 325
- electron microscope, 331–333, 341, 343–351, 358–361, 363, 366, 368, 497
- goniometer, 346
- hierarchy, 364
- orchestra, 327
- product, 341, 365
- repertoire, 347

Britain
- electron microscopy, 289, 307
- electron optics, 290

British instrument, 292

Broadminded director, 240
Brüche
  group, 359
  institute, 321

C

Cambridge Instrument Company, 281, 322, 498
Cambridge microscope, 260
Camera
  cine, 322
  design, 6
  system, 343
Carbonized specimens, 5
Cavendish laboratory, 238, 248, 258, 292, 300, 305
Chief instrument maker, 21
Chromatic aberration, 79, 92, 96, 98, 100, 101, 105, 106, 153, 256, 267, 378, 384, 385, 393
  coefficient, 75
  constant, 152
  magnification, 106, 108
Chromatic field aberrations, 106–108
Cine camera, 322
Complex instrument, 312
Compound
  condenser lens, 103
  lens, 96, 98, 104
Condenser lens, 27, 101, 102, 229, 237
  compound, 103
  strength, 101, 106
Convergent beam, 117, 125
Convergent beam illumination, 106
Cooled specimen container, 420
Crystalline specimen, 79
Cylinder lens, 219, 389
Cylindrical polepieces, 109

D

Darmstadt project, 391, 394, 395, 397, 398
Deflected beam, 124
Delft
  electron microscopy, 166
  microscope, 159, 163
Demagnifying lenses, 255
Demountable emission microscope, 3
Dennis Gabor, 162, 173, 250
Diffraction lens, 235
Disk specimen, 79

E

Earliest instruments, 178, 182
Einzel objective lens, 271
Electrical gear, 296
Electromagnetic lenses, 455
Electron
  beam, 15, 35, 37, 38, 40–42, 77, 79, 84, 110, 117, 122–125, 144, 178, 181, 183, 203–205, 222, 226, 247, 257, 322, 329, 346, 445, 454, 470
    deflection, 124
    interferometry, 205
    lithography, 169
    quality, 37
  biprism, 37, 205, 211, 215, 219, 221, 222
  biprism interference, 203, 204
  bombardment, 28, 76, 179, 180, 182, 206, 413, 415, 455
  detection, 258
  diffraction, 29, 39, 205, 216, 330, 331, 346, 349, 469
    equipment, 146
    experiments, 144, 202
    facility, 341
    instrumentation, 229
    patterns, 203, 234
  emission, 26
  Fresnel biprism, 222
  gun, 73, 117, 122, 123, 127, 130, 167, 178, 266, 300, 306, 329, 333
  holograms, 241
  holographic microscope, 241
  holography, 34, 35, 41, 42, 221, 228, 238, 240, 241
  image, 26, 27, 39, 142
  image intensifier, 246
  intensity variations, 246
  interaction, 204

interference fringes, 38
interference microscope, 38, 39, 219
interferometer, 37, 191, 219
interferometry, 220
irradiation, 34
lens, 25, 74, 95, 143, 178, 228, 239, 240, 250, 378, 379, 393, 398, 401, 451
  astigmatism, 401
  design, 95
  system, 95, 108, 109
line probe, 210
lithography machine, 169
micrographs, 10, 15, 16, 20, 29, 100, 130, 183, 185, 473
microscope, 2, 6, 7, 27–29, 34, 45, 48, 51, 72–74, 76, 141, 144, 147, 150, 158, 173, 175–178, 205, 215, 225, 226, 228–232, 290, 292–294, 298, 312, 314, 317, 320, 324, 377, 378, 381, 388, 407, 408, 423, 443, 444, 448, 453, 469–471, 475–478, 483, 484, 486
  alignment, 164
  development, 497
  for biological specimens, 192
  history, 483, 484
  specimen, 15
microscopists, 9, 15, 19, 239–241, 289, 292, 301, 302
mirror, 42, 169
multipliers, 187
optical
  biprism, 201
  means, 204
  theory, 95
optician, 250
optics, 2, 3, 13, 25, 27, 52, 73, 74, 142, 144, 174–177, 184, 186, 187, 228, 232, 250, 258, 289–292, 314, 317, 345, 377, 379, 380, 382, 391, 455, 458, 478, 480, 486–488
  for electron microscopes, 95
  history, 173
  in Britain, 291
path, 28
penetration, 238
phase microscope, 127
physics, 250, 258
physics team, 236
probe, 167
projection images, 211
rays, 215
scattering, 163
shutter, 182
source, 38, 167, 250
swarm, 240
tube, 142, 176
velocity, 239
wave, 201, 205, 219
wave nature, 312
wave optics, 363
Electron microscopy (EM), 5, 13, 14, 27–31, 45, 48, 50, 53, 72, 74, 78, 84, 160, 169, 178, 181, 182, 203, 228, 232, 234, 289, 298, 300, 301, 317, 320, 321, 324, 377, 384, 408, 409, 424–426, 428, 444, 469, 470, 472, 476–478, 483, 484
  beam, 9
  history, 191, 496
  in
    Berlin, 311, 313, 321
    Britain, 289, 307
    Delft, 166
    Holland, 139, 497
    Tübingen, 361
  laboratory, 16
  techniques, 210
Electronics stabilization, 147
Electrostatic
  cylinder lenses, 210
  einzel lens, 385
  electron lenses, 38, 455
  instrument, 476
  lenses, 27, 80, 163, 249, 262, 271, 317
  microscopes, 359, 377, 475, 476, 487
  reducing lens, 87
  stigmator, 117, 119, 121
Eliminating astigmatisms, 121
EMB
  microscope, 471
  model, 470
Emission microscope, 3, 5, 25, 26, 42, 150
EMU model microscope, 471

F

Field aberrations, 92, 98, 103, 104, 106
Fluorescent screen, 2, 4, 5, 18, 25, 27, 142, 144, 157, 163, 175–177, 183, 211, 226, 233, 246, 247, 257, 445, 453, 454
Focal length, 74, 86, 88, 95, 96, 120, 126, 150, 165, 271, 290, 332, 333, 454, 455, 458
Foil lens, 397
Foolproof instruments, 185
French microscope, 483
Fresnel fringes, 35, 37, 152, 167

G

GEC instrument, 487
Genuine Bosch products, 341
Geometrical electron optics, 378
GOI
  instrument, 478
  microscope, 478

H

Hampstead
  instrument, 305
  meeting, 302
Historical instrument, 352
History
  electron microscope, 483, 484
  electron optics, 173
Hitachi instruments, 488
Holland
  electron microscope, 163
  electron microscopy, 139, 497
Homemade instruments, 300, 475

I

Illuminating beam, 100, 103, 106
Impinging electrons, 179
Imported instruments, 486
Incoherent electron sources, 42
Instrument
  AEG, 476
  Bosch, 348, 359, 360, 366
  complex, 312
  design, 306
  designers, 235
  development, 472
  electrostatic, 476
  maker, 283
  manufacturers, 275
  RCA, 273, 299, 498
  Siemens, 304, 484
Instrumental
  performance, 239, 247
  perturbations, 247
  problem, 226
  resolution, 248
  resolving power, 239
Instrumentation, 19
Ion lithography machine, 169
Iron polepiece lens, 228
Irradiated specimen area, 415
Isolated specimen, 420

J

Japan Electron Optics Laboratory (JEOL), 74

K

Kaiser Wilhelm Institute (KWI), 323, 334, 361
Kossel
  institute, 202, 204
  lines, 202

L

Laboratory of Electron Microscopy (LEM), 361
Leica camera, 4, 147
Lens
  aberration, 52, 80, 84, 230
  aberration coefficients, 85
  action, 384
  anastigmatic, 109
  configuration, 143
  construction, 242
  currents, 142, 156, 165
  dimensions, 385
  strength, 95, 98, 102, 106, 108
  supplies, 237
  supplies electronic stabilization, 233
  systems, 74, 98
Loaned instrument, 351

M

Magnetic
  condenser lens, 27, 38
  electron lenses, 25
  lens, 6, 9, 20, 72, 109, 186, 249, 271, 291, 306, 312, 317, 331–333, 341, 458, 470, 476, 497, 498
  lens anisotropic aberrations, 110
  lens design, 93, 101
  lens for electron microscopes, 95
  microscopes, 5, 475
  stigmator, 120, 121
Magnification
  chromatic aberration, 106, 108
  factor, 455
  range, 428
  steps, 415
Manchester
  area, 293
  houses, 308
  laboratory, 229
  meeting, 166, 303
  university, 249, 250
Mark II microscope, 156
Maximum magnification, 164, 233, 417
Metropolitan Vickers
  company, 228, 232, 240
  electron microscopes, 229
Microbeam, 78
Microbeam illumination, 78
Micrographs, 10, 15, 16, 18, 20, 29, 72, 73, 88, 93, 100, 111, 183, 185, 266, 268, 279, 300, 301, 470, 471
Microscope
  AEG, 317, 368
  alignment electron, 164
  Bosch, 332, 341, 343, 344, 348–351, 358–361, 363, 366, 368, 497
  column, 262, 298
  Delft, 159, 163
  electron, 2, 6, 7, 27–29, 34, 45, 48, 51, 72–74, 76, 141, 144, 147, 150, 158, 173, 175–178, 205, 215, 225, 226, 228–232, 290, 292–294, 298, 312, 314, 317, 320, 324, 377, 378, 381, 388, 407, 408, 423, 443, 444, 448, 453, 469–471, 475–478, 483, 484, 486
  equipment, 408
  family, 438
  performance, 429
  production, 360
  RCA, 232, 273, 350
  room, 410
  Ruska, 455, 456
  Siemens, 72, 149, 307, 366, 470
  systems, 414
Miniature electron lenses, 250
Minimum aberrations, 105
Möllenstedt
  energy analyser, 222
  group, 361
  symposium, 222
Multilens imaging, 235

N

National Physical Laboratory (NPL), 232, 238, 292, 294–297
Nonmetallic specimens, 273
Nuclear physics, 392, 401
Nuclear physics projects, 324

O

Opaque metal specimens, 266

P

Philips Electron Optics, 170
Philips microscope, 163, 164, 305
Physics
  colloquium, 401, 402
  education, 201
  electron, 258
  laboratory, 295, 329, 382
  teacher, 202, 386
  textbooks, 35, 387
Platelets, 157
Polepiece, 89, 91, 110, 122, 125, 143
  lenses, 248, 331, 456
  magnetic lenses, 312
Projector
  lens, 27, 92, 98, 109, 150, 151, 177, 178, 229, 234, 237, 335, 341, 346
  lens focal length, 110
  polepieces, 149
Prototype instrument, 237

R

RCA
instrument, 273, 299, 498
microscope, 232, 273, 303, 350
scientific instrument, 487
type, 14, 16, 187, 296, 298, 300, 304
Recording beam, 449
Reich Mark (RM), 354
Resolution
capability, 458, 462
instrumental, 248
level, 465
limit, 386, 388, 393
range, 415
record, 464
value, 412
Resolving power, 5, 6, 8, 14, 15, 17, 27, 35, 72, 75, 79, 84, 101, 178, 186, 230, 232, 233, 236–238, 240, 242, 243, 408, 411, 416, 417, 432
in electron microscopy, 186
instrumental, 239
Robert Bosch company, 324, 327
Routine instrument, 291
Royal Microscopical Society, 251, 293, 301–303, 305
Ruska microscope, 455, 456

S

Scanning electron microscope (SEM), 322
Scattered
beams, 450
electron, 389
electron waves, 389
Scherzer
anecdotes, 401
defocus, 402
theorem, 378, 384
Scientific instrumentation, 230
Secondary electrons, 257
Seidel aberrations, 84
Sequential projection lenses, 341
Serially produced instrument, 232
Shadowed micrographs, 471
Shirley Institute, 292, 293, 300–302
Siemens
company, 408
electron microscope, 409
Elmiskop I, 243, 351
instrument, 304, 484
list, 356
management, 318
microscope, 72, 149, 307, 366, 470
Solvay physics congresses, 184
Specially coated lenses, 332
Specimen
airlock, 331, 336
area, 412, 430
border, 418
cartridge, 426
chamber, 268, 419
drift, 411, 429
elasticity, 426
grid, 424
holder, 175, 336, 343, 415, 419, 434
lock, 343
materials, 10
natural, 411
preparation techniques, 27, 80, 226, 228, 236, 247, 299, 301, 304–306
stage, 229, 260, 332
support, 433
surface, 262, 264, 421, 429, 430
techniques, 19
temperature, 76
thicknesses, 238
Spherical aberration, 52, 74, 78, 79, 84, 86, 87, 96, 98, 105, 106, 110, 111, 114, 146, 147, 162, 165, 176, 186, 221, 222, 239–242, 384, 385, 388–390, 393, 395, 397
coefficient, 74, 111, 116
constant, 384, 385
integral, 385
Spherically corrected microscope, 396
Stable specimens, 410
Stanford physics department, 189
Stigmator, 20, 117, 119–125
electrostatic, 117, 119, 121
magnetic, 120, 121
radius, 119

Stray electrons, 183
Symmetrical electron lens, 249

T
Telefocal electron gun, 117
Thermionic electron gun, 239
Thick specimens, 148, 396–398
Thinnest specimens, 5, 182
Transmission electron microscope (TEM), 322
Troublesome instruments, 21

U
Ultimate magnification, 409
Unconventional instruments, 476
Unipole magnetic lens, 74, 75
Unresolved specimen, 397
Unsaturated polepieces, 101

V
VAP report, 329–332
Versatile microscope, 158

W
Weak lens, 96
World War II, 27, 72, 186, 187, 191, 289, 293, 301, 316, 321, 328, 332, 356, 359, 444, 456, 457, 470

Z
Zone lens, 167

Kolibri 360 GmbH
Ferdinand-Jühlke-Straße 7
99095 Erfurt, Deutschland
produktsicherheit@kolibri360.de